NANOPAINTS

NANOPAINTS

Characterization, Formulation, and Emerging Applications

Edited by

BHASHA SHARMA
Department of Chemistry, Shivaji College, University of Delhi, India

VIJAY CHAUDHARY
Department of Mechanical Engineering, Amity School of Engineering and Technology, Amity University, Noida, Uttar Pradesh, India

PRAKASH CHANDER THAPLIYAL
Advanced Concrete, Steel and Composites Group, CSIR-Central Building Research Institute, Roorkee, Uttarakhand, India

VIJAY MANNARI
Coatings Research Institute, Eastern Michigan University, Ypsilanti, MI, United States

ELSEVIER

Elsevier
Radarweg 29, PO Box 211, 1000 AE Amsterdam, Netherlands
125 London Wall, London EC2Y 5AS, United Kingdom
50 Hampshire Street, 5th Floor, Cambridge, MA 02139, United States

ISBN: 978-0-443-26512-9

For Information on all Elsevier publications
visit our website at https://www.elsevier.com/books-and-journals

Publisher: Matthew Deans
Acquisitions Editor: Ana Claudia Garcia
Editorial Project Manager: Emily Thomson
Production Project Manager: Anitha Sivaraj
Cover Designer: Greg Harris

Typeset by MPS Limited, Chennai, India

Contents

List of contributors

Hani Nasser Abdelhamid Department of Chemistry, College of Science, Imam Mohammad Ibn Saud Islamic University (IMSIU), Riyadh, Saudi Arabia

Noelia L. Alchapar Institute of Environment, Habitat and Energy (INAHE) Scientific and Technological Centre Mendoza, National Council for Scientific and Technological Research (CONICET), Buenos Aires, Argentina

K.M. Anis-Ul-Haque Department of Chemistry, Jashore University of Science and Technology, Jashore, Bangladesh

Julieta Balter Institute of Environment, Habitat and Energy (INAHE) Scientific and Technological Centre Mendoza, National Council for Scientific and Technological Research (CONICET), Buenos Aires, Argentina

Harish C. Barshilia Surface Engineering Division, CSIR-National Aerospace Laboratories, Bengaluru, Karnataka, India

Parthasarathi Bera Surface Engineering Division, CSIR-National Aerospace Laboratories, Bengaluru, Karnataka, India

Vijay Chaudhary Department of Mechanical Engineering, Amity School of Engineering and Technology Amity University, Noida, Uttar Pradesh, India

Dinesh Chawla Department of Mechanical Engineering, School of Engineering and Technology, Manav Rachna International Institute of Research and Studies, Faridabad, Haryana, India

Partha Pratim Das School of Materials Science and Engineering, Nanyang Technological University, Singapore, Singapore

Shweta Gomey Department of Chemistry, Amity Institute of Applied Sciences, Amity University, Noida, Uttar Pradesh, India

Pallav Gupta Department of Mechanical Engineering, Madan Mohan Malaviya University of Technology, Gorakhpur, Uttar Pradesh, India

Sumit Gupta Department of Mechanical Engineering, Amity School of Engineering and Technology, Amity University, Noida, Uttar Pradesh, India

Tejendra K. Gupta Department of Physics, Amity Institute of Applied Sciences, Amity University Uttar Pradesh, Noida, Uttar Pradesh, India; Department of Chemistry, Amity Institute of Applied Sciences, Amity University, Noida, Uttar Pradesh, India

Yaman Hooda Department of Civil Engineering, School of Engineering and Technology, Manav Rachna International Institute of Research and Studies, Faridabad, Haryana, India

Chinnala Kalyani Department of Chemistry, Amity Institute of Applied Sciences, Amity University, Noida, Uttar Pradesh, India

Kanak Department of Chemistry, Amity Institute of Applied Sciences, Amity University, Noida, Uttar Pradesh, India

Twinkle Kawatra Department of Chemistry, Amity Institute of Applied Sciences, Amity University, Noida, Uttar Pradesh, India

Anand Sawroop Khanna SSPC India, Mumbai, India

Most. Afroza Khatun Department of Chemical Engineering, Jashore University of Science and Technology, Jashore, Bangladesh

Ashish Kumar Department of Science, Technology and Technical Education, Nalanda College of Engineering, Bihar Engineering University Government of Bihar, Patna, Bihar, India

R.V. Lakshmi Surface Engineering Division, CSIR-National Aerospace Laboratories, Bengaluru, Karnataka, India

Vijay Mannari Coatings Research Institute, Eastern Michigan University, Ypsilanti, MI, United States

Jimmy Mehta Department of Mechanical Engineering, School of Engineering and Technology, Manav Rachna International Institute of Research and Studies, Faridabad, Haryana, India

Prateek Mittal Department of Mechanical Engineering, School of Engineering and Technology, Manav Rachna International Institute of Research and Studies, Faridabad, Haryana, India

Manu Satya Prakash Pathariya Department of Mechanical Engineering, Amity School of Engineering and Technology Amity University, Noida, Uttar Pradesh, India

Priti Department of Paint Technology, Mahamaya Polytechnic of Information Technology, Chail, Kaushambi, Uttar Pradesh, India

Md. Wasikur Rahman Department of Chemical Engineering, Jashore University of Science and Technology, Jashore, Bangladesh; Department of Physics and Astronomy, University of Texas Rio Grande Valley, Edinburg, TX, United States; Department of Mechanical Engineering, University of Texas Rio Grande Valley, Edinburg, TX, United States

Rishi Raj Department of Physics, Amity Institute of Applied Sciences, Amity University Uttar Pradesh, Noida, Uttar Pradesh, India

Md. Sohel Rana Department of Chemical Engineering, Jashore University of Science and Technology, Jashore, Bangladesh

Manoj Raula Department of Chemistry, Amity Institute of Applied Sciences, Amity University, Noida, Uttar Pradesh, India

Radha Sachan Department of Paint Technology, School of Chemical Technology, Harcourt Butler Technical University, Kanpur, Uttar Pradesh, India

Ankita Saxena Department of Biotechnology, School of Engineering and Technology, Noida International University, Greater Noida, Uttar Pradesh, India

Sony Setiya Department of Mechanical Engineering, Amity School of Engineering and Technology Amity University, Noida, Uttar Pradesh, India

Gaurav Sharma Department of Chemistry, Amity Institute of Applied Sciences, Amity University, Noida, Uttar Pradesh, India; Department of Physics, Amity Institute of Applied Sciences, Amity University Uttar Pradesh, Noida, Uttar Pradesh, India

Nancy Sharma Department of Physics, Amity Institute of Applied Sciences, Amity University Uttar Pradesh, Noida, Uttar Pradesh, India

Sushant Sharma Department of Mechanical Engineering, School of Engineering and Technology, Manav Rachna International Institute of Research and Studies, Faridabad, Haryana, India

Tushar Sharma Department of Mechanical Engineering, Amity School of Engineering and Technology Amity University, Noida, Uttar Pradesh, India

Neeta Shukla Department of Physics, BGS College of Engineering and Technology (BGSCET), Bengaluru, Karnataka, India

Abhinay Thakur Department of Research Impact and Outcome, Division of Research and Development Cell, Lovely Professional University, Phagwara, Punjab, India

Prakash Chander Thapliyal Advanced Concrete, Steel and Composites Group, CSIR-Central Building Research Institute, Roorkee, Uttarakhand, India

Sheenam Thatai Department of Physics, Amity Institute of Applied Sciences, Amity University Uttar Pradesh, Noida, Uttar Pradesh, India; Department of Chemistry, Amity Institute of Applied Sciences, Amity University, Noida, Uttar Pradesh, India

Rohit Verma Department of Physics, Amity Institute of Applied Sciences, Amity University Uttar Pradesh, Noida, Uttar Pradesh, India

PART

I

Formulation and characterization

1

Pioneering the classification and formulation of paints

Manu Satya Prakash Pathariya and Vijay Chaudhary
Department of Mechanical Engineering, Amity School of Engineering and Technology Amity University, Noida, Uttar Pradesh, India

1.1 Introduction

The push for the intense development of nanomaterials stems from the desire to transcend the limitations of traditional materials and satisfy the demands of the twenty-first century. Nanomaterials have proven indispensable in tackling the numerous issues encountered by various industries as we navigate an increasingly complex and linked environment. Their distinctive features, including increased mechanical strength, electrical conductivity, and catalytic activity, result from their nanoscale dimensions, making them particularly well-suited for a wide range of applications [1].

Nanomaterials allow breakthrough solutions in fields ranging from electronics to energy and healthcare to environmental remediation [2]. They are critical in increasing the efficiency of photovoltaic cells, developing high-capacity batteries, and improving medication delivery methods for personalized treatment. Nanomaterials are the foundation of nanotechnology, supporting their potential to revolutionize many sectors, ensure sustainability, and push the boundaries of what is possible.

The realm of materials science has witnessed a remarkable evolution with the advent of nanotechnology, a field that explores the manipulation and utilization of materials at the nanoscale. Among the many applications of nanomaterials, their integration into paint formulations has opened a new frontier of possibilities in the coatings industry [3]. This chapter explores the utilization of nanomaterials in paints and delves into the realm of nanopaints, elucidating the wide-ranging applications that have emerged from this transformative innovation [4–6].

Nanopaint is one of the many applications of nanomaterials that has received a lot of interest [7,8]. This new product uses the

Nanopaints. DOI: https://doi.org/10.1016/B978-0-443-26512-9.00001-X

unique features of nanoparticles to generate coatings with exceptional properties. Nanopaint not only improves the durability and longevity of surfaces, but it also has a slew of practical advantages, including self-cleaning, UV protection, and the ability to harvest solar energy [9]. The invention and use of nanopaint are examples of nanoparticles' transformative impact, since they alter the possibilities of an ordinary household item [10].

The imperative for integrating nanomaterials into paint formulations arises from the continual quest to enhance the performance, durability, and functionality of coatings. Traditional paints, while effective for many purposes, often face limitations when confronted with the growing demands of modern industries and applications [11]. Nanomaterials, with their unique properties and capabilities at the nanoscale, offer a compelling solution to address these challenges.

In this present chapter, the world of nanomaterials for nanopaints, formulation, and methodology of paints has been explored in depth. Also, shedding light on paint synthesis, characterization, and the diverse range of applications has been discussed in detail. This chapter includes the intricacies of nanopaint and examines the underlying principles that render it a paradigm-shifting innovation in the field of coatings. Through a comprehensive analysis, we aim to highlight the critical role nanomaterials play in reshaping the scientific and technological landscape, driving innovation, and offering solutions to contemporary challenges.

1.1.1 Nanopaints: a revolution in paint technology

In the contemporary world, surface coatings are becoming more prominent, adorning our surroundings, and they offer a variety of different protections and also provide an esthetic appeal to a wide range of objects and structures [12]. They have a wide range of applications, like providing corrosion resistance and antifouling properties to architectural wonders, giving consumer goods additional attractive features, and paints and coatings that are inseparable from our daily lives. Yet, the demands placed on these coatings are continually evolving; hence, there is a need for innovations that transcend the limitations of traditional paint materials. This challenge has given rise to the emergence of nanopaints, a class of materials uniquely positioned to revolutionize the realm of surface coatings [13].

The impetus for nanopaints is grounded in the ever-growing need for advanced, multifunctional coatings that can adapt to the increasingly complex demands of our modern society.

Traditional paints, while serving admirably in their intended purpose, often fall short in meeting the multifaceted requirements of durability, environmental sustainability, and performance across a diverse array of substrates. Nanopaints, a subcategory of nanomaterials, address these shortcomings by harnessing the extraordinary properties that arise at the nanoscale.

Nanopaints, as the name implies, contain tiny components, frequently nanoparticles, in their formulation [14–17]. This infusion of nanoparticles causes a change in the fundamental properties of paints [18,19]. Nanoparticles give nanopaints exceptional properties such as increased mechanical strength, resistance to UV radiation, and the capacity to repel water and pollutants [20,21]. These characteristics enhance the functional capabilities of surface coatings, making them a viable option for a wide range of applications [22].

Nanopaints are used and applied in a wide range of industries and sectors, from construction and automotive to aerospace and healthcare. They enable architects to design self-cleaning and energy-efficient structures, car manufacturers to construct vehicles with durable and scratch-resistant coverings, and even medical device manufacturers to develop biocompatible coatings. Nanopaints' adaptability exemplifies their versatility, making them a vital tool in today's technological scene. One such application explored was the development of zinc oxide (ZnO) nanopaints for the prevention of marine biofouling by Yong et al. [1]. The results of the study have been shown in Fig. 1.1 [1].

In this chapter, we embark on a comprehensive exploration of nanopaints, shedding light on their formulation, unique characteristics, and the diverse spectrum of applications they serve. We delve into the complexities of nanopaint, looking at the underlying concepts that make it a game-changing invention in the world of coatings. By examining the technology and applications of nanopaints, we aim to underscore their pivotal role in the evolution of surface coatings and the promise they hold for reshaping the future of various industries.

1.2 Classification of nanopaints and nanocoatings

Nanocoatings function by incorporating nanostructures into a material to modify it at the molecular level. When these nanostructures come into contact with a substrate, they produce a self-assembling hydrophobic layer a few tens of nanometers thick. These nanoparticles adhere securely to surfaces,

Figure 1.1 In situ marine fouling tests over a period of 30 days in a lagoon connected to Jeju Sea. The figure shows how the substrate reacts to sea water conditions, two substrates are shown before and after being immersed in a lagoon in Jeju Sea for 30 days. The substrate without ZnO nanopaint shows significant biofouling, while the substrate with ZnO nanopaint shows little biofouling. From H.E. Yong, K. Krishnamoorthy, K.T. Hyun, S.J. Kim, Preparation of ZnO nanopaint for marine antifouling applications, Journal of Industrial and Engineering Chemistry 29 (2015) 39–42. https://doi.org/10.1016/j.jiec.2015.04.020.

repelling components such as water and dirt, similar to the natural lotus effect. Almost any material can be nanocoated, and the technique is resistant to temperatures as high as 350°C. It's ideal for fine-textured materials and precision applications like ultrathin nonstick coatings for quicker mold releases or protective layers for often cleaned surfaces like glass and facades.

Selecting the right nanocoating requires careful consideration of desired outcomes, the substrate type, and what the substrate needs protection from since different nanocoatings possess varied properties or combinations of them. represents the classification of nanopaints. The various classifications of nanopaints are shown in Fig. 1.2.

1.2.1 Superhydrophobic coating

At the forefront of surface treatment innovations, superhydrophobic coatings mark a leap beyond the capabilities of conventional waterproof solutions. Rather than merely facilitating

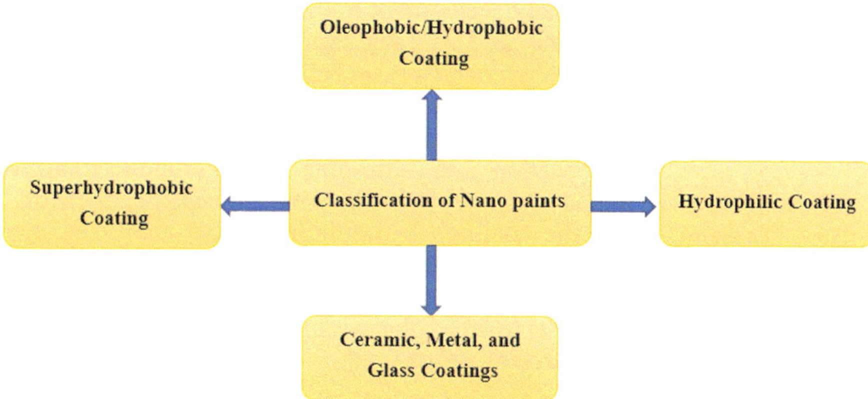

Figure 1.2 Classification of nanopaints based on coating technology.

the runoff of water, these advanced coatings employ a dynamic repulsion strategy against it. The method of application is straightforward: a spray coats the target surface and, upon drying, initiates a level of hydrophobicity that leverages tiny water droplets for self-cleaning purposes. This innovative approach ensures that any contact with water not only repels unwanted substances like dust, dirt, and microbes but also gathers and removes them as water beads up and rolls off the surface.

The implications of such technology are profound, especially in settings prone to moisture-related bacterial growth. By keeping water perpetually moving and thwarting its bonding with surfaces, superhydrophobic coatings inherently curb the likelihood of bacteria taking hold. The versatility of this technology stretches beyond water, effectively warding off various substances such as mud, ketchup, and sirup, thus maintaining cleanliness and dryness across treated surfaces.

Moreover, these coatings significantly mitigate ice formation by preventing water adherence, a benefit that plays a critical role in averting damage across numerous applications, from ensuring the integrity of pedestrian pathways to safeguarding electrical systems. The reduction of ice accumulation also translates into lessened rust and corrosion on metallic surfaces, bolstering their durability, performance, and safety by minimizing water exposure. The utility of superhydrophobic coatings spans a broad spectrum of industrial uses, adaptable to a variety of materials, including plastics, metal alloys, microelectronics, and glass. This adaptability makes them indispensable for applications ranging from architectural exteriors to machinery and high-tech gadgets.

A pivotal development in this domain is the creation of a highly transparent and durable superhydrophobic organic–inorganic composite nanocoating by Feng et al. [2]. This breakthrough was achieved through a meticulous sol-gel dip-coating process, fine-tuning the nanocoating's characteristics by controlling the polymerization and ultrasonic vibration-induced aggregation of inorganic colloidal particles. The innovation lies in the ability to adjust the coating's hydrophobicity and clarity by manipulating the monomer concentration and the size of the aggregates. Remarkably, this coating exhibits superior thermal resistance, retaining its hydrophobic qualities within a wide temperature spectrum, from 20°C to 90°C [2].

1.2.2 Oleophobic/hydrophobic coating

In the realm of material science and engineering, the development of oleophobic (oil-repelling) and hydrophobic (water-repelling) coatings represents a pinnacle of innovation designed to protect and extend the life of surfaces in a multitude of applications. These coatings, which cleverly manipulate the interactions between surfaces and fluids, owe their efficacy to a sophisticated blend of chemical compounds, including the pivotal titanium dioxide (TiO_2), as well as a host of other organic and inorganic substances. The creation of these protective films is not just a matter of chemical composition but also of the ingenious methods applied in their fabrication, which can broadly be categorized into two distinct approaches: the top-down approach and the bottom-up approach.

The top-down approach to fabricating these coatings is characterized by its methodical precision, utilizing techniques such as lithography and particle coating. These methods, intricate and precise, involve the meticulous structuring and application of materials onto surfaces, sculpting at the macro or micro level to achieve the desired oleophobic and hydrophobic properties. This approach mirrors the careful planning and execution of a sculptor, where the final form is chiseled out from the larger mass, revealing the functional beauty beneath.

Conversely, the bottom-up approach embodies the principles of growth and assembly from the molecular or atomic level, utilizing techniques that include electrocoating, vapor deposition, spray coating, spin coating, and dip coating. Here, the process is akin to planting a seed and nurturing it to bloom; the coating grows organically, layer by layer, to form a cohesive, protective barrier on the surface. This method highlights the remarkable ability to construct complex structures from the simplest of beginnings, a testament to human ingenuity in mimicking nature's evolutionary strategies.

The applications of oleophobic and hydrophobic coatings are as varied as they are vital. In an age where clarity and durability are paramount, these coatings offer antifogging properties essential for devices and products that must maintain transparency under diverse environmental conditions, such as lenses and optical instruments. Beyond clarity, these coatings impart surfaces with enhanced scratch resistance, significantly extending the lifespan of everyday items by fortifying them against the wear and tear of daily use. Perhaps most importantly, the corrosion resistance provided by these coatings safeguards the integrity of materials exposed to potentially damaging substances, ensuring both functionality and esthetics are preserved over time.

In crafting this narrative on oleophobic and hydrophobic coatings, it's clear that the interplay between the complexity of chemical compounds and the ingenuity of fabrication techniques—encompassing both the top-down and bottom-up approaches—illuminates a path toward not just preserving the material world around us, but also enhancing our interaction with it. Through the lens of material science, we are reminded that the quest for durability, transparency, and resistance in materials is not just a technical challenge but a reflection of the human desire to innovate and improve upon the natural world.

1.2.3 Hydrophilic coating

Harnessing the power of titanium dioxide (TiO_2), this cutting-edge coating technique unveils a remarkable approach to cleanliness. At its core, TiO_2 acts as a catalyst when acted upon by sunlight, engaging with and dismantling dirt particles in a dance of degradation. This process is beautifully complemented by the coating's affinity for water. Its hydrophilic nature doesn't just attract water; it encourages it to spread into a uniform layer that glides over the surface, sweeping away the remnants of dirt with graceful efficiency.

But the magic of this hydrophilic coating doesn't stop with mere cleanliness. It brings to the table a reduction in surface friction and a boost in lubrication, qualities that expand its realm of possibility. From the sun-drenched faces of solar panels, where it plays a pivotal role in preserving their pristine condition and operational efficacy, to the towering glass facades of skyscrapers, where it bestows the gift of self-cleansing windows, and even into the sterile environments of medical devices that demand impeccably smooth surfaces—its versatility shines through. This innovation not only redefines cleanliness but also opens new avenues for applications that benefit from its unique properties.

1.2.4 Ceramic, metal, and glass coatings

In the vanguard of surface technology, the advent of silicon dioxide-based ceramic coatings has ushered in a new era of protection, blending scratch resistance with an unparalleled hardness, heat resilience, and insulating prowess. These coatings, primarily celebrated in the automotive domain for their finish-enhancing properties, also play a pivotal role in guarding against graffiti and bolstering medical device manufacture with an invincible layer of durability.

The advancement in nanocoatings is ever-evolving. In the year 2000, the study conducted by Jordan et. al. delved into the performance of plasma-sprayed alumina–titania coatings, focusing on a range of characteristics such as porosity, hardness, and resistance to various forms of wear and damage. Notably, the nanostructured versions of these coatings showcased remarkable enhancements in several key areas. Despite their increased porosity and slightly diminished hardness, these nanocoatings demonstrated superior resistance to indentation crack growth, cracking, spallation, and both abrasive and sliding wear. Furthermore, this has led to new improvements to the nanoscale structure of the alumina–titania, which also showed notable gains in adhesive strength. More recently, Jevnikar et al. [3] conducted a study to assess the impact of functionalizing Y-TZP ceramic surfaces with a nanostructured alumina coating on the bond strength of resin-modified glass ionomer dental cement. In their experiment, 160 disc-shaped specimens were divided into two groups of 80, with half receiving an alumina coating created through the hydrolysis of aluminum nitride (AlN) powder. They evaluated the shear bond strengths of both the resin-modified glass ionomer cement and the composite resin luting agent on coated and uncoated surfaces. Their findings, supported by SEM analyses, indicated that alumina-coated Y-TZP ceramics provided a highly retentive surface for bonding, with significantly higher bond strengths observed in the coated groups compared to the uncoated ones.

The sol-gel technique emerges as a beacon of innovation in the realm of protective coatings, celebrated for its chemical fidelity, oxidative governance, and a fortified stand against corrosion, especially on metallic canvases. This environmentally congenial method stands as a testament to an alternative pathway, sidestepping the hazardous byways of conventional corrosion resistance strategies. A review discussed recent developments and applications of sol-gel protective coatings on various metal substrates, including steel, aluminum, copper, magnesium, and their alloys, while also addressing challenges in industrial production and future research directions.

Zhang et al. explored [5] the effects of contamination on the resin bond strength to nanostructured alumina-coated zirconia ceramic. They applied nanostructured alumina coatings to zirconia ceramic disks using the hydrolysis of aluminum nitride powder. The study included scenarios where specimens were either contaminated with saliva or silicone disclosing agent and subsequently cleaned with phosphoric acid or simply rinsed with tap water. Control specimens were left uncontaminated. The tensile bond strength (TBS) was assessed using a universal testing machine, and the topography of the debonded surfaces was examined for fractographic features using optical and scanning electron microscopy. The study concluded that while uncontaminated nanostructured alumina-coated zirconia ceramic demonstrated durable TBS, contamination significantly reduced it. Phosphoric acid was effective in removing saliva contamination, but not the silicone disclosing agent. The alumina coating improved resin bonding to zirconia ceramic, negating the need for air-abrasion before bonding.

In another study, Zhang et al. [6] investigated the influence of various factors such as weathering duration, water pH, rainfall duration, and intensity on the release of SiO_2-ENPs, Ago-ENPs, and TiO_2-ENPs from coated panels. They observed that the concentration of these particles increased with weathering duration and varied under different pH conditions. Their findings indicated that weathering time and rainfall duration play significant roles in controlling the release of particles.

Al-Kattan et al. [7] investigated the compounds released from paint that had aged and contained nano-TiO_2. They studied the colloidal stability of particulate material with various pH levels and ionic compositions. Through gentle agitation in water, they were able to obtain a stable suspension from the old paint powder. This resulted in a suspension with a TiO_2 content of 580 $\mu g\ \ell^{-1}$ and an average particle size of 200–300 nm. While most of the particles in this suspension were tiny pieces of the paint matrix that also contained nano-TiO_2, electron imaging revealed some independent nano-TiO_2 particles. However, the organic paint binder trapped a considerable fraction of these particles.

The researchers [7] concluded that pure nano-TiO_2 demonstrated expected colloidal behavior. Its stability increased as the pH increased, and it agglomerated significantly above the isoelectric point, leading it to settle in the presence of Ca. Even when subjected to 3 mM Ca, the TiO_2 released showed only slight variations in particle size, potential, and colloidal stability. It was noted that the behavior of these newly released engineered nanomaterials (ENMs) may differ from that of the original materials. They proposed that future research should focus on knowing the particles that are truly released, as the toxicity difference between the original and released components could be large.

According to Xiao et al. [8], thermoset nanocomposites (TSNCs) may offer significant advantages over standard thermoset materials, making them an appealing alternative for widespread industrial use, particularly in coatings. They suggest that, while experimental design of TSNCs can be difficult, computational design can provide useful insights and bring out ideal design options. They presented a thorough computational methodology designed specifically for the design, analysis, and assessment of TSNC-based coatings. They emphasized the method's unique ability to relate material formulation, processing circumstances, and coating qualities quantitatively. This connection aids in a holistic analysis for the optimal TSNC coating design, with case studies demonstrating the efficacy of their methodology. Xiao and Huang [9] also investigated the design of sustainable multifunctional nanocoatings, stressing the problems posed by the material's inherent multiscale nature as well as the complexities required in design and manufacturing processes. They wanted to solve sustainability and design difficulties at the same time with their research; thus, they used multiscale systems theory and engineering sustainability concepts. Their main goal was to improve system performance by synchronizing the characterization and improvement of materials, products, and associated production processes at various scales.

Xiao et al. [8] used multiscale modeling and simulation methodologies that were tightly integrated, ranging from delicate molecular modeling to more typical continuum modeling. They were able to establish macroscopic system performance predictions anchored in fundamental molecular processes by integrating multiple approaches and theories from different sizes. They went on to investigate mathematical and methodological approaches for creating a long-term, goal-driven design for material-process-product systems. They said that their proposed methodology, which will be completely demonstrated in a comprehensive study on thermoset nanocomposite coatings, can considerably benefit researchers in the development of novel materials and the discovery of new knowledge. Furthermore, this methodology can provide scientific direction as well as uncover a plethora of novel solutions and prospects for sustainable production.

Nagaraj and Manjushree [10] discussed novel nanopaints that change color in response to voltage changes. They emphasized that when these paints are exposed to voltage changes, they exhibit a specific color spectrum change. While other materials can achieve a comparable effect, nanopaints outperform them in terms of cost efficiency, memory, real-time adaptation, and energy usage. They presented a clear sketch of the framework components and their interaction based on environmental inputs, which was validated by simulation results. They found that delivering voltage to a device designed to interact with these programmable paints, aided by an application programming interface (API), could enable this novel functionality to be realized.

1.3 Methods of formulation of nanopaints

The development and commercialization of nanopaints represents a huge step forward in coating technology, utilizing the unique features of nanoparticles to improve different aspects of conventional paints. This section describes the various approaches, as seen in Fig. 1.3, used in the creation of

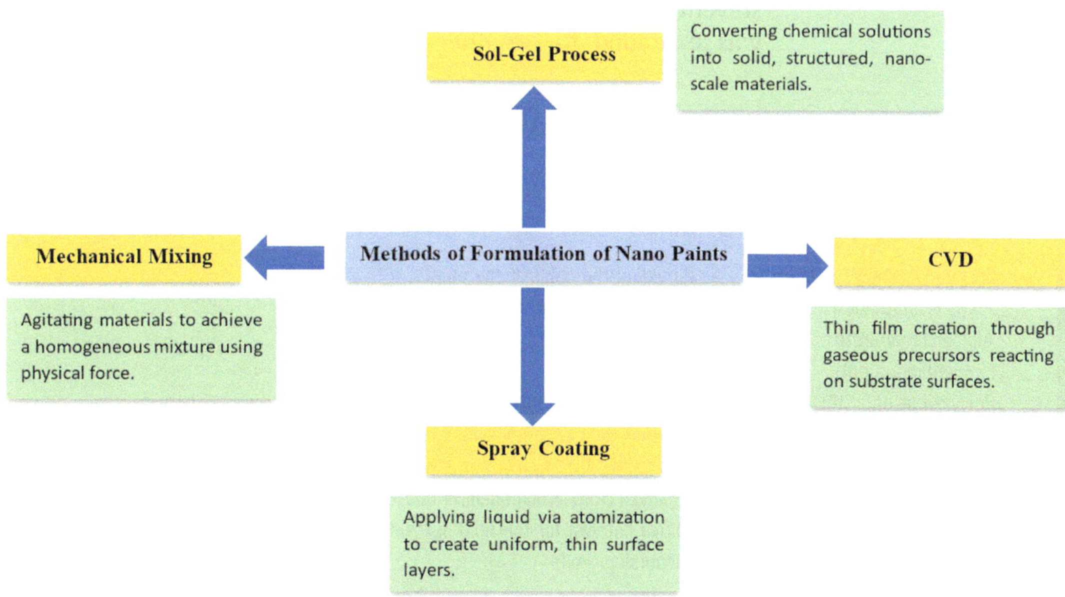

Figure 1.3 Methods of formulation for nanopaints.

nanopaints, each designed to exploit distinct nanomaterial properties for a variety of purposes.

1.3.1 Mechanical mixing

While straightforward, this process can be extended for more complex applications, as illustrated by Yong et al. [1], who published their findings in the Journal of Industrial Engineering Chemistry in 2015. Their study focused on the creation of a zinc oxide (ZnO) nanopaint for antifouling applications in the marine industry; they utilized a high-energy ball milling process, which falls within the larger topic of mechanical mixing.

In their investigation [1], ZnO nanoparticles were used as the pigment and alkyd resin as the binder. The formulation also contained cobalt naphthenate as an outer coat drier, nanosized zirconium oxide (ZrO_2) as an inner coat drier, polyvinyl alcohol, aluminum stearate as an antisettling agent, and soya lecithin as a wetting agent. These components were blended in particular amounts, as shown in Table 1.1 [1].

The mixture was milled for eight hours. This process was carried out utilizing a tungsten carbide bowl and balls, with a ball-to-powder ratio of 10:1 and a milling speed of 300 revolutions per minute (rpm). Following that, toluene was added as a solvent or thinner to the milled result, and the milling process was extended for another 30 minutes. This rigorous method

Table 1.1 Composition of ZnO nanopaint and the role of its additives.

Material used	Role	Weight percentage (%)
ZnO nanoparticles (pigment)	Providing the color of the paint.	20
Alkyd resin (binder)	Film-forming component of paint.	60
Aluminum stearate (antisettling agent)	Preventing pigment and binder settling.	0.5
Polyvinyl alcohol (thickener)	Improving the viscosity and preventing coagulation.	0.2
Soya lecithin (wetting agent)	Wetting the pigment in the binder for uniform dispersion.	0.6
Nanosized zirconia	Chemical cross-linking agent of unsaturated fatty acids.	0.6
Cobalt napthenate (upper coat drier)	Active catalyst for the lipid autoxidation process.	0.6
Toluene (thinner/solvent)	Dispersing agent.	17.5

Source: From H.E. Yong, K. Krishnamoorthy, K.T. Hyun, S.J. Kim, Preparation of ZnO nanopaint for marine antifouling applications, Journal of Industrial and Engineering Chemistry 29 (2015) 39–42. https://doi.org/10.1016/j.jiec.2015.04.020.

yielded a homogeneous, white ZnO paint, demonstrating the adaptability of mechanical mixing in generating nanopaints with specified qualities. The resulting paint was recognized for its capacity to be applied as a coating on a variety of surfaces.

In addition to the study by Yong et al. [1], the research conducted by Krishna Mohan et. al. [12] titled "Al_2O_3 and CuO nanoparticulate-based paints for marine applications" was studied to better understand the usage of mechanical mixing for nanopaint formulation. The high-energy ball milling strategy was used to formulate ZnO nanopaints [1], and the mechanical mixing method was confirmed for its efficacy in the processing of Al_2O_3 and CuO particles [12]. Initially in 100-micron particle form, these materials were reduced to nanosize, specifically 60–80 nm for Al_2O_3 and 65–80 nm for CuO [12], using the usual ball milling procedure. This process was purposefully chosen because of its favorable effect on the forging characteristics of the materials, hence improving the overall excellent properties of the resultant material matrix.

The relevant study by Krishna Mohan et. al. [12] showcased that ball milling is an important step in the creation of nanoparticles. The milling operation lasted 14 hours and used tungsten carbide balls at a constant speed of 300 rpm. The ball-to-powder ratio was rigorously maintained at 10:1 by volume to achieve effective contact and particle size reduction.

Subsequently, the resulting nano powders were integrated into the paint formulation, with 250 grams of nanopaint being prepared by stirring the desired quantities of nano powder into the mixture. The components comprising the paint formulation are linseed alkyd resin (30 g), aluminum stearate (2.5 g), thickener A (0.5 g), soya lecithin (3 g), TiO_2 powder (10 g), mineral turpentine oil (40 g). The mechanical mixing, in this context, underscores the versatility of the method in enhancing the mechanical and structural attributes of nanopaints through precise and controlled milling techniques.

These examples from Yong et al. and Krishna Mohan et al. demonstrate the potential for mechanical mixing, especially when paired with high-energy approaches, to make nanopaints with improved characteristics. It emphasizes the method's adaptability and efficacy in the field of nanopaint formulation, which contributes considerably to the improvement of coating technology.

1.3.2 Sol-gel process

The sol-gel process, a key approach in nanopaint formation, includes transitioning from a colloidal solution (sol) to a solid phase (gel). This approach is useful for achieving uniform dispersion of nanoparticles, such as silica or zirconia, within the paint matrix. The resulting nanopaints have superior properties such as increased chemical resistance, hardness, and scratch resistance, making them excellent for applications that require endurance and toughness.

Gąsiorek et al. [14] used the flexible sol-gel approach to manufacture silica and two unique silica-zirconia hybrid materials, using the reagents listed in Table 1.2 [14]. Organically modified SiO_2-based matrices were created using tetraethoxysilane (TEOS, 98%) and dimethyldiethoxysilane (dMdEOS, 98%) as oxide precursors. These precursors aided in the formation of a base SiO_2 sol, which acted as an interlayer for further alterations.

A thorough approach was used to modify the organically modified silica matrices with zirconia (referred to as SiO_2/ZrO_2). One hour after the reaction began, the base SiO_2 (A) sol was coupled with a ZrO_2 (B) sol, which was produced from zirconium(IV) butoxide (ZrOBu, 80% in *n*-butanol). The obtained sols were homogenized for 0.5 hours.

Furthermore, a distinct type of organically modified silica-zirconia hybrid material (SiO_2/GPTMS/ZrO_2) was created. This technique involved the integration of an organically modified SiO_2 (C) sol with ZrO_2 (B) sol an hour after the reaction began. The modification technique used (3-glycidoxypropyl) trimethoxysilane (GPTMS, 97%) and TEOS as precursors.

All compounds were synthesized under acidic conditions using hydrochloric acid (HCl, 31%–38%) and/or acetic acid (CH_3COOH, 99.5%), as stated in Table 1.2 [14]. To slow the hydrolysis and condensation processes, acetylacetone (AcAc, 99.6%) was used as a complexing agent to stabilize the zirconia sol in a zirconium terakis-acetylacetone configuration. The sol-

Table 1.2 Chemical compounds and volume ratio of substrate used to sol-gel synthesis.

Sol	Chemical compounds
SiO_2	TEOS + dMdEOS + EtOH + HCl (1:1:4:0.02)
SiO_2/ZrO_2	A. TEOS + dMdEOS + EtOH + HCl (1:1:4:0.02)
SiO_2/ZrO_2	B. ZrOBu + BuOH + AcAc + CH_3COOH (1:4:0.5:0.5)
SiO_2/GPTMS/ZrO_2	C. GPTMS + TEOS + EtOH + HCl (1:1:4:0.02)
SiO_2/GPTMS/ZrO_2	B. ZrOBu + BuOH + AcAc + CH_3COOH (1:4:0.5:0.5)

Source: *From J. Gąsiorek, A. Mazur-Nowacka, A. Szczurek, B. Babiarczuk, W.J. Tic, G. Tic, et al. Influence of zirconia and organic additives on mechanical and electrochemical properties of silica sol-gel coatings. Materials. 14 (2021) 2389.*

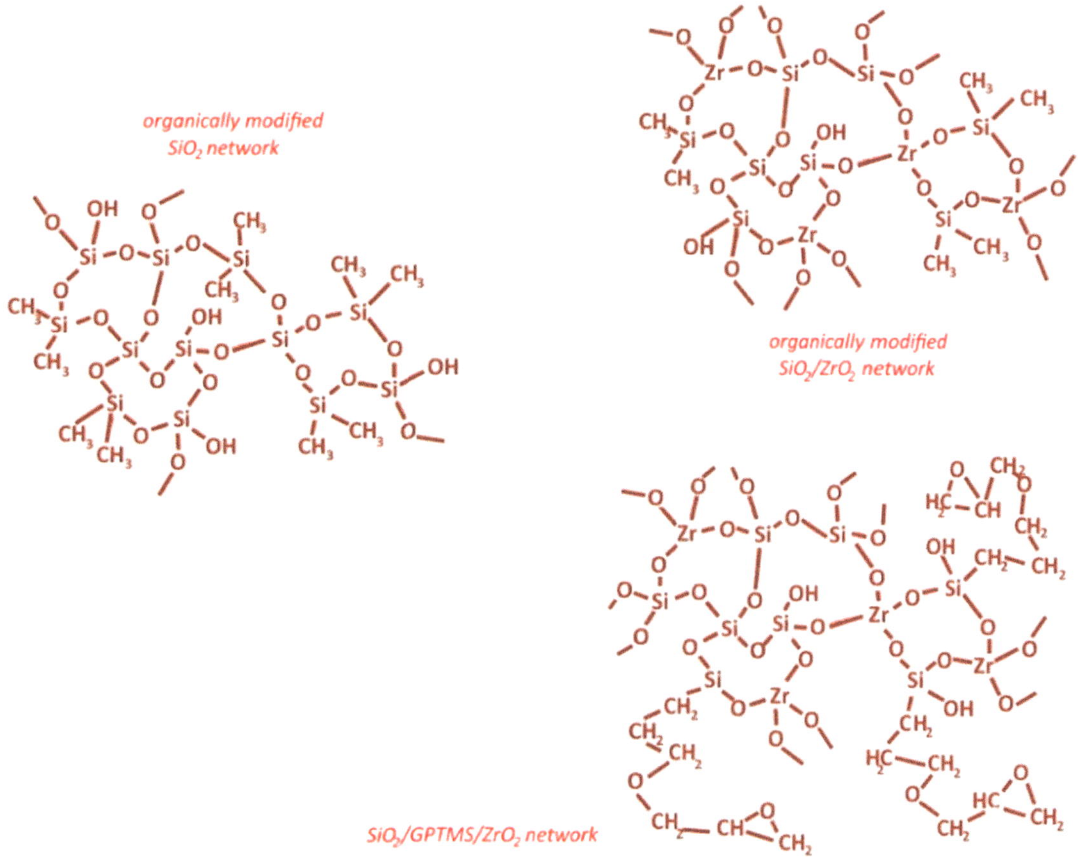

Figure 1.4 Block diagrams for the sol-gel oxide networks. From J. Gąsiorek, A. Mazur-Nowacka, A. Szczurek, B. Babiarczuk, W.J. Tic, G. Tic et al. Influence of zirconia and organic additives on mechanical and electrochemical properties of silica sol-gel coatings. Materials 14 (2021) 2389.

gel procedure used solvents including ethanol (EtOH, 96%, POCH) and butanol (BuOH, 99.9%, POCH) to achieve the needed material characteristics. The block diagrams of the resulting sol-gel oxide networks are shown below in Fig. 1.4 [14].

1.3.3 Chemical vapor deposition

The chemical vapor deposition (CVD) technique is a key step in the production of high-quality thin films, which are essential for sophisticated technological applications. A rigorous approach to precursor synthesis is essential, as demonstrated by the processes used in the research [16]. Gold colloids were produced in toluene utilizing the Brust two-phase chemical reduction process. For example, $HAuCl_4 \cdot 3H_2O$ was dissolved in deionized water and combined with tetraoctylammonium bromide in toluene. The reaction began with the addition of $NaBH_4$, which resulted in a dark red organic layer that was washed and dried over Na_2SO_4, yielding a gold solution with a concentration of 4.3 mmol of atoms per liter. Notably, to avoid contamination of films formed from these colloids, thiol capping groups were deleted, which resulted in colloidal deterioration over time.

Aerosol-assisted CVD, a version of conventional CVD, was used, with depositions occurring in a cold-wall horizontal-bed CVD reactor. To inhibit ion migration into the film, this arrangement used SiO_2 barrier glass, which was properly dimensioned. A carbon heating block with precise temperature control provided ideal deposition conditions. The precursor solution formed an aerosol, which was driven into the reactor by nitrogen gas via PTFE and glass tubing. The deposition procedure took 20–30 minutes, depending on the gas flow rate, and the films were cooled in situ using nitrogen gas flow.

The tungsten phenoxide synthesis was carried out as previously described, by reacting WCl_6 with phenol in toluene. Furthermore, titanium isopropoxide [$Ti(O^iPr)_4$] was used, emphasizing the need to use high-purity precursors for complicated material synthesis. The precision of chemical ratios and the purity of the reaction environment are important in these approaches for efficient thin film deposition. The methods presented here, from precursor synthesis to final deposition, demonstrate the stringent criteria and exact execution required in CVD techniques to generate films of high quality and desired characteristics.

1.3.4 Spray coating

Spray coating is a versatile and prevalent industrial process, wherein a coating material is propelled through a nozzle to form a layer on a substrate. This method accommodates a range of materials including liquid, molten, or softened particles. Its widespread application is evident in the coating of diverse materials, such as metals, glass, wood, textiles, and leather, notably in the application of organic lacquers and binders [18].

In the realm of nanocoatings, spray coating is adeptly employed for applying nanocomposite coatings. These coatings, which may be in solution, dispersion, or molten state, are directly sprayed onto the substrate. This application is typically succeeded by a phase of drying and/or cooling to establish the final coating layer. Thermal spraying is a specialized variant of this technique, where particles are partially melted and then rapidly propelled toward the substrate using a flame or an arc. This method, which encompasses the use of electrical or chemical heating, facilitates the application of metallic, ceramic, or polymeric materials [19,20].

A further advancement within this field is plasma spraying. Here, coating powders are introduced into a plasma jet, which can reach temperatures around 10,000K, and then applied in a molten state over the substrate. The characteristics of the resultant coating are influenced by several parameters including the nature of the coating powders, the composition and flow rate of the plasma gas, energy input, torch geometry, and the distance of the torch from the substrate [21].

While spray-coating techniques are effective for casting nanoparticle-polymer dispersions to form nanocomposite polymers on various substrates, they have limitations. Specifically, these techniques are not suited for nanoscale thin-film coatings, as they typically result in coatings with thicknesses at the micrometer level. Hence, while spray coating offers a broad scope of application in nanocoating formulation, its utility is contingent on the desired scale and specific characteristics of the final coating.

1.4 Applications of nanopaints

1.4.1 Utilization of nanopaints and nanocoatings in the automobile sector

Surface protection is crucial in the high-end automotive industry to protect against UV radiation, wear and tear, and heat; improve adhesion; and reduce engine friction. The automotive coating industry prioritizes several functional characteristics, including (1) self-cleaning with nano-TiO_2 and nano-SiO_2, (2) fingerprint resistance with silica nanoparticles, (3) UV protection provided by nanoparticles such as zinc oxide, cerium oxide, titanium oxide, and iron oxide, (4) chemical resilience with CNTs and graphene, (5) abrasion resistance with silica and aluminum oxide nanoparticles, and (6) scratches specific uses in the automotive area include: (1) scratch-resistant coatings, (2) self-cleaning coatings for glass surfaces, and (3) antibacterial coatings for interior trims and upholstery, (4) hydrophobic and oleophobic coatings to prevent fingerprints on automotive displays, (5) additives for lubrication, (6) coatings to prevent engine corrosion, (7) nanocoatings designed for engine wear resistance, (8) additional anticorrosion coatings for engines, (9) fingerprint-resistant coatings for mirrors and interior surfaces, and (10) coatings that prevent ice formation.

1.4.2 Nanopaints and nanocoatings in the energy domain

An important aspect of nanocoating's impact on photovoltaics is the improvement in light entrapment. This is achieved through the application of nanostructured antireflective layers on the front cover of solar cells, effectively reducing reflection losses. Moreover, these nanostructures facilitate the up conversion or down conversion of light wavelengths, thereby optimizing the use of the solar spectrum.

Furthermore, the application of nanotechnological process technologies, such as plasma-aided procedures and the nanoscale design of surface and layer structures, plays a crucial role in enhancing the efficiency and structure of solar cells. These advancements at the nano level are critical in optimizing the performance of various types of solar cells, showcasing the broad and impactful role of nanotechnology in the evolution of solar energy technologies.

The energy sector is undergoing a remarkable transformation, catalyzed by the integration of nanotechnology into various applications, including vehicular pollution control and filtration systems. The advent of nanoscale materials in catalytic converters has significantly advanced the reduction of toxic emissions from automobiles. The use of nanoparticles in these converters has shown to enhance the redox reactions necessary for converting exhaust gases into less harmful substances, thanks to their increased surface area and the intrinsic reactivity of nanoparticles.

The development of advanced air filtration technologies is another area where nanotechnology is making strides. The creation of filters using nanofibers illustrates a leap forward in air purification methods. These filters boast an expansive surface area and a high aspect ratio, which contributes to their enhanced ability to trap and retain particulate matter. Such filters not only improve air quality but do so without substantially impeding airflow, an essential consideration in maintaining energy efficiency.

Moreover, the use of nanocoatings in the automotive industry reflects a commitment to prolonging the lifespan and appearance of vehicles. These nanocoatings are designed to bolster the durability of surfaces, improving their resistance to scratches and environmental wear-and-tear. The incorporation of nanoparticles into these coatings has been shown to impact the mechanical properties significantly, enhancing the hardness and resilience of the coated surfaces.

Collectively, these innovations underscore the growing role of nanotechnology in promoting environmental sustainability and operational efficiency within the energy sector. The continuous refinement and application of nanomaterials promise not only to improve existing processes but also to pave the way for novel solutions to current environmental challenges.

1.4.3 Role of nanopaints in the aerospace industry

In the aerospace and aviation sectors, the application of nanocoatings has become pivotal in protecting aircraft structures from the rigors of operation within extreme environments. The rigorous demands for resilience to corrosion, abrasion, wear, and extreme temperatures in these sectors have catalyzed the advancement of nanocoatings designed to meet these challenges.

Nanocoatings are engineered to enhance the durability, performance, and reliability of various aerospace components. They serve as a barrier against erosion and wear, thus improving the surface quality of aircraft parts. Additionally, these coatings are essential for shielding against various forms of corrosion, such as pitting, exfoliation, and oxidation, which are prevalent in the harsh conditions encountered during flight.

Advancements in multifunctional nanocoatings have led to the development of coatings that not only protect against corrosion using environmentally benign materials but also possess the ability to detect mechanical damage to the aircraft's exterior. These smart coatings can initiate a chemical or physical response to damage, adhere optimally to substrates, and enhance fatigue resistance. Furthermore, they can change color as required, adding to their multifunctional capabilities.

Nanomaterials employed in aircraft engines are prized for their high strength-to-weight ratio and thermal stability, qualities that are paramount in the high-stress environment of aerospace applications. Aerospace entities, including companies and agencies, are increasingly exploring the use of nanocoatings to imbue aircraft frames, interiors, and engines with special properties. These include enhancements in hardness, fuel efficiency, self-cleaning capabilities, wear and corrosion resistance, thermal performance, and flame retardancy.

The benefits of nanocoatings extend beyond mere surface protection to encompass a reduction in carbon footprint, lower maintenance and cleaning costs, and mitigation of erosion and ice accumulation. In aviation, particularly within military applications, maintenance costs are a significant concern; the introduction of anticorrosion nanocoatings has the potential to reduce these costs substantially.

The industry has seen a rising trend in the use of multilayered nanocoatings that are resistant to thermal shock, extreme temperatures, and wear caused by corrosion and erosion. Such coatings have proven especially beneficial in turbo engine applications, significantly extending the service life of these critical components.

Moreover, nanoparticle coatings have been noted for their crack-healing properties, which contribute to improved performance at high temperatures, increased creep resistance, and enhanced material strength. The scope of nanocoating applications in aerospace is broad, ranging from silicon dioxide and carbon nanotubes to aluminum oxide and graphene, each material bringing unique properties and benefits suited to specific applications within anticorrosion, thermal barrier, flame retardancy, and antiicing domains.

1.4.4 Role of nanopaints in marine vessel protection

In the realm of maritime engineering, nanotechnology-based coatings have emerged as a significant innovation, offering multifaceted performance enhancements for marine vessels [22]. These nanocoatings are engineered to revitalize the esthetic and functional aspects of ships' exteriors, effectively reversing the effects of oxidation. The application of such coatings reinstates the vessel's original hue, imparts a high degree of gloss, and fortifies the surface against mechanical degradation and ultraviolet radiation.

Addressing the pervasive challenge of biofouling, which encompasses the undesirable accrual of biological entities on submerged surfaces, nanocoatings present an ecologically considerate approach. Traditional antifouling agents, characterized by their biocidal activity, are now subject to stringent environmental regulations to mitigate their ecological impact. Nanotechnology facilitates the development of coatings that minimize bio-adhesion through molecular design, thus preventing the attachment and growth of fouling organisms without the deleterious environmental effects associated with conventional biocide-laden paints.

The application of nonstick nanocoatings, specifically designed to resist the adhesion of marine life, has been adopted across naval and commercial maritime sectors. These coatings are applied to an array of structural materials, including steel, aluminum, and fiberglass hulls, thereby obviating the enduring fixation of fouling agents such as barnacles and aquatic flora.

The U.S. Navy's utilization of nanostructured cemented carbide coatings is a testament to the enhanced durability offered by these advanced materials. Furthermore, the incorporation of carbon nanotubes into marine coatings has yielded significant improvements in anticorrosive properties and mechanical integrity. Such coatings exhibit an enhanced modulus under stress, augmenting their resistance to impact and abrasion—a critical attribute in the preservation of maritime vessels.

Graphene, noted for its exceptional barrier properties, is currently under investigation for its potential in developing impermeable, rustproof coatings. This line of research, pursued by several entities, aims to exploit the unique characteristics of graphene to provide robust, long-term protection against corrosive maritime elements.

1.4.5 The role of nanopaints in biomedical applications

Nanocoatings, in the realm of biomedical applications, have garnered escalating interest for their ability to endow materials with enhanced and specialized properties. These nanoscale coatings, when applied to bone implants or scaffolds, serve multifaceted roles—protective or functional—contingent upon their specific application within the musculoskeletal system. Protective nanocoatings are akin to armor, providing a barrier against chemical degradation, wear from impact or abrasion, and the weakening effects of metal fatigue. In contrast, functional nanocoatings are akin to performance enhancers, primarily elevating the biological interaction, such as fostering osseointegration and thwarting infections, at the crucial juncture where material and bone converge. Current research ventures delve into both the protective and functional virtues of orthopedic nanocoatings, with an overarching goal to augment both the longevity and effectiveness of orthopedic devices.

AI disclosure

During the preparation of this work, the author(s) used ChatGPT in order to increase the document's readability. After using this tool/service, the author(s) reviewed and edited the content as needed and take(s) full responsibility for the content of the publication.

References

[1] H.E. Yong, K. Krishnamoorthy, K.T. Hyun, S.J. Kim, Preparation of ZnO nanopaint for marine antifouling applications, Journal of Industrial and Engineering Chemistry 29 (2015) 39–42. Available from: https://doi.org/10.1016/j.jiec.2015.04.020.

[2] Q. Feng Xu, J. Nong Wang, K.D. Sanderson, Organic-inorganic composite nanocoatings with superhydrophobicity, good transparency, and thermal stability, ACS Nano 4 (4) (2010) 2201–2209. Available from: https://doi.org/10.1021/nn901581j.

[3] P. Jevnikar, K. Krnel, A. Kocjan, N. Funduk, T. Kosmač, The effect of nano-structured alumina coating on resin-bond strength to zirconia ceramics, Dental Materials 26 (7) (2010) 688–696. Available from: https://doi.org/10.1016/j.dental.2010.03.013.

[4] D. Wang, G.P. Bierwagen, Sol–gel coatings on metals for corrosion protection, Progress in Organic Coatings 64 (4) (2009) 327–338. Available from: https://doi.org/10.1016/j.porgcoat.2008.08.010.

[5] S. Zhang, A. Kocjan, F. Lehmann, T. Kosmac, M. Kern, Influence of contamination on resin bond strength to nano-structured alumina-coated zirconia ceramic, European Journal of Oral Sciences 118 (4) (2010) 396–403. Available from: https://doi.org/10.1111/j.1600-0722.2010.00752.x.

[6] X. Zhang, M. Wang, S. Guo, Z. Zhang, H. Li, Effects of weathering and rainfall conditions on the release of SiO_2, Ag, and TiO_2 engineered nanoparticles from paints, Journal of Nanoparticle Research 19 (10) (2017) 338. Available from: https://doi.org/10.1007/s11051-017-4022-4.

[7] A. Al-Kattan, A. Wichser, S. Zuin, et al., Behavior of TiO_2 released from nano-TiO_2-containing paint and comparison to pristine nano-TiO_2, Environmental Science and Technology 48 (12) (2014) 6710–6718. Available from: https://doi.org/10.1021/es5006219.

[8] J. Xiao, Y. Huang, C. Manke, Computational design of thermoset nanocomposite coatings: methodological study on coating development and testing, Chemical Engineering Science 65 (2) (2010) 753–771. Available from: https://doi.org/10.1016/j.ces.2009.09.028.

[9] J. Xiao, Y. Huang, Design of sustainable multifunctional nanocoatings: a goal-driven multiscale systems approach, Chinese Journal of Chemical Engineering 19 (4) (2011) 666–673. Available from: https://doi.org/10.1016/S1004-9541(11)60039-6.

[10] S. Nagaraju, B.M. Manjushree, DynaSurface: a framework for device notifications using nano coatings in connected devices. Proceedings of the 9th IEEE International Conference on Collaborative Computing: Networking, Applications and Worksharing, COLLABORATECOM 2013. December 2013, pp. 505–514. Available from: https://doi.org/10.4108/icst.collaboratecom.2013.254080.

[11] Z. Ranjbar, S. Rastegar, Nano mechanical properties of an automotive clear-coats containing nano silica particles with different surface chemistries, Progress in Organic Coatings 72 (1-2) (2011) 40–43. Available from: https://doi.org/10.1016/j.porgcoat.2010.11.001.

[12] M.V.K. Mohan, Al_2O_3 and CuO nano particulate-based paints for marine applications, Engineering Research Express 4 (3) (2022) 035056.

[13] A. Ealias, M.P. Saravanakumar, A review on the classification, characterization, synthesis of nanoparticles and their application, IOP Conference Series: Materials Science and Engineering 263 (2017) 032019.

[14] J. Ga̧siorek, A. Mazur-Nowacka, A. Szczurek, et al., Influence of zirconia and organic additives on mechanical and electrochemical properties of silica sol-gel coatings, Materials 14 (9) (2021) 2389.
[15] M. Yu, M. Liang, J. Liu, S. Li, B. Xue, H. Zhao, Effect of chelating agent acetylacetone on corrosion protection properties of silane-zirconium sol-gel coatings, Applied Surface Science 363 (2016) 229–239. Available from: https://doi.org/10.1016/j.apsusc.2015.12.013.
[16] R.G. Palgrave, I.P. Parkin, Aerosol assisted chemical vapor deposition using nanoparticle precursors: a route to nanocomposite thin films, Journal of the American Chemical Society 128 (5) (2006) 1587–1597.
[17] What are oleophobic thin film coatings? https://www.azom.com/article.aspx?ArticleID = 22157, (accessed 20.01.24).
[18] S. Kuroda, J. Kawakita, M. Watanabe, H. Katanoda, Warm spraying—a novel coating process based on high-velocity impact of solid particles, Science and Technology of Advanced Materials 9 (3) (2008) 033002.
[19] S. Paulussen, R. Rego, O. Goossens, D. Vangeneugden, K. Rose, Plasma polymerization of hybrid organic-inorganic monomers in an atmospheric pressure dielectric barrier discharge, Surface and Coatings Technology 200 (1-4) (2005) 672–675. Available from: https://doi.org/10.1016/j.surfcoat.2005.02.134.
[20] F. Leroux, C. Campagne, A. Perwuelz, L. Gengembre, Fluorocarbon nano-coating of polyester fabrics by atmospheric air plasma with aerosol, Applied Surface Science 254 (13) (2008) 3902–3908. Available from: https://doi.org/10.1016/j.apsusc.2007.12.037.
[21] A.S.H. Makhlouf, Current and advanced coating technologies for industrial applications, Nanocoatings and Ultra-Thin Films: Technologies and Applications, Elsevier Ltd., Germany, 2011, pp. 3–23. Available from: https://doi.org/10.1016/B978-1-84569-812-6.50001-4.
[22] Nanocoatings in the marine industry. https://www.nanotechmag.com/issue-30-article-nanocoatings-in-the-marine-industry/, (accessed January 22.01.24).

2

Broad spectrum of nanomaterials and their properties into paints

Vijay Chaudhary, Tushar Sharma and Sony Setiya
Department of Mechanical Engineering, Amity School of Engineering and Technology Amity University, Noida, Uttar Pradesh, India

2.1 Introduction

Nanotechnology is the science of developing materials by controlling the individual atoms and molecules to create devices that are thousands of times smaller than the current technology. They are the structures between 1 and 100 nm in size [1]. These materials can impart remarkable benefits such as improved durability, UV resistance, antimicrobial properties, self-cleaning capabilities, and enhanced mechanical strength to paints. Nanotechnology may improve the primary properties of traditional materials and add functionalities to existing materials (e.g., paints/coatings or glass may gain self-cleaning, antimicrobial, and pollution-reducing properties) [2]. Nanotechnology offers ideas and solutions for many fields of activity, such as medicine, industry, information, communication or agriculture. Nanotechnology represents a direct manipulation of materials at the nanoscale, or an ambitious set of procedures that offers almost complete control of the deep structure of matter [3]. The use of nanoparticles in coatings has been demonstrated to enhance coating performances as well as to bring new functions to the system allowing the development of multipurpose coatings. A coating is described as "nanocoating" if it contains a nanocomponent.

Nanofillers help to develop synergy as well as other properties of material such as mechanical, electrical, and thermal [4]. Generally, the size of a single particle of nanofiller is defined as below 100 mm, but it may also differ sometimes [5]. Natural biopolymer nanofibrils, which are composed of various biopolymer molecules—that is, cellulose, collagen, and chitin—are

Nanopaints. DOI: https://doi.org/10.1016/B978-0-443-26512-9.00002-1

often used due to their biocompatibility, biodegradability, durability, availability, and unique mechanical properties [5]. The cellulose-based organic nanofillers were introduced with the polymers for the fabrication of transparent composite film. The main characteristics of nanocellulose organic fillers imposed on the composite films are lower cytotoxicity, good thermal and mechanical stability, blood compatibility, and a good antibacterial effect. The selection of inorganic fillers is primarily oriented on different characteristics such as structural, electrical, or electronic properties. Both metal and metal oxide particles are reinforced with matrix and they enhance the overall performance of the resulting biofiber composite films [6]. Metal oxide nanoparticles such as zinc oxide (ZnO NPs), titanium oxide (TiO_2 NPs), silica (SiO_2 NPs), aluminum oxide (Al_2O_3 NPs), cerium oxide (CeO_2 NPs), iron oxide (Fe_2O_3 NPs), and copper oxide (CuO NPs) have been added to biopolymer films as active nanofillers [5]. Nanofillers can significantly improve or adjust the different properties of the materials into which they are incorporated, such as optical, electrical, mechanical, thermal properties, or fire-retardant properties, sometimes in synergy with conventional fillers [7].

2.2 Importance of nanomaterial in paints

Nanomaterials have evolved the paint industry by significantly enhancing the performance and durability of coatings. In the last decade the development of nanostructured coatings has generated great interest. Different types of coatings are being used in the industry. From them, paints are widely employed due to its application process is relatively simple, robust and inexpensive regarding other coating techniques. Currently, this field is focused on finding new formulations with improved properties over environmental and microbial degradation, mechanical strength, hydrophobicity, etc., considering the use of nanostructured materials [8]. Various common nanomaterials (NMs), including TiO_2, SiO_2, Ag, ZnO, graphene, carbon nanotubes (CNTs), and some others, are used in paints and coatings [9]. Nanomaterials are expected to bring breakthrough development to the leading edge of nanotechnology [10]. Incorporating nanomaterial into conventional paint materials showed a significant improvement in the paint characteristics such as biocompatibility, mechanical strength, thermal stability, corrosion protection, long-term stability, and antibacterial behavior [11]. Furthermore, nanomaterials enhance water and stain resistance, making painted surfaces more durable and easier to maintain (Table 2.1).

2.3 Nanomaterial and nanotechnology

Materials are defined as nanomaterials if their size or one of their dimensions is in the range of 1–100 nm [32]. At this scale, materials often exhibit unique physical and chemical properties that differ significantly from their bulk counterparts. The recent rapid development of nanoscience is the result of a new-found ability to observe and control structure at small length and time scales, coupled with the development of computational capabilities that are most effective at small scales [33]. Nanotechnology applications will give rise to cleaner energy production, lighter and more durable materials, inexpensive clean water production [34]. These properties include increased strength, lighter weight, enhanced chemical reactivity, and improved electrical conductivity.

Nanomaterials like titanium dioxide and zinc oxide are established photocatalysts that are well-known in a variety of functions. They can be applied in paint and coating as pigment while providing functionalities like antimicrobial [35]. The nanopaint has mechanical, thermal, electrical, chemical, and physical properties. Nanopaint acts as a coolant, is anticorrosive, is self-healing, and has less viscosity in nature with a high Young's modulus [36]. Nanoparticles like titanium dioxide (TiO_2) and zinc oxide (ZnO) are used in paints for their excellent UV-blocking capabilities, which help prevent fading and degradation of painted surfaces under sunlight. Additionally, some nanomaterials impart self-cleaning properties to paints, where photocatalytic nanoparticles help break down organic contaminants, making it easier for rain to wash them away. Nanomaterials also enhance the antimicrobial properties of paints, with nanoparticles like silver (Ag) and copper oxide (CuO) inhibiting the growth of bacteria and mold, thus maintaining cleaner surfaces. Moreover, nanomaterials can improve the water and stain resistance of paints, making them more durable and easier to maintain. In industrial applications, they contribute to enhanced corrosion resistance, protecting metal surfaces from rust and degradation. Overall, the incorporation of nanomaterials in paints leads to high-performance coatings that meet the demands of modern applications, offering both functional and esthetic benefits. The use of nanoparticles in coatings has been demonstrated to enhance coating

Table 2.1 Summary of different fibers and nanofillers for polymeric composite materials.

S. No.	Fiber	Nanofiller	Remark	References
1.	Ramie	Carbon nanotubes	Improvement in bending force and flexural module. The integration of CNT's increased the fibers and matrix load transfer.	[12]
2.	Kenaf	Silica nanoparticles	Improvement in tensile and flexural strength	[13]
3.	*Coccinea indica*	Nanoclay	Rise in the weight of nanoclay increased the bending, tensile, impact and compression properties of composite.	[14]
4.	Flax/*Sisal*	SiC nanopowder	Distribution of SiC nanopowder into the fiber has been shown to increase the tensile and impact strength by 50%.	[15]
5.	Cane grass	Montmorillonite	By the addition of MMT the mechanical properties such as young's modulus and flexural strength increased significantly.	[16]
6.	Bamboo	Nano-SiO_2	With increase in nano-SiO_2 content the tensile and flexural strength increases.	[17]
7.	Jute	SiO_2	The tensile strength, young's modulus and flexural strength of jute increases by the addition of SiO_2.	[18]
8.	Sugar palm	Silica aerogel nanoparticles	Enhancement of tensile strength and modulus of elasticity by the addition of silica aerogel nanoparticles.	[19]
9.	Bagasse	Multi walled carbon nanotubes	Addition of multi walled carbon nanotubes yielded positive effect on tensile and flexural strength improves significantly.	[20]
10.	Wood	Montmorillonite nanoclay	Modulus of elasticity increased with the addition of MMT nanoclay and it also improves the interfacial adhesion between the fiber and matrix.	[21]
11.	Napier	Nanoclay	It has a significant impact on the flexural properties of the composite.	[22]
12.	Clay	Epoxy	A well dispersed silica in silica/epoxy composite consistently enhanced the tensile and compressive strength.	[23]
13.	Hemp	Cloisite 30B	Decrease in ultimate tensile strength by 20% and moisture absorption reduced by 8%.	[24]
14.	Bamboo	Mont morillonite clay filler	Addition of clay led to negative effect on mechanical properties.	[25]
15.	*Sisal*	Nanoclay (Garamite)	Substantial increase in mechanical properties and decrease in water absorption.	[26]
16.	Bagasse	Nano-SiO_2	5% Nano-SiO_2 added samples show 71.46% increase in tensile strength.	[27]
17.	Wood	Nano-SiO_2	Most of the properties have been improved.	[28]

(Continued)

Table 2.1 (Continued)

S. No.	Fiber	Nanofiller	Remark	References
18.	Sugar cane	TiO_2	Increase in tensile strength after addition of TiO_2.	[29]
19.	Kenaf	Montmorillonite	By addition of oil palm nanofiller 24.9% increase in tensile strength and 28.3% increase in impact strength.	[30]
20.	Jute	Graphene	Graphene as nanofiller enhanced the machinability of composite material.	[31]

performances as well as to bring new functions to the system, allowing the development of multipurpose coatings [37].

Nanotechnology, on the other hand, is the field of science and engineering dedicated to studying, manipulating, and utilizing materials at the nanoscale. It involves the design, production, and application of nanomaterials and devices. The meaning of the term nanotechnology is any technology operating on the nanoscale that has applications in the real world, that is, to employ single atoms and molecules to form functional structures [38]. Nanotechnology is connected with systems and materials, the components and structures of which represent novel, significantly improved chemical, physical, and biological properties, processes, and phenomena because of their nanoscale size [39]. Nanotechnology's rapid development worldwide is a testimony to the transformative power of identifying a concept or trend and laying out a vision at the synergistic confluence of diverse scientific research areas [40]. The field also plays a crucial role in environmental sustainability, enabling the development of more efficient catalysts for chemical reactions and materials for energy-efficient solar cells.

2.4 Types of nanoparticles

Nanoparticle manufacturing is an essential component of nanotechnology because the specific properties are realized at the nanoparticle, nanocrystal, or nanolayer level, and the assembling of precursor particles and related structures is the most generic route to generate nanostructured materials [41]. A nanoparticle can be either a zero dimensional where the length, breadth and height is fixed at a single point for example nanodots, one dimensional where it can possess only one parameter for example graphene, two dimensional where it has length and breadth for example carbon nanotubes or three dimensional where it has all the parameters such as length, breadth and height for example gold nanoparticles. The nanoparticles are of different shapes, sizes, and structures [42].

2.5 Organic nanoparticles

The development of nanotechnology using organic materials is one of the most intellectually and commercially exciting stories of our times [43]. Nature provides a wide range of examples of organic NPs, such as protein aggregates, lipid bodies, milk emulsions, or more complex organized structures such as viruses, to name a few. One of the most important features of organic NPs is that they offer relatively simple routes for the encapsulation of materials. This, together with the fact that the molecules used for the fabrication of the organic NPs can be biodegradable [44]. On the nanoscale, organic compounds are converted into organic nanomaterials. Some examples of organic nanoparticles or polymers are liposomes, dendrimers, micelles, and ferritin. These types of nanoparticles have hollow interiors, and they are also sensitive from light and heat. Dendrimers are branched, tree-like molecules with functional groups that can be used for drug delivery, sensing, or catalysis. By integrating organic nanoparticles, nanopaints become more resistant to mechanical damage, wear, and abrasion, while also offering self-cleaning properties through hydrophobic surfaces that repel water and dirt. Additionally, these nanoparticles provide protection against UV radiation and chemical degradation, enhancing the longevity of the paint in harsh environments. In industrial applications, nanopaints can also offer anticorrosion and antimicrobial benefits, making them ideal for protecting metal surfaces. Furthermore, organic nanoparticles improve pigment dispersion, resulting in more vibrant colors and better color stability over time. This combination of features makes organic nanoparticle-based paints highly effective for a wide range of applications, from construction to automotive industries.

2.6 Inorganic nanoparticles

Inorganic nanoparticles are particles that are not made up of carbon. Metal and metal oxide-based nanoparticles are generally categorized as inorganic nanoparticles [45]. Magnetic nanoparticles (mNPs) are one of the most significant inorganic nanomaterials. A magnetic core (e.g., maghemite (g-Fe_2O_3) or magnetite (Fe_3O_4)) is generally present. Other metals, such as nickel and cobalt, are also employed, although their applications are limited due to their toxicity and oxidation vulnerability [46]. In nanopaints, commonly used inorganic nanoparticles include titanium dioxide (TiO_2), zinc oxide (ZnO), silica (SiO_2), aluminum oxide (Al_2O_3), silver (Ag), and cerium oxide (CeO_2). TiO_2 is often used for its strong UV-blocking and photocatalytic abilities, protecting surfaces from UV damage while improving brightness and opacity. ZnO nanoparticles add antimicrobial and UV-filtering properties, contributing to self-cleaning effects and preventing mold. Silica improves hardness and scratch resistance, frequently resulting in hydrophobic or self-cleaning surfaces. Aluminum oxide enhances thermal stability and mechanical strength, making the coatings more resistant to high temperatures and wear. Silver nanoparticles provide antimicrobial protection, which is especially useful in environments like healthcare and the food industry, where they help prevent bacterial and fungal growth. Cerium oxide adds UV protection and anticorrosive benefits, valuable in sectors such as automotive and aerospace. Overall, the integration of inorganic nanoparticles significantly enhances the durability and resistance of nanopaints for diverse industrial and consumer applications.

2.7 Properties of nanomaterials used in nanopaints

The use of nanotechnology in paints, known as nanopaints, has provided these coatings with enhanced properties such as improved mechanical strength, durability, corrosion resistance, and the ability to respond to environmental changes. These properties enable the creation of coatings that are stronger, more resistant to wear, and have better functional properties than traditional paints. The resulting nanopaint and coating possess excellent qualities, such as increased scratch-resistant, increased hardness, mold and bacteria resistant, hence yielding a coating product that is highly durable with high performance characteristics [47]. This section explores the various properties that nanomaterials impart to nanopaints, focusing on the following key categories: mechanical properties, wear resistance, thermal properties, magnetic properties, and optical properties. Nanomaterials can be produced with outstanding magnetic, electrical, optical, mechanical, and catalytic properties that are substantially different from their bulk counterparts [48].

2.8 Mechanical properties

Mechanical properties of nanomaterials refer to the mechanical characteristics of nanomaterials under different environments and various external loads [49]. Nanomaterials have excellent mechanical properties due to the volume, surface and quantum effects of nanoparticles [50]. As nanoparticles are added to a common material, these particles will refine the grain to a certain extent, forming an intragranular structure or an intergranular structure, thereby improving the grain boundary and promoting the mechanical properties of materials [51,52]. When integrated into paint formulations, they enhance characteristics such as tensile strength, toughness, and elasticity, making the coatings more resistant to external forces and deformation. Nanoparticles, such as titanium dioxide (TiO_2), silica (SiO_2), and carbon nanotubes (CNTs), act as reinforcement agents in paint matrices. Their nanoscale dimensions allow them to fill the microscopic voids in the polymer matrix, leading to a denser and more compact structure. This densification significantly improves the tensile strength and hardness of the coatings, making them more durable and able to withstand mechanical stress. In addition to strength, nanopaints exhibit improved flexibility and elasticity due to the incorporation of nanomaterials. For instance, the addition of polymer-based nanoparticles can enhance the ability of coatings to flex and expand without cracking. This is especially important for surfaces that experience frequent thermal cycling, such as outdoor structures exposed to varying temperatures. Nanomaterials also provide resistance to crack propagation. When a coating develops micro-cracks due to external stress, nanoparticles at the crack tips can slow down or prevent further growth of the cracks by dispersing the stress, thus prolonging the life of the coating.

2.9 Wear resistance

Nanopaints are known for their superior wear resistance, which refers to the ability of a material to withstand mechanical action such as abrasion, erosion, or friction. This property is particularly important in industries where coatings are subjected to constant physical contact or movement. Nanoparticles, such as alumina (Al_2O_3) and silica (SiO_2), greatly enhance the abrasion resistance of nanopaints. These hard inorganic materials are capable of withstanding significant surface wear, making the coatings ideal for high-traffic areas or surfaces that come into regular contact with abrasive elements. The nanoparticles form a protective barrier that prevents the paint from being worn down by friction or impact. The erosion resistance of nanopaints is also improved through the inclusion of nanomaterials. Coatings used on surfaces exposed to harsh environments—such as wind, rain, or particulate matter—benefit from the increased toughness and protective capabilities provided by nanomaterials. Erosion-resistant nanopaints are commonly used in sectors like aerospace and automotive, where surfaces face constant exposure to environmental stressors.

2.10 Thermal properties

The thermal properties of nanopaints, including thermal conductivity and resistance to temperature fluctuations, are critical in applications where temperature control or stability is required. Nanoparticles enable the creation of paints that either resist heat transfer or dissipate heat efficiently, depending on the intended use. Nanoparticles such as alumina and silica can increase the heat resistance of nanopaints. By acting as thermal barriers, these nanoparticles reduce the thermal conductivity of the coatings, making them suitable for applications where heat insulation is required. For example, nanopaints with high heat resistance are used in industries such as automotive manufacturing, where surfaces are exposed to extreme temperatures during operations like welding or engine performance. In other cases, nanomaterials such as graphene or carbon nanotubes are added to paints to improve thermal conductivity. In terms of thermal deformation reduction, nanocoatings were considered as an effective technical approach [53]. These materials have excellent heat dissipation properties, allowing the coated surface to cool more effectively. This is particularly useful for electronic devices or mechanical systems that generate significant amounts of heat and require efficient heat management. According to the nature of the nanomaterial, extraordinary thermal and electrical conductivity can be exhibited at the nanoscale level compared to bulk counterparts. One example of this is graphene attained from graphite [54]. Nanopaints also exhibit improved resistance to thermal expansion. Coatings that experience large temperature fluctuations can expand and contract, which may lead to cracking or delamination. Nanomaterials help to stabilize these expansions, maintaining the integrity of the paint layer even under extreme temperature changes.

2.11 Magnetic properties

Nanomaterials in paints can also be engineered to possess magnetic properties, allowing for specialized applications where magnetic responsiveness is required. These magnetic nanopaints are useful in areas such as data storage, sensing technologies, and medical applications. The incorporation of magnetic nanoparticles, such as iron oxide (Fe_3O_4), into nanopaints enables the coatings to respond to magnetic fields. These coatings can be used in applications that require controlled magnetic behavior, such as in electromagnetic interference (EMI) shielding, which protects electronic devices from external electromagnetic fields. Magnetic nanomaterials also have potential uses in sensor technologies. For example, nanopaints containing magnetic particles can be applied to surfaces for detecting the changes in magnetic fields. The magnetic behavior of elements can change at the nanoscale. A nonmagnetic element can become magnetic at the nanoscale level [55]. This property is particularly useful in industrial settings for nondestructive testing of structural integrity or for security purposes. The production of nanoparticles has been fast-tracked in the past decade and has led to the development of various magnetic nanoformulations such as liposomes, metallic/nonmetallic, and polymeric nanoparticles [56]. Magnetic nanopaints can also be applied in biomedical fields, where magnetic nanoparticles are used for targeted drug delivery or imaging techniques. Although this is a specialized area of research, the potential for magnetic nanomaterials in functional coatings is vast.

2.12 Optical properties

Nanomaterials in paints can manipulate the optical properties of coatings, allowing for enhanced color, gloss, reflectivity, and UV resistance. These characteristics are especially important in industries like automotive, architecture, and consumer electronics, where the appearance and light management of surfaces are critical. Noble metal NPs have size-dependent optical properties and exhibit a strong UV–visible extinction band that is not present in the spectrum of the bulk metal [57]. Nanoparticles such as titanium dioxide (TiO_2) and zinc oxide (ZnO) can improve the brightness and opacity of paints by enhancing light scattering. This results in more vibrant colors and increased gloss, which is particularly desirable for decorative coatings or automotive finishes. TiO_2 and ZnO nanoparticles also provide excellent UV-blocking properties, protecting surfaces from UV-induced degradation. These nanoparticles absorb and scatter harmful UV radiation, preventing damage to the underlying material and extending the life of the paint. UV-resistant nanopaints are widely used in outdoor applications to protect surfaces from fading, chalking, or cracking. Certain nanomaterials, such as silica and titanium dioxide, can create self-cleaning coatings with hydrophobic or super hydrophilic properties. Optical layers for reflection reduction on dashboards or hydrophobic and dirt-repellent "easy-to-clean" surfaces on car mirrors are further examples of applications of nanotechnology in automobiles [58]. By providing additional functionalities, such as the stimuli-responsive colloidal stability, which in turn can influence the optical properties of gold nanoparticles under certain environmental conditions [59]. These coatings can repel water, dirt, and other contaminants, keeping surfaces clean and clear. Additionally, antireflective nanopaints are useful in optical devices, such as lenses or screens, where the control of light reflection is essential for performance.

Nanomaterials have fundamentally transformed the properties of paints, creating nanopaints that are far superior to conventional coatings. By enhancing mechanical strength, wear resistance, thermal stability, magnetic responsiveness, optical behavior, and catalytic activity, nanomaterials offer multifunctional coatings that address a wide range of industrial and consumer needs. These advanced properties make nanopaints suitable for applications in automotive, aerospace, architecture, healthcare, and electronics, among others.

2.13 Utilization of nanoparticles for nanopaints

Paints are a mixture of pigments and binders generally available in a liquid state. Paints form a solid film when applied on a surface. This film protects the surface from many dangers like corrosion, weathering, chemical attacks, etc. Timber or metal structures can extend their life by coating them with paint. They also provide an esthetic appearance to the surfaces. Painting is done to protect the surface from the effects of weathering, to prevent the wood from decaying and metal from corroding, to provide a decorative finish, and to obtain a clean, hygienic and healthy living atmosphere. It is not surprising that those coatings can suffer from premature failure and/or exhibit defects that may or may not result in failure. Most of the examples of failure and defects are abrasion removal of a portion of the surface of the coating, adhesion failure, surface contamination or condensation, bloom paint film exposed to condensation or moisture during curing, bridging failure to brush paint into corners and over welds, and bubbles or bubbling trapped air/solvent within the coating that is not released before the surface dries.

The different effects described can be used to overcome several drawbacks in paints that are currently available on the market. Some of the properties of paints are enhanced by adding nanoparticles into them.

1. Antimicrobial coating: Fighting bacteria and other microbes is commonly achieved by adding certain chemical agents. Two inorganic approaches can be used for sanitizing surfaces. The first is based on the photocatalytic activity of titanium dioxide. Photocatalytic surfaces have an antibacterial side effect due to their ability to break down organic substances in dirt, with the help of silver nanoparticles [60]. The second exploits the toxicity of certain metallic cations such as silver. Silver has long been known for its excellent antimicrobial effect due to the release of silver ions, which are taken up by microbes and exert a toxic effect [61].
2. Antifouling surface: By adding nanoparticles in paints, they could improve their performance, for example, by making them lighter and giving them different properties. Thinner paint coatings, used, for example, on aircraft, would reduce their weight, which could be beneficial to the environment. It may also be possible to substantially reduce the solvent content of paints, with resulting environmental benefits. Antifouling surface treatment is also valuable in process applications such as heat exchange, where it could lead to energy savings.
3. Antifog: When a cold surface is introduced into a warmer environment, fogging occurs due to the formation of tiny droplets on the surface, which scatter light and distort reflections. This phenomenon is unavoidable unless the surface is heated. However, a superhydrophilic coating can help minimize droplet formation. Instead of forming individual droplets, the water merges into a thin, continuous layer, reducing its impact on the surface's reflectivity. One effective solution is a photocatalytic TiO_2 coating, which becomes super hydrophilic under sufficient exposure to UV light, helping to maintain clarity on surfaces like mirrors [62].
4. Corrosion prevention: In automotive manufacturing, steel components are often heat-treated to attain the required

shape. During this process, which can reach temperatures as high as 1000°C, the steel is susceptible to corrosion. To safeguard the steel from corrosion at these elevated temperatures, a coating composed of nanoparticles can be applied, offering effective protection [63].

5. UV protection: Coatings are essential in managing the impact of electromagnetic radiation, including UV, visible light, and infrared, on different surfaces. A variety of organic and inorganic pigments are chosen for this purpose. Titanium dioxide (TiO_2), with a refractive index of 2.5–2.7, is the most efficient white pigment. Its optimal hiding power is achieved at a particle size of approximately 250 nm.
6. Strength enhancement: Adding nanoscale components into composites will enhance their tensile and impact strength. Carbon nanotubes offer the highest tensile strength ever observed. In this sense, carbon nanotubes are expected to be of great importance in the future.

The addition of nanosilicon dioxide to paints can improve the macro and micro hardness and abrasion. Adding nanosilicon dioxide to polymeric resins creates paints with excellent abrasion benefits when using a nanopaint, depending much on its actual composition and functionality, and can therefore not be generalized. Equally, the production of nanomaterials and the nanopaints has a strong impact on the environment. Due to the high surface area of the particles, even a small amount of addition can bring a big change in the performance of the coating. Perfect distribution of nanoparticles in the polymer matrix reflects the best possible outcome. This problem can be overcome by using dispersing agents and methods like ultrasonication or functionalization with silane and fluoro-compounds. Among all the pigments, nanoinorganic materials are considered to be the most advanced to improve mechanical and performance properties. Nano-(Tio2), ZnO, (silica) SiO_2, (alumina) Al_2O_3, carbon black, etc. Each kind can perform more than one function. For example, nano-ZnO can be used as a UV blocker, but it can also improve the scratch and abrasion resistance.

2.14 Application and performance of nanopaints in different industries

Research on the development of conductive nanopaints has been increasing over the years. Specifically, in the automotive field, nanopaints are extremely important because this type of coating can enhance the structural, thermal, and electrical properties of the substrate and contribute to imitating the metallic-like appearance when replacing the chrome-plated paint [64]. Nanoparticle additives provide a better corrosion rate as compared to the original basecoat. The optimum anticorrosion behavior for both TiO_2 and SiO_2 nanopaints were achieved at weight percentages of 1.5 and 1.0 wt.%, respectively. Therefore, the nanopaint has the potential to provide better corrosion performance for automotive surface applications [65]. The application of nanotechnology in the field of coating can facelift the coating industry. Properties like corrosion resistance, flame resistance, UV stability, gloss retention, and chemical and mechanical properties are improved significantly using nanoparticles [66].

The modification with a low concentration of TiO_2 and ZnO nanoparticles offers an enhancement in the physicochemical properties of paints, allowing them to be applied in the construction field as a promising nanocomposite [67]. A range of nanomaterials are also being used to add new functionalities, such as self-cleaning properties, to traditional construction industry products, for example, paint and cement [68]. Paints are a class of materials used for processing and protecting surfaces. They have numerous applications in multiple areas, in marine industries, and these are especially used as corrosion and fouling-protecting coatings. The use of Fe_2O_3 and ZnO as a pigment in the formulation of nanopaints gives both the antifouling and anticorrosion properties, additionally giving other mechanical properties, hydrophobicity, surface morphology, and roughness, among others [69]. Automotive clear coats are subjected to different kinds of environmental damage, for instance, UV radiation, acid rain, tree sap, bird droppings, stone chipping, and marring. Marring is mainly due to car wash brushes, decreases the gloss of automotive paint, and alters the color [70]. TiO_2 nanoparticles when mixed with basecoat paint lead to better surface appearance and structure, such as increased gloss unit value, reduced surface roughness, and improved adhesion to the substrate-primer surface [71].

2.15 Conclusion

In summary, the integration of nanomaterials in paint formulations has transformed the coatings industry by significantly improving their performance and adding new functionalities. The use of nano-sized particles enhances mechanical strength and offers innovative solutions for surface protection and improved esthetics. The key points below outline the major advancements and future potential of nanopaints:

1. Nanoinorganic pigments such as TiO_2, ZnO, SiO_2, Al_2O_3, and carbon black are highly effective in boosting the mechanical and performance characteristics of paints.
2. Nanopaints are highly promising for improving corrosion resistance, especially for automotive surface applications.
3. Various nanomaterials are being incorporated to introduce advanced functionalities, such as self-cleaning capabilities.
4. The addition of TiO_2 nanoparticles in basecoat paints leads to enhanced surface smoothness and a more appealing finish.

References

[1] V. Parkavi, V.S. Benitha, Behavioural studies of nano formulated paints, International Journal of Innovative Research in Science, Engineering and Technology 3 (3) (2014). https://www.ijirset.com.

[2] M.J. Hanus, A.T. Harris, Nanotechnology innovations for the construction industry, Progress in Materials Science 58 (7) (2013) 1056–1102. Available from: https://doi.org/10.1016/j.pmatsci.2013.04.001.

[3] C.-M. Gavrilescu, C. Paraschiv, P. Horjinec, D.-M. Sotropa, R.-M. Barbu, The advantages and disadvantages of nanotechnology, Romanian Journal of Oral Rehabilitation 10 (2) (2018) 153–159.

[4] D.M. Marquis, É. Guillaume, C. Chivas-Joly, Properties of nanofillers in polymer, Nanocomposites and Polymers with Analytical Methods, In Tech, Croatia, 2011.

[5] E. Jamróz, P. Kulawik, P. Kopel, The effect of nanofillers on the functional properties of biopolymer-based films: a review, Polymers 11 (4) (2019) 675. Available from: https://doi.org/10.3390/polym11040675.

[6] P. Jagadeesh, M. Puttegowda, S. Mavinkere Rangappa, S. Siengchin, Influence of nanofillers on biodegradable composites: a comprehensive review, Polymer Composites 42 (11) (2021) 5691–5711. Available from: https://doi.org/10.1002/pc.26291.

[7] D.M. Marquis, É. Guillaume, C. Chivas-Joly, Properties of nanofillers in polymer, Nanocomposites and Polymers with, Analytical Methods 59 (2011) 261. Available from: https://www.intechopen.com.

[8] Á. Yedra, G. Gutiérrez-Somavilla, C. Manteca-Martínez, M. González-Barriuso, L. Soriano, Conductive paints development through nanotechnology, Progress in Organic Coatings 95 (2016) 85–90. Available from: https://doi.org/10.1016/J.PORGCOAT.2016.03.001.

[9] M. Azani, A. Hassanpour, Nanotechnology in the fabrication of advanced paints and coatings: dispersion and stabilization mechanisms for enhanced performance, ChemistrySelect 9 (19) (2024). Available from: https://doi.org/10.1002/slct.202400844.

[10] L.Y. Jun, et al., Importance of nanomaterials in engineering application, Contemporary Nanomaterials in Material Engineering Applications (2021) 1–20. Available from: https://doi.org/10.1007/978-3-030-62761-4_1.

[11] V.S. Benitha, K. Jeyasubramanian, V.S. Prabhin, S. Dhanabalan, A. Thirumurugan, Nanomaterials in paints, Handbook of Nanomaterials,

Elsevier - Health Sciences Division, 2024, pp. 693–720. Available from: https://10.1016/B978-0-323-95511-9.00024-X. Vol.1.
[12] S. Nam, A.N. Netravali, Green composites. I. Physical properties of ramie fibers for environment-friendly green composites, Fibers and Polymers 7 (4) (2006) 372–379. Available from: https://doi.org/10.1007/BF02875769.
[13] F. Bajuri, N. Mazlan, M.R. Ishak, J. Imatomi, Flexural and compressive properties of hybrid kenaf/silica nanoparticles in epoxy composite, Procedia Chemistry 19 (2016) 955–960. Available from: https://doi.org/10.1016/j.proche.2016.03.141.
[14] B. Mylsamy, S.K. Palaniappan, S. Pavayee Subramani, S.K. Pal, K. Aruchamy, Impact of nanoclay on mechanical and structural properties of treated *Coccinia indica* fibre reinforced epoxy composites, Journal of Materials Research and Technology 8 (6) (2019) 6021–6028. Available from: https://doi.org/10.1016/j.jmrt.2019.09.076.
[15] G.R. Kalagi, R. Patil, S.K. Shetty, M. Nayak, Effect of SiC nano powder on multiaxial woven and chopped randomly oriented flax/sisal fiber reinforced composites, MATEC Web of Conferences 144 (2018) 02005. Available from: https://doi.org/10.1051/matecconf/201814402005.
[16] A.V. Ratna Prasad, K.B. Rao, K.M. Rao, K. Ramanaiah, S.P.K. Gudapati, Influence of nanoclay on the mechanical performance of wild cane grass fiber-reinforced polyester nanocomposites, International Journal of Polymer Analysis and Characterization 20 (6) (2015) 541–556. Available from: https://doi.org/10.1080/1023666X.2015.1053335.
[17] R.K. Sinha, K. Sridhar, R. Purohit, R.K. Malviya, Effect of nano SiO_2 on properties of natural fiber reinforced epoxy hybrid composite: a review, Materials Today: Proceedings, Elsevier Ltd., 2019, pp. 3183–3186. Available from: https://doi.org/10.1016/j.matpr.2020.02.657.
[18] K.P. Ashik, R.S. Sharma, N. Raghavendra, Effect of filler on mechanical properties of natural fiber reinforced composites, Asian Journal of Chemistry 29 (8) (2017) 1697–1701. Available from: https://doi.org/10.14233/ajchem.2017.20551.
[19] R.M. Shahroze, M.R. Ishak, S.M. Sapuan, Z. Leman, M. Chandrasekar, M. Asim, Effect of silica aerogel additive on mechanical properties of the sugar palm fiber-reinforced polyester composites, International Journal of Polymer Science 2019 (2019). Available from: https://doi.org/10.1155/2019/3978047.
[20] H.Y. Kordkheili, S.E. Shehni, G. Niyatzade, Effect of carbon nanotube on physical and mechanical properties of natural fiber/glass fiber/cement composites, Journal of Forestry Research (Harbin) 26 (1) (2015) 247–251. Available from: https://doi.org/10.1007/s11676-014-0003-y.
[21] M.S. Islam, et al., Hybrid nanocomposites, BioResources 10 (1) (2015) 1394–1407.
[22] K.H. Lim, M.S.A. Majid, M.J.M. Ridzuan, K.S. Basaruddin, M. Afendi, Effect of nano-clay fillers on mechanical and morphological properties of napier/epoxy composites, Journal of Physics: Conference Series 908 (2017) 012010. Available from: https://doi.org/10.1088/1742-6596/908/1/012010.
[23] N. Sapiai, A. Jumahat, N. Manap, Full Paper effect of nanofillers dispersion on mechanical properties of clay/epoxy and silica/epoxy nanocomposites, Jurnal Teknologi 76 (9) (2015) 107–111. Available from: https://www.jurnalteknologi.utm.my.
[24] M. Haq, R. Burgueño, A.K. Mohanty, M. Misra, Hybrid bio-based composites from blends of unsaturated polyester and soybean oil reinforced with nanoclay and natural fibers, Composites Science and

Technology 68 (15–16) (2008) 3344–3351. Available from: https://doi.org/10.1016/j.compscitech.2008.09.007.
[25] G. Han, Y. Lei, Q. Wu, Y. Kojima, S. Suzuki, Bamboo–fiber filled high density polyethylene composites: effect of coupling treatment and nanoclay, Journal of Polymers and the Environment 16 (2) (Apr. 2008) 123–130. Available from: https://doi.org/10.1007/s10924-008-0094-7.
[26] B. Venkatram, C. Kailasanathan, P. Seenikannan, S. Paramasamy, Study on the evaluation of mechanical and thermal properties of natural sisal fiber/general polymer composites reinforced with nanoclay, International Journal of Polymer Analysis and Characterization 21 (7) (2016) 647–656. Available from: https://doi.org/10.1080/1023666X.2016.1194616.
[27] S.B. Hosseini, S. Hedjazi, L. Jamalirad, A. Sukhtesaraie, Effect of nano-SiO_2 on physical and mechanical properties of fiber reinforced composites (FRCs), Journal of the Indian Academy of Wood Science 11 (2) (2014) 116–121. Available from: https://doi.org/10.1007/s13196-014-0126-y.
[28] G.L. Devnani, S. Sinha, Effect of nanofillers on the properties of natural fiber reinforced polymer composites, Materials Today: Proceedings 18 (3) (2019) 647–654. Available from: www.sciencedirect.comwww.materialstoday.com/proceedings2214-7853.
[29] G.D. Vilakati, A.K. Mishra, S.B. Mishra, B.B. Mamba, J.M. Thwala, Influence of TiO_2-modification on the mechanical and thermal properties of sugarcane bagasse–EVA composites, The Journal of Inorganic and Organometallic Polymers and Materials 20 (4) (2010) 802–808. Available from: https://doi.org/10.1007/s10904-010-9398-x.
[30] N. Saba, M.T. Paridah, K. Abdan, N.A. Ibrahim, Effect of oil palm nano filler on mechanical and morphological properties of kenaf reinforced epoxy composites, Construction and Building Materials 123 (2016) 15–26. Available from: https://doi.org/10.1016/j.conbuildmat.2016.06.131.
[31] V. Sridharan, T. Raja, N. Muthukrishnan, Study of the effect of matrix, fibre treatment and graphene on delamination by drilling jute/epoxy nanohybrid composite, Arabian Journal for Science and Engineering 41 (5) (2016) 1883–1894. Available from: https://doi.org/10.1007/s13369-015-2005-2.
[32] N. Baig, I. Kammakakam, W. Falath, I. Kammakakam, Nanomaterials: A review of synthesis methods, properties, recent progress, and challenges, Materials Advances 2 (2021) 1821–1871. Available from: https://doi.org/10.1039/d0ma00807a.
[33] K.T. Ramesh, Nanomaterials, Nanomaterials, Springer US, Boston, MA, 2009, pp. 1–20. Available from: 10.1007/978-0-387-09783-1_1.
[34] R.A. McIntyre, Common nano-materials and their use in real world applications, Science Progress 95 (1) (2012) 1–22. Available from: https://doi.org/10.3184/003685012X13294715456431.
[35] M.C. Chen, P.W. Koh, V.K. Ponnusamy, S.L. Lee, Titanium dioxide and other nanomaterials based antimicrobial additives in functional paints and coatings: review, Progress in Organic Coatings 163 (2022) 106660. Available from: https://doi.org/10.1016/j.porgcoat.2021.106660.
[36] R.V.S.K. Jakkula, P. Sethuramalingam, Analysis of coatings based on carbon-based nanomaterials for paint industries–a review, Australian Journal of Mechanical Engineering 21 (3) (2023) 1008–1036. Available from: https://doi.org/10.1080/14484846.2021.1938953.
[37] J. Vidales-Herrera, I. López, Nanomaterials in coatings: an industrial point of view, Handbook of Nanomaterials for Manufacturing Applications,

Elsevier, 2020, pp. 51–77. Available from: https://doi.org/10.1016/B978-0-12-821381-0.00003-X.
[38] T. Kaehler, Nanotechnology: basic concepts and definitions, Clinical Chemistry 40 (9) (1994) 1797–1799. Available from: https://doi.org/10.1093/clinchem/40.9.1797b.
[39] M. Nasrollahzadeh, S.M. Sajadi, M. Sajjadi, Z. Issaabadi, An introduction to nanotechnology, Interface Science and Technology, Elsevier B.V, The Netherlands, 2019, pp. 1–27. Available from: https://doi.org/10.1016/B978-0-12-813586-0.00001-8. Vol. 28.
[40] M.C. Roco, The long view of nanotechnology development: the National Nanotechnology Initiative at 10 years, Journal of Nanoparticle Research 13 (2) (2011) 427–445. Available from: https://doi.org/10.1007/s11051-010-0192-z.
[41] M.C. Roco, Nanoparticles and nanotechnology research, Journal of Nanoparticle Research 1 (1999) 1–6.
[42] A.M. Ealias, M.P. Saravanakumar, A review on the classification, characterisation, synthesis of nanoparticles and their application, IOP Conference Series: Materials Science and Engineering, Institute of Physics Publishing, Berlin, London, New York, 2017. Available from: https://doi.org/10.1088/1757-899X/263/3/032019.
[43] A.C. Grimsdale, K. Müllen, The chemistry of organic nanomaterials, Angewandte Chemie International Edition 44 (35) (2005) 5592–5629. Available from: https://doi.org/10.1002/anie.200500805.
[44] G. Romero, S.E. Moya, Synthesis of organic nanoparticles, Frontiers of Nanoscience 4 (1) (2012) 115–141. Available from: https://doi.org/10.1016/B978-0-12-415769-9.00004-2. Elsevier Ltd.
[45] A.M. Ealias, M.P. Saravanakumar, A review on the classification, characterisation, synthesis of nanoparticles and their application, IOP Conference Series: Materials Science and Engineering 263 (2017) 032019. Available from: https://doi.org/10.1088/1757-899X/263/3/032019.
[46] B.H. Alshammari, et al., Organic and inorganic nanomaterials: fabrication, properties and applications, RSC Advances 13 (20) (2023) 13735–13785. Available from: https://doi.org/10.1039/D3RA01421E.
[47] L. Li, Q. Yang, Part II Coating Materials Nanotechnology, John Wiley & Sons, Bristol, UK, 2019.
[48] N. Baig, I. Kammakakam, W. Falath, Nanomaterials: a review of synthesis methods, properties, recent progress, and challenges, Materials Advances 2 (6) (2021) 1821–1871. Available from: https://doi.org/10.1039/D0MA00807A.
[49] Q. Wu, W. Miao, Y. Zhang, H. Gao, D. Hui, Mechanical properties of nanomaterials: a review, Nanotechnology Reviews 9 (1) (2020) 259–273. Available from: https://doi.org/10.1515/ntrev-2020-0021.
[50] B. Zou, C.Z. Huang, J. Wang, B.Q. Liu, Effect of nano-scale TiN on the mechanical properties and microstructure of Si_3N_4 based ceramic tool materials, Key Engineering Materials 315–316 (2006) 154–158. Available from: https://doi.org/10.4028/www.scientific.net/KEM.315-316.154.
[51] J.L. Wang, L.J. Meng, Influence of carbon nano-fiber on mechanical property of PALC, Applied Mechanics and Materials 535 (2014) 785–787. Available from: https://doi.org/10.4028/www.scientific.net/AMM.535.785.
[52] X.H. Wang, C.H. Hu, M.D. Yi, H.F. Zhang, Effects of nano-ZrO_2 on the microstructure and mechanical properties of Ti(C,N)-based cermet die materials, Advanced Materials Research 154–155 (2010) 1319–1323.

Available from: https://doi.org/10.4028/www.scientific.net/AMR.154-155.1319.

[53] X. Gao, K. Zhang, M. Wang, T. Zan, P. Gao, C. Liu, Optimization of nano coating to reduce the thermal deformation of ball screws, Nanotechnology Reviews 11 (1) (2022) 438–451. Available from: https://doi.org/10.1515/ntrev-2022-0029.

[54] S.K. Krishnan, E. Singh, P. Singh, M. Meyyappan, H.S. Nalwa, A review on graphene-based nanocomposites for electrochemical and fluorescent biosensors, RSC Advances 9 (16) (2019) 8778–8881. Available from: https://doi.org/10.1039/C8RA09577A.

[55] E. Roduner, Size matters: why nanomaterials are different, Chemical Society Reviews 35 (7) (2006) 583. Available from: https://doi.org/10.1039/b502142c.

[56] V.V. Mody, A. Singh, B. Wesley, Basics of magnetic nanoparticles for their application in the field of magnetic fluid hyperthermia, European Journal of Nanomedicine 5 (1) (2013). Available from: https://doi.org/10.1515/ejnm-2012-0008.

[57] I. Khan, K. Saeed, I. Khan, Nanoparticles: properties, applications and toxicities, Arabian Journal of Chemistry 12 (7) (2019) 908–931. Available from: https://doi.org/10.1016/j.arabjc.2017.05.011.

[58] M. Werner, W. Wondrak, C. Johnston, 15. Nanotechnology and transport: applications in the automotive industry, Nanoscience and Nanotechnology, De Gruyter, Hoboken, New Jersey, 2018, pp. 260–282. Available from: https://doi.org/10.1515/9783110547221-015.

[59] M. Chanana, L.M. Liz-Marzán, Coating matters: The influence of coating materials on the optical properties of gold nanoparticles, Nanophotonics 1 (3–4) (2012) 199–220. Available from: https://doi.org/10.1515/nanoph-2012-0008.

[60] I.M. Hosny, I. Anous, Nanomaterials and their applications in interior design, Materials Science, Environmental Science, Engineering (2014). Available from: https://www.semanticscholar.org/paper/Nanomaterials-and-their-applications-in-interior-Hosny-Anous/6399273e199597739a39a94117c8b8a3fd5c849c.

[61] S. Leydecker, Nano Materials, Birkhäuser, Basel, 2008, p. 192. Available from: https://doi.org/10.1007/978-3-7643-8321-3.

[62] S.S. Kumar, Advances in high performance polymer nano composite coatings for better resistance to corrosion and ultraviolet rays, International Journal of Engineering and Applied Sciences 2 (10) (2015) 5. Available from: www.ijeas.org.

[63] S.S. Pathak, A.S. Khanna, T.J.M. Sinha, HMMM cured corrosion resistance waterborne ormosil coating for aluminum alloy, Progress in Organic Coatings 60 (3) (2007) 211–218. Available from: https://doi.org/10.1016/j.porgcoat.2007.07.014.

[64] M.N. de Queiroz, et al., Conductive nanopaints: a remarkable coating, Nanotechnology in the Automotive Industry, Elsevier, Amsterdam, The Netherlands, 2022, pp. 429–449. Available from: https://doi.org/10.1016/B978-0-323-90524-4.00020-7.

[65] S.Z.A. Sakinah, W.A. Wan Hamzah, J. Alias, Corrosion performance of nanopaint for automotive application, Proceedings of the 2nd Energy Security and Chemical Engineering Congress (2023) 299–308. Available from: https://doi.org/10.1007/978-981-19-4425-3_27.

[66] A.S. Khanna, Nanotechnology in high performance paint coatings, Asian Journal of Experimental Sciences 21 (2) (2008) 25–32.

[67] R. Solano, D. Patiño-Ruiz, A. Herrera, Preparation of modified paints with nano-structured additives and its potential applications, Nanomaterials and Nanotechnology 10 (2020). Available from: https://doi.org/10.1177/1847980420909188.
[68] M.J. Hanus, A.T. Harris, Nanotechnology Innovations for the Construction Industry, Elsevier Ltd., Amsterdam, The Netherlands, 2013. Available from: https://doi.org/10.1016/j.pmatsci.2013.04.001.
[69] M.V.K. Mohan, T.V.K.B. Prakash, A. Mukherjee, Antifouling paint formulations with ZnO and Fe_2O_3 nano-paints for marine applications, Sustainable Chemistry and Pharmacy 30 (2022) 100858. Available from: https://doi.org/10.1016/j.scp.2022.100858.
[70] P. Bertrand-Lambotte, J.L. Loubet, C. Verpy, S. Pavan, Nano-indentation, scratching and atomic force microscopy for evaluating the mar resistance of automotive clearcoats: study of the ductile scratches, Thin Solid Films 398–399 (Nov. 2001) 306–312. Available from: https://doi.org/10.1016/S0040-6090(01)01377-3.
[71] S.Z.A. Sakinah, W.H. Azmi, J. Alias, Characterization of TiO_2 nanopaint for automotive application, IOP Conference Series: Materials Science and Engineering 863 (2020) 012053. https://doi.org/10.1088/1757-899X/863/1/012053.

3

Classification of natural nanopaints and their modern aspects

Jimmy Mehta[1], Prateek Mittal[1], Yaman Hooda[2] and Pallav Gupta[3]

[1]Department of Mechanical Engineering, School of Engineering and Technology, Manav Rachna International Institute of Research and Studies, Faridabad, Haryana, India [2]Department of Civil Engineering, School of Engineering and Technology, Manav Rachna International Institute of Research and Studies, Faridabad, Haryana, India [3]Department of Mechanical Engineering, Madan Mohan Malaviya University of Technology, Gorakhpur, Uttar Pradesh, India

3.1 Introduction

The distinction between paint and coating lies within the broader category of "coatings," which encompasses paints. Nano ingredients are frequently incorporated into coatings, although the coating itself typically lacks a nano-structured composition except in specialized applications such as electronics. Nanocoatings serve various purposes, including antigraffiti, antistatic, or UV-blocking functionalities. In contrast, paint acts as a coating applied to a substrate for decorative or protective purposes against oxidation, corrosion, or degradation. Coatings can also modify surface properties such as corrosion resistance or electrical conductivity, with precise thickness control being crucial. Ultimately, the primary role of paint is to shield the substrate, forming an insulating layer against external factors like humidity and UV radiation. Different paints cater to specific protection needs, incorporating additives for corrosion inhibition, peel prevention, or UV absorption. Matching the type of paint to the substrate material is vital for effective protection.

Nanoparticles with diameters less than 100 nanometers offer unique properties due to their high surface-area-to-weight ratio, leading to enhanced performance in coatings and paints. Nanomaterials such as carbon dots, graphene, and titanium dioxide nanoparticles find applications in photovoltaic devices, biosensing, drug delivery systems, and corrosion-resistant

Nanopaints. DOI: https://doi.org/10.1016/B978-0-443-26512-9.00003-3

coatings. These materials facilitate the development of smart coatings capable of responding to environmental stimuli and enhancing surface properties like scratch resistance and antimicrobial activity. In the medical field, nanotechnology has led to significant advancements, while in the maritime industry, nanocoatings have addressed challenges such as biofouling and reducing the environmental impact of copper-based antifouling paints. Nanotechnology also provides solutions for energy efficiency, with nanocomposite insulation coatings improving thermal resistance and energy generation in food manufacturing equipment. Furthermore, nanotechnology plays a crucial role in enhancing the efficiency of solar panels by creating nanostructured surfaces and modifying topcoats to inhibit dirt buildup, resulting in increased sunlight conversion and prolonged panel lifespan. Overall, nanomaterials have become integral to enhancing coating characteristics, offering benefits such as enhanced durability, corrosion resistance, and thermal efficiency. These advancements have expanded applications across industries, showcasing the versatility and potential of nanotechnology in addressing diverse challenges in our evolving world. Nanopaint, a liquid coating applied by spraying onto surfaces and then heated to harden the topcoat, offers exceptional durability despite its thinness, making it an ideal choice for various home decor and furnishing applications. Ongoing research and development continue to refine nanopaint formulations, allowing for tailored properties based on user-defined parameters. Like traditional paint, nanopaint consists of a liquid containing microscopic particles, such as nanotubes, which can alter their behavior in response to changing conditions or specific commands. The resulting coating, known as nanocoating, forms an ultrathin, optically transparent polymer film on the treated substrate, often measuring only a single molecule in thickness. A coating is a cohesive layer applied to a substrate, while a nanocoating is an ultrathin layer as thin as one molecule. Research on nanopaints is advancing, with formulations tailored to specific requirements. Nanopaints typically contain nanotubes that adjust their behavior based on external conditions. Nanocoatings, invisible and measured in nanometers, offer long-lasting protection, especially in electronics and automotive applications. Nanopaints, utilizing nanocarbon structures, enhance durability and performance.

Fig. 3.1 illustrates the evolution of science, the development and implementation of technology, and the future prospects of developed nanomaterials, hence, nanopaints. Electronics, automotive, and biomedical industries benefit from nanopaints and

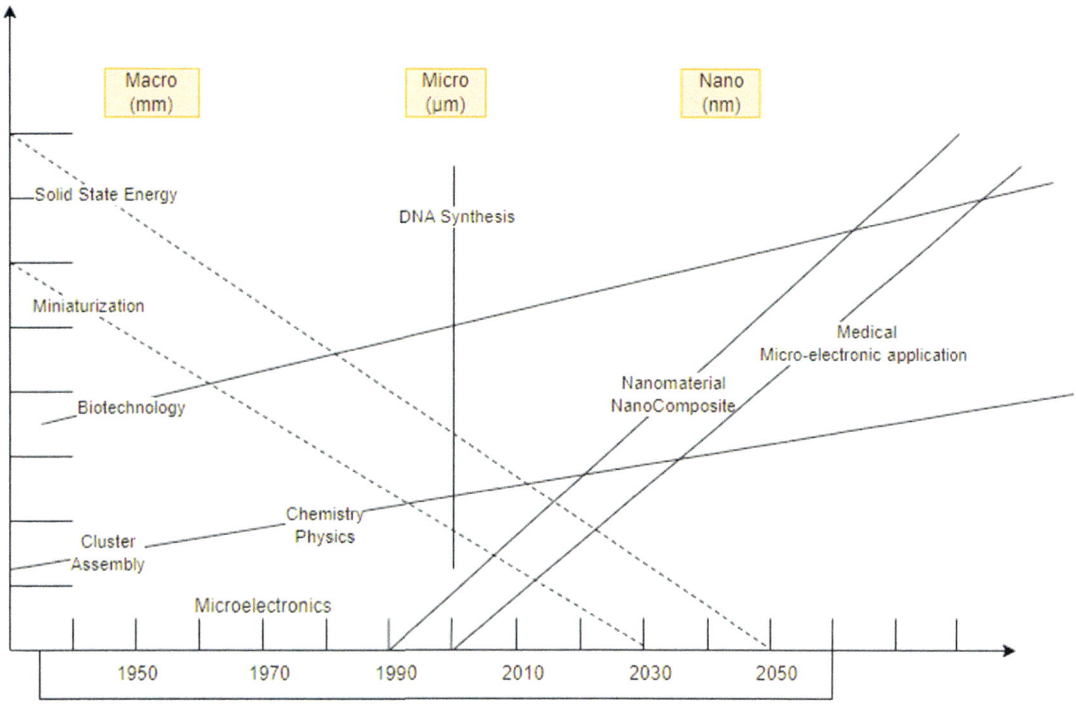

Figure 3.1 Evaluation of science, technology, and future.

coatings. Application methods include physical vapor deposition and chemical vapor deposition. Nanotechnology, exemplified by carbon nanotubes, yields coatings with superior qualities, reducing costs and enhancing performance in marine environments.

3.1.1 Working of nanopaint

When nanopaint is applied, it adopts the properties of conventional paints but forms an extremely thin and nearly invisible surface coating. This top layer is invisible to the naked eye but can be observed under high magnification. Despite its thinness, nanocoating is exceptionally durable, measured on the nanometer scale. It offers longevity, with applications lasting for months on electronics and years on automotive sprays and waxes. The application of nanopaint combines paint and coating technologies to enhance substrate qualities, including hardness, scratch resistance, mold and bacteria resistance, and ease of cleaning, resulting in longer-lasting products (Fig. 3.2).

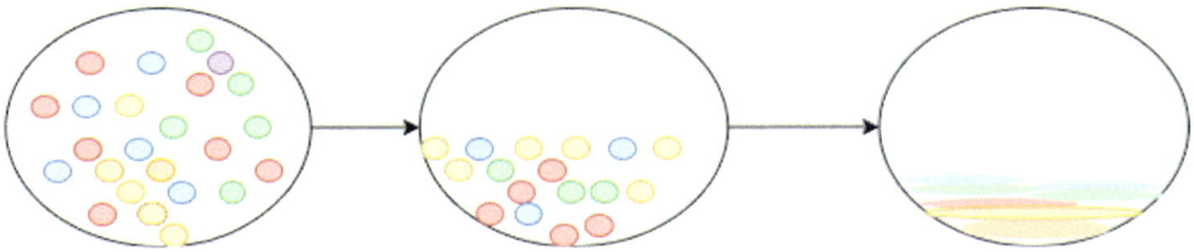

Figure 3.2 Nanopaint being applied.

3.1.2 Materials used for nanopaints

Structures densely packed with molecules, such as nanocarbon, nanotubes, nanofiber, and graphite, serve as the foundation for producing nanopaints and coatings. These nano-based coatings offer exceptional qualities, including enhanced scratch resistance, hardness, and resistance to mold and bacteria, resulting in highly durable and high-performance coatings. Nanopaints find applications across various sectors, including electronics, optics, aerospace, marine, automotive, oil and gas, building and civil engineering, and biomedical fields. Different methods, such as physical vapor deposition, atomic layer deposition, chemical vapor deposition, electrospray, electrospinning, sol–gel, and aerosol coating are employed for applying nanopaints and coatings, tailored to specific market requirements. Carbon nanotubes (CNTs) are crucial in elevating paints' mechanical and physical characteristics. They enhance durability, strength, and hardness, thereby rendering the paint film more resilient to abrasion and wear. Moreover, they improve adhesion to substrates, mitigating issues such as peeling and cracking. Additionally, CNTs facilitate the development of conductive coatings suitable for electronics, sensors, and applications requiring electromagnetic interference (EMI) shielding. CNT-infused paints can also offer self-cleaning and antimicrobial properties, proving invaluable in medical and food processing environments. Furthermore, CNTs contribute to creating paints with superior thermal insulation properties, outperforming traditional insulation materials like fiberglass. Nanopaints have diverse applications, including safeguarding building facades and structures with waterproof coatings, enhancing oils and wood drying oils with nano additives, providing waterproof solutions for plastic and smooth facades, and developing antigraffiti coatings. Table 3.1 shows the list of various nanoparticles being used for different applications.

Table 3.1 Nanoparticles for surface coating applications.

Nanoparticle	Application
SiO_2	Corrosion and scratch resistance
TiO_2	Self-cleaning, optical effect, antifogging, antibacterial
Al_2O	Scratch resistance and corrosion protection
ZnO	UV absorber
FeO_2	Medical applicants
Ag colloids	Antibacterial
Metal colloids	UV absorber
Carbon nanotubes	Conductive coatings
Antimony tin oxide (ATO) and indium tin oxide (ITO)	IR/UV absorber

As per the available literature, there are many advantages of nanopaints when employed for various applications but a few of them are as mentioned as follows:

- *Beneficial for Both Porous and Nonporous Surfaces*: Preserving industrial surfaces often involves treating them to reduce their porosity. Decreased porosity shields surfaces from moisture infiltration and simplifies the process of fortifying them against corrosion, particularly rust and salt erosion. Nanocoatings, due to their minute size, can be applied to traditional porous surfaces as well as those previously treated.
- *Microscopic Homogeneity for Enhanced Dirt Resistance*: Nanocoatings offer an unparalleled level of uniformity to treated substrates, surpassing any other available option. By occupying minuscule surface pores, a properly administered nanocoating can render materials nearly impervious to dirt. A simple wipe with a cloth can restore many treated surfaces to their original appearance and functionality.
- *Exceptional Hydrophobic and Weather-Resistant Properties*: Nanocoatings exhibit remarkable water-repellent capabilities, initially developed and tested for high-performance vehicles. Their ability to deter pooling and water damage, proven effective on the racetrack, is equally advantageous in industrial environments.
- *Resistance to Chemicals, Oils, and Salt*: Industrial workplaces often employ harsh and corrosive chemicals, while sites near water bodies must guard against salt damage. Selecting the appropriate nanocoating can significantly enhance resistance to these corrosive elements.

- *Ultraviolet Protection*: Ultraviolet radiation significantly contributes to the degradation of outdoor facilities and equipment. Nanocoatings filter out a portion of UV light, preventing its reach onto the surface. However, a purpose-designed anti-UV coating remains superior in UV protection.
- *Unmatched Durability and Longevity*: Nanocoatings can be engineered for extreme hardness, with the toughest coatings reaching a rating of 9 on the Mohs scale of mineral hardness. Only materials harder than this can cause scratches. For instance, a diamond, rated 10 on the scale, can potentially damage the coating.
- *Versatile Temperature Resilience*: Nanocoatings exhibit superior temperature resistance compared to standard paint coatings. Formulations designed for temperature resilience can provide protection even in temperatures exceeding −50°F. Moreover, nanocoatings can be tailored to withstand extreme heat conditions.

3.2 Traditional classification methods (historical context and limitations)

In 1959 Richard Feynman, often revered as the pioneer of modern nanotechnology, introduced the concept of manipulating matter at the atomic level in his renowned lecture at Cal Tech, titled "There's Plenty of Room at the Bottom." This groundbreaking idea laid the foundation for the emerging field of nanotechnology. In 1974 Norio Taniguchi provided a clearer definition of nanotechnology as manipulating materials at the atomic or molecular scale. Building upon Taniguchi's work, Eric Drexler expanded the concept in the 1980s, introducing molecular nanotechnology with the revolutionary idea of nanoscale assemblers capable of orchestrating chemical reactions with atomic precision. A significant milestone came in 1991 with Sumio Iijima's discovery of CNTs, propelling nanotechnology to the forefront of scientific exploration. Since then, nanotechnology has experienced exponential growth, influencing various industries, including energy conservation, computer science, biomedicine, electronics, and coatings. The term "nanotechnology" was coined by Norio Taniguchi in 1974 to emphasize the exacting precision required for machining and refining materials. However, it was not until 1981 when K. E. Drexler proposed a groundbreaking "bottom-up" approach, departing from the conventional "top-down" methodologies discussed by Feynman and Taniguchi. Drexler's innovative approach involved manipulating molecules and engineering at the molecular level to fabricate molecular machines and devices with atomic precision. The publication of Drexler's influential book, "Engines of Creation," in 1986 played a crucial role in popularizing the term "nanotechnology." The prefix "nano" is derived from the Greek word for "dwarf," signifying a billionth of a unit, such as a meter. Thus a nanometer (nm) equals one billionth of a meter. Despite the widespread use of the term "nanotechnology," it has occasionally led to misunderstandings. Some technologies categorized as nanotechnology operate on the micrometer scale rather than the nanometer scale (1–100 nm). Additionally, a significant portion of research focuses on examining nanostructures rather than directly applying nanotechnology. Nanomaterials are not entirely novel substances but rather minute variations of familiar materials such as gold, silver, and platinum. Moreover, historical practices like using carbon nanoparticles for tire reinforcement and natural phenomena such as photosynthesis now fall under the umbrella of nanotechnology. The history of nanomaterials extends beyond human exploration, with nanostructures emerging after the Big Bang and appearing in ancient meteorites, seashells, and other natural phenomena. Even early human activities, such as the use of fire, contributed to the creation of nanoscale smoke particles. Scientific exploration into nanomaterials began much later, with investigations into nanostructured catalysts dating back over 70 years. By the early 1940s, production and utilization of silica nanoparticles had commenced in various consumer goods. Significant progress in nanomaterial production occurred in the 1960s and 1970s through the development of metallic nanopowders for magnetic recording tapes. The synthesis of nanocrystals using inert-gas evaporation techniques in 1976 marked a significant milestone, paving the way for further advancements in nanotechnology applications. Recent discoveries, such as those related to Maya blue paint, illustrate the intricate nature of nanostructured materials. Despite centuries of use, the precise mechanisms underlying its unique properties remain elusive. Studies of authentic samples have revealed that Maya blue paint comprises needle-shaped palygorskite crystals forming a superlattice with intercalated metal nanoparticles, resulting in its characteristic blue hue and resistance to acids and bio-corrosion. Currently, nanophase engineering continues to advance, encompassing a broad range of structural and functional materials across both inorganic and organic domains. This progress enables precise manipulation of mechanical, catalytic, electric, magnetic, optical, and electronic functionalities. Techniques such as cluster assembly and doping of conventional materials with nanoscale semiconductor colloids are opening up new horizons in various industries, from semiconductor manufacturing to optical computing. The potential of nanotechnology to revolutionize technology and enhance the performance of everyday materials holds promise for driving innovation across numerous disciplines.

3.3 Modern approaches to classification

Nanopaint signifies a groundbreaking coating engineered to revolutionize surface or material characteristics based on user-defined specifications. Similar to traditional paint, nanopaint is initially applied in liquid form before solidifying. This liquid comprises a suspension of minute particles known as nanotubes, which dynamically adjust their behavior in response to external changes or specific commands. Currently, nanopaint is undergoing extensive research and development.

A groundbreaking prototype of nanopaint has been developed, capable of shielding against RF (radio frequency) fields akin to metals like copper. When applied to the internal walls of a structure, this substance can selectively permit or hinder signals to and from cell phones, portable radios, and other wireless devices. The potential applications of nanopaint are vast. Notably, its application on building exteriors could regulate their infrared (IR) reflectivity or absorption based on external conditions. This innovation could significantly enhance energy efficiency by enabling structures to absorb thermal energy during chilly yet sunny days, reflect it during hot spells, conserve it on cold nights, and dissipate it on warmer evenings. Specialized nanopaint formulations could serve a diverse array of functions, including:

- Blocking cell phone signals in inappropriate environments such as theaters, hospitals, or funerals.
- Providing glass with adjustable opacity.
- Endowing surfaces of motor vehicles or industrial machinery with self-repair capabilities.
- Facilitating the alteration of surface textures as desired.
- Deterring the proliferation of pathogens such as bacteria or viruses.
- Repelling or neutralizing toxic chemicals, acids, or other corrosive substances.

3.4 Classification criteria

3.4.1 Chemical composition

Nanopaints, a revolutionary class of coatings, can be categorized based on their chemical composition, which significantly influences their properties and applications. The chemical composition of nanopaints typically involves the combination of various materials at the nanoscale, offering unique functionalities and performance characteristics. Here are some common classifications based on chemical composition:

- *Polymer-Based Nanopaints*: Polymer-based nanopaints utilize polymers as the primary matrix or binder material. These polymers can be organic or synthetic and provide structural integrity and adhesion properties to the coating. Nanoparticles, such as metal oxides or CNTs, are often incorporated into the polymer matrix to impart specific functionalities, such as enhanced strength, flexibility, or electrical conductivity.
- *Sol–Gel Nanopaints*: Sol–gel nanopaints are synthesized through the sol–gel process, involving the conversion of a colloidal suspension (sol) into a solid network (gel). These nanopaints typically consist of inorganic precursors, such as metal alkoxides or metal salts, which undergo hydrolysis and condensation reactions to form a three-dimensional network of nanoparticles embedded in a gel-like matrix. Sol–gel nanopaints are known for their excellent adhesion, durability, and optical transparency, making them suitable for applications requiring high-performance coatings, such as antireflective coatings on glass or ceramics.
- *Hybrid Nanocomposite Nanopaints*: Hybrid nanocomposite nanopaints combine organic and inorganic components to create a synergistic material with enhanced properties. These nanopaints often incorporate both polymer matrices and inorganic nanoparticles, such as metal oxides, quantum dots, or carbon nanomaterials. By carefully controlling the composition and structure of the hybrid nanocomposites, they can tailor their mechanical, electrical, thermal, and optical properties for specific applications, including corrosion protection, electromagnetic shielding, or optical sensing.
- *Functionalized Nanopaints*: Functionalized nanopaints are designed with specific functional groups or chemical moieties attached to the nanoparticle surface. These functional groups can impart desired properties, such as hydrophobicity, antimicrobial activity, or chemical reactivity, to the coating. Functionalized nanopaints are particularly useful in applications requiring surface modification or tailored surface interactions, such as biomedical coatings, self-cleaning surfaces, or sensors.
- *Bio-Based Nanopaints*: Bio-based nanopaints utilize renewable and biodegradable materials as key components in their chemical composition. These nanopaints may incorporate

natural polymers, such as cellulose, chitosan, or proteins, as well as nanoparticles derived from biocompatible sources. Bio-based nanopaints offer environmental sustainability and biocompatibility advantages, making them suitable for eco-friendly coatings in various industries, including packaging, construction, and healthcare.

3.4.2 Particle size

Nanopaints, distinguished by their capacity to modify surface properties at the nanoscale, can be categorized based on the size of the nanoparticles they incorporate. Particle size significantly impacts the behavior and functionality of nanopaints, thus serving as a fundamental criterion for classification. Here are some common classifications based on particle size:

- *Nanoemulsions*: Nanoemulsion nanopaints encompass nanoparticles typically sized between 10 and 100 nanometers suspended in a liquid medium. These nanopaints demonstrate outstanding stability and uniformity owing to the diminutive size of the dispersed phase particles. Nanoemulsions find application in scenarios requiring smooth and uniform coatings, such as inks, coatings for electronic devices, or biomedical applications.
- *Nanoparticle Dispersions*: Nanoparticle dispersion nanopaints comprise larger nanoparticles, typically ranging from 100 to 500 nanometers, dispersed within a liquid medium. These nanopaints strike a balance between stability and ease of handling, rendering them suitable for a diverse range of applications, including surface coatings, paints, and functional coatings for industrial use.
- *Nanopowders*: Nanopowder nanopaints entail nanoparticles ranging from 1 to 100 micrometers, often presented in the form of dry powders. These nanopaints are commonly employed in applications necessitating high solid content and durability, such as automotive coatings, aerospace coatings, or protective coatings for industrial machinery.
- *Nanocomposites:* Nanocomposite nanopaints comprise nanoparticles dispersed within a polymer matrix, featuring particle sizes spanning from a few nanometers to several micrometers. These nanopaints offer enhanced mechanical, thermal, and electrical properties compared to conventional polymer coatings, making them suitable for demanding applications such as structural materials, electronic devices, or aerospace components.

- *Quantum Dot Nanopaints*: Quantum dot nanopaints encompass semiconductor nanoparticles typically sized between 2 and 10 nanometers. These nanopaints showcase distinctive optical properties, including tunable emission wavelengths and high quantum yields, rendering them ideal for applications such as displays, lighting, or biological imaging.
- *Ultrafine Nanopaints*: Ultrafine nanopaints encompass nanoparticles sized on the order of a few nanometers to tens of nanometers. These nanopaints offer exceptional coverage and adhesion properties, making them suitable for high-performance coatings, adhesives, or sealants across industries such as automotive, aerospace, and electronics.

3.4.3 Surface properties

Nanopaints can be categorized based on their surface properties into several classifications, each offering unique characteristics and applications. Here are some common classifications:

- *Hydrophobic Nanopaints*: Engineered to repel water and other liquids, hydrophobic nanopaints incorporate hydrophobic nanoparticles that create a nonporous surface. This prevents water penetration and damage, making them ideal for waterproofing surfaces such as building exteriors, roofs, and marine vessels.
- *Hydrophilic Nanopaints*: In contrast to hydrophobic nanopaints, hydrophilic nanopaints are formulated to attract and absorb water. These nanopaints contain nanoparticles with functional groups that promote water absorption, enhancing surface wetting and adhesion. They find applications in coatings for biomedical devices, drug delivery systems, and moisture-sensitive surfaces.
- *Antifouling Nanopaints*: Designed to inhibit the adhesion of biological organisms to submerged surfaces, antifouling nanopaints utilize nanoparticles with unique surface structures or chemical compositions. These nanoparticles deter marine fouling, reducing drag and enhancing the performance of submerged structures like ship hulls and underwater equipment.
- *Self-Cleaning Nanopaints*: Incorporating nanoparticles with photocatalytic properties, such as titanium dioxide (TiO_2), self-cleaning nanopaints break down organic contaminants and pollutants when exposed to light. This creates surfaces resistant to dirt, grime, and stains, making them easier to clean and maintain. They are commonly used for building exteriors, solar panels, and automotive coatings.

- *Anticorrosion Nanopaints*: Formulated to prevent corrosion, anticorrosion nanopaints contain nanoparticles that act as corrosion inhibitors, forming protective barriers on metal surfaces. These nanoparticles passivate the metal surface or release corrosion-inhibiting compounds over time, prolonging service life and reducing maintenance costs for infrastructure, pipelines, and industrial equipment.
- *UV-Resistant Nanopaints*: With nanoparticles that absorb or reflect ultraviolet (UV) radiation, UV-resistant nanopaints protect surfaces from UV-induced degradation, discoloration, and fading. They are commonly used for outdoor applications such as automotive coatings, outdoor furniture, and architectural finishes to maintain aesthetic appeal and durability.
- *Thermal Insulating Nanopaints*: Thermal insulating nanopaints incorporate nanoparticles with low thermal conductivity to reduce heat transfer through surfaces. These nanopaints provide thermal insulation, regulating indoor temperatures, improving energy efficiency, and reducing heating and cooling costs. They are applied to building exteriors, roofs, and industrial equipment exposed to high temperatures.

3.4.4 Functionalization

Nanopaints can be categorized based on their functionalization, which refers to the specific chemical or physical properties imparted to the nanoparticles within the paint formulation. Here are some common classifications:

- *Functionalized Nanopaints*: These nanopaints contain nanoparticles that have undergone chemical modification or functionalization with organic or inorganic compounds to impart specific properties. Functionalization involves attaching functional groups, such as amino or carboxyl groups, onto the nanoparticle surface to enhance compatibility, stability, or reactivity with the paint matrix or target surface. They offer tailored functionalities for various applications, including enhanced adhesion, controlled release, or targeted surface interactions.
- *Polymer Nanocomposite Nanopaints*: Polymer nanocomposite nanopaints incorporate nanoparticles dispersed within a polymer matrix to impart specific properties to the coating. Nanoparticles can be functionalized to improve their dispersion within the polymer matrix and enhance compatibility with the paint formulation. These nanopaints offer advantages such as improved mechanical strength, thermal

stability, barrier properties, and surface functionalities compared to conventional polymer coatings.

- *Photoactive Nanopaints*: Photoactive nanopaints contain nanoparticles with photoactive properties, such as photocatalytic or photochromic nanoparticles, that respond to light stimuli. Photocatalytic nanoparticles, like titanium dioxide (TiO_2) or zinc oxide (ZnO), can degrade organic pollutants or inhibit microbial growth when exposed to light, suitable for self-cleaning or antimicrobial coatings. Photochromic nanoparticles change color or transparency in response to light, applied in smart coatings for sensing or display purposes.
- *Magnetic Nanopaints*: Magnetic nanopaints include magnetic nanoparticles, such as iron oxide (Fe_3O_4) or cobalt ferrite ($CoFe_2O_4$), exhibiting magnetic properties. Functionalization with surfactants or polymers improves dispersion and stability within the paint matrix. They enable applications like magnetic resonance imaging (MRI) contrast agents, magnetic targeting for drug delivery, or magnetic separation of contaminants in wastewater treatment.
- *Conductive Nanopaints*: These contain nanoparticles with high electrical conductivity, like CNTs or graphene, dispersed within the paint matrix. Forming conductive networks within the coating enables applications like EMI shielding, electrostatic discharge protection, or flexible electronics. Functionalization improves dispersion, adhesion, or compatibility with the paint formulation.
- *Luminescent Nanopaints*: Luminescent nanopaints contain nanoparticles emitting light, such as quantum dots or rare-earth nanoparticles, dispersed within the paint matrix. Functionalization with ligands or surface coatings controls emission properties, wavelength, or stability. Applied in fluorescence imaging, display technologies, security inks, or decorative coatings.

3.5 Case studies

Currently, TiO_2 nanoparticles find extensive application in various construction materials and structures, including cement paints, special cement, cement mortars, road pavements, self-cleaning materials for both concrete and bitumen, air purification systems, antibacterial compositions, and finishing materials for external and internal works [1]. These photosensitive catalysts are particularly favored for creating self-cleaning surfaces in concrete due to their remarkable superhydrophobic properties, which help maintain the aesthetic appearance of structures over extended periods. The inception of large-scale applications of photocatalytic self-cleaning materials dates back to 1996 when Italcementi participated in the construction of the Dives in Misericordia Church in Rome. This architectural marvel, envisioned as three massive sails constructed from precast reinforced concrete, required a unique concrete formulation. The desired concrete needed to exhibit high strength and durability, and retain its white color owing to the self-cleaning properties of the surface. Subsequently, photocatalytic cements found their way into prestigious European architectural projects, including the Cité de la Musique and the Hotel de Police in Bordeaux, France, as well as the Saint John Court in Monte Carlo, Monaco. Furthermore, schools in Mortara, Italy (1999), and multistorey residential buildings in Ostend, Belgium, benefited from the incorporation of photocatalytic cements.

In addition to architectural applications, cement paints and plasters formulated with photocatalysts are extensively utilized in Italy for constructing residential buildings in urban environments, enhancing the eco-friendly attributes of tunnels, underground parkings, petrol stations, and other infrastructural projects.

Beyond their self-cleaning properties, building materials containing TiO_2 offer promising prospects for monitoring urban pollution. Photocatalytic systems employing "TiO_2/cement" can effectively mitigate various pollutants, including NOx, SOx, NH_3, CO, volatile organic hydrocarbons (e.g., benzene, toluene), organic chlorides, aldehydes, and condensed aromatic compounds. Furthermore, doping TiO_2 mesoporous films with small amounts of nanosilver enhances their antibacterial efficacy, even in the absence of ultraviolet radiation. Both hydrophilic and hydrophobic coatings find applications primarily on flat surfaces and fundamental construction materials such as concrete, reinforced concrete, stone, and wood.

Major applications include:

- ***Antibacterial Nanopaint for Healthcare Facilities:***

  Healthcare-associated infections (HAIs) represent a pressing concern within hospitals and medical facilities globally, posing significant challenges to patient health and safety. Traditional paints utilized on hospital walls and surfaces have inadvertently fostered environments conducive to bacterial growth, exacerbating the transmission of infections among patients and healthcare workers [2–4]. In response to this issue, the development and application of antibacterial nanopaint containing silver nanoparticles have emerged as a promising solution. Hospitals and healthcare facilities have begun implementing this innovative nanopaint on various surfaces, including walls, door handles, and other frequently touched areas. The introduction of this antibacterial nanopaint has yielded tangible results, with its antimicrobial properties significantly reducing the incidence of HAIs. Consequently, this proactive measure has not only enhanced patient outcomes but has also contributed to the reduction of healthcare costs associated with treating such infections [5,6].

- ***Self-Cleaning Nanopaint for Urban Infrastructure:***
  Urban pollution and environmental degradation pose significant challenges to maintaining clean and aesthetically pleasing infrastructure in cities. Graffiti and environmental contaminants further exacerbate these issues, detracting from the appearance of buildings and public spaces and necessitating frequent and costly cleaning and upkeep efforts [7–9]. In response to this problem, using self-cleaning nanopaint with photocatalytic properties, often incorporating TiO_2 nanoparticles, has emerged as a viable solution. Municipalities and urban planners are now actively implementing this innovative nanopaint on various urban structures, including bridges, buildings, and public art installations. The results of adopting self-cleaning nanopaint have been promising, with notable decreases in maintenance expenses and significant improvements in the aesthetics of urban infrastructure. This advanced paint effectively breaks down organic pollutants and inhibits graffiti adhesion, contributing to the overall cleanliness and visual appeal of urban environments [10,11].
- ***Thermal Insulating Nanopaint for Energy Efficiency:***
  Heating and cooling buildings represent major contributors to carbon emissions and energy costs, highlighting the urgency for energy-efficient solutions. Inadequate insulation in buildings exacerbates this issue, resulting in heat loss during winter and heat gain in summer, ultimately leading to heightened energy consumption. To address this challenge, the development and application of thermal insulating nanopaint containing ceramic or aerogel nanoparticles have emerged as a promising solution [12–14]. Commercial and residential buildings are now integrating thermal insulating nanopaint into their infrastructure, applying it to walls, roofs, and windows. Implementing thermal insulating nanopaint has yielded significant results, notably enhancing energy efficiency and reducing heating and cooling expenses. By providing superior insulation properties, this

innovative paint ensures optimal indoor comfort levels year-round, contributing to environmental sustainability and cost savings for building owners and occupants [15–18].

- ***Nanopaints in Computer Industry:***

  Nanotechnology, stemming from micro-engineering concepts in physics and material sciences, has revolutionized modern industries, particularly computing and information technology. This evolution has led to the development of nanoscaled materials and devices, with nanotubes replacing traditional silicon chips in computer technologies, offering enhanced efficiency in minimal space. Inspired by the works of Feynman and Drexler, computer scientists are exploring the potential of revolutionary nanocomputers [19–21], facilitating advancements in various sectors. In urban infrastructure, self-cleaning nanopaints, leveraging photocatalytic properties with TiO_2 nanoparticles, combat environmental pollutants and graffiti, reducing maintenance costs and enhancing aesthetics. Similarly, thermal insulating nanopaints contribute to building energy efficiency by providing superior insulation properties and maintaining comfort levels year-round [22–26]. Integrating nanotechnology and information and communication technology has yielded remarkable improvements in interconnected communities, economic competitiveness, and environmental stability. Novel technologies like smart sensors, nanochips, and quantum computing are enhancing computational capabilities and fostering environmental sustainability. Furthermore, nanotechnology intersects with artificial intelligence, machine learning, and big data platforms, driving the Fourth Industrial Revolution. These technologies enable data collection, analysis, and utilization for various applications, from disease management in healthcare to market analysis in business [27–30]. The convergence of nanotechnology, robotics, and drones, termed the internet of robotics and nano things, enhances automation and precision in industries. Nanosensors equip drones with improved detection capabilities, facilitating tasks in maintenance, inspections, and security surveys. Additionally, nanotechnology-based 3D printing revolutionizes production management, minimizing waste generation and enabling customized, sustainable practices across industries. Overall, nanocomputer applications permeate modern industries, driving efficiency, sustainability, and innovation, with continuous collaboration among scientists expected to further explore its vast potential [31].

- ***Nanotechnology and Bioprocessing Industry:***

  Scientific and engineering endeavors are bridging nanotechnology with the bioprocessing industry, aiming to enhance the manufacturing of food and biomaterials with superior quality. Key nanotechnological applications include pathogenic identification, food monitoring, biosensor devices, smart packaging materials, and nanoencapsulation of active food compounds. Successful integration of societal, cultural, and ethical considerations is crucial for the widespread acceptance and adoption of nano-based bio-processed products in mainstream markets. With global population growth, ensuring safe, healthy, and sustainable food production is paramount. Nanotechnology facilitates improved sensitivity in food quality monitoring through sensors and diagnostic devices, while also aiding in brand protection against counterfeit food products [32,33]. Furthermore, nanotechnology exhibits significant promise in diminishing energy consumption in food production, effectively tackling environmental concerns. It also assumes a pivotal role in fermentation processes by enhancing biomolecule production at a reduced cost compared to conventional methodologies. This integration has propelled the advancement of biocompatible substances and nanostructures with wide-ranging applications across the food, bioprocessing, and winemaking sectors. Nanodevices and nanomaterials are indispensable for the progression of food monitoring and supply chain management, augmenting analytical capabilities and substituting traditional monitoring methods with cutting-edge nano-based technologies. Additionally, nanotechnology has revolutionized bio-composting materials, developing biodegradable nanocomposites that boast superior mechanical and functional characteristics. These materials extend beyond the food industry, finding utility in biomedical, therapeutic, packaging, sensor, and cosmetic domains.

While nanocomposites offer numerous benefits such as biocompatibility and antimicrobial activity, rigorous toxicity and safety assessments are imperative to ensure consumer satisfaction and public safety [34]. Incorporating nanotechnology into the food industry has led to the creation of functional food products with improved bioactivity and solubility profiles. Nano-encapsulation methods have been particularly effective in enhancing the delivery of bioactive compounds, thereby maximizing consumer health benefits. However, rigorous safety regulations and comprehensive profiling are essential to guarantee the safe consumption of nano foods. Despite these challenges, the extensive utilization of nanotechnology in food processing, manufacturing, and monitoring shows significant promise in addressing global food security issues and enhancing public health outcomes.

3.6 Applications

3.6.1 Biomedical applications

Nanopaints demonstrate significant promise in a wide array of biomedical applications, harnessing the distinctive attributes of nanoparticles to tackle challenges in drug delivery, imaging, diagnostics, and therapy. A notable application in the biomedical field is their contribution to targeted drug delivery systems. In these systems, nanoparticles integrated within paints serve as carriers for therapeutic agents, enabling precise delivery to specific tissues or cells within the body. Through functionalization, these nanoparticles enable controlled drug release, thereby enhancing therapeutic effectiveness while minimizing adverse effects. Additionally, nanopaints can be engineered to respond to external stimuli such as pH, temperature, or enzymes, facilitating triggered drug release at specific sites. In medical imaging, nanopaints play a pivotal role in enhancing contrast for various imaging techniques like MRI, computed tomography, and optical imaging. Nanoparticles with intrinsic or modified surface properties improve the visibility of tissues and organs, enabling more accurate diagnosis and disease monitoring. Moreover, nanopaints find applications in biosensing and diagnostics, where functionalized nanoparticles can detect specific biomarkers or analytes in biological samples, enabling early disease detection and monitoring. These biosensors offer rapid, sensitive, and cost-effective diagnostic solutions for various medical conditions. Furthermore, nanopaints are currently under investigation for therapeutic purposes such as photothermal therapy and photodynamic therapy. Nanoparticles with photothermal or photosensitizing properties, when incorporated into paints, can selectively target and eradicate cancer cells or pathogens upon exposure to light, while preserving healthy tissues.

3.6.2 Environmental applications

Nanopaints, leveraging nanotechnology, offer innovative solutions for diverse environmental challenges owing to their unique properties. These paints comprise nanoparticles ranging in size from 1 to 100 nanometers, affording precise control over their properties and functionalities. In environmental applications, nanopaints offer significant benefits, particularly in areas such as pollution reduction, energy conservation, and sustainability. A key application lies in mitigating air pollution through the development of photocatalytic nanopaints. These paints incorporate photocatalytic nanoparticles like TiO_2 or zinc oxide (ZnO), which actively degrade air pollutants such as volatile organic compounds and nitrogen oxides (NOx) upon exposure to sunlight or artificial UV radiation. This photocatalytic activity helps purify the air, improving overall air quality and mitigating the harmful effects of pollutants on human health and the environment [35–37].

Additionally, nanopaints contribute to energy conservation by incorporating nanomaterials with thermal insulating properties. These paints form a barrier that reduces heat transfer through walls and roofs, thereby minimizing the need for heating and cooling in buildings. By improving energy efficiency, nanopaints help reduce greenhouse gas emissions and alleviate the environmental impact associated with energy consumption. Moreover, nanopaints offer sustainable solutions by enhancing the durability and longevity of coated surfaces. Nanoparticles reinforce the paint matrix, superiorly resisting weathering, corrosion, and microbial growth. This longevity reduces the frequency of maintenance and repainting, resulting in lower resource consumption and waste generation over the lifespan of painted surfaces [37].

3.6.3 Industrial applications

Industrial applications of nanopaints encompass diverse sectors, including automotive, aerospace, construction, and electronics. Nano-enhanced coatings offer superior properties such as enhanced durability, corrosion resistance, and scratch resistance, making them ideal for protecting surfaces in harsh environments. Additionally, nanopaints can provide functional properties such as self-cleaning, antifouling, and antimicrobial characteristics, meeting the stringent requirements of various industrial applications. Table 3.2 summarizes the properties and applications as per their applications.

3.6.4 Automotive applications

In the luxury car market, ensuring surface protection is essential to safeguard against UV damage, wear, and heat and to enhance adhesion while reducing engine friction. Within the automotive coatings industry, a range of functional properties are highly sought after. These include self-cleaning attributes facilitated by nanoparticles like Nano TiO_2 and nano SiO_2, antifingerprint capabilities provided by silica nanoparticles, UV resistance typically achieved through zinc oxide, cerium oxide,

Table 3.2 Properties and applications of nanopaints [1,35,38–40].

Surface properties	Applications
Mechanical properties like tribology, hardness, and scratch resistance	Wear protection of machines and equipment, mechanical protection of soft materials like leather, polymers, and textiles
Wetting properties like antiadhesive, hydrophobic, hydrophilic	Antigraffiti, antifouling, lotus-effect, self-cleaning surface, ceramics, etc.
Thermal and chemical properties like heat resistance and insulation, corrosion resistance	Corrosion protection for machinery and equipment, heat resistance for turbines and engines, and thermal insulation. Building materials, etc.
Biological properties like biocompatibility and antiinfective	Biocompatible implants, antibacterial surfaces in medical tools, etc.
Electronic and magnetic properties like magnetoresistance and dielectric	Ultrathin dielectrics for field-effect transistors, magnetoresistive sensors, and data memory
Optical properties like antireflection, photo and electro-chromatic	Photo and electro-chromatic windows, antireflective screens, solar cells, etc.

titanium oxide, and iron oxide nanoparticles, chemical resistance bolstered by CNTs and graphene, as well as abrasion and scratch resistance offered by silica, aluminum oxide, and alumina nanoparticles.

Applications for these coatings are diverse, spanning from scratch-proof coatings to self-cleaning treatments for glass, antibacterial coatings for interior trim and upholstery, hydrophobic and oleophobic antifingerprint coatings for automotive [41] displays, lubricant additives, anticorrosion coatings for engines, wear-resistant nanocoatings for engine components, and antiicing coatings.

In both the aviation and aerospace sectors, protecting surfaces is paramount to withstand harsh environmental conditions. Stringent requirements for corrosion resistance, extreme temperature endurance, abrasion resistance, and wear resistance in engine components have led to a surge in demand for high-performance coatings. These coatings play a critical role in enhancing the durability, reliability, and performance of aerospace components while addressing challenges such as erosion, sliding wear, fretting wear, and various corrosion-related issues like pitting, exfoliation, oxidation, and hot corrosion. Tailored multifunctional nanocoatings designed for aerospace applications offer corrosion protection using environmentally friendly materials. They also provide corrosion and mechanical damage sensing capabilities, optimal adhesion properties, and enhanced fatigue resistance. Desired attributes for nanomaterials in aircraft engines include high strength, lightweight properties, and thermal stability. Many aerospace companies, airline operators, and agencies are exploring the integration of nanocoatings to impart specialized characteristics to aircraft frames, interiors, engine parts, and other component surfaces. These coatings offer benefits such as increased hardness, improved fuel efficiency, self-cleaning capabilities, resistance to wear and corrosion, enhanced thermal performance, flame retardancy, reduced environmental impact, lower maintenance expenses, erosion protection, and drag reduction for fuel-saving advantages. Advancements in nanotechnology have enabled the development of multifunctional aerospace coatings utilizing nanodimensional materials, which provide notable property advantages, potentially enhancing the performance and functionality of aerospace structures. However, careful considerations regarding stability, mechanical properties, and optical properties are crucial, as they may be affected when utilizing nanodimensional materials in aerospace coatings.

Taylor and Sieradzki [42] have extensively studied the development of multifunctional aerospace coatings, emphasizing the advantages and potential disadvantages of nanodimensional materials. Their research underscores the importance of stability, mechanical properties, and optical properties in optimizing the functionality of aerospace coatings, particularly in addressing the unique challenges faced in the aerospace sector.

3.7 Challenges and future directions

The primary challenge in utilizing nanoparticles for coating purposes is achieving dispersion and stability, as nanoparticles tend to agglomerate due to their high surface energy and large surface area. Additionally, reducing pigment size to the nano level may cause a loss of color and opacity. To address this, stable binders are necessary to prevent the photocatalytic activities of nanoparticles like nano TiO_2. Hardening issues may arise with ultrafine powders, and extensive nanoparticle usage could lead to new environmental concerns, such as the generation of toxic materials and hazards, including the catalysis of chemical reactions inside the body, posing potential dangers. The promising future of nanocoatings lies in advancements in nanotechnology, offering enhanced properties. The automotive industry stands to gain significantly from nanocoatings, to formulate systems where the interior and exterior possess distinct properties. Creating coating systems with highly soluble nanoparticles, even at high concentrations without viscosity increase, and developing hybrid coatings could ensure effective protection against environmental factors.

3.8 Conclusion

Nanopaints offer significant benefits for asthma patients and have wide-ranging applications in various settings including hospitals, homes, industrial locations, and storage rooms. They demonstrate effectiveness in mitigating unpleasant odors and preventing issues stemming from dampness, such as the detachment of plastering materials from structures. Furthermore, nanopaints and coatings play a crucial role in asset maintenance and function as waterproofing agents. Through the integration of multifunctional materials, they enhance anticorrosive properties, forming protective surface layers that mitigate pollution damage. By employing hybrid technology, nanopaints exhibit a range of properties, including oil repellency, which proves advantageous in surface maintenance. Surface preparation techniques, such as optimizing surface energy and roughness, contribute to the efficacy of nanopaint applications. TiO_2 present in nanopaints undergoes oxidation, enhancing safety for human exposure and acting as a catalytic agent to improve formulations in nanopaint solutions. Nanopaints distinguish themselves from traditional conversion paints due to the higher concentration of atoms on the surface of nanoparticles, resulting in enhanced performance and broader applicability.

3.8.1 Implications for research and industry

The market size of nanopaints and coatings is projected to reach $13.2 billion by 2026, reflecting a compound annual growth rate of 4.8% during the period 2021–2026. The primary catalyst for this growth is the escalating demand for nanotechnology in the paints and coatings sector. Key nanotechnology resins, such as graphene, CNTs, and nanosilver, enjoy widespread adoption in this industry. The increasing interest in nanopaints and coatings is driven by their multifaceted applications across diverse end-use industries. These applications target the enhancement of product attributes such as abrasion and wear resistance, chemical and corrosion resistance, stealth/IR shielding effects in defense applications, antireflection properties, and flame retardancy. The rapid expansion of the automotive and marine industries has additionally stimulated demand for superhydrophobic coating nanopaints and coatings, thus further propelling the market growth. Furthermore, the flourishing aerospace industry is anticipated to significantly contribute to the expansion of the nanopaints and coatings market in the forecast period. COVID-19 Impact: However, the nanopaints and coatings materials market has been adversely affected by the COVID-19 pandemic, primarily due to a reduction in the production of new aircraft across various regions. The Middle East aviation industry and economies have been particularly affected by the suspension of air traffic. According to the International Air Transport Association, global air traffic for the full year of 2020 is expected to decline by 51% compared to 2019.

The Welsh Government services and information indicate that the demand for production of short-haul aircraft is anticipated to decrease by 15% globally (resulting in approximately 1500 fewer aircraft out of 10,000) and by 30%–65% for long-haul aircraft by the end of 2025 (equating to around 600–1300 fewer aircraft out of 2000). This decline in aerospace manufacturing is projected to impede the growth of the nanopaints and coatings market across various regions. Nanocoatings are materials engineered through the reduction of materials at the molecular level, resulting in a denser product. The introduction of nanoparticles brings forth numerous advantages and opportunities for the paint and coating industry, which has been quick to harness the potential of nanotechnology. The incorporation of nanoparticles into coatings can enhance various properties of coating systems, yielding multipurpose coatings with minimal cost differences. These coatings, often comprised of self-assembling monolayers, find applications ranging from scratch-resistant coatings on glass to self-cleaning surfaces, moisture absorption, antigraffiti coatings, and corrosion resistance. Moreover, the unique composition, improved strength and flexibility, as well as excellent gloss and transparency render nanocoatings highly effective. Many nanoparticles, such as nano-ZnO, are nontoxic, providing an additional advantage to the coating industry. Nanocoatings can be applied through various methods, including chemical vapor phase deposition, physical vapor phase deposition, electrochemical deposition, sol–gel methods, electro-spark deposition, and laser beam surface treatment.

Key Advantages of Nanocoating:

- Enhanced surface appearance.
- Superior chemical resistance.

- Reduction in permeability to corrosive environments, thereby improving corrosion properties.
- Optical clarity.
- Increased modulus and thermal stability.
- Easy-to-clean surfaces.
- Antiskid, antifogging, antifouling, and antigraffiti properties.
- Improved thermal and electrical conductivity.
- Retention of gloss and other mechanical properties like scratch resistance.
- Antireflective properties.
- Free from chromate and lead.
- Good adherence to various types of materials.

The primary challenge in utilizing nanoparticles for coating purposes lies in achieving dispersion and stability. Nanoparticles tend to agglomerate due to their high surface energy resulting from their large surface area. Additionally:

- Reducing pigments to nano size may lead to loss of color and opacity.
- A stable binder is necessary to inhibit the photocatalytic activities of nano TiO_2.
- Ultrafine powder may encounter hardening issues.
- Extensive nanoparticle usage may introduce new environmental concerns, such as the creation of toxic materials and other hazards. Ultrafine particles have the potential to catalyze chemical reactions within the body, posing potential dangers.

The future of nanocoating appears promising, with advancements in nanotechnology expected to further enhance coating properties. The automotive sector stands to benefit significantly from nanocoating advancements. The future goal for nanocoating is to develop systems where the nanoparticles remain soluble even at high concentrations without increasing viscosity. Hybrid coatings incorporating nanoparticles may offer the required properties for effective coating systems capable of withstanding various environmental impacts.

References

[1] Nano Materials in architecture, Interior architecture and Design. Leydecker, Sylvia (2008). Nano forum.org, European Nano technology Gateway, Nano forum report.

[2] J. Hulla, S. Sahu, A. Hayes, Nanotechnology: history and future, Human & Experimental Toxicology 34 (2015) 1318–1321.

[3] Z. Ullah, Nanotechnology and its impact on modern computer, Global Journals of Research in Engineering 12 (2013) 34–38.

[4] G. Tsaramirsis, A. Kantaros, I. Al-Darraji, D.P.C. Apostolopoulos, A. Pavlopoulou, M. Alrammal, et al., A modern approach towards an industry 4.0 model: from driving technologies to management, Journal of Sensor 2022 (2022) 5023011.
[5] A. Singh, S. Dubey, H.K. Dubey, Nanotechnology: the future engineering, Nanotechnology 6 (2019) 230–233.
[6] M. Roco, B. Müller, E. Wagner, G. Borchard, T. Di Francesco, K. Jurczyk, et al., Nanoscience and Nanotechnology: Advances and Developments in Nano-Sized Materials, De Gruyter, Berlin, Germany, 2018.
[7] G.A. Silva, A new frontier: the convergence of nanotechnology, brain machine interfaces, and artificial intelligence, Frontiers in Neuroscience 12 (2018) 843.
[8] S. Talebian, T. Rodrigues, J. Neves, B. Sarmento, R. Langer, J. Conde, Facts and figures on materials science and nanotechnology progress and investment, ACS Nano 15 (2021) 15940–15952.
[9] J.-C. Jeon, Designing nanotechnology QCA–multiplexer using majority function-based NAND for quantum computing, Journal of Supercomputing 77 (2021) 1562–1578.
[10] O. Bongomin, G.G. Ocen, E.O. Nganyi, A. Musinguzi, T. Omara, Exponential disruptive technologies and the required skills of industry 4.0, Journal of Engineering 2020 (2020) 4280156.
[11] N. Kumar, A. Dixit, Role of nanotechnology in futuristic warfare, Nanotechnology for Defence Applications, Springer: Berlin/Heidelberg, Germany, 2019, pp. 301–329.
[12] G. Opait, Nanotechnology and drones, boom-art in the advanced tech, Economics and Applied Informatics 1 (2020) 93–104.
[13] K.D. Sattler, Handbook of Nanophysics: Nanomedicine and Nanorobotics, CRC Press, Boca Raton, FL, USA, 2010.
[14] J. Santos, R.S. de Oliveira, T.V. de Oliveira, M.C. Velho, M.V. Konrad, G.S. da Silva, et al., 3D printing and nanotechnology: a multiscale alliance in personalized medicine, Advanced Functional Materials 31 (2021) 2009691.
[15] B.E. Yamamoto, A.Z. Trimble, B. Minei, M.N. Ghasemi Nejhad, Development of multifunctional nanocomposites with 3-D printing additive manufacturing and low graphene loading, Journal of Thermoplastic Composite Materials 32 (2019) 383–408.
[16] S. Becker, Nanotechnology in the marketplace: how the nanotechnology industry views risk, Journal of Nanoparticle Research 15 (2013) 1426.
[17] R.K. Goyal, The role and applications of nanomaterials in the automotive industry, Nanomaterials and Nanocomposites, CRC Press, Boca Raton, FL, USA, 2021, pp. 51–59.
[18] I.E. Ekengwu, O.G. Utu, C.E. Okafor, Nanotechnology in automotive industry: the potential of graphene, Nanotechnology 9 (2019) 31–37.
[19] P. Zhang, L. Wang, H. Wei, J. Wang, A Critical Review on Effect of Nanomaterials on workability and mechanical properties of high-performance concrete, Advances in Civil Engineering 2021 (2021) 1–24.
[20] V.K. Bajpai, M. Kamle, S. Shukla, D.K. Mahato, P. Chandra, S. KyuHwang, et al., Prospects of using nanotechnology for food preservation, safety, and security, Journal of Food and Drug Analysis 26 (2018) 1201–1214.
[21] S. Palit, Recent advances in the application of nanotechnology in food industry and the vast vision for the future, Nanoengineering in the Beverage Industry 20 (2020) 1–34.
[22] C. Yadav, Food and bioprocessing industry, Handbook of Polymer Nanocomposites for Industrial Applications, Elsevier, Amsterdam, The Netherlands, 2021, pp. 295–324.

[23] L. Anmol, G. Veena, Nanotechnology approach in food science: a review, International Journal of Food Sciences and Nutrition 3 (2018) 183–186.
[24] T. Garde-Cerdán, B.S. Costa, P. Rubio-Bretón, E. Pérez-Álvarez, Nanotechnology: recent advances in viticulture and enology, Journal of the Science of Food and Agriculture 101 (2021) 6156–6166.
[25] T. Ghosh, G.B. Raj, K.K. Dash, A comprehensive review on nanotechnology based sensors for monitoring quality and shelf life of food products, Measurement: Food (2022) 100049.
[26] D. Thangadurai, J. Sangeetha, R. Prasad, Nanotechnology for Food, Agriculture, and Environment, Springer, Berlin/Heidelberg, Germany, 2020.
[27] X. Hea, H. Deng, H.-M. Hwang, The current application of nanotechnology in food and agriculture, Journal of Food and Drug Analysis 27 (2019) 1–21.
[28] D. Kalita, S. Baruah, The impact of nanotechnology on food, Nanomaterials Applications for Environmental Matrices, Elsevier, Amsterdam, The Netherlands, 2019, pp. 369–379.
[29] A. Bhardwaj, A. Bhardwaj, A. Misuriya, S. Maroli, S. Manjula, A.K. Singh, Nanotechnology in dentistry: present and future, Journal of International Oral Health 6 (2014) 21.
[30] A.E. Fadiji, D.M.N. Mthiyane, D.C. Onwudiwe, O.O. Babalola, Harnessing the known and unknown impact of nanotechnology on enhancing food security and reducing postharvest losses: constraints and future prospects, Agronomy 12 (2022) 1657.
[31] G. Pandey, Challenges and future prospects of agri-nanotechnology for sustainable agriculture in India, Environmental Technology & Innovation 11 (2018) 299–307.
[32] J. Xue, Y. Luo, Sustainable food and agriculture system: a nanotechnology perspective, ES Food & Agroforestry 5 (2021) 1–3.
[33] K. Vishwakarma, N. Upadhyay, N. Kumar, D.K. Tripathi, D.K. Chauhan, S. Sharma, et al., Potential applications and avenues of nanotechnology in sustainable agriculture, Nanomaterials in Plants, Algae, and Microorganisms, Academic Press, Cambridge, MA, USA, 2018, pp. 473–500.
[34] J. Schulte, Nanotechnology: Global Strategies, Industry Trends and Applications, John Wiley & Sons, Hoboken, NJ, USA, 2005.
[35] S. Leydecker, Nano Materials in Architecture, Interior Architecture and Design, first ed., BirkhauserVerlag AG, Berlin, 2008.
[36] E. Karana, O. Pedgley, V. Rognoli, Materials Experience: Fundamentals of Materials and Design, 1st ed., Butterworth-Heinemann, USA, 2013.
[37] R. Lumley, A. Morton, Nanoengineering of metallic materials, in: R.H.J. Hannink, A.J. Hill (Eds.), Nanostructure Control of Materials, first ed., Woodhead Publishing Ltd, 2006, pp. 219–247. Ch. 9.
[38] Nanomaterials, Nanotechnologies and Design: An Introduction for Engineers and Architects, in: M.F. Ashby, P.J. Ferreira, D.L. Schodek (Eds.), Elsevier Ltd., 2009.
[39] H. Hosono, Y. Mishima, H. Takezoe, K.J.D. MacKenzie, Smart coatings. Multilayered and multifunctional in situ ultrahigh-temperature coatings, Nanomaterials: From Research to Applications, Elsevier, Amsterdam, 2007.
[40] Nanotechnology and Construction. November 2006.
[41] Z. Ranjbar, S. Rastegar, Nanomechanical properties of an automotive clear coats containing nano silica particles with different surface chemistries, Progress in Organic Coatings 72 (2010) 40–43.
[42] S.R. Taylor, K. Sieradzki, The development of a multi-functional aerospace coating: considerations in the use of nano-dimensioned materials, Progress in Organic Coatings 47 (2003) 169–173.

4

Conventional methodology and advanced techniques employed in nanopaints

Anand Sawroop Khanna
SSPC India, Mumbai, India

4.1 Introduction

Basically, a paint coating consists of a resin in a solvent, to which one or more pigments are added along with different additives, which help in providing several micro-level benefits to the coating properties. Resin is the main component that provides thickness and gives barrier properties; pigments give color, strength, and corrosion resistance, while several additives improve the properties of the coatings, such as fast drying, UV resistance, dispersion, etc. There is a broad classification of coatings in terms of single and two components, single layer versus multilayer coating systems (Fig. 4.1), conventional versus high-performance coating, and functional and smart coatings. Fig. 4.2 summarizes this classification [1].

The process of paint manufacturing is quite simple. It involves the choice of a suitable resin, which depends upon its chemistry: alkyd, vinyl, epoxy, polyurethane, etc., which are chosen depending upon the severity of the environment.

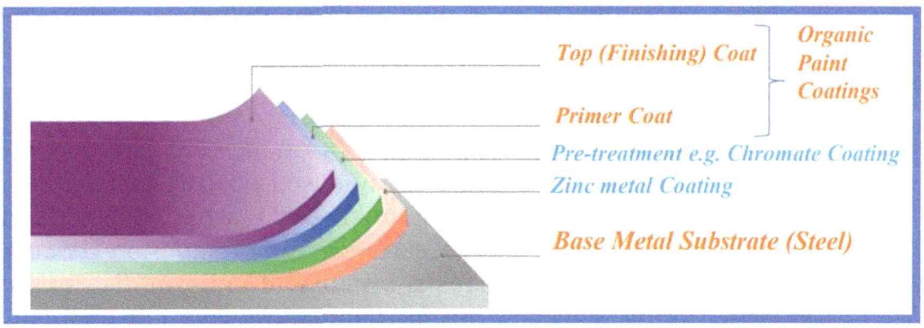

Figure 4.1 A typical multicoat paint system usually practiced in many industrial applications.

Nanopaints. DOI: https://doi.org/10.1016/B978-0-443-26512-9.00004-5

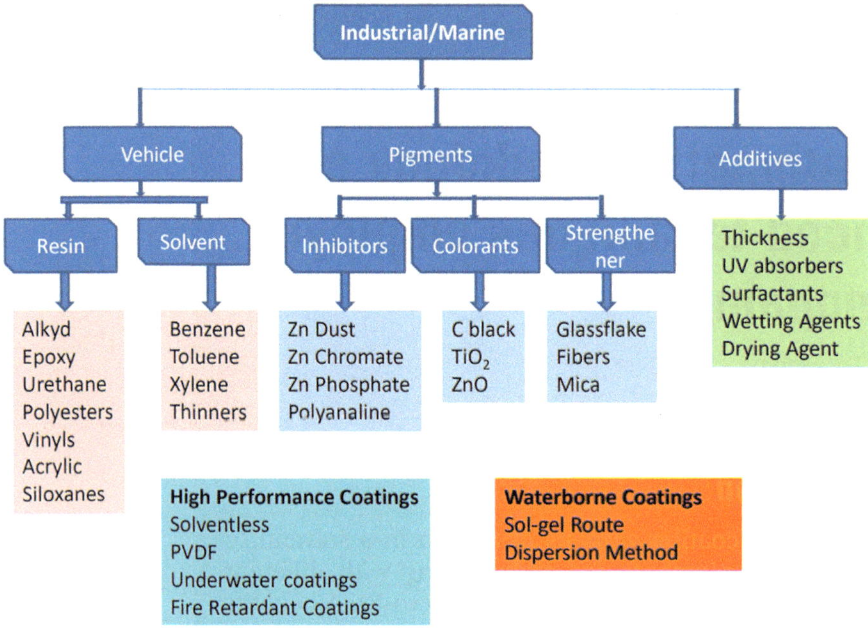

Figure 4.2 Classification of various paint coatings.

The chosen resin is added to a suitable solvent, usually benzene, xylene, and toluene. To this are added the pigments and additives, which are well dispersed to get a uniform paint composition.

In most of the conventional paints, the pigments and additives used are usually in micro size. However, if the size of these is decreased, it has been found that the performance of one or more properties of the paint starts improving. And when the size of particles is reduced to nanodimensions (1 to 100 nm), then the properties of the paint are found to be the most superior. This is shown in Fig. 4.3, where it has been shown that when the size of one or more pigments and/or additives is decreased to nanosize, the paint performance properties reach a peak [2]. This happens due to very tiny particles, dispersed uniformly and having a large surface area, resulting in enhanced paint coating performance. To cite an example, the UV resistance of an epoxy-based coating can be enhanced by adding about 10% of micro-ZnO, while the addition of just 0.03% of nano-ZnO can give much superior UV resistance to the same paint. This is because of uniform distribution at optimum concentration for nano-sized ZnO particles compared to a random distribution with relatively less surface area as shown

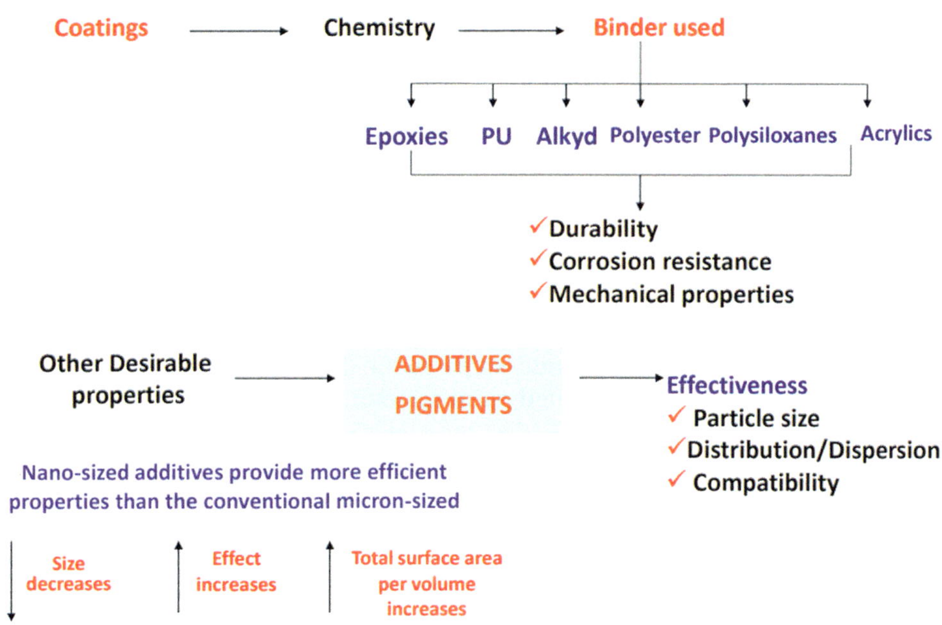

Figure 4.3 Schematic of the choice of raw materials for paint and their size effect.

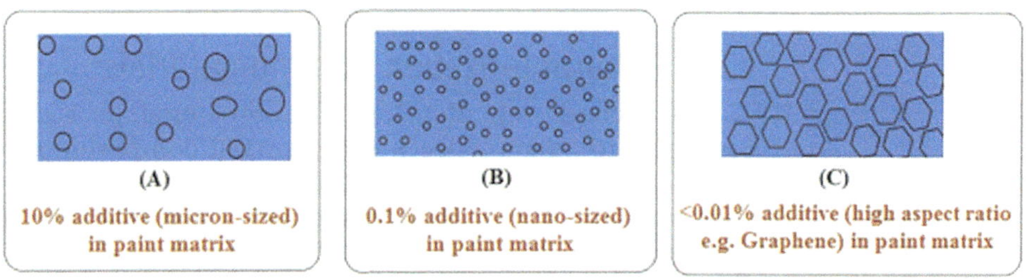

Figure 4.4 Showing the (A) distribution of 10% micro-ZnO particles in epoxy matrix, (B) 0.03% nano ZnO particles in epoxy matrix, and (C) 0.1% nanographene particles in Epoxy matrix.

in Fig. 4.4(A and B), where it can be seen how micro-sized ZnO particles are randomly distributed, leaving lots of epoxy matrix space where UV light hits and deteriorates the paint system, while Fig. 4.4(B) shows a very even distribution of nano ZnO particles in the resin matrix, which does not allow the epoxy resin to be attacked by UV light. In the same way, it can be seen that a 2D nanographene distribution in an epoxy matrix, as shown in Fig. 4.4(C), helps in enhancing the performance of the epoxy coating.

4.2 Synthesis of nano-modified paints

A conventional paint or coating is called a nanopaint or coating when a nano-sized pigment or additive is added or incorporated into it. As a result, it provides enhanced performance of one or more of its properties compared to its properties when the same pigment or additive is added in macro/micro size. As discussed above, nanomaterial, by virtue of its very high surface area, occupies a given surface at a significantly lower dosage of incorporation, thereby reducing the material consumption in coating formulation. Given below are the systematic steps to be followed in order to synthesize a nano-modified paint system:

4.2.1 Identification of nanomaterial

The first step in developing or formulating a nanopaint or coating is to identify the nanomaterial in an appropriate physical state (powder/dispersion/formulated additive). Nanomaterial identified thereby shall also be known in literature for providing the intended performance property. A few such examples are given in Table 4.1.

4.2.2 Characterization of nanomaterials

The second step of using a nanomaterial in a paint and coating is to characterize it for chemical composition, particle shape, particle size, etc., using one or more or a combination of advanced characterization techniques such as scanning electron microscopy (SEM), TEM, XRF, XRD, and many other electron-optical techniques.

It is found that it is possible to synthesize a nanoparticle based upon the method of synthesis. For example, ZnO can be synthesized in spherical form or in flake form. Fig. 4.5 shows nanoparticles of ZnO, TiO_2, and Fe_2O_3 in different sizes and shapes. Based upon their size and shape, each of these nanoparticle has a different effect on the properties of the coating. The effect would be different depending upon the same shape having different concentrations or the same concentration with different shapes. This is shown in Figs. 4.6–4.8 for various nanoparticles on different paint properties. For example, in Fig. 4.6, Yellowing effect of a white epoxy based coating was studied as a function of the concentration of nano-ZnO with spherical and flake shapes with sizes varying from 30 to 90 nm.

Table 4.1 A tentative list of nanoparticles and their effects on coatings [1–4].

Function	Nanomaterial (examples)	Advantage/effect	Application
UV protection, IR reflective or IR absorbing.	TiO_2; ZnO, CeO_2, iron oxide pigments (transparent iron oxide; needle-shaped particles with a length of 50–100 nm and width of 2 nm)	Enhanced UV resistance, blocking of IR and visible light, indoor climate control.	Construction (facades), wood preservation, glass, plastics.
Scratch resistance.	Oxide (synthetic amorphous silica), SiO_2, Al_2O_3)	Improved scratch resistance.	Automotive, information and communication, parquet flooring, consumer goods (furniture), optics (lenses).
Color brilliance, shade, color effects.	Carbon black; Oxides (TiO_2, Fe_2O_3, Fe_3O_4, SiO_2, Cr_2O_3) on mica flakes or SiO_2 spheres, with metal pigments, ZnO	Intensify effects of metal pigments; stabilize pigments and fillers; positive effects in dispersion paints; prevent crack formation (phyllosilicates/sheet silicates); improve resistance to fading.	Automotive, consumer goods (furniture), construction.
Self-cleaning (easy-to-clean).	Organic–inorganic hybrid polymers (organically modified ceramics), nanosilica/colloidal silica embedded in resin particles following polymerization; silanes (silicon-based mixtures with other chemicals, e.g., fluorine compounds); TiO_2	Dirt and water repellent, Protection against algae and fungi; antigraffiti protection: easy removal of unwanted paint.	Automotive, construction (facades), glass.
Switchable (electrochromic, photochromic, thermochromic).	Tungsten oxide (WO_3) (electrochromic)	Color effects.	Automotive.
Self-assembly.	Polymer gel, specific organic–inorganic hybrid polymers	Self-healing surfaces.	Automotive, cosmetics.
Monolayer adhesive films.	Polymers	Ultra-thin layers.	Automotive, consumer goods.
Corrosion protection, wood preservation.	Zinc or aluminum coated with nano-TiO_2, nano-clay (like hydrotalcite $Mg_4Al_2(OH)_{12}CO_3xH_2O$)	Nanoclay coatings delay the fading of wood (which is a result of the bleeding of complex chemicals like tannins).	Construction, automotive, wood preservation.

(Continued)

Table 4.1 (Continued)

Function	Nanomaterial (examples)	Advantage/effect	Application
Optimized flow characteristics.	Oxide (synthetic amorphous silica)	Generate new rheological properties (elasticity, flow characteristics, thixotropy).	Various.
Conductive coatings for electro-static paint spraying.	Carbon: Fullerenes, carbon nanotubes (CNT), graphene	Enhanced spraying processes.	Automotive.
Photocatalytic effect, antimicrobial effect.	TiO_2, ZnO, Ag	Removal of grease, dirt, algae, bacteria, fungi, odourants and pollutants, transformation of NO_x and ozone from the atmosphere into harmless compounds.	Construction (facades, noise barriers, tiles), road surface, vehicles, wood preservation, glass.
Fire retardant.	SiO_2	When a certain temperature is exceeded, a heat-insulating carbon foam layer is created on the wood surface followed by a flame-resistant ceramic layer.	Construction, protection of wood against fire.

(A) ZnO 30nm Granular (B) ZnO 90nm Granular (C) Nano ZnO flakes

(D) Rutile 130nm (E) Fe_2O_3 75nm Sphere (F) Fe_2O_3 Rods 180nm

Figure 4.5 Nano particles with different shapes and sizes.

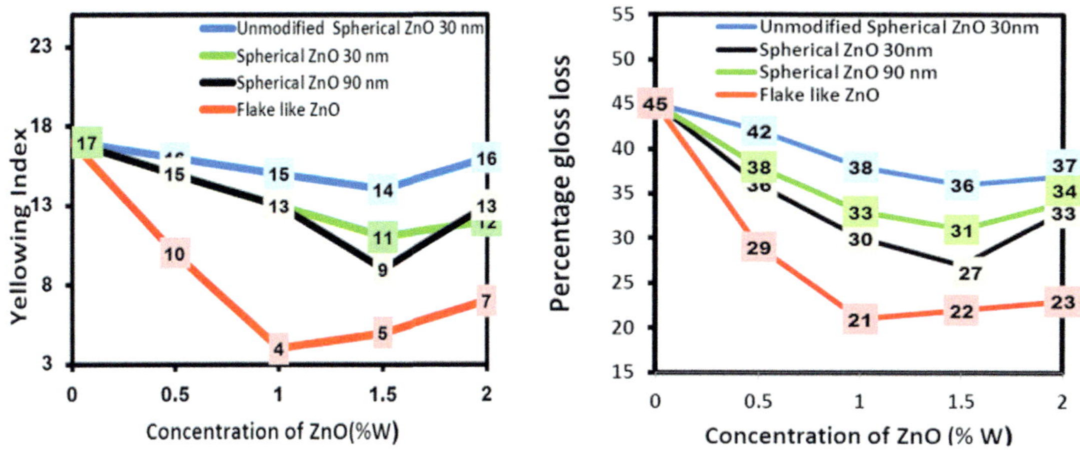

Figure 4.6 Effect of concentration and shape and size of nano ZnO on the yellowing effect and gloss of epoxy based coating.

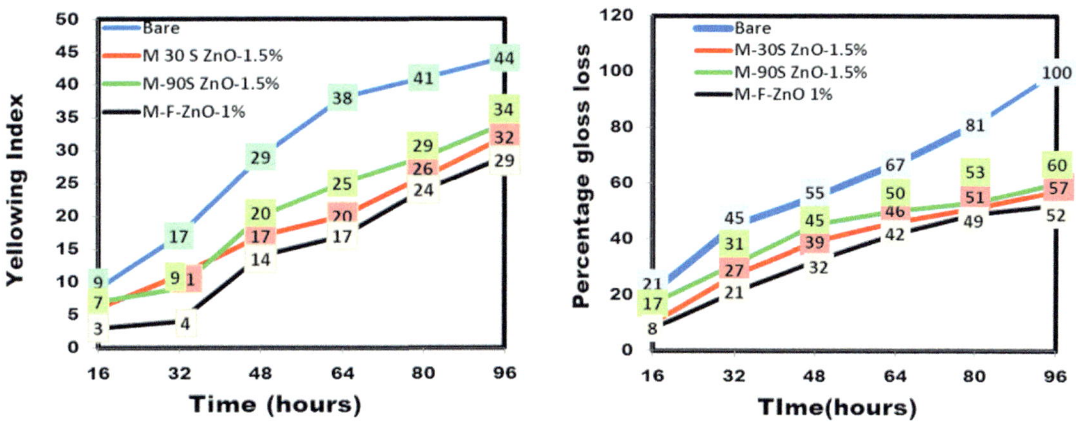

Figure 4.7 Effect of time of exposure during accelerated weathering test for various shapes and sizes of nano ZnO particles in an epoxy based coating.

In Fig. 4.9, the same two effects were studied at various exposure times in an accelerated weathering test. It shows that nano-ZnO, in flake shape, gives the best results in both tests. Also, the lower size of nanoparticles, gives the lesser yellowing and gloss change [5].

Fig. 4.8 displays the effect of spherical and rod-shaped nano-Fe_2O_3 on the tensile modulus and Tabor abrasion of an epoxy-based coating containing nano Fe_2O_3. Here, it is clear that the rod-shape Fe_2O_3 gives superior modulus and Tabor abrasion.

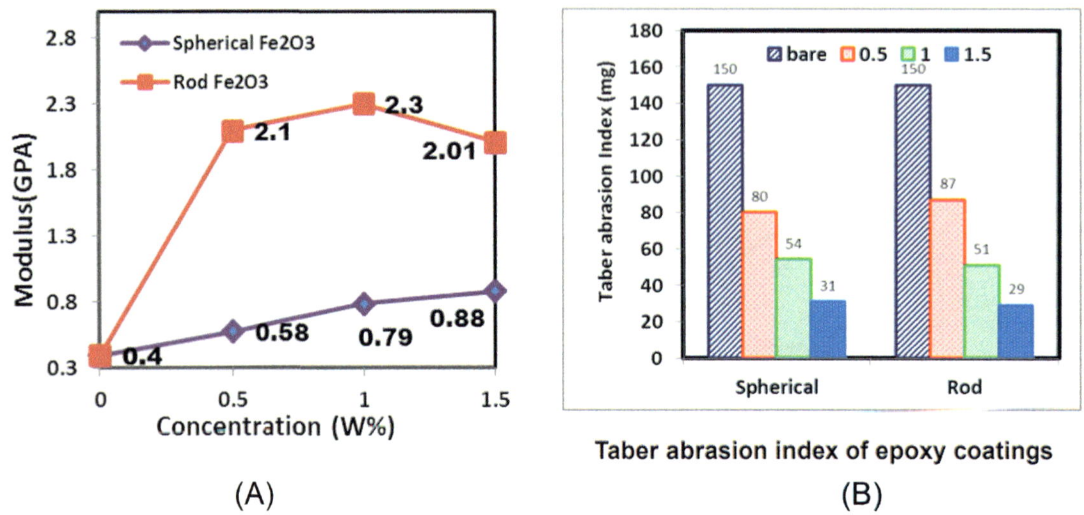

Figure 4.8 Effect of concentration and shape of nano Fe_2O_3 on the (A) Modulus epoxy based coating, and (B) tabor abrasion of epoxy based coatings.

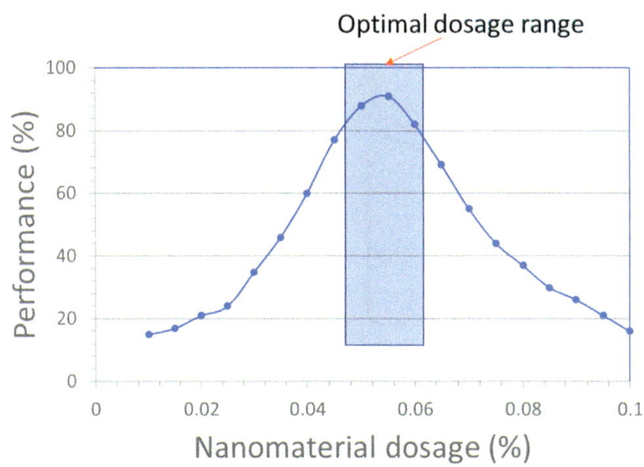

Figure 4.9 Representative graph showing plot of nanomaterial dosage against performance for determining optimal dosage range.

4.2.3 Optimization of dosage of nanomaterial

The last and the most critical step of incorporating nanomaterial into a nano-modified paint or coating is to identify the optimum concentration of the nano additive/pigment, which imparts maximum performance to the paint and coating.

The general expectation from the incorporation of a nanomaterial into a paint coating is that the intended property increases with increasing the nanomaterial concentration or dosage. However, several studies in the literature clarify that the relationship between nanomaterial concentration/dosage and performance property is not linear [6,7]. Factors like shape, size, and type of nanomaterial and its interaction with other components in the system affect the optimum dosage where the performance reaches its peak. For example, in a study aimed at identifying the effect of the concentration of silane-treated TiO_2 nanoparticles on the weathering performance of polyurethane coatings, authors incorporated nanoparticles at different dosages ranging from 0 to 2.0 wt.% and the photocatalytic activity of the coating was examined as a function of nanomaterial concentration. The optimum concentration/dosage of silane-treated nano TiO_2 was somewhere between 0.5 and 1.0 wt.%. On either side of this range, weathering performance was found to be inferior. Therefore, it is critical to run a ladder experiment and clearly identify the optimum dosage of nanomaterial where intended performance is at peak.

Optimal dosage of nanomaterial can be determined through measuring the performance of nanomaterial paint against varying nanomaterial doses. For clarity and better understanding, this has been illustrated here with an example. To develop or formulate a nano-modified two-component (2K) epoxy coating with a good weathering resistance, one can follow the steps detailed below to identify the optimum dosage of additive/nanomaterial.

1. For weathering resistance, choose a nanomaterial that has UV absorption properties, such as ZnO, and synthesize nano-ZnO using a suitable method or purchase it from a supplier.
2. Characterize the synthesized/purchased nanomaterial using SEM, TEM, XRD, UV-visible spectroscopy, etc., to confirm its size, structure, and properties.
3. Usually, the concentration of these nanomaterials is in the range lower than 0.1%. For optimizing the dosage, prepare resin with nanomaterial concentration ranging from 0.01% to 0.1%. Make sure that ZnO is uniformly distributed.
4. On mixing with hardener, coat it on steel panels and allow it to cure.
5. Expose the coated panel to the UV chamber to assess nanomaterial performance.
6. Measure the percentage improvement of the performance.

In order to assess the effect of the shape of ZnO nanomaterial, another set of experiments can be carried out with ZnO with different morphology.

If needed, measure other properties of the film, such as corrosion resistance by salt spray, humidity, mechanical properties, abrasion resistance, etc., so as to gauge the effect of nanomaterials on them.

Plot a graph of the dosage versus performance to determine optimum dosage (a representative graph is depicted in (Fig. 4.9).

Fig. 4.10 shows a narrative of how particle arrangement changes with the concentration of nanomaterial. At a dosage below optimum level, generally particles are not well connected with each other, leading to discontinuous dispersion of particles (Fig. 4.10A). In this case, the intended performance is not optimum. As the concentration of nanomaterial increases, performance improves in a nonlinear fashion and reaches a stage where there are just sufficient well-dispersed nanoparticles that influence each other, leading to optimum performance (Fig. 4.10B). Further increase in concentration of nanomaterial leads to agglomeration and poor dispersion, resulting in reduced performance (Fig. 4.10C). This is explained by ***continuum theory***, according to which each nanoparticle has its range of influence, which reduces to almost nil beyond this range. Therefore, the optimum effect of each nanoparticle would reach maximum when there is the best distribution, as shown in Fig. 4.10(A). In case nanoparticle is overcrowded (Fig. 4.10C) or nonuniformly distributed (Fig. 4.10B), the range of influence gets disturbed and the optimum property is lost.

As nanomaterial properties depend on many factors, their dosage in paint and coatings too varies significantly from one another. Primary parameters for deciding the dosage are their activity and target application. Other factors like cost, health, and environmental impact also need to be considered. Table 4.2

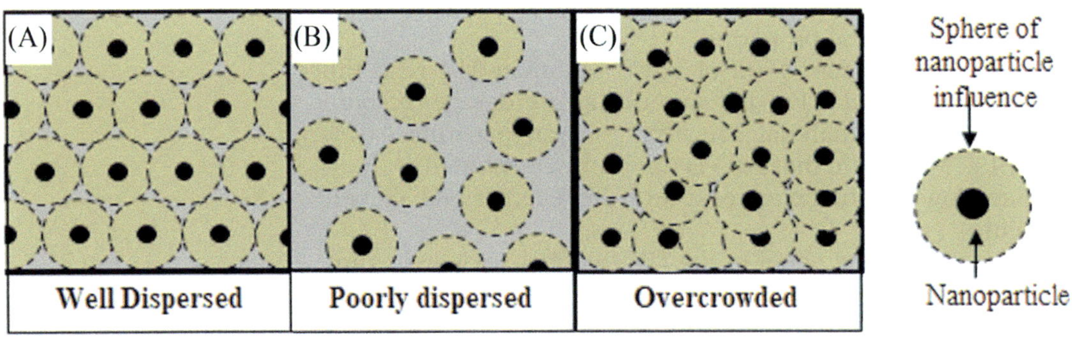

Figure 4.10 Distribution of nanoparticles (A) the best distribution which gives optimum properties, (B) distribution when there are not enough particles, and (C) when there are large number of nanoparticles, which leads to agglomeration thus reducing the effect on the property.

Table 4.2 Typical doses of a few nano particle addition in paint coatings with application areas.

Nanomaterial	Application area	Dosage range (wt.%)
Silver	Antimicrobial	0.002–0.01
ZnO	UV absorption	0.05– 0.5
	Antimicrobial	0.2–1
Graphene	Protection performance	0.2–1
Silica	Mechanical, scratch resistance	Up to 7
Clays	Gas barrier	Up to 10

gives the doses for a few typical nanomaterials for various application areas.

Since nanomaterial dosages are generally very low, it is very important to make sure they are uniformly dispersed across the paint film postapplication. Uneven dispersion leads to reduced activity of nanomaterials, which demands higher dosage. The most important step is to disperse them with a high shear process such as high-speed mixers or preferably with sonication. Wherever possible, attempts should be made to minimize the interaction of nanomaterials with paint ingredients. Unwanted interaction would reduce the activity of nanomaterial.

4.2.4 Health impacts

Handling of nanomaterials during the synthesis of paint coating is very important. Nanomaterials can enter the human body through inhalation during synthesis, processing, grinding, milling operations, or spraying of nanomaterial-loaded paint, which poses potential health risks. Fine droplets of the size range of a few to 100 nm generated from aerosol spray can reach lungs easily, which may lead to alveolar collapse [8]. Studies on chronic inhalation toxicity of nanomaterials show clear effects in rats as inflammatory reactions and sometimes generate tumors. Apart from inhalation, nanomaterial scan also enter body through dermal absorption. Healthy skin is known to be a good barrier against nanomaterials (TiO_2, ZnO) found in sunscreens. However, the studies conducted showed that few nanomaterials, like gold nanoparticles and quantum dots, can enter deeper layers of skin in mice [9] and pigs [10]. Most of these nanomaterials studied are in the size range of 10–15 nm. Under normal working conditions, such small-sized nanomaterials are unlikely to be released as free particles from paint or coatings. A new electromagnetic separation technique has been discovered to separate nanoparticles from the waste paint solutions [11].

4.3 Types of nano-modified paint systems

This can be divided into two sections. Section 4.1 will deal with the formation of nano-sized polymers with and without the addition of nanoparticles. Here, the paint coatings are made by special techniques such as the sol-gel technique, and the coatings formed are very thin, ranging from 1 to 5 microns. These coatings are highly corrosion resistant, scratch resistant, and can be applied on very sensitive components such as glass and mobile covers. The application areas that they cover are self-cleaning coatings, self-healing coatings, antifog, antiglow, scratch resistance, etc.

In the second section, we would deal with thick coatings where typical paint properties such as corrosion resistance, strength, UV resistance, etc., are modified simply by the addition of nanoparticles with proper dispersion in the resin matrix. Tough and hard coatings with high corrosion resistance can be formed as a result.

4.3.1 Section 1

One of the advantages of the addition of nano particles to coatings is that it makes them smart; that is, in addition to their protection of the substrate, they would also trigger an action, such as self-cleaning coatings, self-healing, conducting, etc. The best way to make such coatings is by the sol-gel method.

The sol-gel process is a wet-chemical technique used for making thin and protective coatings. In this process, the sol (or solution) evolves gradually toward the formation of a gel-like network containing both a liquid phase and a solid phase. Typical precursors are metal alkoxides and metal chlorides, which undergo hydrolysis and polycondensation reactions to form a colloidal solution.

The most common sol-gel process is to start with a silane such as tetra-methoxy silane $Si(OCH_3)_4$, which in water hydrolyzes, and gives silicon tetrahydroxyl. This, when treated with an organic resin, for example, epoxy resin, having at least one –OH group in its long chain, such as GPTMS, which has at least one –OH group in the chain, undergoes a polycondensation reaction leading to the formation of a precursor, which basically is an inorganic–organic hybrid. –Si-O– bond with metal gives inorganic properties, such as a chemical bond with a steel substrate, while organic resin gives barrier properties, and the molecule formed has nano dimensions. Let us now take the example of generating an epoxy-based precursor. For this we choose a GPTMS (3-Glycidyloxypropyltrimethoxysilane), which is a bifunctional organosilane with three methoxy groups on one side and an epoxy ring on the other. To create a strong epoxy precursor, this can be treated with MTMS, which would give a primary sol as shown in Fig. 4.11. To this now, if treated with an amine hardener such as HMMM, results in a coating, which has much superior properties than a normal 2K epoxy coating.

Now if we want to create a hydrophobic coating using the above coating, it is required to react with a fluoro-based resin, which in turn would result in producing a trifluoro resin that has excellent hydrophobicity, as shown in Fig. 4.12 (Fig. 4.13).

Such a coating is very stable and inert and has almost no reactivity. That is why even water is unable to spread on it. It makes very little contact with the coated surface. The water droplets, when added on the surface of this coating, show hydrophobicity; that is, the water droplets do not spread but form a tiny contact with the coated surface [12–14]. On measuring the contact angle of these water droplets, it was found that it was between 107 and 110 degrees, confirming the surface as hydrophobic in nature. The way the water droplets are stationary on the surface confirms the water droplets have been

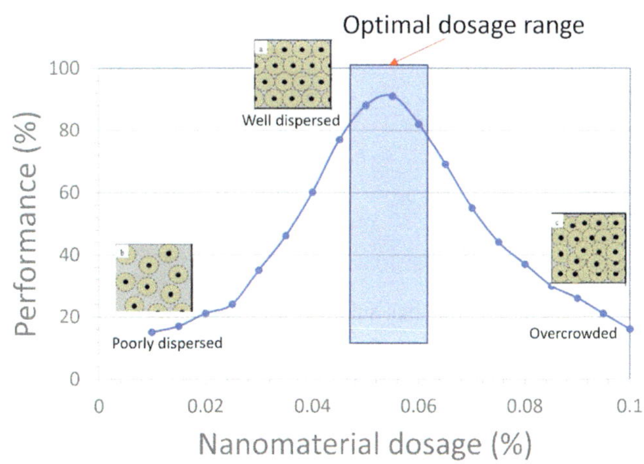

Figure 4.11 Schematic of arrangements of nanoparticles in the paint coating system based upon whether there is optimum less or more loading of nanomaterials.

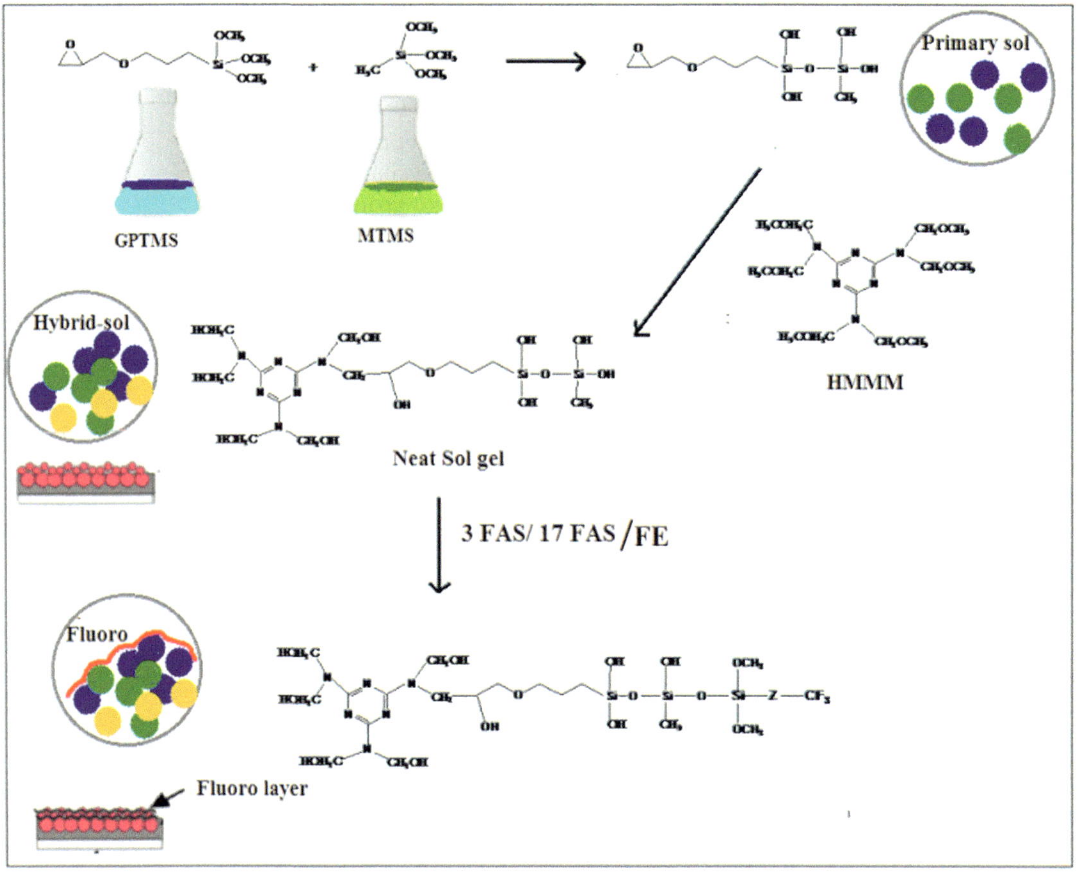

Figure 4.12 Schematic of sol-gel process leading to the formation of tri-fluro resin.

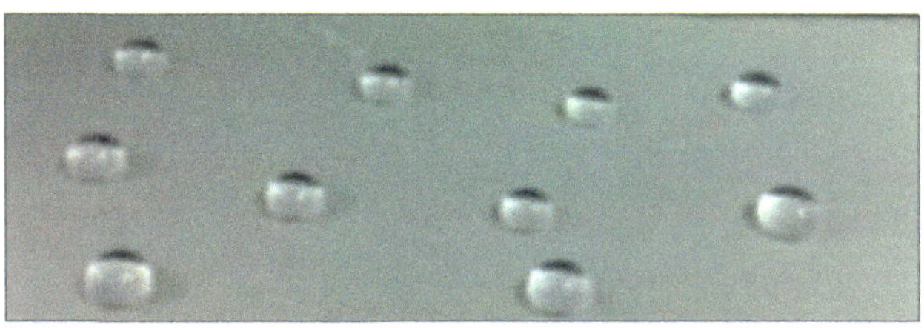

Figure 4.13 Confirms the sol-gel coating formed has suitable hydrophobicity.

struck on the large surface roughness of the coated surface. It was estimated it was in the range of 75–80 nm, as shown in Fig. 4.14(A). Now, in order to release these drops, it is required to reduce the surface roughness lower than 30 nm, which

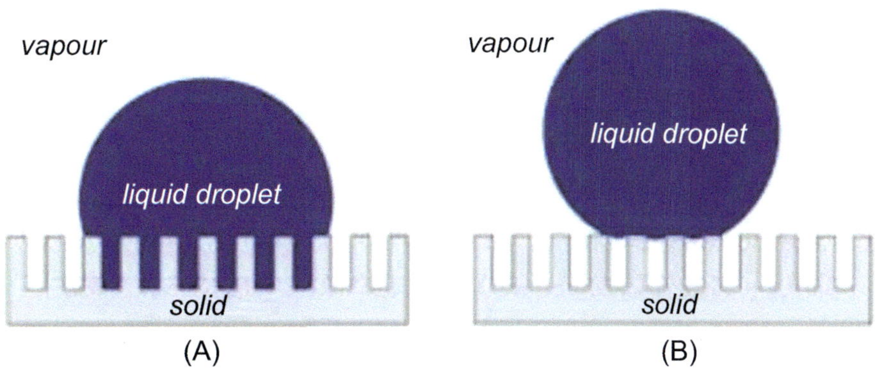

Figure 4.14 Showing the effect of nano particles such as nano ZnO on reducing the surface rougness of the coating resulting in the release of water drop which now slides on a small tile.

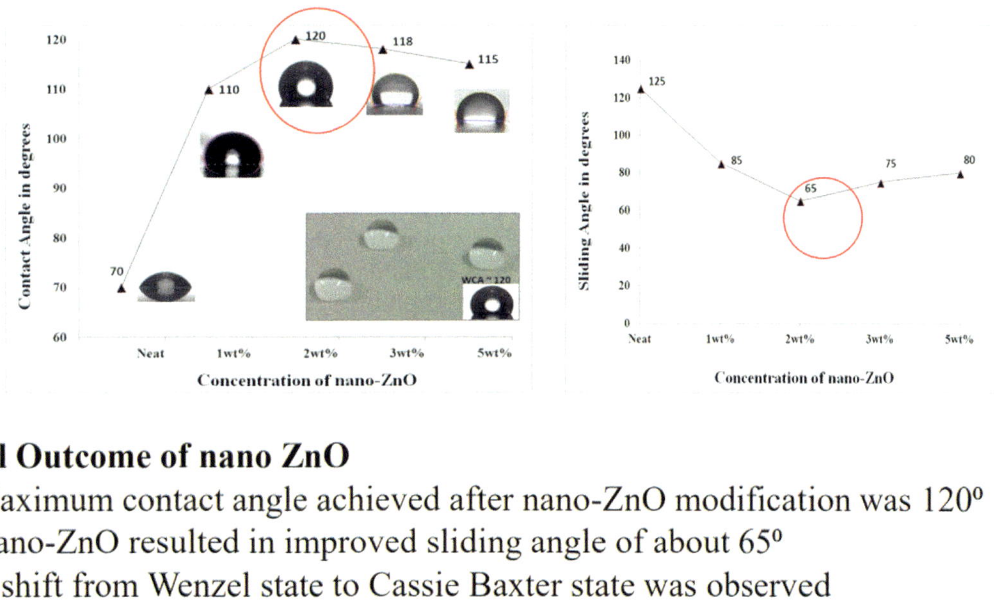

Figure 4.15 Showing the effect of 30 nm ZnO nanoparticles addition, which reduced the surface roughness below 35 nm and the sliding angle to 65 degrees resulting in self-cleaning activity.

results in making these water drops lose on the surface, and on slight tilting the surface, the water drops move and result in self-cleaning of the surface. This is shown in Fig. 4.15, where the flluronated epoxy coating was further treated with nanoparticles such as nano-ZnO or nano-SiO_2 with a size of about 30 nm. This results in reducing the surface roughness, which

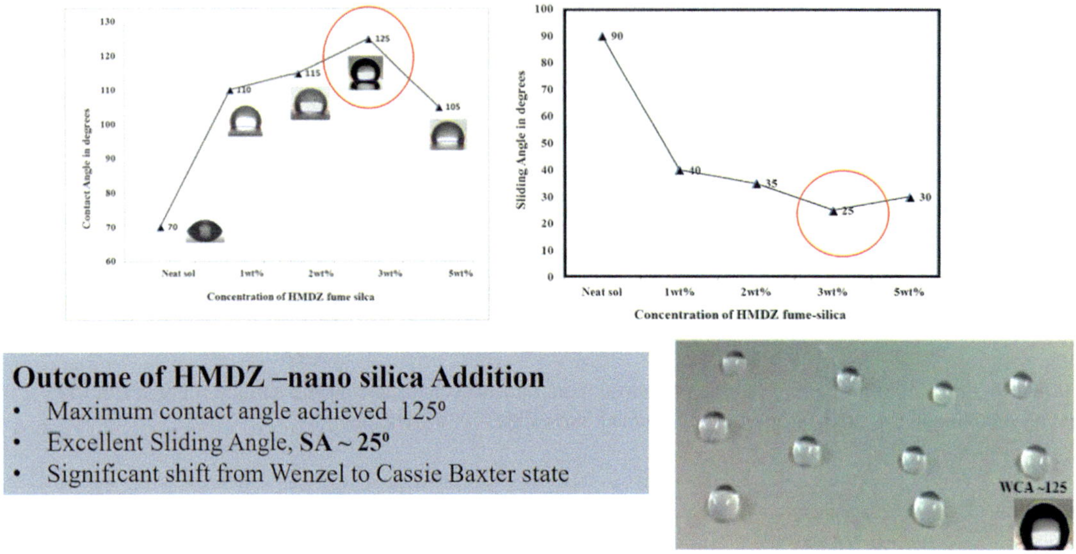

Figure 4.16 Effect of HMDZ-coated silica nanoparticles to enhance the contact angle and reduce the sliding angle of the water droplets to give more efficient self-cleaning ability.

makes the water droplets lose and start sliding at a lower sliding angle. Depending upon the shape, size, and concentration of nanoparticles, it is possible to get various surface roughnesses of the surface and thereby several sliding angles. The lowest sliding angle would be the best. Fig. 4.15 shows that by adding nano ZnO particles of 30 nm, it was possible to achieve a contact angle of 120 degrees and a surface roughness less than 32 nm and a sliding angle of 65 degrees [15].

A similar work using silica nanoparticles, encapsulaled in polymer such as HMDZ, resulted in a better contact angle of 125 degrees and superior sliding angle of 25 degrees as shown in Fig. 4.16.

4.3.2 Self healing

Self-healing provides re-healing of an old coating in case it is damaged by a small scratch or it delaminates or fails during service. The process is very similar to a cut in a human body, which cures after some time. This happens due to movement of healing cells in our blood which reach the damaged portion of skin healing the damage. In the same way, self-healing coatings require the physical presence of healing materials in the coating and a suitable mechanism by which these cells reach the damaged area and heal the coating.

Thus, the main challenges of the self-healing coatings are:

1. Synthesis of healing materials in some form, which allows them to remain in the coating without any deterioration, and when damage occurs in the coating, they reach and heal the damaged area.
2. Some mechanism that allows these healing materials to reach selectively at the damaged location of the coating to begin the healing activity.

Two of the most common strategies are available to tackle this problem: (a) by creating spherical cells, called microcapsules, which hold the healing materials and are sensitive to opening up during heavy physical activity such as scratching, impact, or abrasion of the coating. The traditional concept of a self-healing mechanism using a double-capsule/catalyst system [16] suggests the healing material remain intact until rupture triggers the release of encapsulated content, allowing the healing agent to flow in the crack plane, with further reaction to form a barrier film. Postrelease, the healing agent may undergo a reaction with the functional group of the matrix, oxidation or polymerization of the monomer, chain entanglement, etc., to form a film along the disrupted surface. The stepwise mechanism of self-healing is shown in Fig. 4.17. The efficiency of the

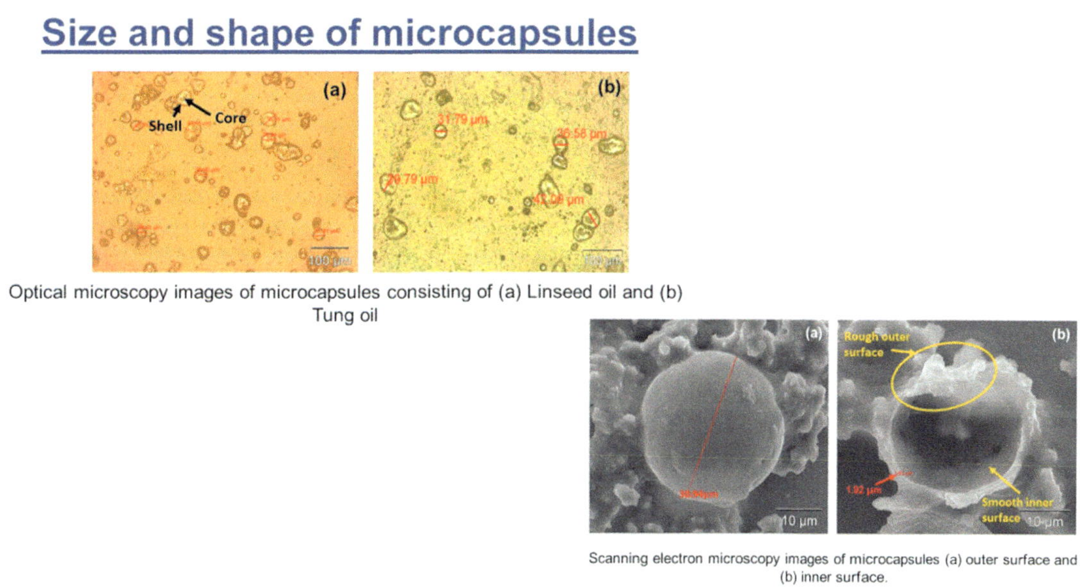

Figure 4.17 (A) Description of the microcappsule formation using formaldehyde and healing material linseed oil and (B) SEM micrograph of the microcapsule containing healing material.

self-healing process is not only influenced by the capillary forces generated due to the cracks but also by the flow properties of the healing agent, that is, viscosity, and the ability of the healing agent to form sufficiently strong bonds with the substrate.

The most challenging step was to create micro-containers, similar to the blood cells, which would retain the healing agent intact, without any deterioration in its chemical structure. As reported, encapsulation of linseed oil and tung oil using a one-step in-situ polymerization technique possesses numerous challenges, that is, gelation, poor control over the mechanism of encapsulation, and unnecessary losses of ingredients. Thus, a modified two-step in-situ polymerization technique was employed, which assisted in better control over tailoring the characteristic properties of microcapsules, such as their size, shape, morphology of the outer shell, and thickness of the shell. Important process parameters such as type of emulsifier, concentration of emulsifier, pH, temperature, stirring speed, and reaction time of encapsulation were optimized [17–19]. The SEM images of the microcapsules are shown in Fig. 4.17.

Further, the microcapsules synthesized at optimal conditions were incorporated in the epoxy coating. As seen in Fig. 4.17, the healing ability of the coating was examined for 48 hours in a laboratory environment using an optical microscopic technique. The coating loaded an optimized concentration of 3 wt.% microcapsules (coating system B) healed seamlessly within 24 hours (Fig. 4.18A–F). The self-healing ability in the coating is provided by linseed oil that is released from the rupturing of the microcapsules; the oil thus released fills the cracks, forming a film due to the oxidative reaction of linseed oil in the presence of atmospheric oxygen and thus dries. For the same reason, coating system A, which is in devoid of a healing agent, that is, linseed oil encapsulated in the synthesized microcapsules, no healing was observed, as can be seen from Fig. 4.18(A–C). The reason can be attributed to the absence of a healing agent (linseed oil) encapsulated in the synthesized microcapsules.

The use of halloysite nanotubes, which are bi-functional; the outer layer is chemically similar to silica, and the inner layer is chemically similar to alumina. This bi-functional nature allows the outer silica layer and inner alumina layer to behave in two distinct manners at the same time, where the surface potentials are governed by the changes in the pH of the medium or surroundings. This unique dual-functional capability allows the use of halloysite nanotubes for potential applications where sustained or controlled release of active material on demand is necessary. Often halloysite nanotubes are employed for

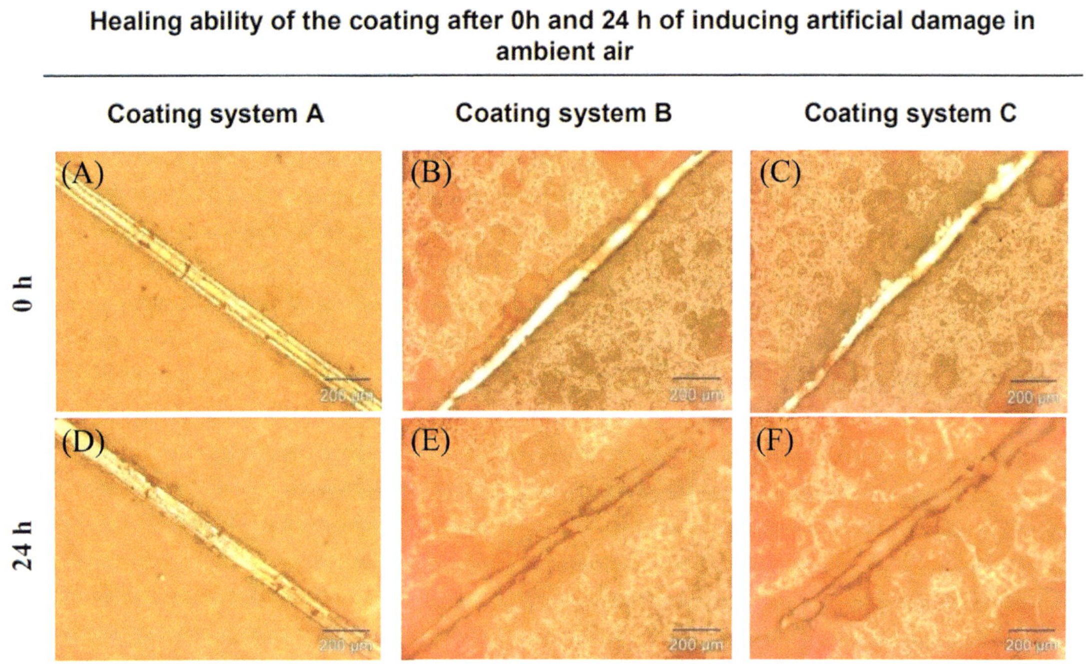

Figure 4.18 Depicting the self-healing process of self-healing using three different concentrations of healing materials.

applications such as drug delivery or release of corrosion inhibitors at the site of corrosion, where their release is triggered due to the changes in the pH occurring in the defect region [16–18].

In acidic medium, the inner alumina layer has positive surface potential (Fig. 4.19), which assists the negative ions of the corrosion inhibitor to chemisorb on the lumen surface, known as the loading of the inhibitor in the halloysite nanotube. The loading steps are based on vacuum cycling of halloysite nanotube suspension in a saturated solution containing a corrosion inhibitor, as schematically represented in Fig. 4.19. However, in alkaline medium, the inner alumina layer has negative surface potential (Fig. 4.19), which results in repelling the negatively charged corrosion inhibitor adsorbed on the lumen surface, assisting in the release of corrosion inhibitor from the lumen surface, schematically represented in Fig. 4.19.

4.3.3 Graphene-based nanocoatings

Graphene is one of the most important and useful nanomaterial ever discovered. It has an umpteen number of applications,

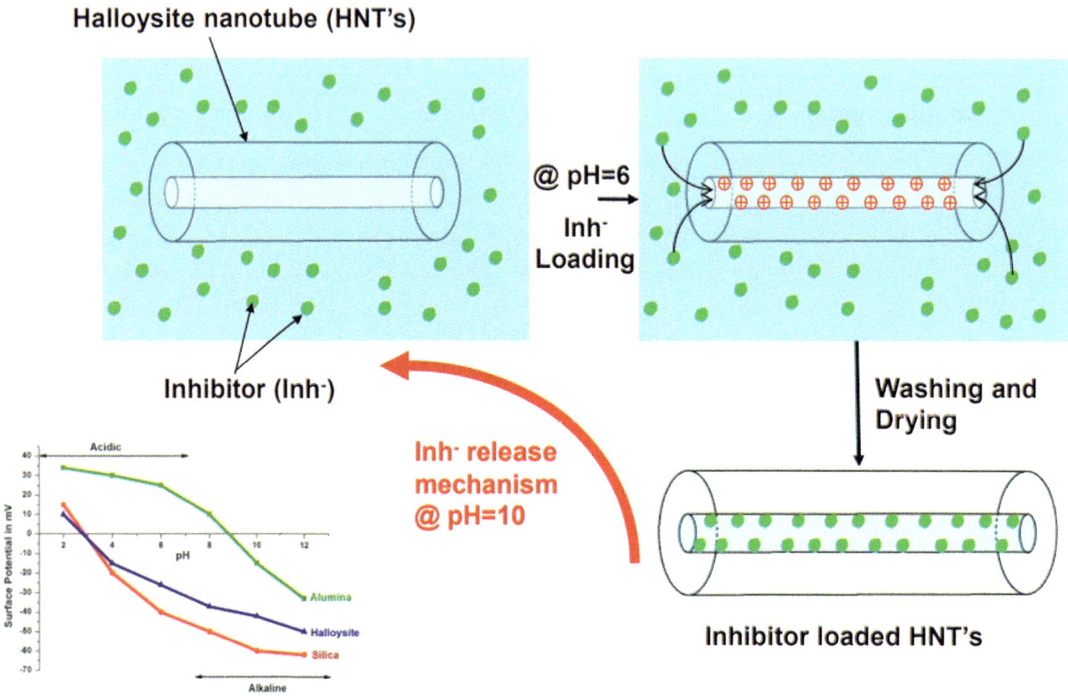

Figure 4.19 Halloycite nano tubes which release healing materials by changing pH of the system.

ranging from medical, construction, electronics, chemical sensors, hydrogen storage, and paint coatings. It has the simplest 2D structure, which imparts several useful properties to materials. Though there are several methods to make graphene, the top-down approach of synthesizing graphene from natural graphite appears to be the best. It can be done using a mechanical method called mechanical exfoliation or chemical exfoliation using strong chemicals. The best results are obtained when using natural graphite sources where the graphene layers are arranged in a lamellar structure, exactly like paper sheets are arranged in a stack of paper, as shown in Fig. 4.20. On blowing the stack, one gets single or multi sheets depending upon how tightly the sheets are struck and what pressure they get from blowing. Separation of of various graphene layers is shown using this principle in Fig. 4.20(B), indicating how a 2D single graphene or multilayer graphene is obtained.

The graphene so obtained is in the colloidal form, which powder graphene can be obtained by evaporation of the water. This is a cumbersome process and requires huge energy and time. So an alternative method of collecting graphene faster is

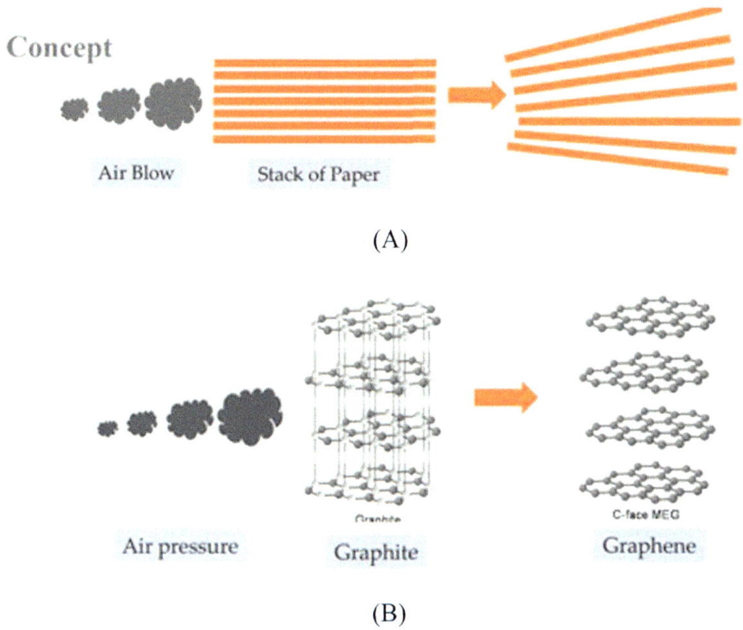

Figure 4.20 Concept of graphene extraction using mechanical exfoliation method.

to functionalize the colloidal solution, which turns graphene into micron-level granules that can be filtered easily using nano and micro filters. These functionalized graphene particles are encapsulated in a polymer layer, which releases pure graphene when added to a paint formulation.

Let us now discuss the graphene-based coating formulations. Graphene can be used to make pure graphene on metal substrates as well as it can be added as a corrosion pigment or additive to enhance the corrosion resistance as well as other properties of the coatings. It must be remembered that graphene is a pure carbon and has no affinity to bind to a metallic surface. Therefore, in order to make a pure graphene-based coating, it must be first made active so that it can bond to a metallic surface. For this, the best method is to make use of the −C-N− bond to bind graphene to an active entity. The sol-gel technique is used to bind graphene to an N-antity, an amino group here. Graphene is treated with an amino silane such as tri-methoxy amino silane, which forms a bond with the amino group to bind graphene and the silane group with its −OH entity after hydration binds to the metallic surface. This way, a strong graphene-based coating is formed with enough corrosion resistance as well as strong resistance to the diffusion of moisture. This is illustrated in Fig. 4.21. The coating formed is less

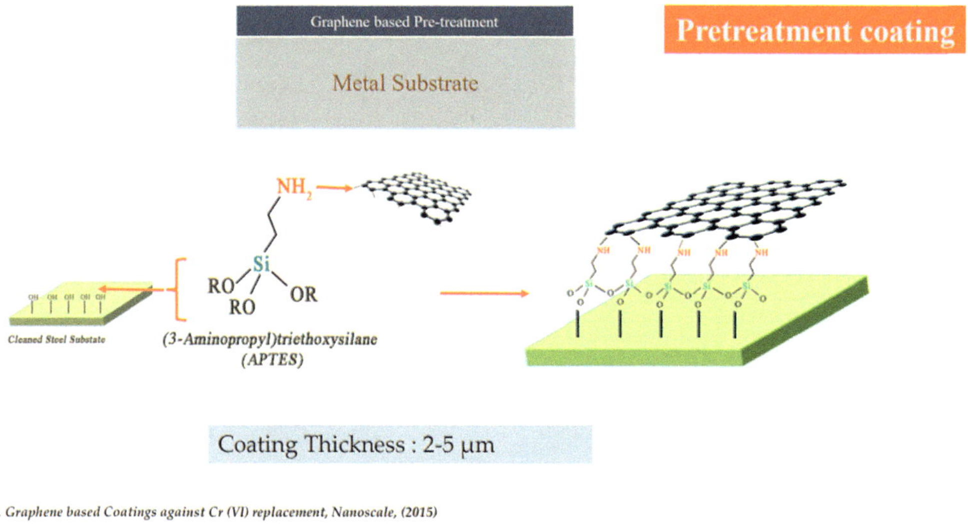

Figure 4.21 Sol-gel method to synthesize graphene based coating on metal.

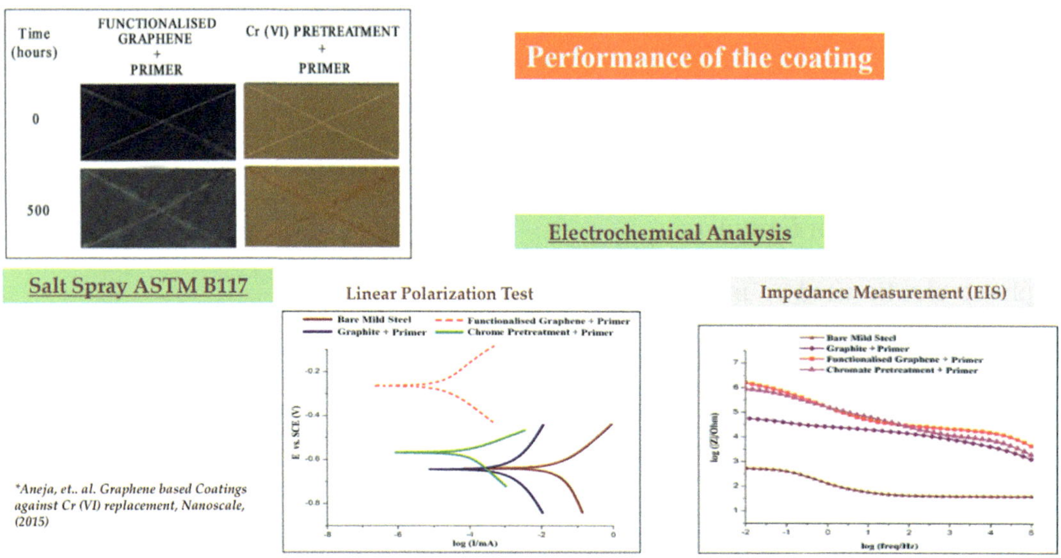

Figure 4.22 Showing the morphology of the graphene coating (A) and the corresponding corrosion rate (B), and impedance results (C).

than 10 microns thick and is uniform with good adherence. Its corrosion rate is very low Fig. 4.22.

Let us now consider the modification of an epoxy coating with the addition of graphene. Several formulations were made starting from 0.5% of graphene to 5% of graphene. It was found that the best results were obtained when graphene concentration was 1%. Higher concentration of graphene resulted in poor properties. The properties of the 1% graphene-based epoxy paint were compared with the zinc-based primer: an organic zinc-rich primer containing 72% of zinc and inorganic zinc primer containing 82% of zinc. It was found that the 1% graphene-based epoxy was about twice as good as the inorganic zinc-rich primer, and it was almost 10 times better than the organic–zinc rich primer. Diffusion data in Fig. 4.23 confirms it.

It is also possible to make antistatic coatings using graphene by just adding 0.5% of graphene in epoxy. As you can see in Fig. 4.24, the charge gets evenly distributed in the graphene-based coating while it remains concentrated at one point in the epoxy coating having no graphene [19,20].

4.4 Practical examples of some nanoadditives modified coatings

4.4.1 Nanomodified waterproofing compound

Waterproofing is one of the severe problems in the old buildings. Lots of waterproofing formulations available in the market are based upon the selection of some resins, mostly acrylic, which is mixed with cement to enhance its flexibility and reduce porosity. This results in reducing leaking and also gives

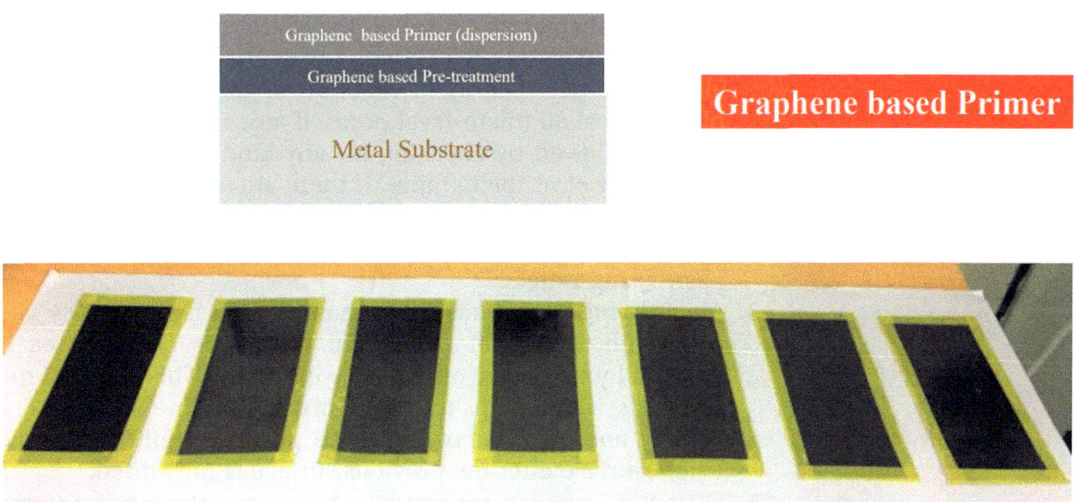

Figure 4.23 Graphene based epoxy coating staring from 0.5% to 5% graphene.

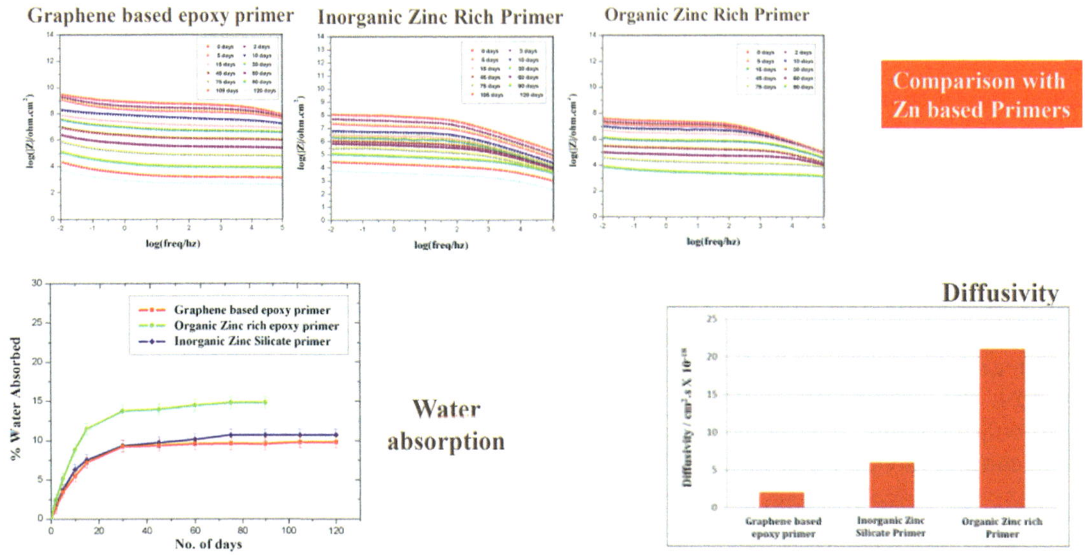

Figure 4.24 Comparison of graphene based coating with organic zinc-based and inorganic zinc-rich primers.

enough strength and flexibility to the coating in case of sudden impact or temperature effect. It has been found that modification of the resin with nanoparticles gives higher strength and no leaking. A product was made with a very small amount of nanoparticle addition in the resin. This nanomodified compound is now available with a name called Thermacool WP-51. Its application works in three steps:

1. Apply an absorption coat of this product by diluting it 4–5 times with water and a small amount of cement, about 20% by weight. This is nicely spread over the whole area and is allowed to dry. The main function of this application is to blunt or seal all micro-level pores, if any.
2. This is followed by checking for any kind of cracks or holes in the roof or at the parapet. If there are any holes or cracks, they are further widened and filled with a paste made from 50% of WP-51 and 50% of cement, nicely converted into a paste using water.
3. After drying these cracks, the same paste is made less viscous with additional water, and the whole surface area is covered by this paste and is allowed to dry. This is called the first coat, which helps to seal all cracks and fine pores.
4. Finally, another coat of the same paste is applied over the whole surface after the previous surface is fully dried.
5. Over this, one can seal a waterproofing surface using Coba, tiles, or any surface reflective coating for a cooling effect.

4.4.2 Fire retardant nanocoating

A transparent nanomodified fire retardant coating was developed to either extinguish the fire or retard its rate of burning on any substrate, such as wood, plastics, or metal. As shown in Fig. 4.25, after 7 minutes of burning the coated surface, there was no burning on the part. In the same way, a PE floor mat uncoated burned immediately compared to the same one coated with the nanomodified fire retardant coating DLFR-26. The technical data sheet (TDS) for this is given in Fig. 4.25(B).

4.4.3 Acid resistant nanomodified epoxy coating

A highly corrosion-resistant, brushable epoxy-based coating, DL-AR, was developed for such applications where very strong corrosion resistance from acids, alkalis, or other strong mixed media is required. This is especially made for battery rooms in ships and submarines, pickling side walls that constantly get acid vapors from acid tanks in a seven-tank process, sewage treatment plants, refineries, and chemical plants where SO_2 and H_2S vapors keep coming. The epoxy is modified with special nanoparticles. Coated panels resist attack at any pH ranging from 1 to 14 and in various organic solvents (Fig. 4.26) and TDS in Table 4.3.

Before we conclude the chapter, it is important to mention a few important applications of nanotechnology in paint coatings

(A)

Product Data-Sheet

Sr.no	Type	Property
1.	Component	Single Pack
2.	Composition	Proprietary Formulation
3.	% Vol Solid	60%
4.	Density	1.13 Kg/Litre
5.	Viscosity	Viscous
6.	Recommended Thickness	50 -60 Microns
7.	Theoretical Spreading rate	6-8 Sq Meter Per Litre (For All The Three Coats)
8.	Recoating Interval	3 to 12 hrs After Every Coat
9.	Drying Time	24 H After All The Three Coats
10.	Flexibility	Good
11.	Scratch/Abrasion resistance	Moderate
12.	Resistance to acids /alkalies	Good
13.	Resistance to Solvents	Poor
14.	Resistance to salts	Good
15.	Weatherability	Moderate

Flame Resistance

DL FR – 26 coated HDPE PIPE does not catch fire even after 12 minutes of continuous heating using a candle flame. There is also no propagation of fire.

(B)

Figure 4.25 (A) Showing that even after 7 min of burning, there is no effect on the coated wood surface and (B) datasheet of DLAR26.

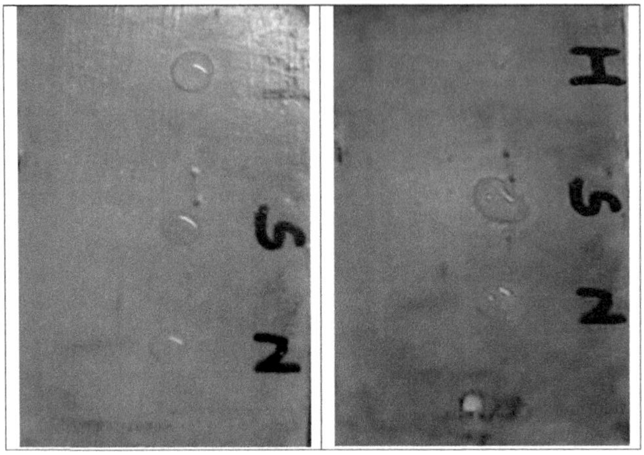

Figure 4.26 Depicts that nanomodified 2K epoxy-polyamide coating is not affected by acids like H_2SO_4 or HCl.

by other researchers. One very important application is the use of nanosilver to make antimicrobial coatings. Nanosilver has been considered the most effective in creating an antimicrobial effect in coatings [8,21].

Bacteria and microbe control in hospitals requires a very high level of cleanliness to prevent infection. To achieve this in day-to-day work, surfaces need to be cleaned frequently and sterilized. This ensures safety for doctors and patients while preventing the spread of disease.

Antimicrobial coatings can make the task of keeping surfaces sterile much simpler since microbes are killed or destroyed on contact. Nanosilver powder mixed with paint provides an antimicrobial coating solution to hospitals, healthcare providers, and other organizations that need it. Silver is a well-documented antimicrobial. It can inhibit crucial cellular functions in microbes and is known to kill bacteria, fungi, and even certain viruses [8].

A new approach to synthesizing silver nanopowder, developed by a team of Indian researchers at the Mumbai-based Bhabha Atomic Research Centre (BARC), uses an environmentally friendly process that minimizes the need for harsh chemicals. The method also prevents the release of silver nanoparticles into the environment by attaching them to a silica substrate with the use of gum arabic. This is an eco-friendly biopolymer derived from the sap of the acacia tree. In addition to providing paints with antimicrobial properties, the powder could also be used in waste management or to clean water, according to the researchers.

Table 4.3 TDS for nano modified epoxy polyamide coating DL-AR.

TDS of DL-AR Coating

Product Description:

Product Data -Sheet

Type	Two Pack		
Mixing Ratio	Base: Hardener- 4:1		
Induction time	10 Minutes		
Pot Life	4-5 hours (@30 °C & 65% RH)		
Composition	Epoxy resin with suitable hardener fortified with nano particles		
Volume solids	80% ± 5%		
Application		Nozzle orifice	Nozzle pressure
	Brush	---------	------------
	Conventional spray	1.5-3 mm	3-4 kg/cm^2
	Airless spray	0.53-0.68mm	Not less than176 kg/cm^2
Recommended DFT	150-1000 microns (based upon various applications)		
Theoretical Spreading Rate	@ 100 microns DFT 8 Sqmt per Lt		
Drying time	Surface: 4-5 Hrs (@30 °C & 65% RH) Hard Dry: Overnight (@30 °C & 65% RH)		
Over coating	Min 24 hr at max 7 days (@30 °C & 65% RH)		
Color	Grey and any inert shade / aluminum		
Finish	Semi Glossy to eggshell		
Thinner (MP Thinner)	Brush: 0-5% Conventional Spray: 8-10% Air less Spray: 0-3%		

Chemical Resistance against fumes and spillages:	Acid	Alkalis	Solvents	Salt water	Fresh water
	Excellent	Excellent	Excellent	Excellent	Excellent
Temperature Resistance:	up to 250°C				
Weatherability:	Excellent with a suitable top coat				
Flexibility:	Very Good				

Dust deposition on surfaces, painted glass, aluminum composite panels, and solar panels is the biggest source of a surface looking ugly. And in the case of solar panels, it affects the dampening of collection efficiency. Thus, any effort to create a dust-free surface can look good and clean and would not affect the efficiency of solar panels. It has been reported that TiO_2-based nanocoatings can be effective in controlling dust and dirt when the coated substrate is exposed to UV radiation. One new company has developed a nanoparticle delivery system that can be used to create dust-repellent coatings for solar panels, improving their efficiency and minimizing needed maintenance [22]. With time, the solar panels can become less efficient due to the accumulation of dust, dirt, and other debris on the panel surface. On a long enough timescale, soiling can reduce solar panel efficiency by up to 30%. These particles catalyze UV light to decompose airborne dirt and dust particles, helping to slow or even prevent soiling. In practice, the nanocoating may improve the energy production of a solar panel by 3% each year, all without increasing the cost of manufacturing. The technology is very useful for a large number of solar plants having thousands of solar panels that are difficult to clean by conventional washing with water and soap. Hence, dust removal coatings have a great future.

Finally, it can be said that nanomaterials can enhance several properties of paints and coatings, such as antimicrobial, anticorrosive, and dust-repellent properties, that can offer benefits to a wide range of industries. The use of nanomaterials is still mostly experimental, but several early commercial applications have shown how nanotechnology may enable new kinds of industrial paint and coatings. The benefits of adding nanoparticles far outweigh the costs and can lead to longer-lasting, more-efficient end products.

4.5 Conclusion

Nanotechnology is the most important invention of this decade. It has applications almost in every field. In paint coatings, the use of nanotechnology helps in reducing the quantity of raw materials and also enhances the properties of the coating. To make use of nanotechnology in paint coatings, it is important to identify the proper material, which can affect the focused property of the paint, synthesize it and characterize it, and then optimize its concentration. Another precaution is to properly mix it in resin using a high shear process or sonication and create a uniform distribution. It is possible to use conventional paint systems and add nanoadditives or pigments in that or use sol-gel technology to create nanosized paint coatings that are thin and very effective. Another important area where nanotechnology has made good paint systems is self-cleaning, self-healing, antimicrobial paint systems, waterproofing and fire-resistant coatings.

References

[1] A. Khanna, Nanotechnology in high performance paint coatings, Asian Journal of Experimental Sciences 21 (2) (2008) 25–32.
[2] U.A. Samad, M.A. Alam, E.M. Sherif, et al., Coatings 9 (2019) 1–4.
[3] https://www.fortunebusinessinsights.com/industry-reports/paints-and-coatings-market-101947.
[4] https://www.mordorintelligence.com/industry-reports/india-paints-and-coatings-market.
[5] Use of nanomaterials in coatings. Fact Sheet published by Federal Environment Agency, Germany, https://www.umweltbundesamt.de/sites/default/files/medien/378/publikationen/use_of_nanomaterials_in_coatings_0.pdf, 2014.
[6] Y. Huang, F. Yu, Y.S. Park, et al., Co-administration of protein drugs with gold nanoparticles to enable percutaneous delivery, Biomaterials 31 (34) (2010) 9086–9091.
[7] J.P. Ryman-Rasmussen, J.E. Riviere, N.A. Monteiro-Riviere, Penetration of intact skin by quantum dots with diverse physicochemical properties, Toxicological Sciences: An Official Journal of the Society of Toxicology 91 (1) (2006) 159–165.
[8] L. Guo, et al., Polymer/nanosilver composite coatings for antibacterial applications, Colloids and Surfaces A: Physicochemical and Engineering Aspects 439 (2013) 69–83.
[9] H. Paschen, C. Coenen, T. Fleischer, et al. TA-Projekt Nanotechnologie Endbericht. Büro für Technikfolgen-Abschätzungbeim Deutschen Bundestag (TAB), 2003.
[10] Deutsches Lackinstitut, Brandschutz-Beschichtungen – 40 Minutenkönnen Leben retten. Lack imGespräch, InformationsdienstDeutschesLackinstitut Nr 112 (5) (2012) Seite: http://www.lacke-und-farben.de/fileadmin/templates/img/pdf/LIG_112.pdf.
[11] https://cordis.europa.eu/article/id/36226-ensuring-the-safety-of-nanoparticles-in-paints-and-coatings.
[12] R.G. Wankhede, A.S. Khanna, Development of water-repellent organic–inorganic hybrid sol–gel coatings on aluminum using short chain perfluoro polymer emulsion, Applied Surface Science 283 (2013) 1051–1059.
[13] N. Rajagopalan, A. Khanna, Effect of methyltrimethoxy silane modification on yellowing of epoxy coating on UV (B) exposure, Journal of Coatings 2014 (2014).
[14] S. Pathak, A.S. Khanna, T. Sinha, Sol gel derived organic-inorganic hybrid coating: a new era in corrosion protection of material, Corrosion Reviews 24 (5-6) (2006) 281–306.

[15] R. Grover Wankhede, A.S. Khanna, N. Birbilis, AFM studies to understand the mechanism of fume-silica particles incorporated in epoxy silane sol-gel coatings, Journal of Surface Engineered Materials and Advanced Technology (2013). Available from: http://www.scirp.org/journal/jsemat.
[16] B.M. Gurumurthy, Y.M. Shivaprakash, A. Hiremath, M.C. Gowrishankar, P.K. Jayashree, Self healing materials: a new era in material technology: a review, International Journal of Applied Engineering Research 11 (2016) 1373–1378.
[17] K. Thanawala, N. Mutneja, A.S. Khanna, R.K.S. Raman, Development of self-healing coatings based on linseed oil as autonomous repairing agent for corrosion resistance, Materials 7 (2014) 7324–7338.
[18] K. Thanawala, A.S. Khanna, R.K. Singh Raman, Development of self-healing coatings using encapsulated linseed oil and tung oil as healing agents, NACE Corrosion Conf. paper No. 7310 (2016) 1–9.
[19] K. Thanawala, A.S. Khanna, R.K.S. Raman, Tung oil-urea formaldehyde microcapsules for anti-corrosive self-healing epoxy coatings, Materials Science & Surface Engineering 3 (2015) 151–156.
[20] K.S. Anejaa, H.L.M. Böhmc, A.S. Khannaa, S. Böhma, Functionalised graphene as a barrier against corrosion, FlatChem 1 (2017) 11–19.
[21] P. Ganguli, S. Chaudhuri, Nanomaterials in antimicrobial paints and coatings to prevent biodegradation of man-made surfaces: A review, Materials today 45 (3) (2021) 3769–3777.
[22] P. Adapala, et al., Development and evaluation of nano-silica dispersed polyurethane based coatings for improved anti-graffiti and scratch resistance, Open Journal of Applied Sciences 5 (12) (2015) 808.

5

Elementary discussion on characterization techniques to analyze nanopaints

Partha Pratim Das
School of Materials Science and Engineering, Nanyang Technological University, Singapore, Singapore

5.1 Introduction

Nanopaints, situated at the nexus of cutting-edge nanotechnology and traditional coatings, signify a transformative leap in material science. This class of innovative materials harnesses the unique and often extraordinary properties of nanoparticles—particles typically ranging in size from 1 to 100 nm to redefine the capabilities and applications of conventional paints [1,2]. In essence, nanopaints represent a paradigm shift in the realm of surface coatings, promising a spectrum of functionalities that extend far beyond the esthetic, permeating various industries with their potential. The distinctive feature of nanopaints lies in the intentional integration of nanomaterials, which could include materials, such as metal oxides, carbon nanotubes (CNTs), or engineered polymers, into the paint matrix. This deliberate inclusion of nanoscale components imparts nanopaints with exceptional characteristics, elevating their performance to meet the demands of modern applications [3–5]. Industries, such as automotive, construction, healthcare, and beyond are increasingly turning to nanopaints to overcome longstanding challenges and to pioneer new frontiers of performance. The allure of nanopaints is not limited to their visual appeal; it extends to their multifunctionality. By capitalizing on the unique attributes of nanoparticles, these paints transcend traditional roles [6–9]. They become dynamic materials capable of self-healing, exhibiting antimicrobial properties, and providing enhanced ultraviolet (UV) protection [10,11]. Nanopaints, therefore emerge not only as a means of enhancing the esthetics of surfaces but also as intelligent and responsive coatings that adapt to environmental

Nanopaints. DOI: https://doi.org/10.1016/B978-0-443-26512-9.00005-7

conditions and specific application requirements. Crucial to the understanding and optimization of nanopaints is an exploration of their composition and structure. The intricate interplay between nanoparticles and the paint matrix governs the ultimate performance of these materials. As researchers delve into the nuances of this interplay, new possibilities for innovation and customization emerge, offering tailored solutions to the diverse challenges faced by industries and consumers alike [12,13].

In essence, nanopaints represent a convergence of art and science, where the canvas is not merely adorned but transformed into a dynamic and responsive interface. The exploration of nanopaints encapsulates not just a technological advancement in coatings but a journey into the future of materials that redefine the boundaries of what is achievable in surface functionalities and esthetics. Fig. 5.1 shows the broad classification of nanomaterials.

Nanomaterials, particularly, nanoscale titanium dioxide (TiO_2) and silicon dioxide (SiO_2), are anticipated to enhance paint properties due to their unique structural characteristics, including size, shape, and increased surface area. Other nanomaterials, such as silver (Ag), zinc oxide (ZnO), aluminum oxide, cerium dioxide, copper oxide, and magnesium oxide are also subjects of ongoing investigation in the paint industry [14,15].

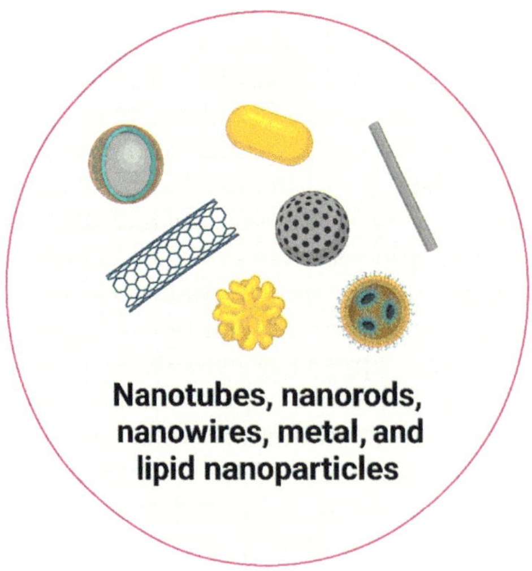

Figure 5.1 Broad classification of nanomaterials (Created in *biorender.com*).

5.2 Nanomaterials used in paints

Nanomaterials are increasingly utilized in paints to enhance various properties and functionalities. Some common nanomaterials employed in paints include [7,16–20]:

- Nano titanium dioxide (TiO_2): TiO_2 nanoparticles are known for their photocatalytic activity and UV-blocking capabilities. They are used to provide UV protection, enhance durability, and contribute to a self-cleaning effect in paints.
- Nano silicon dioxide (SiO_2): SiO_2 nanoparticles improve macro- and micro-hardness, abrasion resistance, and scratch resistance in paints. However, their addition may reduce paint elasticity, impacting the ability to withstand dimensional changes caused by temperature and humidity fluctuations.
- Nano silver (Ag): nano Ag particles exhibit excellent antimicrobial properties, making them valuable in paints for inhibiting the growth of bacteria and pathogens. However, their effectiveness outdoors may be limited due to exposure to external conditions.
- Nano zinc oxide (ZnO): ZnO nanoparticles contribute to UV-blocking properties in paints, protecting against UV radiation. They are used in sun-protective coatings for outdoor applications.
- Nano cerium dioxide (CeO_2): CeO_2 nanoparticles are known for their catalytic properties and are employed in certain paints to enhance oxidation resistance and durability.
- Carbon nanotubes (CNTs): CNTs can improve the mechanical strength and conductivity of paints. They find applications in antistatic coatings and those requiring enhanced mechanical properties.
- Nano alumina (Al_2O_3): Al_2O_3 nanoparticles contribute to improved hardness and wear resistance in paints. They are used in coatings for surfaces that require enhanced durability.
- Nano clays: nano-sized clay particles, such as montmorillonite, can be incorporated into paints to enhance barrier properties, mechanical strength, and fire resistance.
- Nano iron oxide (Fe_2O_3): nano-sized Fe_2O_3 particles are used as pigments in paints, providing color, corrosion resistance, and UV-blocking properties.
- Nano zinc phosphate: nano zinc phosphate is employed in certain paints as a corrosion inhibitor, enhancing the protective properties of coatings against rust.

5.3 Characterization techniques to analyze nanopaints

Characterization techniques play a pivotal role in analyzing nanopaints, providing insights into their composition and performance. Spectroscopic methods, such as infrared (IR) and Raman spectroscopy unravel molecular structures, while imaging techniques like scanning electron microscopy (SEM) and atomic force microscopy (AFM) offer a closer look at morphological features. Thermal analysis methods, including differential scanning calorimetry (DSC) and thermogravimetric analysis (TGA), assess thermal stability. These techniques collectively enable researchers to tailor nanopaints formulations for optimal functionality, ensuring their successful application across various industries [21,22]. Fig. 5.2 shows the broad spectrum of characterization techniques.

5.3.1 Spectroscopy techniques

Spectroscopy is a scientific method that involves the interaction of electromagnetic radiation with matter to gain insights into the composition, structure, and properties of materials.

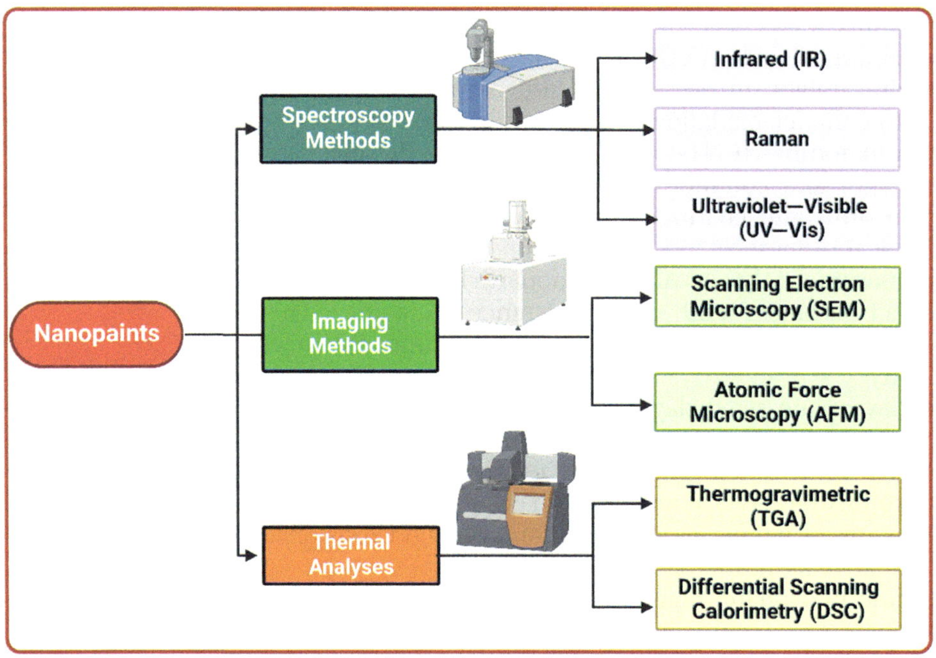

Figure 5.2 Spectrum of characterization techniques (Created in *biorender.com*).

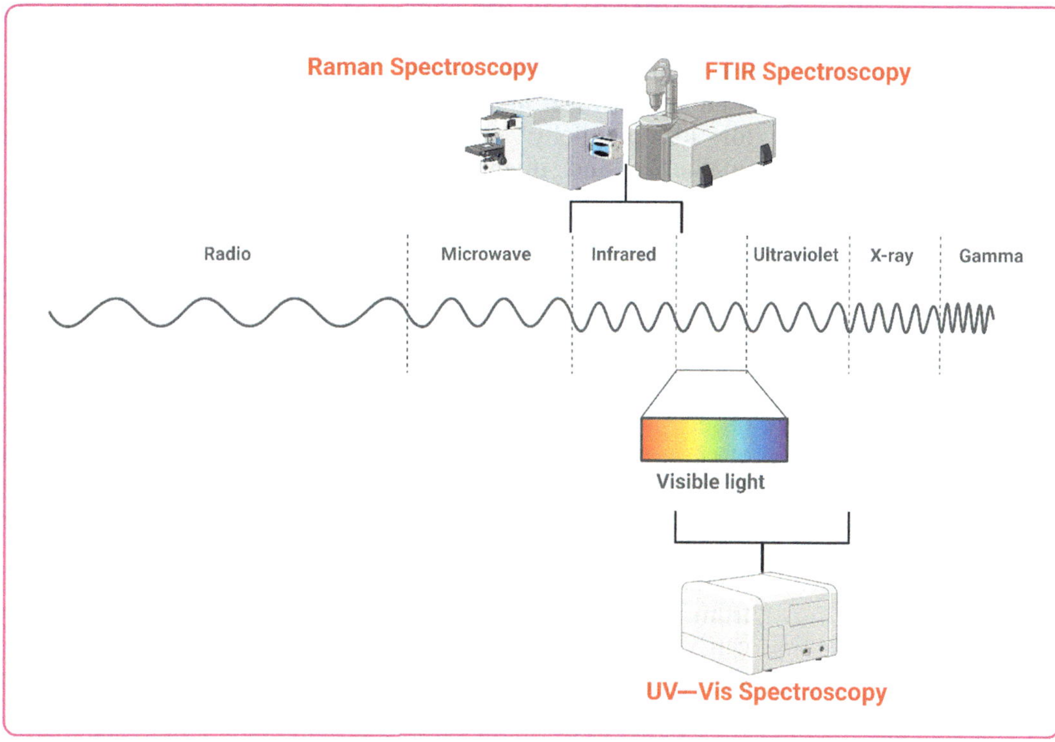

Figure 5.3 Spectroscopy techniques working range in the electromagnetic spectrum (Created in *biorender.com*).

This analytical technique exploits the principle that different materials absorb, emit, or scatter light at characteristic wavelengths, allowing scientists to identify and quantify various components within a sample. Spectroscopy technique's working range in the electromagnetic spectrum is shown in Fig. 5.3.

5.3.1.1 Ultraviolet–visible spectroscopy

Ultraviolet–visible (UV–vis) spectroscopy operates based on the fundamental principle of measuring the absorption of UV and vis light by a sample. The technique involves passing a beam of UV or vis light through a sample and detecting the amount of light absorbed at different wavelengths. The sample absorbs specific wavelengths of light corresponding to electronic transitions within its molecules. The absorption spectrum, generated by plotting the absorbance against wavelength, provides valuable information about the sample's composition, concentration, and electronic structure [23,24]. According to the

Beer–Lambert law, absorbance is directly proportional to the concentration of the absorbing species and the path length of the light through the sample. UV–vis spectroscopy is widely used in chemical analysis, biochemistry, and material science for its simplicity, sensitivity, and versatility in providing qualitative and quantitative insights into the properties of various substances [25,26].

This technique plays a crucial role in the in-depth characterization of nanopaints due to its ability to provide a wealth of information about their properties. When examining the optical properties, the absorption spectrum reveals details about electronic transitions within the nanocoating material, offering insights into its electronic structure. The interference fringes observed in the UV–vis spectra enable precise measurements of coating thickness, with changes indicating alterations in the coating structure. In the case of metallic nanopaints, such as those made of gold or Ag, the technique is particularly powerful in detecting surface plasmon resonance (SPR). The SPR phenomenon is manifested as a distinct peak in the UV–vis spectrum, and monitoring variations in its position and intensity can yield valuable information about the coating environment and interactions with surrounding media [27,28]. Moreover, UV–vis spectroscopy allows for quantitative analysis, as calibration curves can be established for determining concentrations of specific components within the nanocoating. This analytical method is also instrumental in studying the stability and degradation of nanopaints over time, providing critical insights into their long-term performance. Integration with other characterization techniques, such as microscopy and spectroscopy, enhances the overall understanding of the properties of nanopaint, and meticulous sample preparation and calibration are essential for ensuring the accuracy and reliability of the obtained results [29].

5.3.1.2 Fourier transform infrared spectroscopy

Fourier transform infrared spectroscopy (FTIR) operates on the principle of measuring the absorption of IR radiation by a sample. In FTIR, a broadband IR light source is used to generate an interferogram by passing the light through the sample. The interferogram is then Fourier-transformed to obtain the IR spectrum, which represents the absorption of IR radiation at different frequencies. The sample interacts with the IR light, causing specific vibrational transitions in its molecular bonds. These transitions are unique to the chemical composition of the sample, allowing for the identification of functional groups and molecular structures. The resulting FTIR spectrum is a plot of absorbance or transmittance versus wavenumber, providing detailed information about the sample's chemical composition and structure. FTIR is widely employed in various fields, including chemistry, biology, and materials science specially in the field of nanopaints, for qualitative and quantitative analysis of organic and inorganic compounds based on their IR absorption characteristics [30,31].

5.3.1.3 Raman spectroscopy

Raman spectroscopy operates on the fundamental principle of inelastic scattering of monochromatic light when interacting with a sample. In this technique, a laser light source is directed onto the sample, and the incident photons undergo scattering, producing both elastic (Rayleigh) and inelastic (Raman) scattered photons. In Raman scattering, a small fraction of the photons experiences an energy exchange corresponding to the vibrational modes of the molecular bonds in the sample. This results in the generation of Stokes lines (indicative of photons gaining energy) and anti-Stokes lines (indicative of photons losing energy). The Raman spectrum is then created by detecting the intensity of the scattered light at different wavelengths. Peaks in the spectrum represent specific vibrational modes of the molecules in the sample, offering insights into their chemical composition and molecular structure. Raman spectroscopy is nondestructive, requires minimal sample preparation, and is widely employed across various scientific disciplines for its ability to provide detailed information about materials in a noninvasive manner [32,33].

This technique is a powerful analytical tool for the in-depth analysis of nanopaints, providing crucial insights into their chemical composition and molecular characteristics. By leveraging the inelastic scattering of monochromatic light, Raman spectroscopy facilitates the identification of individual components within nanopaints, enabling the discrimination between different nanoparticles, polymers, and additives present in the coating formulation. The technique's chemical mapping capabilities allow for the visualization of spatial distribution, aiding in the assessment of nanopaints homogeneity and identifying potential variations [34,35]. Moreover, Raman spectroscopy is sensitive to molecular vibrations, making it particularly effective in detecting molecular changes, such as cross-linking reactions or chemical transformations during the curing or aging processes of nanopaints. Its quantitative analysis capabilities contribute to quality control, providing valuable information about the concentration of specific components. The nondestructive nature of Raman spectroscopy allows for the assessment of nanocoating properties without compromising their integrity, and its in situ application enables real-time monitoring of dynamic processes. Complemented by other analytical techniques, such as microscopy, Raman spectroscopy offers a comprehensive approach to unraveling the intricate details of nanopaints, essential for their development, optimization, and application in diverse industrial settings [36,37].

5.3.2 Imaging techniques

5.3.2.1 Atomic force microscopy

AFM operates on a principle rooted in the interaction between a sharp-tip probe and a sample surface at the atomic or nanoscale level. In AFM, a cantilever with a sharp tip is brought into proximity to the sample surface without direct contact. The interaction forces between the tip and the sample induce deflections in the cantilever, and these deflections are precisely measured using a laser beam and photodetector system. The resulting topographical information is then used to generate a high-resolution, three-dimensional image of the sample surface. By scanning the tip across the sample in a systematic manner, AFM can produce detailed images with subnanometer resolution. This technique allows for the visualization and characterization of various materials, including biological samples, polymers, and nanoparticles, providing insights into surface morphology, mechanical properties, and interactions at the nanoscale. AFM's nondestructive nature and capability to operate in various environments, including air and liquids, make it a powerful tool for nanoscale imaging and analysis in both research and industry [38].

This technique helps in the analysis of nanopaints, offering high-resolution imaging and characterization at the nanoscale. In the context of nanocoating analysis, AFM is employed to investigate the surface morphology, thickness, and mechanical properties of the nanopaints. The technique involves scanning a sharp-tip cantilever over the nanocoating surface without direct contact, detecting the interaction forces between the tip and the coating. This allows for the generation of detailed topographical maps, providing information on features, such as surface roughness, grain size, and the uniformity of the nanocoating. Additionally, AFM can be used to measure the mechanical properties of nanopaints, including adhesion, stiffness, and elasticity. AFM plays a crucial role in the comprehensive characterization of nanopaints, aiding in the optimization of coating processes and the assessment of their performance in diverse applications, including electronics, optics, and biomedical devices [39–42].

5.3.2.2 Scanning electron microscopy

SEM operates on the principle of using a focused beam of electrons to scan the surface of a specimen, providing high-resolution images with exceptional depth of field. In SEM, electrons are emitted from a heated filament and accelerated towards the specimen. The interaction between the electron beam and the specimen's surface generates various signals, including secondary electrons, backscattered electrons, and characteristic X-rays. Secondary electrons, emitted from the specimen's surface, are commonly used for imaging. The electron beam scans the specimen point by point, and the signals are detected and used to create detailed topographical and compositional images. SEM is known for its high magnification capabilities, yielding images with subnanometer resolution. This technique is widely employed in materials science, biology, and other fields for the detailed examination of surface structures, morphology, and composition at the micro- and nanoscale. The versatility of SEM makes it invaluable for both qualitative and quantitative analysis, offering insights into the fine details of materials and biological specimens [43].

In nanopaint analysis using SEM, a focused electron beam is directed at the coating surface, and the interaction of electrons with the sample generates various signals, including secondary and backscattered electrons. Secondary electrons are particularly useful for imaging, offering high-resolution topographical information about the nanocoating surface. SEM allows for the visualization of nanoscale features, such as the thickness, roughness, and uniformity of the coating. Additionally, the technique can provide information about the distribution of nanoparticles within the coating matrix. Energy-dispersive X-ray spectroscopy (EDS) can be coupled with SEM to obtain elemental composition data, offering insights into the chemical composition of the nanocoating. SEM's capability to operate in various imaging modes, such as secondary electron imaging, backscattered electron imaging, and EDS, makes it a versatile tool for comprehensive nanocoating analysis, aiding in quality control, optimization of coating processes, and understanding the structure–property relationships crucial for the performance of nanocoated materials [44–46].

5.3.3 Thermal analysis techniques

Thermal analysis techniques are valuable tools for the characterization of nanopaints providing insights into their thermal properties, stability, and behavior under different temperature conditions.

5.3.3.1 Thermogravimetric analysis

TGA is a powerful thermal analysis technique employed to investigate changes in the mass of a sample as a function of temperature or time. The fundamental principle of TGA lies in subjecting a precisely weighed sample to a controlled temperature program within a specified atmosphere. The sample, typically a few milligrams, undergoes physical and chemical transformations, such as evaporation, decomposition, or oxidation, as the temperature increases. Simultaneously, an empty sample pan or a reference material is subjected to the same conditions, serving as a baseline for comparison. The instrument continuously measures and records the weight changes of both the sample and reference. The resulting TGA curve, displaying the weight loss or gain as a function of temperature or time, provides critical insights into the thermal stability, decomposition kinetics, and composition of the sample. Analyzing features, such as onset temperature, peak temperatures, and completion temperatures aids in understanding the thermal behavior of the material [47,48]. TGA is widely utilized across diverse scientific disciplines for its ability to characterize the thermal properties of materials, ranging from polymers and pharmaceuticals to minerals and biomaterials.

TGA is a valuable tool for the analysis of nanopaints, offering insights into their thermal stability and decomposition behavior. In the context of nanopaints, TGA provides crucial information about the weight changes as a function of temperature, helping to characterize the thermal properties of the coating material. The procedure involves placing a small amount of the nanocoating onto the sample pan and subjecting it to a controlled temperature ramp. As the temperature increases, the nanocoating undergoes thermal processes, such as the release of volatile components, decomposition, or other reactions. The

TGA instrument continuously monitors and records the mass changes in the nanocoating, generating a thermogram that illustrates the weight loss or gain over the temperature range. Analyzing the thermogram allows researchers to identify key thermal events, such as onset temperatures, peak temperatures, and completion temperatures, which are indicative of the coating's thermal stability and composition. TGA serves as a valuable technique for quality control, optimization of processing parameters, and understanding the thermal behavior of nanopaints in various applications, including corrosion resistance, biomedical coatings, and advanced functional nanopaints [21,49–51].

5.3.3.2 *Differential scanning calorimetry*

DSC is a thermal analysis technique that measures the heat flow associated with thermal transitions in materials as a function of temperature. The basic principle of DSC revolves around comparing the heat flow into or out of a sample with that of a reference material while both are subjected to a controlled temperature program. The sample and reference materials are placed in separate pans within a calorimeter, and both pans are heated or cooled simultaneously. As the sample undergoes physical or chemical changes, such as phase transitions, melting, crystallization, or chemical reactions, it either absorbs or releases heat. This results in a temperature difference between the sample and reference pans, which is detected by a differential thermocouple. The recorded heat flow is plotted against temperature, generating a DSC curve. Peaks or troughs in the curve correspond to specific thermal events, and the area under the curve provides quantitative information about the heat absorbed or released during these transitions. DSC is widely employed across various scientific disciplines to investigate the thermal properties of materials, aiding in the characterization of polymer coatings, pharmaceuticals, alloys, and numerous other substances [47,48].

This technique is extensively utilized for the analysis of nanopaints, providing valuable insights into their thermal properties and behavior. In the realm of nanocoating analysis, DSC is employed to investigate the heat flow associated with thermal transitions, such as curing, melting, crystallization, or degradation, as a function of temperature. To conduct a DSC analysis on nanocoatings, a small amount of the coating material is accurately weighed and placed in a sample pan, while an empty reference pan serves as a baseline. The pans are simultaneously heated or cooled at a controlled rate, and the resulting DSC curve depicts the heat flow difference between the sample and reference. For nanocoatings, DSC can reveal crucial information about curing kinetics, glass transition temperatures, and thermal stability. Peaks or troughs in the DSC curve correspond to specific thermal events, aiding in the identification of critical temperature points relevant to the nanocoating's performance. The area under these peaks provides quantitative data on the heat absorbed or released during these transitions. DSC analysis proves valuable in optimizing nanopaint formulations, ensuring proper curing, and assessing the overall thermal behavior critical for various applications, including corrosion resistance, biomedical coatings, and functional surface modifications [52–54].

5.4 Effect of nanomaterials on paints

Nano TiO_2 is incorporated into paint for its exceptional properties, including photocatalytic activity and UV protection. The synergistic effect of photocatalysis and hydrophilic properties creates a self-cleaning feature, reducing the need for regular maintenance. However, research indicates that the use of photocatalytic nano TiO_2 in organic paints can lead to the deterioration of the binder under UV irradiation. To address this, the rutile type is preferred in organic facade coatings for enhanced UV protection. The introduction of nano SiO_2 to paints enhances macro- and micro-hardness, abrasion resistance, scratch resistance, and weather resistance. Despite these advantages, the inclusion of nano SiO_2 reduces paint elasticity, which is essential for resisting dimensional changes caused by temperature and humidity fluctuations. Paints containing nano Ag exhibit potent antimicrobial properties against bacteria and pathogens. Yet, outdoor application poses challenges as studies reveal that both nano Ag and nano TiO_2 struggle to completely prevent microbial and algal growth on test surfaces and provide limited defense against fungal colonization [3,55,56]. The overall summary of the characterization techniques is highlighted in Table 5.1.

5.5 Applications of nanopaints

Nanopaints, which incorporate nanomaterials into traditional paint formulations, offer a wide range of applications across various industries. The unique properties conferred by nanoparticles contribute to enhanced performance and

Table 5.1 Overall summary of characterization techniques.

Sl. no	Techniques	Advantages	Disadvantages
1.	Ultraviolet–visible (UV–vis) spectroscopy	Can be applied to a wide range of chemical species	Only be used to analyze the substance that are absorb in the UV or visible region of the electromagnetic spectrum.
2.	Fourier transform infrared spectroscopy (FTIR)	Identifies organic compounds and functional groups Nondestructive Rapid and qualitative analysis	Limited sensitivity for some materials
3.	Raman spectroscopy	Provides high-resolution spectra Differentiates between chemical species	Fluorescence interference in some samples
4.	Scanning electron microscopy (SEM)	High-resolution imaging Depth of focus for 3D visualization	Sample preparation may alter surface features
5.	Atomic force microscopy (AFM)	High-resolution topographic mapping Operates in various environments	Limited imaging speed
6.	Differential scanning calorimetry (DSC)	Measures heat flow during transitions Provides thermal behavior insights	No direct structural information
7.	Thermogravimetric analysis (TGA)	Analyzes weight loss for decomposition patterns	Mass loss of volatiles is not equivalent to formation of degradation

functionalities. Some key applications of nanopaints include as discussed below.

- ***Antibacterial coatings:*** nanopaints containing antimicrobial nanoparticles, such as Ag or copper nanoparticles, can be applied to surfaces in hospitals, public spaces, or household environments to inhibit the growth of bacteria, fungi, and other microorganisms.
- ***Anticorrosive coatings:*** nanopaints with corrosion-resistant nanoparticles, such as ZnO or cerium oxide nanoparticles, are used to protect metal surfaces from corrosion, extending

the lifespan of structures and equipment in various industries, including automotive and marine.

- ***Self-cleaning surfaces:*** nanopaints with photocatalytic nanoparticles, like TiO_2, enable self-cleaning surfaces. When exposed to light, these nanoparticles break down organic compounds, preventing the accumulation of dirt and pollutants on surfaces.
- ***UV-blocking coatings:*** nanopaints incorporating nanoparticles with UV-blocking properties, such as ZnO or TiO_2, are utilized in protective coatings for outdoor structures, vehicles, and equipment to prevent UV-induced degradation and color fading.
- ***Thermal insulation coatings:*** nanopaints containing insulating nanoparticles, such as aerogels or silica nanoparticles, provide thermal insulation when applied to surfaces. This is particularly useful in construction, reducing energy consumption for heating and cooling.
- ***Transparent conductive coatings:*** nanopaints incorporating conductive nanoparticles, like indium tin oxide nanoparticles, can be applied to create transparent conductive coatings. These coatings find applications in electronics, solar cells, and touchscreens.
- ***Antifog coatings:*** nanopaints with hydrophilic nanoparticles, such as silica nanoparticles, can be used to create antifog coatings on glass surfaces. These coatings promote the even distribution of water droplets, preventing fog formation.
- ***Scratch-resistant coatings:*** nanopaints containing nanoparticles with high hardness, such as nanoparticles of ceramics or polymers, enhance the scratch resistance of surfaces. This is valuable in automotive applications and electronic devices.
- ***Color-shifting coatings:*** nanopaints incorporating nanoparticles that exhibit structural color effects can create coatings with color-shifting properties. These coatings find applications in decorative and esthetic finishes.
- ***Biomedical coatings:*** nanopaints are used in biomedical applications for coatings on medical devices, implants, or drug delivery systems. The controlled release of drugs or the addition of diagnostic nanoparticles can enhance the functionality of these medical coatings.
- ***Flexible and stretchable coatings:*** nanopaints with nanoparticles like CNTs or graphene oxide can impart flexibility and stretchability to coatings. These coatings find applications in flexible electronics and wearable devices.

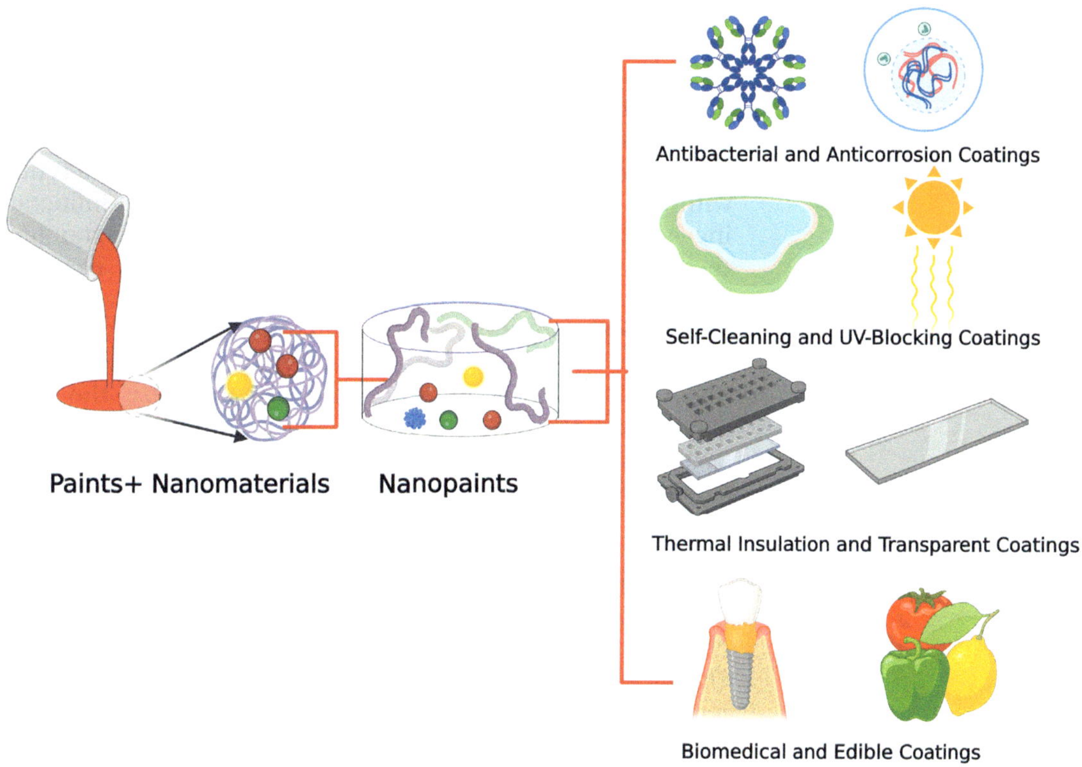

Figure 5.4 Application of nanopaints.

The versatility of nanopaints and their ability to impart unique properties make them increasingly valuable in diverse industries as shown in Fig. 5.4 contributing to advancements in materials science and technology.

5.6 Challenges in the characterization of nanopaints

Characterizing nanopaints poses unique challenges due to the complexity introduced by the incorporation of nanomaterials into traditional paint matrices. Some of the key challenges in nanopaints characterization include:

Small size of nanoparticles: nanopaints typically contain nanoparticles with dimensions in the nanometer range. The small size poses challenges in accurately measuring and visualizing the nanoparticles using conventional techniques.

Uniform dispersion: achieving and maintaining a uniform dispersion of nanomaterials within the paint matrix is crucial for optimal performance. Challenges arise due to the tendency of nanoparticles to agglomerate, settle, or exhibit uneven distribution, impacting the coating's properties.
Surface interactions: nanoparticles may interact differently with the paint matrix compared to larger particles, influencing surface properties and performance. Understanding these interactions is crucial for tailoring the coating's functionalities.
Instrument resolution limits: traditional microscopy techniques may have limitations in resolving nanoscale features, making it challenging to precisely characterize the size, shape, and distribution of nanoparticles within the paint matrix.
Chemical analysis: nanopaints often contain complex formulations, and characterizing the chemical composition of nanomaterials within the coating matrix is challenging. Techniques, such as X-ray photoelectron spectroscopy or nuclear magnetic resonance may be required for detailed chemical analysis.
Quantitative analysis: quantifying the amount of nanomaterials in nanopaints accurately is challenging, particularly when dealing with low concentrations. Specialized techniques like inductively coupled plasma mass spectrometry or gravimetric methods may be necessary.
Dynamic behavior: nanopaints may exhibit dynamic behavior, and characterizing changes over time or in response to environmental conditions requires specialized techniques capable of real-time or in situ analysis.
Multicomponent systems: nanopaints are often multicomponent systems comprising nanoparticles, polymers, solvents, and additives. Characterizing each component individually and understanding their interactions pose additional challenges.
Safety concerns: characterizing potential health and safety risks associated with the use of nanomaterials in nanopaints is essential. Robust methodologies are needed to assess the release of nanoparticles during application and use.

5.7 Future directions in the characterization of nanopaints

The future outlook in the area of nanopaints characterization holds promising developments driven by advancements in nanotechnology, materials science, and analytical techniques. Several key trends and directions can be anticipated:

- ***Advanced imaging techniques:*** continued advancements in nanoscale imaging techniques, such as high-resolution electron microscopy, will enable more detailed visualization of nanopaint structures. Techniques like correlative microscopy, combining multiple imaging modalities, may become more prevalent for a comprehensive understanding.
- ***In situ characterization:*** the development of in situ characterization methods will allow researchers to monitor changes in nanopaints in real-time and under dynamic conditions. In situ spectroscopy, microscopy, and sensing techniques will play a crucial role in capturing the dynamic behavior of nanopaints during application and use.
- ***Machine learning and data analytics:*** the integration of machine learning and data analytics into nanopaints characterization will enhance the processing and interpretation of large datasets. These technologies can aid in identifying patterns, correlations, and predicting performance based on complex nanopaint formulations.
- ***Multimodal characterization:*** integration of multiple characterization techniques into a single platform, such as combining spectroscopy with imaging techniques, will provide a more comprehensive understanding of nanopaint properties. This multimodal approach can offer synergistic insights into both the chemical and physical aspects of nanopaints.
- ***Nondestructive techniques:*** the development of nondestructive characterization techniques will become increasingly important. Techniques that allow for the analysis of nanopaints without altering their structure or properties, such as nondestructive spectroscopy or imaging, will gain significance for quality control and postapplication assessments.
- ***Environmental and health impact assessment:*** as awareness of environmental and health concerns related to nanomaterials grows, there will be an increased focus on developing characterization techniques specifically tailored for assessing the release and impact of nanoparticles from nanopaints in different environments.
- ***Smart nanopaints:*** the emergence of smart or responsive nanopaints with functionalities that can be altered in response to external stimuli may necessitate the development of specialized characterization methods to monitor and control these dynamic properties.
- ***Standardization and regulation:*** with the increasing use of nanopaints in various industries, there will likely be a push

toward standardization of characterization methods. Regulatory bodies may establish guidelines for nanopaint characterization to ensure consistency and safety in applications.

- ***Biosafety and biocompatibility assessments:*** given the potential use of nanopaints in biomedical applications or consumer products, there may be an increased emphasis on characterizing their biosafety and biocompatibility. Techniques assessing cytotoxicity, biodegradability, and interactions with biological systems will be crucial.
- ***Industry collaboration:*** collaborative efforts between academia, industry, and regulatory bodies will be essential for addressing challenges and advancing nanopaint characterization. Sharing knowledge, resources, and establishing common standards will contribute to the development of effective and safe nanopaint technologies.

In summary, the future outlook for nanopaints characterization is characterized by a convergence of advanced techniques, interdisciplinary collaboration, and a focus on addressing environmental and safety considerations. This dynamic field will likely witness transformative developments that contribute to the optimization and responsible use of nanopaints in various applications.

5.8 Conclusions

This chapter underscores the significance of diverse characterization techniques in comprehending the intricacies of nanopaints. Spectroscopy methods unveil chemical compositions, microscopy techniques offer visual insights into morphology, and thermal analysis elucidates thermal stability. The amalgamation of these techniques empowers researchers to optimize nanopaint formulations, ensuring their tailored application across industries. As we move forward, the continual evolution of characterization methods promises to unlock new dimensions in nanopaint research, fueling advancements in nanomaterial science and high-performance coatings.

References

[1] P.P. Das, V. Chaudhary, F. Ahmad, A. Manral, Effect of nanotoxicity and enhancement in performance of polymer composites using nanofillers: a state-of-the-art review, Polymer Composites 42 (5) (2021) 2152–2170. Available from: https://doi.org/10.1002/pc.25968.

[2] S.J. Ahn, S.J. Lee, J.K. Kook, B.S. Lim, Experimental antimicrobial orthodontic adhesives using nanofillers and silver nanoparticles, Dental Materials 25 (2) (2009) 206–213. Available from: https://doi.org/10.1016/j.dental.2008.06.002.
[3] Parna Ganguli, Surabhi Chaudhuri, Nanomaterials in antimicrobial paints and coatings to prevent biodegradation of man-made surfaces: a review, Materials Today: Proceedings 45 (2021) 3769–3777. Available from: https://doi.org/10.1016/j.matpr.2021.01.275.
[4] J. Vidales-Herrera, I. López, Nanomaterials in coatings: an industrial point of view, Handbook of Nanomaterials for Manufacturing Applications, Elsevier, United Kingdom, 2020, pp. 51–77. Available from: https://doi.org/10.1016/B978-0-12-821381-0.00003-X.
[5] K. Virmani, C. Deepak, S. Sharma, U. Chadha, S.K. Selvaraj, Nanomaterials for automotive outer panel components: a review, European Physical Journal Plus 136 (9) (2021). Available from: https://doi.org/10.1140/epjp/s13360-021-01931-w.
[6] Hala S. Hussein, The state of the art of nanomaterials and its applications in energy saving, Bulletin of the National Research Centre 47 (1) (2023). Available from: https://doi.org/10.1186/s42269-023-00984-4.
[7] K.L. Arun, M. Udhayakumar, N. Radhika, A comprehensive review on various ceramic nanomaterial coatings over metallic substrates: applications, challenges and future trends, Journal of Bio- and Tribo-Corrosion 9 (1) (2023). Available from: https://doi.org/10.1007/s40735-022-00717-6.
[8] Maria Teresa Ferreira, Eliana Soldado, Giovanni Borsoi, Maria Paula Mendes, Inês Flores-Colen, Nanomaterials applied in the construction sector: environmental, human health, and economic indicators, Applied Sciences 13 (23) (2023). Available from: https://doi.org/10.3390/app132312896.
[9] A.A. Keller, A. Ehrens, Y. Zheng, B. Nowack, Developing trends in nanomaterials and their environmental implications, Nature Nanotechnology 18 (8) (2023) 834–837. Available from: https://doi.org/10.1038/s41565-023-01409-z.
[10] Huimeng Feng, Tong Wang, Lin Cao, Yanan Pu, Zhipeng Zhao, Shougang Chen, Recent achievements and applications of photothermal self-healing coatings: a review, Progress in Organic Coatings 187 (2024). Available from: https://doi.org/10.1016/j.porgcoat.2023.108153.
[11] Semin Kim, Hyeonyeol Jeon, Jun Koo, Mo, Dongyeop X. Oh, Jeyoung Park, Practical applications of self-healing polymers beyond mechanical and electrical recovery, Advanced Science (2024). Available from: https://doi.org/10.1002/advs.202302463.
[12] Amina Othmani, Overview of the Major Types of Nanomaterials Used for Environmental and Energy Applications: Challenges and Prospects, Springer Science and Business Media LLC, 2023, pp. 1–13. Available from: https://doi.org/10.1007/978-981-99-6924-1-1.
[13] S. Malik, K. Muhammad, Y. Waheed, Nanotechnology: a revolution in modern industry, Molecules 28 (2) (2023). Available from: https://doi.org/10.3390/molecules28020661.
[14] M.C. Chen, P.W. Koh, V.K. Ponnusamy, S.L. Lee, Titanium dioxide and other nanomaterials based antimicrobial additives in functional paints and coatings: review, Progress in Organic Coatings 163 (2022). Available from: https://doi.org/10.1016/j.porgcoat.2021.106660.

[15] Arkan Alali, Afrouzossadat Hosseini-Abari, Abbas Bahrami, Maryam Yazdan Mehr, Biosynthesis of copper oxide and silver nanoparticles by bacillus spores and evaluation of the feasibility of their use in antimicrobial paints, Materials 16 (13) (2023). Available from: https://doi.org/10.3390/ma16134670.
[16] Hyunjoo Hong, Claudia Som, Bernd Nowack, Development of a benefit assessment matrix for nanomaterials and nano-enabled products—toward safe and sustainable by design, Sustainability 15 (3) (2023). Available from: https://doi.org/10.3390/su15032321.
[17] G.M. Patel, V. Shah, J. Bhaliya, P. Pathan, K.M. Nikita, Polymer-based nanomaterials: an introduction, Smart Polymer Nanocomposites: Design, Synthesis, Functionalization, Properties, and Applications, Elsevier, India, 2022, pp. 27–59. Available from: https://doi.org/10.1016/B978-0-323-91611-0.00018-9.
[18] H. Huang, X. Sheng, Y. Tian, L. Zhang, Y. Chen, X. Zhang, Two-dimensional nanomaterials for anticorrosive polymeric coatings: a review, Industrial and Engineering Chemistry Research 59 (35) (2020) 15424–15446. Available from: https://doi.org/10.1021/acs.iecr.0c02876.
[19] E.O. Ogunsona, R. Muthuraj, E. Ojogbo, O. Valerio, T.H. Mekonnen, Engineered nanomaterials for antimicrobial applications: a review, Applied Materials Today 18 (2020). Available from: https://doi.org/10.1016/j.apmt.2019.100473.
[20] A. Tiwari, A.K. Yadav, V. Bagaria, Nanomaterials-based antimicrobial coatings for medical devices, *Antiviral and Antimicrobial Coatings Based on Functionalized Nanomaterials: Design, Applications, and Devices*, Elsevier, India, 2023, pp. 545–568. Available from: https://doi.org/10.1016/B978-0-323-91783-4.00008-5.
[21] R.V.S.K. Jakkula, P. Sethuramalingam, Analysis of coatings based on carbon-based nanomaterials for paint industries – a review, Australian Journal of Mechanical Engineering 21 (3) (2023) 1008–1036. Available from: https://doi.org/10.1080/14484846.2021.1938953.
[22] Abhinay Thakur, Savaş. Kaya, Ashish Kumar, Recent trends in the characterization and application progress of nano-modified coatings in corrosion mitigation of metals and alloys, Applied Sciences 13 (2) (2023). Available from: https://doi.org/10.3390/app13020730.
[23] Istie Rahayu, Wayan Darmawan, Deded Sarip Nawawi, et al., Surface modification of fast-growing wood with a titanium-dioxide-based nanocoating to improve weathering resistance, Coatings 13 (11) (2023). Available from: https://doi.org/10.3390/coatings13111924.
[24] S.A. Afolalu, O.M. Ikumapayi, T.S. Ogedengbe, J.F. Kayode, A.T. Ogundipe, T. C. Jen, A review on emerging trends in the scientific application of nano-coatings and nanoparticle synthesis, Materials Today: Proceedings 2024 (2023). Available from: https://doi.org/10.1016/j.matpr.2023.06.376.
[25] A.V. Pipa, A. Puth, J. Böcker, et al., Laser absorption spectroscopy for plasma-assisted thermochemical treatment. Part I.: applicability of the Beer–Lambert law and interpretation of spectroscopic data, Plasma Sources Science and Technology 32 (8) (2023). Available from: https://doi.org/10.1088/1361-6595/ace9f9.
[26] A. Chen, J. Zhang, Infrared spectroscopy analysis of (3-Aminopropyl) triethoxysilane (APTES) coating, (2023). Accessed on Feb. 21, 2024. [Online].
[27] Y.Z. Qian, Q. Wang, D.Y. Zhang, Y.Y. Wang, B. Li, H.Y. Wang, A high-performance long-range surface plasmon resonance sensor based on the co-modification of carbon nanotubes and gold nanorods, Optical Fiber

Technology 80 (2023). Available from: https://doi.org/10.1016/j.yofte.2023.103460.
[28] S.T. Mhaske, P.Y. Borse, S.U. Mestry, Characterization of Antimicrobial Nanocoatings, Elsevier BV, 2024, pp. 37–59. Available from: https://doi.org/10.1016/b978-0-323-95756-4.00006-3.
[29] A. Behera, P. Mallick, S.S. Mohapatra, Nanocoatings for anticorrosion, Corrosion Protection at the Nanoscale, Elsevier, India, 2020, pp. 227–243. Available from: https://doi.org/10.1016/B978-0-12-819359-4.00013-1.
[30] M. El-Azazy, A.S. El-Shafie, K. Al-Saad, Introductory chapter: Infrared spectroscopy - principles and applications, IntechOpen, 2023. Available from: https://doi.org/10.5772/intechopen.109139.
[31] S.S. Panicker, S.P. Rajeev, Spectroscopy for material characterization-review, AIP Conference Proceedings (2023) 2752. Available from: https://doi.org/10.1063/5.0136076.
[32] W.P. Evangeline, E. Saranya, E. Rajalakshmi, et al., Advancements of Raman spectroscopy in cosmetics and dermatology, South African Journal of Botany 167 (2024) 122–129. Available from: https://doi.org/10.1016/j.sajb.2024.02.011.
[33] C. Chen, Applications of Raman spectroscopy in the diagnosis and monitoring of neurodegenerative diseases, Frontiers in Neuroscience 18 (2024) 1301107.
[34] U. Mahajan, M. Dhonde, K. Sahu, P. Ghosh, P.M. Shirage, Titanium nitride (TiN) as a promising alternative to plasmonic metals: a comprehensive review of synthesis and applications, Materials Advances 5 (3) (2024) 846–895. Available from: https://doi.org/10.1039/d3ma00965c.
[35] N. Thakur, N. Thakur, A. Kumar, et al., A critical review on the recent trends of photocatalytic, antibacterial, antioxidant and nanohybrid applications of anatase and rutile TiO_2 nanoparticles, Science of The Total Environment 914 (2024). Available from: https://doi.org/10.1016/j.scitotenv.2023.169815.
[36] S. Jakhmola, S. Das, K. Dutta, Emerging research trends in the field of polyurethane and its nanocomposites: chemistry, synthesis, characterization, application in coatings and future perspectives, Journal of Coatings Technology and Research 21 (1) (2024) 137–172. Available from: https://doi.org/10.1007/s11998-023-00841-z.
[37] A.B. Rashid, M. Haque, S.M.M. Islam, K.M.R. Uddin Labib, Nanotechnology-enhanced fiber-reinforced polymer composites: recent advancements on processing techniques and applications, Heliyon 10 (2) (2024). Available from: https://doi.org/10.1016/j.heliyon.2024.e24692.
[38] K.W. Shinato, F. Huang, Y. Jin, Principle and application of atomic force microscopy (AFM) for nanoscale investigation of metal corrosion, Corrosion Reviews 38 (5) (2020) 423–432. Available from: https://doi.org/10.1515/corrrev-2019-0113.
[39] J.R.V. Sai Kiran, S. Prabhu, Robot nano spray painting – a review, IOP Conference Series: Materials Science and Engineering 912 (3) (2020). Available from: https://doi.org/10.1088/1757-899x/912/3/032044.
[40] L. Giraud, A. Tourrette, E. Flahaut, Carbon nanomaterials-based polymer-matrix nanocomposites for antimicrobial applications: a review, Carbon 182 (2021) 463–483. Available from: https://doi.org/10.1016/j.carbon.2021.06.002.
[41] M.S. Selim, S.A. Higazy, Z. Hao, X. Chen, Ceramic polymer nanocomposites as eco-friendly marine antifouling coatings, Advances in Nanotechnology for Marine Antifouling, Elsevier, China, 2023, pp. 117–140. Available from: https://doi.org/10.1016/B978-0-323-91762-9.00010-1.

[42] M.S. Selim, S.A. El-Safty, M.A. Shenashen, S.A. Higazy, A. Elmarakbi, Progress in biomimetic leverages for marine antifouling using nanocomposite coatings, Journal of Materials Chemistry B 8 (17) (2020) 3701–3732. Available from: https://doi.org/10.1039/c9tb02119a.
[43] S.A. Othman, N.N. Musa, Characterization of analgesics: a preliminary review, Journal of Health and Quality of Life 1 (2024) 16–20.
[44] Zahra H. Mohammad, Faizan Ahmad, Nanocoating and its application as antimicrobials in the food industry: a review, International Journal of Biological Macromolecules 254 (2024). Available from: https://doi.org/10.1016/j.ijbiomac.2023.127906.
[45] Abhilasha Mishra, Pawan Kumar Mishra, Gagan Bansal, Recent Advances in Coating Characterization Techniques, Elsevier BV, 2024, pp. 461–484. Available from: https://doi.org/10.1016/b978-0-443-19009-4.00024-2.
[46] R. Bansal, H.C. Barshilia, K.K. Pandey, Nanotechnology in wood science: innovations and applications, International Journal of Biological Macromolecules 262 (2024). Available from: https://doi.org/10.1016/j.ijbiomac.2024.130025.
[47] P.K. Gallagher, M.E. Brown, Handbook of Thermal Analysis and Calorimetry, Elsevier, 2003.
[48] M.E. Brown, P.K. Gallagher, Handbook of Thermal Analysis and Calorimetry: Recent Advances, Techniques and Applications, Elsevier, 2011.
[49] K.V.S. Anirudh, M.V. Bottu, D. Sarvamangala, Production of TiO_2 nanoparticles by green and chemical synthesis—a short review, International Journal of Scientific Engineering and Research 9 (2018) 1633–1648.
[50] F.A. Shilar, S.V. Ganachari, V.B. Patil, Advancement of nano-based construction materials—a review, Construction and Building Materials 359 (2022). Available from: https://doi.org/10.1016/j.conbuildmat.2022.129535.
[51] D.G. Goodwin, A.S. Adeleye, L. Sung, K.T. Ho, R.M. Burgess, E.J. Petersen, Detection and quantification of graphene-family nanomaterials in the environment, Environmental Science and Technology 52 (8) (2018) 4491–4513. Available from: https://doi.org/10.1021/acs.est.7b04938.
[52] I. Gurrappa, L. Binder, Electrodeposition of nanostructured coatings and their characterization – a review, Science and Technology of Advanced Materials 9 (4) (2008). Available from: https://doi.org/10.1088/1468-6996/9/4/043001.
[53] S.S. Patil, A. Rane, V.K. Abitha, S. Jadhav, S. Narute, Jayaja, Nanofillers in surface coatings: a review, Moroccan Journal of Chemistry 3 (2015) 3–4.
[54] M. Armstrong, M. Sivasubramanian, N. Selva Palam, M. Adam Khan, C. Rajaganapathy, A recent examination on the nano coating techniques in heat transfer applications, Materials Today: Proceedings 46 (2021) 7942–7947. Available from: https://doi.org/10.1016/j.matpr.2021.02.660.
[55] I. Serafini, A. Ciccola, Nanotechnologies and nanomaterials, Nanotechnologies and Nanomaterials for Diagnostic, Conservation and Restoration of Cultural Heritage, Elsevier, Italy, 2019, pp. 325–380. Available from: https://doi.org/10.1016/B978-0-12-813910-3.00014-8.
[56] J. Rawat, S. Ray, P.V.C. Rao, N.V. Choudary, Recent developments of nanomaterial doped paints for the minimization of biofouling in submerged structures, Materials Science Forum 657 (2010) 75–82. Available from: https://doi.org/10.4028/www.scientific.net/MSF.657.75.

PART II

Properties and performance evaluation

Asset maintenance and enhancement of cyclability by nanopaints

Ankita Saxena[1], Jimmy Mehta[2] and Pallav Gupta[3]

[1]Department of Biotechnology, School of Engineering and Technology, Noida International University, Greater Noida, Uttar Pradesh, India [2]Department of Mechanical Engineering, School of Engineering and Technology, Manav Rachna International Institute of Research and Studies, Faridabad, Haryana, India [3]Department of Mechanical Engineering, Madan Mohan Malaviya University of Technology, Gorakhpur, Uttar Pradesh, India

Methodologically, a blend of laboratory experiments and field trials is employed to assess the performance of nanopaint-coated assets under realistic conditions. Key factors such as durability, adhesion, and resilience to environmental elements are meticulously evaluated to gauge the efficacy of nanopaints in asset upkeep. The findings of this study underscore the substantial potential of nanopaints to transform asset management practices, providing economical solutions to lengthen asset lifespan and curtail life cycle expenses. Moreover, the adoption of nanopaints shows promise in advancing sustainability efforts by reducing resource consumption and environmental impact associated with frequent maintenance routines. This research yields valuable insights into asset maintenance and highlights the transformative capacity of nanotechnology in infrastructure management.

6.1 Introduction

Materials possessing at least one external dimension are called nanomaterials. They have a size ranging between 1 and 100 nm. Nanomaterials can occur naturally. They can be created as the by-products of combustion reactions or produced through engineering to perform a specialized function. Nanomaterials can have distinct physical and chemical properties. The most

Nanopaints. DOI: https://doi.org/10.1016/B978-0-443-26512-9.00006-9

used nanomaterials for the paint industry are nanoscale silicon dioxide (SiO_2) and titanium dioxide (TiO_2). Nanomaterials such as zinc oxide (ZnO), silver oxide (Ag_2O), cerium dioxide (CeO_2), aluminum oxide (Al_2O_3), copper oxide (CuO), and magnesium oxide (MgO) are also undergoing research.

Depending on the type of paint and the desired functionality, nanomaterials can be integrated as free powders, stabilized particles in a suspension/dispersion, or incorporated into masterbatches or granulates, resulting in the nanomaterial firmly embedded in the paint matrix [1,2].

There also exist paints or products that have a nanostructured surface, contain nanoporous materials, or temporarily contain nanostructured particles (e.g., produced from water glass while drying the paint). As these end products do not contain any nanomaterial, they are considered to pose no nano-specific risk to humans or the environment.

Nanotubes, nanocarbon, graphite structures, and nanofiber are compactly packed with molecules that have a penetrative nature. In addition, they can produce functional nanopaints and nanocoatings. This nanopaint and nanocoating possess terrific qualities, such as increased hardness and scratch resistance, and they are antimold and antibacterial [3]. Therefore, a coating product that is highly durable and has high-performance characteristics is yielded (Fig. 6.1).

When a substance or surface can be modified according to a user, it is called nanopaint. The coating of the nanopaint is liquid in state and hardens upon application, the same as ordinary paint. Nanopaints contain a heterogeneous mixture of minute or microscopic particles called nanotubes. These nanotubes alter the behavior of nanopaints when there is a change in the external conditions. Globally, engineers are trying to figure out the wonders of nanopaints and the beautiful results that modifications can bring. Fig. 6.2 shows the detailed steps for preparing composite nanopaints. These steps need to be followed carefully and knowingly.

Researchers have produced a model nanopaint that can obstruct the radio frequency fields. This is quite similar to metals, such as copper. When the nanopaints are used in the interior walls of a building, the material can impede signals to and from portable radios, wireless devices, or cell phones.

Modifying the infrared (IR) reflecting or even absorbing properties depending on the external conditions would be a great application of nanopaint. This, in turn, could ameliorate energy efficiency by absorbing thermal energy on cool but sunny days and reflecting it on hot days [4,5]. This improvisation could

Figure 6.1 Key materials for preparation of nanopaints.

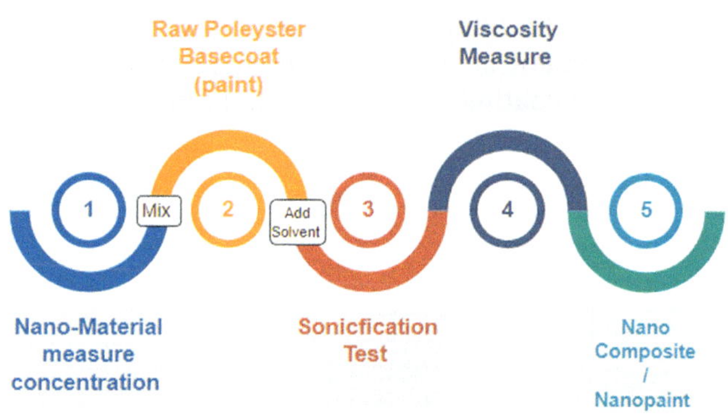

Figure 6.2 Preparation of sample for composite nanopaint.

also retain thermal energy on cold nights and radiate it away on warm nights. Altered or specialized nanopaints might perform a list of functions, such as:

1. Block cell phone signals in unsuitable environments like theaters, hospitals, and funerals.
2. Give glass the ability to become more or less opaque.

3. When damaged, the surfaces of the motor vehicles or industrial machines could repair themselves.
4. Allow the textures of surfaces to be altered at will.
5. Curb the growth of pathogenic bacteria or viruses.
6. Repel or neutralize toxic chemicals, acids, or other corrosive agents.

Over the past few years, research on developing and producing conductive nanopaints has increased. The conductive nanopaints present several exciting properties and valuable applications in automotive, electronics, aerospace, and electrodes. In the automobile industry, nanopaints are very useful since the substrate's structural, thermal, and electrical properties could be enhanced. This further leads to emulating the metallic-like appearance in replacement of the chrome-plated paints. The nanoparticles act as fillers in the conductive nanopaints added to a polymeric matrix [6,7]. Two classes of fillers are mainly used today in conductive nanopaints formulations: metallic nanoparticles and nanostructured carbons. Depending on the type of nanomaterial used, the conductive nanopaints properties are resistance to scratch, electric conductivity, and corrosion inhibition. Fig. 6.3 depicts how the nanopaints act on the dirty surface and the mechanism being followed.

End-user sectors of nanopaints and coatings include electronics and optics, food and packaging, aerospace, marine coatings, building and civil engineering, automobile, oil and gas, and biomedical fields. The methods used to apply the nanopaints and coating processes include physical vapor deposition (PVD),

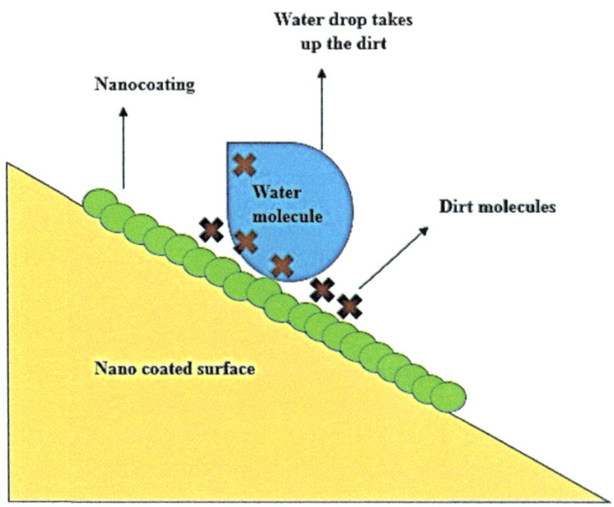

Figure 6.3 Mechanism of nanopaint on the dirt surface.

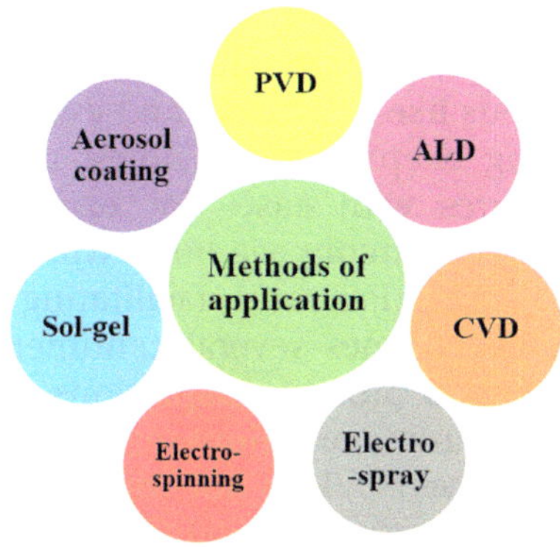

Figure 6.4 Methods of application of nanopaints: Physical vapor deposition (PVD); Atomic layer deposition (ALD); Chemical vapor deposition (CVD).

atomic layer deposition (ALD), chemical vapor deposition (CVD), electrospray and electrospinning, and sol-gel, aerosol coating, depending on individual/market sector. Fig. 6.4 explains the various techniques applied by nanopaints and how they act on different surfaces.

CVD is a process wherein gaseous or vaporous materials dissociate through chemical reactions near a heated substrate surface, leading to the deposition of a coating onto said surface. This method is commonly employed for coating metallic or ceramic compositions, typically on substrates that are also metallic or ceramic. The temperature of the substrate is a critical factor influencing the adhesion of the coating to the substrate. Notably, the multidirectional nature of the CVD method allows for the deposition of coating material on substrates with complex geometries and varying sizes. Its industrial applications span various sectors, including the deposition of refractory materials on turbine blades. However, the CVD approach demands higher substrate temperatures and a slow precipitation rate, making coating large surfaces costly [8–10]. CVD is when a solid substance deposits onto a heated surface through a chemical reaction in the vapor phase. It falls within the vapor-transfer processes category, characterized by atomistic behavior where the deposition entities are atoms, molecules, or combinations. Alongside CVD, this category includes various PVD processes like evaporation, sputtering, molecular-beam epitaxy, and ion plating. While CVD competes directly with PVD processes in many aspects, it is also used alongside them, with newer techniques often being hybrids, such as plasma-enhanced CVD or activated sputtering. Understanding PVD processes is crucial for a comprehensive grasp of CVD.

The advantages of CVD include its nonline-of-sight deposition capability and high throwing power, unlike typical PVD methods like sputtering and evaporation. This makes it adept at coating deep recesses, holes, and complex three-dimensional structures with ease. For example, CVD can completely fill integrated circuits via holes with high aspect ratios. CVD boasts high deposition rates, facilitating the production of thick coatings, sometimes several centimeters thick. Its flexibility allows for various process adaptations and does not necessitate ultra-high vacuum conditions, enhancing its applicability.

However, CVD is not without limitations. Its optimal operating temperature of 600°C and above may pose challenges as many substrates lack thermal stability at such temperatures. Nonetheless, advancements like plasma-CVD and metallo-organic CVD somewhat mitigate this issue. Another drawback is the requirement for chemical precursors with high purity.

The roots of CVD trace back to the late 19th century, when it was initially used to strengthen filaments in incandescent lamps and later for producing pure nickel. Over time, it found applications in various industries, notably in coatings and freestanding shapes post-World War II. Key milestones in its development include introducing terms like CVD and PVD, adopting CVD in semiconductor fabrication, and emerging technologies like plasma CVD and CVD diamond coatings [11].

Despite rapid technological advancements, CVD still faces challenges, such as accurately predicting composition and properties. Despite significant progress in understanding its theory and mechanisms, CVD remains partly an art, relying heavily on experimental innovations for further advancements.

PVD operates differently, involving the conversion of target materials into a gaseous phase before depositing them onto a substrate surface. Unlike CVD, where chemical reactions occur between the gaseous substance and the substrate, PVD relies on thermal energy, laser energy, high-energy particles or ions, electron beam, and resistive heating to achieve this conversion. The steps in PVD encompass target material evaporation, transportation of the evaporated material to the substrate surface, and deposition of the target material as a coating. Various deposition methods fall under PVD, including thermal evaporation, cathodic arc deposition, magnetron sputtering, and laser-pulsed deposition. Particularly noteworthy are novel nanostructured coatings like fullerene, carbon nitride (FL-CNx) and fullerene-like carbon fluoride (FL-CFx) coatings formed via magnetron sputtering, which exhibit low water absorption and are suitable for protecting magnetic storage devices. The structure of

Table 6.1 The commonly used nanomaterials and their applications.

Nanomaterial	Applications
Silicon dioxide (SiO_2)	Corrosion and scratch resistant
Titanium dioxide (TiO_2)	Self-cleaning; antifogging; scratch resistant; antibacterial; UV absorber; and has an optical effect
Aluminum oxide (Al_2O)	Corrosion protection and scratch resistant
Zinc oxide (ZnO)	UV absorber
Ferric oxide (Fe_2O)	Has medical applications
Silver colloids	Antibacterial
Metal colloids	UV absorber
Carbon nanotubes	Conductive coatings
Antimony tin oxide	Infrared and UV absorber
Indium tin oxide	Infrared and UV absorber

nanostructured coatings is influenced by experimental factors affecting mechanical, wear, and corrosion properties. If vapor generation occurs through physical methods without a chemical reaction, the process falls under the category of PVD. PVD is fundamentally a technique for coating via vaporization, operating at an atomic level. It is an alternative to electroplating. It is a well-established technology widely employed for depositing thin films to meet various demands, including enhancing tribological behavior, optical properties, esthetic appeal, and numerous other applications across diverse fields (Table 6.1).

Different types of PVD processes include:

1. Cathodic arc deposition: Here, a high-power electric arc discharges at the target material, causing the material to be ejected into a highly ionized vapor, which is then deposited onto the workpiece.
2. Electron beam PVD: This process involves heating the material to be deposited to a high vapor pressure using electron bombardment in a high vacuum, then transporting it through diffusion and condensation onto the cooler workpieces.
3. Evaporative deposition: In this method, the material to be deposited is heated to a high vapor pressure through resistive heating in a low vacuum.
4. Pulsed laser deposition: This technique employs a high-power laser to ablate material from the target into a vapor.

PVD processes, often called thin film processes, involve atomistic deposition, where the material is transported in vapor form through a vacuum or low-pressure gas or plasma environment to the substrate, where it condenses. Typical deposition rates for PVD range from 10 to 100 Angstroms (1–10 nm) per second. PVD processes enable the deposition of films comprising elements, alloys, and compounds, including reactive deposition processes. The three basic PVD processes are vacuum or thermal evaporation, ion plating, and sputtering. During evaporation, atoms are injected from the source due to thermal effects, ion plating involves ion bombardment of the growing film, and sputtering entails atoms being ejected from the solid target surface by the impact of gas ions, which are then deposited onto the substrate surface.

On the other hand, *spray coating* involves passing a target material through a nozzle to deposit it onto a substrate surface via impact. The sprayed substance may be in liquid, molten, or softened particle form. Thermal spraying, a prominent method within spray coating, melts or heats the coating material before spraying it onto the substrate. Cold spraying, a variation, allows for lower application temperatures but requires high velocities for adequate bonding [12]. Plasma spraying, high-velocity oxy-fuel spraying (HVOF), and high-velocity air fuel (HVAF) are other spray coating methods. The final properties of these processes depend on powder size distribution, feed rate, distance between spray guns and substrates, and spray velocity (Tables 6.2 and 6.3).

The *sol-gel process* entails using a solvent containing a chemically active component precursor, followed by hydrolysis and polycondensation to form a sol. This sol is then polymerized to produce a gel, which is dried and heated to form a coating. While the sol-gel method generates coatings with uniform properties, factors like catalyst nature, initial material, temperature, thickness, and pH influence the coating's characteristics.

Table 6.2 Various deposition conditions along with their values used in CVD technique.

Deposition conditions	CVD
Vacuum requirement	10.2–10.3 Torr
Deposition thickness range	10 nm to 3 μm
Uniformity control	nm
Deposition rate	10–1000 nm/min
Conformality	0%–100% at 4:1 aspect ratio
Contamination particles	Particles can exist due to gas supply, CVD chamber, sample substrate and other sources.

Table 6.3 Various deposition conditions along with their values used in PVD technique.

Deposition conditions	PVD
Vacuum requirement	<10.2 Torr
Deposition thickness range	5 nm > 10 μm
Uniformity control	nm
Deposition rate	10–1000 nm/min
Conformality	15% at 4:1 aspect ratio
Contamination/particles	Indirectly only (flaking)

Although the sol-gel process can produce wear-resistant and nanocomposite coatings, it is time-consuming and costly for large-scale production.

Electrodeposition involves depositing a metal or alloy coating material on a conductive substrate surface using an electric current (electrolysis) in an electrolyte bath containing a salt of the metal or alloy to be coated. The coating's composition, morphology, and texture can be adjusted by varying parameters such as applied potential, current density, deposition time, and plating solution composition [13]. *Laser cladding* utilizes a high-energy laser to melt powdered or wire feedstock target material on the substrate surface, offering advantages such as high cooling rates and ease of automation but suffering from drawbacks like poor metallurgical quality and non-uniform composition.

With the rapid advancement of nanotechnology, integrating nanomaterials into our daily lives and workplaces is increasingly prevalent. Consequently, workers engaged in maintenance tasks are at greater risk of encountering nanomaterials. Despite ongoing research efforts, the pace of nanotechnology development outstrips the generation of knowledge concerning nanomaterial exposure's health and safety implications.

This article elucidates the potential scenarios where maintenance workers may encounter nanomaterials and this inturn can guide them to mitigate the potential exposures.

Effective maintenance is crucial for ensuring the safety and reliability of equipment, machinery, structures (such as bridges or tunnels), and the work environment. Maintenance encompasses various tasks spanning various sectors and work environments, including servicing, repairing, inspecting, testing, and replacing components [14]. Activities may involve accessing closed production systems, replacing filters, surface treatments like paint removal or blasting, and infrastructure repairs such as power grid or water supply maintenance. Given the ubiquitous nature of maintenance across industries, maintenance workers are particularly susceptible to various occupational hazards.

Maintenance can be either proactive, aimed at preventing failures and hazardous conditions, or reactive, focused on repairing malfunctioned equipment. Consequently, maintenance tasks may be integrated into a worker's daily routine or require specialized attention when addressing equipment's malfunctioning. Maintenance activities may constitute construction workers' primary duties.

6.1.1 Nanomaterials in maintenance work

Despite being a nascent industry, nanomaterials are already utilized in numerous applications due to their unique properties. Consequently, potential exposure to nanomaterials during maintenance tasks must be considered across an expanding spectrum of sectors and workplaces.

As the prevalence of products containing nanomaterials rises, maintenance workers are increasingly tasked with servicing such products, potentially exposing them to nanomaterials. Such products include automobiles, energy-efficient tires, electronics, and energy-generation equipment like rechargeable batteries or smart solar panels. Furthermore, building materials themselves may incorporate nanomaterials [15].

Moreover, various maintenance products containing manufactured nanomaterials, such as lubricants and coatings, are commercially available. Without proper prevention measures, these products may pose risks to workers.

However, specific applications of manufactured nanomaterials offer significant benefits for maintenance workers. For instance, intelligent paints embedded with carbon nanotubes can detect microscopic defects like surface cracks or corrosion. By facilitating remote detection of structural issues, such paints minimize the need for hazardous work at heights, particularly in infrastructure inspection scenarios.

While nanomaterials offer several benefits, certain types pose hazards to human health and safety, endangering maintenance workers.

Hazards and exposure routes: Safety risks arise from the explosive, flammable, and catalytic properties of certain nanopowders, particularly metal nanopowders.

Nanomaterials exhibit a wide array of potential toxic effects, often differing from their macro-scale counterparts due to their size, shape, chemical composition, surface characteristics, and aggregation state. Despite forming larger agglomerates or aggregates under normal conditions, nanomaterials can disperse again, potentially entering the body after inhalation. The internal exposure mechanism involves absorption, distribution, and metabolism, with nanomaterials detected in various organs and tissues. Additionally, nanomaterials may carry more toxic substances, exacerbating health risks [16].

The lungs are particularly susceptible to nanomaterial effects, including inflammation, tissue damage, oxidative stress, cytotoxicity, fibrosis, and tumor formation. Cardiovascular impacts are also possible, warranting ongoing research.

Examples of nanomaterials relevant to maintenance workers include those found in paints, disinfectants, and cleaning agents.

Routes of exposure: Workers can be exposed through inhalation, ingestion, or dermal contact, with inhalation being the most common route. Nanoparticles can deposit in the respiratory tract and potentially reach other organs, including the brain. Ingestion may occur directly or via particles cleared from the respiratory tract. Dermal penetration, although less common, remains under investigation.

Exposure risk depends on the likelihood of nanomaterials becoming airborne, with powder form or sprays presenting higher risks than liquid suspensions or bound nanostructures. Before any maintenance work, parameters need to be studied carefully, as shown in Fig. 6.5.

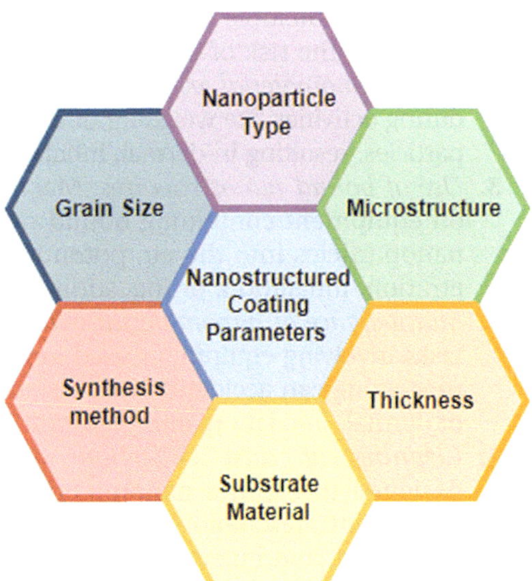

Figure 6.5 Parameters affecting structure of nanopaints.

Maintenance tasks present various scenarios where workers may encounter manufactured nanomaterials:

1. Use of maintenance products: Workers are at risk of exposure while handling maintenance products containing nanomaterials. This includes lubricants, coatings, adhesives, and other products utilized during maintenance activities.
2. Maintenance of installations involving nanomaterials: Workers may encounter nanomaterial exposure while maintaining installations where nanomaterials are utilized or processed. This could include production lines or machinery with nanomaterials or products containing nanomaterials. Workers may come into contact with deposited nanomaterials on surfaces during maintenance tasks [17].
3. Generation of nanomaterials during maintenance: Certain maintenance processes, such as grinding or polishing, can generate nanomaterials as by-products. Workers involved in these activities are at risk of exposure to newly generated nanomaterials during the maintenance process.

Various maintenance activities carry the risk of exposing workers to nanomaterials:

1. *Use of liquid products:* Handling liquid products like lubricants, paints, or adhesives containing nanomaterials can lead to dermal exposure, especially during spill cleanup. Additionally, activities involving high agitation, such as pouring or mixing, may generate aerosols that can be inhaled or deposited onto the skin, increasing exposure risk. Spraying liquid combustible nanomaterials poses additional hazards, including the risk of explosion or fire.
2. *Using nanomaterial powders*: Handling nanomaterial powders during activities like weighing or blending can generate airborne particles, resulting in dermal, inhalation, and exposure risks.
3. *Use of bound nanostructures:* Machining or drilling activities on equipment containing bound nanostructures may release nanoparticles into the air, potentially leading to dermal penetration, inhalation, or ingestion.
4. *Maintenance of nanomaterial-related equipment:* Maintenance tasks involving equipment used in nanomaterial production or processing can accidentally release nanomaterials, posing risks of dermal, inhalation, and ingestion exposure.
5. *Cleaning of dust collection systems:* Cleaning systems designed to capture nanomaterial dust can expose workers to concentrated nanomaterials, increasing the risk of dermal, inhalation, and ingestion exposure.
6. *Spill cleanup:* Spills of nanomaterials during maintenance activities present risks of dermal, inhalation, and ingestion exposure.

7. *Transportation and disposal:* Handling waste materials containing nanomaterials can lead to dermal exposure, inhalation, and ingestion.

6.2 Types of nanopaints

6.2.1 Anticorrosion paints

Anticorrosion paint is generally used to protect solid metal surfaces and sometimes non-metals. Research shows that environmental exposure has a long-term and everlasting impact on the suitability and effectiveness of anticorrosion paint. In corrosion prevention, there are four primary strategies: material selection, environmental adjustment, cathodic protection, and surface modification. Coating is the dominant method due to its versatility in addressing corrosion across diverse applications and conditions. Coatings typically function by providing either passive or active protection. Passive protection occurs when the coating forms a physical barrier of oxides between the substrate and the environment, while active protection involves the introduction of inhibitors to thwart corrosion [18]. These inhibitors may chemically adhere to the material's surface, forming a protective layer, or react with corrosive elements. Typical coating materials include polymers, chromium, and zirconium-based nanocoatings. Nonetheless, continuous advancements are necessary to bolster anticorrosive capabilities and adapt to evolving technological requirements, a challenge that nanotechnology effectively tackles. All the factors for enhanced and effective knowledge are shown clearly in Fig. 6.6.

Nanotechnology offers significant advantages in enhancing anticorrosive properties and addressing the limitations of conventional coatings. For example, traditional polymer coatings often necessitate multiple layers, whereas chromium coatings pose toxicity concerns. Nanocoatings excel due to their minute particle size, which enables efficient filling and sealing of metal surfaces, impeding the diffusion of corrosive agents. Moreover, the heightened surface area of nanomaterials improves adhesion, thus extending the coating lifespan. Additionally, nanomaterials exhibit superior mechanical, chemical, and electronic properties, rendering nanocoatings more durable and resistant to corrosion and wear. For instance, while conventional zinc coatings suffer from poor weldability at the microscale, zinc nanocoatings overcome this issue. Furthermore, the development of multifunctional nanocoatings with self-healing, self-cleaning,

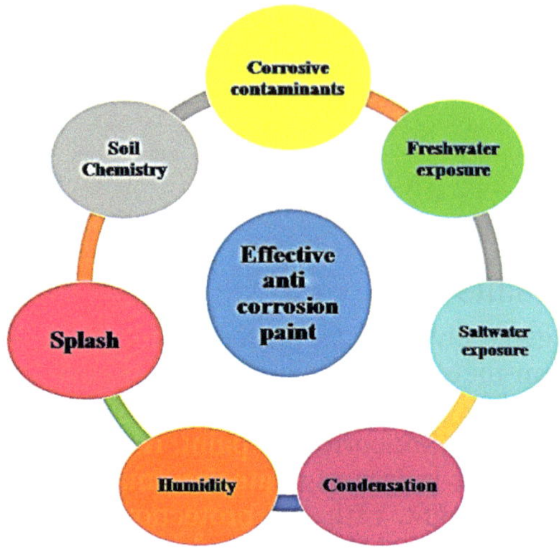

Figure 6.6 Key factors for an effective anticorrosion paint.

corrosion-sensing, and wear-resistant capabilities is feasible. Nanotechnology also enables the creation of intelligent coatings that respond dynamically to external stimuli such as pH, humidity, heat, and mechanical stress, releasing inhibitors as necessary to rectify defects and mitigate damage.

The benefits of anticorrosive nanocoatings are pervasive across various sectors. Everyday consumer products like electronic devices and eyewear incorporate these coatings to enhance longevity. In industrial contexts, construction materials such as scaffolds, windows, air filters, machinery, and automotive components rely on anticorrosive coatings to ensure prolonged service life. Moreover, biomedical materials like bone implants necessitate bio-functional nanocoatings to resist corrosion. Even wood materials can benefit from protection against corrosion by applying nanocoatings [19].

For the development of effective anticorrosion paints certian factors are considered i.e. adhesion to substrate material, barrier properties, corrosion inhibitors, environmental resistance, and durability.

Factors considered are:

1. Reactivity of product
2. Proximity to high-voltage lines
3. Range and peak levels of temperature
4. UV radiation exposure, solvents, and sunlight
5. Exposure to reactive alkaline and acidic substances
6. Projected lifespan of project operation

6.2.2 Waterproof paints

Creating water-resistant or waterproof surfaces is one of the primary uses of nanopaints. These paints make clothing, furniture, electronics, and glass products waterproof. Not only do nanopaints make the surface waterproof, but they also keep the surfaces unstained. This is less time-consuming and pocket-friendly [20]. Nanopaints have shown good oil resistance and resistance to fluid and liquid material (https://www.westerncanadacoatings.ca/nanocoatings/).

6.2.3 Antibacterial paints

Antibacterial nanopaints help avert or at least curtail the growth of microorganisms on the surfaces. These nanopiants are commonly and most widely used in the healthcare sector, where they try to curb the spread of infections. Antibacterial nanopaints have now been found to be used in public places and on surfaces that are in ceaseless contact with people's hands. These include guardrails or handles of the doors, and they are of broad utility in buses and subways. Besides these applications, antibacterial nanopaints are widely used in kitchens and air conditioners [21]. Food factories and food packaging plants also utilize antibacterial nanopaints (https://www.westerncanadacoatings.ca/nanocoatings/).

6.2.4 Thermal insulation paints

The aviation industry and aircraft use nanopaints with thermal insulation. This type of nanopaint with antiheat properties is generally used on metal surfaces exposed to excessively high temperatures. They can be used in gas turbines and parts of spacecraft engines. Thermal insulation paints signify a state-of-the-art approach towards optimization of energy efficiency and controlling temperatures. By leveraging advanced nanotechnology, these paints are engineered with nano-sized particles that act as formidable barriers against heat transfer. With exceptional thermal properties, including low thermal conductivity and high reflectivity, these nanoparticles form a thin yet highly efficient insulating layer upon application, mitigating heat transfer via conduction and radiation. A significant advantage lies in their capacity to reflect thermal radiation, effectively deflecting a substantial portion of infrared radiation emitted by heat sources like the sun or indoor heating systems [21–23]. This feature stabilizes indoor temperatures, curbing the need for excessive heating or cooling and reducing energy consumption. Additionally, their versatility allows for application on diverse surfaces such as walls, roofs, and industrial equipment, rendering them adaptable to various residential and industrial settings. The durability and longevity of nanopaints stand out against conventional insulating materials, as they form resilient coatings capable of enduring harsh environmental conditions, ensuring sustained thermal comfort and energy efficiency over time. Furthermore, their contribution to environmental sustainability is notable, as they aid in diminishing carbon emissions associated with heating and cooling systems, thus fostering a reduced reliance on fossil fuels (https://www.westerncanadacoatings.ca/nanocoatings/).

6.2.5 Antiscratch paints

When the parts are used regularly and continuously, scratches and abrasions are likely to occur. Antiscratch paints eradicate or reduce the total wear and lengthen the life of parts. Antiscratch nanopaints are suitable for use in environments that are not lubricated and are for appliances. Antiscratch nanopaints also increase the mechanical properties of other applications, such as stiffness and hardness (https://www.westerncanadacoatings.ca/nanocoatings/).

6.2.6 Self-healing nanopaints

Self-healing nanopaints are said to be an excellent alternative to protect against corrosion. These self-healing nanopaints have microcapsules or nanocapsules containing unique material released when the paint is damaged and needs healing. Such a self-healing system is designed for metal structures under cathodic protection and is sensitive to electric fields and acidity.

Over the past five decades, new automated systems have significantly influenced the economy and daily human life. This impact has been primarily fueled by the rapid advancements in computing technologies, which have fundamentally reshaped the information landscape. However, the development of intelligent materials with self-responsive properties at the micro and nanoscale emerged in the 1980s. This article highlights a universal approach for creating autonomous materials, focusing on encapsulating active agents into micro and nano-sized containers with controlled release properties. This encapsulation technique is applicable across various innovative materials, offering multiple responsive functionalities. Emphasizing nanocontainer-based self-healing coatings, this concept integrates passive and active components within the coating matrix [24–26]. This approach enables coatings to demonstrate classic passive functionalities like barrier or coloration while exhibiting active responses to internal and external stimuli such as cracks, pH changes, electrochemical potential, temperature, and humidity. By utilizing nanocontainers as coating pigments, significant advancements in coating technologies are expected, including autonomous crack repair to enhance corrosion protection; they have increased durability, extended maintenance intervals, and multifunctionality with additional active features like sensing and antifouling. The success of this approach depends on developing nanocontainer shells with specific permeability tailored to multiple triggers while ensuring stability during formulation, application, and curing processes. Achieving uniform distribution of nanocontainers within the coating is crucial to maintaining integrity and performance, preventing aggregation that could compromise self-healing abilities. Experimental findings suggest an optimal nanocontainer concentration of around 5%, striking a balance between performance and coating characteristics [24]. Various external triggers, such as mechanical impact, pH shifts, electromagnetic irradiation, ultrasonic treatment, temperature, ionic strength, and electrochemical potential, have been explored to activate self-healing mechanisms in these coatings, showing promising potential for corrosion protection and beyond. Moving forward, efforts are aimed at industrial-scale implementation of nanocontainer-based coatings, focusing on cost-effectiveness and expanding functionality to respond to multiple triggering mechanisms, representing significant progress from laboratory-scale demonstrations to real-world applications (https://www.westerncanadacoatings.ca/nanocoatings/).

6.2.7 Antireflective paints

Antireflective nanopaint is transparent and is widely used in making glasses. This type of nanopaint reduces the reflection on the radiant side of the surface. Antireflective nanopaints increase the productivity of solar panels. Owing to the colorless nature of this type of nanopaint, they can be dyed by adding pigments as desired. Various nanostructures and nanomaterials are utilized as antireflective coating layers, with significant interest in materials such as ZnO, Al_2O_3, and TiO_2 due to their distinct optical properties. Spherical-shaped nanoparticles are particularly scrutinized for their potential as antireflective coatings, as their optical characteristics remain consistent regardless of the incident angle of light. ZnO, known for its strong light scattering, appropriate refraction index, and minimal light reflection, is widely applied as an antireflective coating across various solar cell types, bolstered by its high UV resistance. Conversely, TiO_2 nanoparticles, valued for their electrical, optical, and self-cleaning properties, emerge as promising candidates for antireflective coatings. Additionally, due to their favorable optical and electrical properties, Al_2O_3 thin films exhibit exceptional promise for use in single or multilayer coatings, especially in Si solar cells [27,28].

6.3 Significance of asset maintenance

Industrial manufacturing has become increasingly complex with the introduction of artificial intelligence, robotics, machine learning, and augmented reality, to name a few. This has now led to increased complications in maintenance and reliability. Now, asset management strategies have gained momentum and weight. These are explained in Fig. 6.7.

Asset management includes line equipment, packaging machines, transportation fleets, or cutting-edge autonomous robotic vehicles. A proper asset management program can help a manufacturer ensure optimal performance, minimize unscheduled downtime, reduce operating costs, and extend the life of critical equipment.

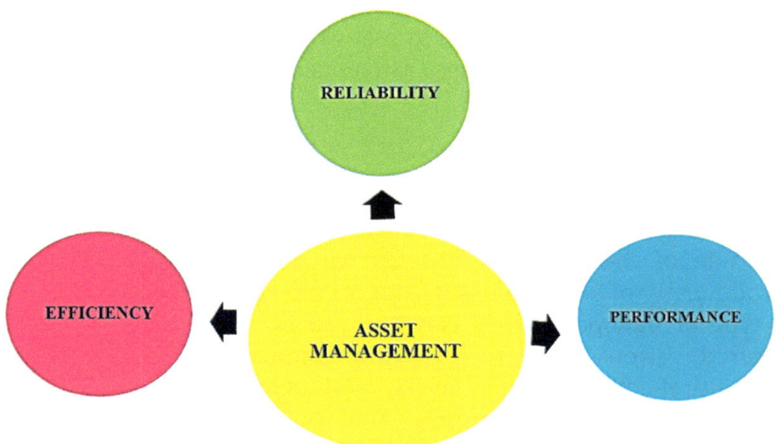

Figure 6.7 Factors affecting asset management.

Asset management is the systematic approach of organizing, operating, maintaining, upgrading, and disposing of assets throughout their entire life cycle. It begins with the strategic planning and purchasing of the asset and moves into efficient implementation and staff training. It includes continuously monitoring that asset's performance with predictive maintenance tools such as sensors to maximize its value while minimizing potential risks of failure and downtime [25].

Asset management includes appropriate and timely actions that follow real-time sensor alerts to ensure maintenance which is performed quickly. In the long run, it includes tracking an asset's overall performance using reliable data to assess risk, evaluate reliability, and make smarter business decisions to impact a company's bottom line positively.

6.4 Conceptualization of life cycle assessment

The life cycle assessment revolves around evaluating environmental burdens, which in turn are associated with a product, process, or activity, by identifying energy and materials used (ISO, 1999). The interrelation can also be understood in Fig. 6.8.

There are four phases of life cycle assessment (ISO 14040 and 14044):

1. Goal
2. Scope definition
3. Inventory analysis
4. Impact assessment

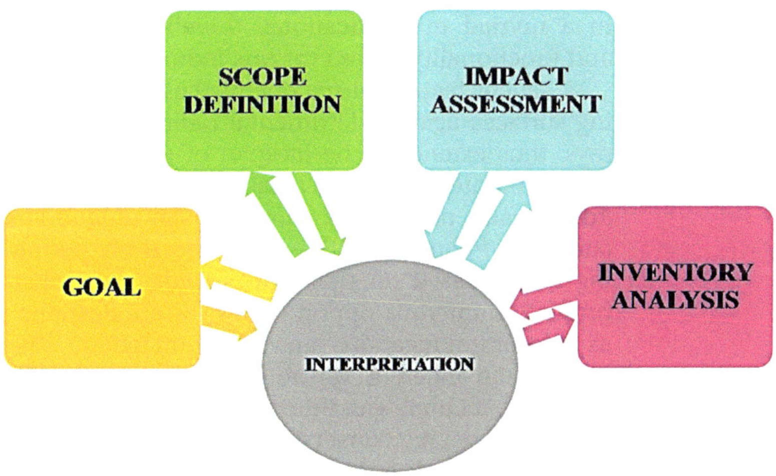

Figure 6.8 Framework of life cycle assessment.

There are many applications of life cycle assessment [29]:

1. Strategic planning
2. Public policymaking
3. Product development
4. Product improvement
5. Marketing
6. Significance of nanopaints

Nanopaints represent a pivotal advancement, particularly in automotive applications, where they serve as critical agents in augmenting substrates' structural, thermal, and electrical characteristics, often replicating the esthetic appeal of chrome-plated paint [9]. Characterized by their nanoparticle formulation, nanopaints encapsulate minute particles ranging from 1 to 100 nm. These nanoparticles endow nanopaints with exceptional properties, including remarkable scratch resistance, durability, and UV protection, surpassing the performance metrics of conventional paints significantly. Integrating nanotechnology in nanopaints marks a significant milestone in materials engineering, offering transformative solutions across diverse sectors, from construction to automotive manufacturing.

Within material science and coatings, nanopaints emerge as revolutionary solutions, boasting multifunctional capabilities. Leveraging the inherent properties of nanoparticles, nanopaints facilitate the establishment of a homogenous surface with unparalleled adhesion and durability compared to conventional counterparts. Engineered to cater to a myriad of user-specific demands, nanopaints can exhibit a spectrum of properties such as heat resistance, electrical conductivity, or hydrophobicity, among others. This inherent versatility renders them adaptable to a myriad of applications, thereby enhancing performance and functionality across various industries.

A salient feature of nanopaints lies in their efficacy in shielding surfaces against detrimental radiation, including UV and IR rays, thus extending the lifespan of painted surfaces while curtailing maintenance costs attributed to weather-induced damage. Additionally, nanopaints showcase the unique ability to modulate radio frequency (RF) waves, selectively permitting or blocking waves based on their wavelength. This inherent capability presents numerous opportunities for applications in fields such as telecommunications and electromagnetic shielding, underscoring the inherent versatility and potential of nanopaints in advanced technological domains. Moreover, nanopaints exhibit promise in energy conservation by harnessing particle nanotechnology to absorb heat energy from sunlight [20]. By mitigating heat transfer through surfaces, nanopaints facilitate reduced energy consumption for cooling purposes across architectural and vehicular contexts. In summary, nanopaints emerge as a profound innovation in coatings and surface protection, offering pioneering solutions for bolstered performance, heightened energy efficiency, and effective radiation management. Ongoing research and development endeavors are poised to refine their properties further and expand their applications, positioning nanopaints as indispensable components in driving technological progress and addressing emergent challenges across diverse industries (https://www.devx.com/terms/nanopaint/).

Benefits of nanopaints:

The primary functional properties and advantages of nanocoatings are manifold:

- *Versatile applications:* Nanocoatings find utility in painting and insulating diverse surfaces within construction projects.
- *Energy efficiency:* They contribute to reducing energy consumption in buildings.
- *Eco-friendly composition*: Being solvent-free and water-based, nanocoatings align with environmentally conscious practices.
- *Water resistance*: Nanocoatings exhibit robust resistance to rain and ambient humidity, safeguarding surfaces.
- *Broad usage spectrum:* They are applicable across various building components, including facades and roofs, across different weather conditions.
- *Strong adhesion:* Nanocoatings boast high adhesion strength, enabling application on various surfaces.
- *Minimal adhesion:* They require little cleaning but offer convenience and ease of upkeep.
- *Antibacterial properties*: Nanocoatings deter the growth of bacteria and fungi, promoting hygienic environments.
- *Customizable esthetics:* By adding different pigments, nanocoatings facilitate the creation of diverse color schemes.
- *UV protection:* They possess anti-UV properties, shielding surfaces from harmful ultraviolet radiation.
- *Simplicity of application:* Nanocoatings are straightforward and do not require specialized equipment.
- *Longevity and particle resistance*: With high durability, they exhibit resilience against airborne particles, ensuring lasting performance.

6.5 Applications of nanopaints

Nanopaint capitalizes on the unique characteristics of nanomaterials to create coatings that outshine traditional paint formulations in terms of effectiveness, durability, and environmental impact. By integrating nanoparticles or nanostructures into the paint mixture, nanotechnology-enhanced coatings demonstrate superior performance across various metrics compared to conventional coatings. For instance, certain nanopaints utilize titanium dioxide nanoparticles to enhance durability and resistance against scratches, UV radiation, and environmental degradation. Additionally, titanium dioxide aids in breaking down air pollutants, resulting in self-cleaning painted surfaces. Conversely, silver nanoparticles in other nanopaints offer antimicrobial properties, inhibiting bacterial growth on treated surfaces. Incorporating these nanoscale elements enhances paint adhesion, corrosion resistance, and other essential properties, prolonging its effectiveness and lifespan beyond traditional paints [30].

Nanopaint forms a resilient surface coating as its particles bond with the substrate upon application, ensuring long-lasting protection. The durability of nanopaint depends on various factors, including paint quality, nanoparticle characteristics, and substrate properties. Generally, nanopaint outperforms traditional paint in durability, with a lifespan of up to a decade or more, depending on environmental conditions [31,32].

Key advantages of nanopaint over traditional paint include increased scratch resistance, enhanced weather durability against elements like rain, snow, and UV radiation, smoother surfaces that repel dirt, and improved adhesion, reducing the risk of peeling or flaking.

Operation of nanopaint: Upon application, nanopaint inherits properties from conventional paints, forming an exceptionally thin yet durable surface layer. This nanocoating is visible only under high magnification, exemplifies the remarkable feature of nanopaint-remarkable thinness combined with exceptional durability. Nanocoating, measured on the nanometer scale, remains invisible to the naked eye and nearly undetectable to touch. Its longevity varies depending on the substrate, with applications on electronics lasting several months and those on automotive sprays and waxes enduring for years. By amalgamating paints and coatings, nanopaint application enhances substrate qualities, bolstering hardness, scratch and mold resistance, and simplifying cleaning processes while ensuring extended product life.

Advanced nanopaint materials: Nanopaints and coatings leverage densely packed molecular structures such as nanocarbon, nanotubes, nanofibers, and graphite to imbue coatings with exceptional properties. These coatings exhibit heightened scratch resistance, increased hardness, and resistance against mold and bacteria, resulting in highly durable, high-performance products. End-user sectors encompass electronics, optics, food and packaging, aerospace, marine coatings, construction, automotive, oil and gas, and biomedical. Application methods include PVD, ALD, CVD, electrospray and electrospinning, sol-gel, and aerosol coating, tailored to specific market requirements [33].

Carbon nanotubes (CNTs) play a significant role in paint formulations, enhancing mechanical and physical properties while improving adhesion and providing electrical conductivity. Due to their self-cleaning and antimicrobial properties, CNT-infused paints find applications in electronics, sensors, electromagnetic interference shielding, and hygiene-sensitive environments. Additionally, CNT-based paints offer enhanced thermal insulation, surpassing traditional materials like fiberglass.

6.5.1 Applications of nanopaint in various materials

Protection of building facades and constructions through waterproof nanopaints. Utilization of waterproof nanopaints for antigraffiti purposes. Incorporation of nano additives in oils, paints, and wood drying oils.

Nanopaint coatings blend the characteristics of traditional paints and coatings, enhancing their quality through increased hardness, scratch resistance, and protection against mold and bacteria. These advancements result in products that are easier to clean and longer lasting. Nanopaints have gained popularity in various applications due to their unique benefits for human life. Today, nanopaints find application in diverse fields, reflecting their growing significance in modern industries.

6.5.2 Self-cleaning and dirt-resistant surfaces

Nanopaint has been developed and used to coat various exterior surfaces like buildings, vehicles, and solar panels, making them resistant to dirt, grime, and pollutants. The nanoparticles present in the paint, such as titanium dioxide, have self-cleaning properties. Sunlight breaks down organic substances (like dirt) and makes the surface hydrophilic, which causes water to spread evenly and wash away the dirt.

6.5.3 Anticorrosion protection

Nanopaint has been used in industries to protect metal structures and components against corrosion. The specially formulated paint contains nano-sized ceramic particles and a corrosion inhibitor, creating a thin, dense, and durable protective layer on the metal surface.

6.5.4 Antibacterial nanopaint

This coating contains antibacterial nanoparticles, typically 30 ppm of nanosilver particles, effectively inhibiting bacterial growth on surfaces. Due to its inorganic nature, the antibacterial properties of these paints are long-lasting, making them highly suitable for maintaining hygiene in environments like hospitals, pharmaceutical facilities, or clean rooms [34].

6.5.5 Thermal control and insulation

Nanopaints can provide efficient temperature regulation and insulation for buildings and vehicles, reducing energy consumption and costs.

6.5.6 Antifog nanopaint coating

Explicitly designed for windows, this coating prevents the formation of mist on surfaces, ensuring clear visibility and protecting against vapor condensation on the interior surface of windows.

6.5.7 Nanopaint for signal blocking

Applied to interior walls, this nanopaint selectively allows or impedes signals from cell phones, portable radios, or other wireless devices. Like copper, this material blocks RF signals, providing passive blocking capabilities for RF energy. Development efforts include creating a spray coating embedded with copper-filled halloysite nanotubes, which, when applied to walls, serve as effective blockers of RF signals.

6.5.8 Nanopaints for building and construction

Nanomaterials offer exceptional protection for construction materials like glass, concrete, sand limestone, or marble against environmental factors such as water staining, moss, algae, soot, and oil stains. Additionally, they serve as effective corrosion inhibitors for reinforced steel. Commercially available paints and coatings create a low-energy surface that makes building exteriors highly hydrophobic and oleophobic, extending maintenance intervals and reducing cleaning requirements. Dirt-repelling protective paints and photocatalytic coatings are widely used in the construction industry to combat the accumulation of dirt on building exteriors, often leading to challenging maintenance issues. Traditional cleaning methods involving chemical detergents, high-energy consumption, and labor-intensive processes result in high maintenance costs. Therefore, there is a growing demand for self-cleaning coatings. Several major multinational companies, including Evonik Degussa, Dupont, Schott, 3M, and Corning, produce antifouling and easy-to-clean coatings for various markets. In recent years, self-cleaning coatings utilizing photocatalytic titanium dioxide (TiO_2) have gained significant attention in the industry. TiO_2 coatings exhibit strong oxidation power and superhydrophilicity, allowing them to kill bacteria, remove foul odors, and easily remove dirt and stains with water or rainfall. Applying nanoparticle TiO_2 self-cleaning coatings significantly reduces the need for costly surface cleaning, particularly for tall buildings like skyscrapers [35–37]. Furthermore, photocatalyst coatings are employed indoors to enhance air quality by reducing exposure to volatile organic compounds and other harmful chemicals in various settings such as hotels, restaurants, commercial buildings, laboratories, hospitals, and residences. The antistick properties of nanocoatings minimize staining and simplify cleaning for interior and exterior surfaces. Moreover, their antigraffiti properties enable easy removal of stubborn stains like graffiti with a high-pressure hose, eliminating the need for intensive cleaning. Water-based, volatile organic compound (VOC)-free, clear-impregnating nanoparticle wood coatings containing nanoscale UV absorbers have been developed for application on masonry and concrete surfaces. These coatings offer superior water repellency, reduce efflorescence, and provide significantly enhanced abrasion resistance.

Major market players in nanopaints: The nanopaints market is fiercely contested, with numerous prominent contenders asserting dominance. Among these industry leaders are BASF SE, Akzo Nobel, Beckers Group, PPG Industries, Valspar, Evonik, Buhler GmbH, Wacker Chemie, DuPont, Henkel Corporation, Yung Chi Paint & Varnish, Dura Coat Products, Nanovere Technologies, I-CanNano, CG2 NanoCoatings, and Tesla NanoCoatings. BASF SE, a formidable force in the chemical sector hailing from Germany, distinguishes itself as a pivotal player in this market. Renowned for its expansive footprint across diverse sectors, including automotive, construction, and aerospace, BASF boasts a legacy of pioneering innovations in nanotechnology, offering nanopaint solutions renowned for their high performance and environmental sustainability.

Similarly, Akzo Nobel, a multinational conglomerate specializing in coatings and specialty chemicals, commands a significant market share in nanopaints. The company showcases various nanopaint offerings spanning architectural, industrial, and marine applications. Akzo Nobel strongly emphasizes research and development initiatives to enhance its product portfolio and extend its market penetration continually. PPG Industries, another heavyweight in the global paints, coatings, and specialty materials arena, emerges as a critical contender in the nanopaints market. With a robust presence in the automotive, industrial, and aerospace sectors, PPG Industries delivers a broad spectrum of nanopaint solutions renowned for their excellence and ingenuity. Propelled by escalating demand for eco-friendly and high-performance coatings, the nanopaints market is witnessing robust expansion. Factors such as heightened requirements for enhanced durability, corrosion resistance, and superior esthetics across various industries drive market growth. Projections indicate that the nanopaints industry is poised to attain a market size of USD 13 billion by 2025, with a compound annual growth rate (CAGR) of approximately 12% over the forecast period. While specific sales revenue figures for individual companies remain undisclosed, industry stalwarts like BASF SE, PPG Industries, and Akzo Nobel are acknowledged for their substantial market presence and commendable financial performance within the nanopaints sector. Overall, the nanopaints market epitomizes a dynamic and fiercely competitive landscape marked by continual investments in research and development aimed at delivering innovative and sustainable solutions. With escalating demand for advanced coatings spanning diverse industries, significant growth prospects lie on the horizon.

6.6 Concluding remarks

Nanopaints present a promising avenue for bolstering asset maintenance and improving cyclability across diverse industries. By harnessing the power of nanoparticles within paint formulations, these coatings offer heightened protection, thus prolonging the operational lifespan of assets while reducing the need for frequent maintenance interventions. Leveraging cutting-edge nanotechnology, nanopaints exhibit remarkable resistance to corrosion, abrasion, UV radiation, and chemical exposure, ensuring enduring durability even in the harshest environmental conditions. Moreover, the integration of nanomaterials enhances the mechanical characteristics of paints, including hardness, flexibility, and adhesion, resulting in resilient coatings capable of withstanding various operational challenges.

In addition to their protective attributes, nanopaints showcase self-cleaning properties, mitigating the accumulation of contaminants and simplifying maintenance tasks, especially in outdoor settings. By harnessing the potential of nanotechnology, nanopaints contribute to sustainable asset management practices by minimizing resource consumption and reducing waste generation. This is achieved through prolonged asset lifespans and diminished maintenance needs, ultimately leading to a more efficient and environmentally conscious asset maintenance and management approach.

Nanopaints represent a groundbreaking solution for enhancing asset maintenance and bolstering cyclability, offering superior protection, durability, and sustainability compared to conventional coatings. Continued research and development endeavors in nanomaterials and paint formulations hold significant promise for further optimizing the performance and applicability of nanopaints across a wide array of industries. This ongoing innovation paves the way for adopting more resilient and efficient asset management strategies, ensuring the longevity and performance of critical assets well into the future.

References

[1] ISO 14040: Environmental Management—Life Cycle Assessment, Principles and Framework (Report), Geneve, CH: International Organization for Standardization (ISO). Retrieved 11 December 2019.

[2] ISO 14044: Environmental Management—life Cycle Assessment, Requirements And Guidelines (Report), Geneve, CH: International Organization for Standardization (ISO). Retrieved 11 December 2019.

[3] M.N. de Queiroz, A.M. de Oliveira Lima, M.E. Gomez Winkler, V.H. Fragal, A. F. Rubira, T. Sequinel, et al., Conductive nanopaints: a remarkable coating, Nanotechnology and Automotive Industry (2022) 439–449. Available from: https://doi.org/10.1016/B978-0-323-90524-4.00020-7.

[4] I.V. Shishkovsky, P.N. Lebedev, Chemical and physical vapor deposition methods for nanocoatings, Nanocoatings and Ultra-Thin Films, Woodhead Publishing, 2011, pp. 55–77.

[5] H. van Lente, J.I. van Til, Articulation of sustainability in the emerging field of nanocoatings, Journal of Cleaner Production 16 (2008) 967–976.

[6] A. Baptista, G. Silva FJ, J. Porteiro, J.L. Míguez, G. Pinto, L. Fernandes, On the physical vapour deposition (PVD): evolution of magnetron sputtering processes for industrial applications, Procedia Manufacturing 17 (2018) 746–757.

[7] P. Nguyen-Tri, T.A. Nguyen, P. Carriere, C. Ngo Xuan, Nanocomposite coatings: preparation, characterization, properties, and applications, International Journal of Corrosion 2018 (2018) 4749501.
[8] G.K. Gueorguiev, C. Goyenola, S. Schmidt, L. Hultman, CFx: a first-principles study of structural patterns arising during synthetic growth, Chemical Physics Letters 516 (2011) 62–67.
[9] N. Hellgren, T. Berlind, G.K. Gueorguiev, M.P. Johansson, S. Stafström, L. Hultman, Fullerene-like BCN thin films: a computational and experimental study, Materials Science and Engineering: B 113 (2004) 242–247.
[10] E. Broitman, G.K. Gueorguiev, A. Furlan, N.T. Son, A.J. Gellman, S. Stafström, et al., Water adsorption on fullerene-like carbon nitride overcoats, Thin Solid Films 517 (2008) 1106–1110.
[11] K. Chu, P.W. Shum, Y.G. Shen, Substrate bias effects on mechanical and tribological properties of substitutional solid solution (Ti, Al) N films prepared by reactive magnetron sputtering, Materials Science and Engineering: B 131 (2006) 62–71.
[12] P.R.T. Avila, R.C. Apolinário, A.M. Rodrigues, J.V. Fernandes, R.R. Menezes, G.D.A. Neves, et al., On improving wear resistance of Cr-Al-N coatings using dynamic glancing angle DC magnetron sputtering, Nanomaterials 11 (2021) 2187.
[13] O. Yilmaz, A. Yorgancioglu, Nanocoatings: preparation, properties, and biomedical applications, Polymeric Nanomaterials in Nanotherapeutics, Elsevier, 2019, pp. 299–331.
[14] X. Wang, Preparation, Synthesis and Application of Sol-gel Method, Vidyasirimedhi Institute of Science and Technology, Rayong, Thailand, 2020. C. Lavollee, M. Gressier, J. Garcia, J.M. Sobrino, J. Reby, M.J. Menu, S. Rossi, M. Fedel, New architectured hybrid sol-gel coatings for wear and corrosion protection of low-carbon steel. Progress in Organic Coatings 99 (2016) 337–345.
[15] Y. Jiang, K. Shi, H. Tang, Y. Wang, Enhanced wettability and wear resistance on TiO_2/PDA thin films prepared by sol-gel dip coating, Surface and Coatings Technology 375 (2019) 334–340.
[16] M. Schlesinger, M. Paunovic, Modern Electroplating, John Wiley & Sons, 2011. Vol. 55.
[17] F. Basile, P. Benito, G. Fornasari, M. Monti, E. Scavetta, D. Tonelli, et al., A novel electrochemical route for the catalytic coating of metallic supports, Studies in Surface Science and Catalysis 175 (2010) 51–58.
[18] E. Toyserkani, A. Khajepour, S.F. Corbin, Laser Cladding, CRC Press, 2004.
[19] The National Institute for Occupational Safety and Health (NIOSH), "Approaches to Safe Nanotechnology"—"Managing the Health and Safety Concerns Associated with Engineered Nanomaterials," Department of Health and Human Services, Centers for Disease Control and Prevention, Publication No. 2009–125, 2009.
[20] European Commission (EC), "Commission Staff Working Paper: Types and Uses of Nanomaterials, Including Safety Aspects. Accompanying the Communication from the Commission to the European Parliament, the Council, and the European Economic and Social Committee on the Second Regulatory Review on Nanomaterials", SWD (2012) 288 final, Brussels, 3 October 2012. Available from: http://eur-lex.europa.eu/LexUriServ/LexUriServ.do?uri = SWD:2012:0288:FIN:EN:PDF.
[21] European Agency for Safety and Health at Work (EU–OSHA), "Workplace Exposure to Nanoparticles", European Risk Observatory, Literature review,

2009. Available from: http://osha.europa.eu/en/publications/literature_reviews/workplace_exposure_to_nanoparticles.
[22] R. Senjen, Nanomaterials - Health and Environmental Concerns. Nanotechnologies in the 21st century, European Environmental Bureau, Issue 2, July 2009. Available from: http://www.eeb.org/?LinkServID = 540E4DA2-D449-3BEB-90855B4AE64E8CE6&showMeta = 0.
[23] C. Lauterwasser, Small size that matter: opportunities and risks of nanotechnologies. Report of Allianz Center for Technology and OECD, (n.d.). Available from: http://www.oecd.org/dataoecd/32/1/44108334.pdf.
[24] M.K. Abdolbaqi, R. Mamat, N.A.C. Sidik, W.H. Azmi, P. Selvakumar, Experimental investigation and development of new correlations for heat transfer enhancement and friction factor of BioGlycol/water based TiO_2 nanofluids in flat tubes, International Journal of Heat and Mass Transfer 108 (2017) 1026–1035.
[25] W.H. Azmi, N.A. Usri, R. Mamat, K.V. Sharma, M.M. Noor, Force convection heat transfer of Al_2O_3 nanofluids for different based ratio of water: ethylene glycol mixture, Applied Thermal Engineering 112 (2017) 707–719.
[26] N. Zawawi, W. Azmi, A. Redhwan, M. Sharif, Coefficient of friction and wear rate effects of different composite nanolubricant concentrations on Aluminium 2024 plate, IOP Conference Series: Materials Science and Engineering 257 (2017) 012065.
[27] K.A. Hamid, W. Azmi, M. Nabil, R. Mamat, Experimental investigation of nanoparticle mixture ratios on TiO_2–SiO_2 nanofluids heat transfer performance under turbulent flow, International Journal of Heat and Mass Transfer 118 (2018) 617–627.
[28] Y. Shao, J. Xiao, W. Wang, M. Engelhard, X. Chen, Z. Nie, et al., Sodium ion insertion in hollow carbon nanowires for battery applications, Nano Letters 13 (2013) 3909–3914.
[29] A. Mahmood, A.A. Ali, M. Nazam, M. Nazim, Developing an interplay among the psychological barriers for the adoption of Industry 4.0 phenomenon, PLOS ONE 16 (2021) e0255115.
[30] X. Liu, M. Antonietti, Molten salt activation for synthesis of porous carbon nanostructures and carbon sheets, Carbon 69 (2014) 460–466.
[31] S. Wang, C. Xiao, Y. Xing, H. Xu, S. Zhang, Carbon nanofibers/nanosheets hybrid derived from cornstalks as a sustainable anode for Li-ion batteries, Journal of Materials Chemistry A 3 (2015) 6742–6746.
[32] N.U.M. Nizam, M.M. Hanafiah, K.S. Woon, A content review of life cycle assessment of nanomaterials: current practices, challenges, and future prospects, Nanomaterials 11 (12) (2021). https://www.mdpi.com/2079-4991/11/12/3324#.
[33] V. Chandrakala, V. Aruna, G. Angajala, Review on metal nanoparticles as nanocarriers: current challenges and perspectives in drug delivery systems, Emergent Materials 5 (6) (2022) 1593–1615. Available from: https://doi.org/10.1007/s42247-021-00335-x.
[34] C. Crunchy, M. Barros, Polymeric nanoparticles: a study on the preparation variables and characterization methods, Materials Science and Engineering: C 80 (2017) 771–784. Available from: https://doi.org/10.1016/j.msec.2017.06.004.
[35] I. Khan, K. Saeed, I. Khan, Nanoparticles: properties, applications and toxicities, Arabian Journal of Chemistry 12 (7) (2019) 908–931. Available from: https://www.sciencedirect.com/science/article/pii/S1878535217300990.

[36] E. Musielak, A. Feliczak-Guzik, I. Nowak, Synthesis and potential applications of lipid nanoparticles in medicine, Materials 15 (2) (2022) 682. Available from: https://doi.org/10.3390/ma15020682.

[37] S. Thomas, B. Harshita, P. Mishra, S. Talegaonkar, Ceramic nanoparticles: fabrication methods and applications in drug delivery, Current Pharmaceutical Design 21 (42) (2015) 6165–6188. Available from: https://doi.org/10.2174/1381612821666151027153246.

7

Antibacterial and antifouling nanopaints

Twinkle Kawatra[1], Sheenam Thatai[1], Rohit Verma[2], Gaurav Sharma[1,2] and Tejendra K. Gupta[1]

[1]Department of Chemistry, Amity Institute of Applied Sciences, Amity University, Noida, Uttar Pradesh, India [2]Department of Physics, Amity Institute of Applied Sciences, Amity University Uttar Pradesh, Noida, Uttar Pradesh, India

7.1 Introduction

The method of adding new functionality to a surface by depositing a thin layer (100 nm) on it is known as nanocoating. Nanocoatings have improved abrasion resistance, increased flexibility, and appropriate adhesion when compared to traditional coatings. A variety of nanocoating techniques are available, including sol–gel, spin coating, self-assembling, electrochemical placement, magnetron-based sputtering, vapor placement, pulsed laser deposition, sonochemical, plasma or ion-beam aided, dip coating, and layer-by-layer (LBL) [1].

Essentially, there are two types of antibacterial nanocoatings: bactericidal coatings and bacteria-resistant coatings. By immobilizing molecules like polyethylene glycol, polymers containing zwitterion, and poly(*N*-vinylpyrrolidone) that can counteract protein adsorption, or by using the physical characteristics of a rough nanosurface to deter bacterial adhesion, bacteria-resistant coatings inhibit the formation of biofilms [1,2].

Bactericidal coatings can also be classified as release-based coatings, which include charged particles or nanoparticles (NPs) of metals, such as silver zinc, copper, or antimicrobial agents, chloride, as well phytochemicals, iodide ammonium, or oxides of nitrogen, and different biocides which are already loaded or encased in the coatings, and contact-killing compounds, which act as soon as they come into contact with the coating, such as quaternary ammonium polymers, synthetic cationic compounds, or naturally occurring molecules of peptides, chitosan, and enzymes; and light-activated

Nanopaints. DOI: https://doi.org/10.1016/B978-0-443-26512-9.00007-0

photocatalysts, like TiO_2, or photosensitizers, which are activated by light irradiation at the proper wavelength to kill the bacteria. The primary process involves the generation of oxygen species that are reactive (ROS) under light irradiation, including the superoxide anion, hydroxyl compounds radical, along with phototoxic singlet oxygen, which have antibacterial qualities [3].

Many antibacterial substances have been created in recent years to create antibacterial coatings. To incorporate the antibacterial agents, a number of delivery methods have also been developed. The most recent and effective method involves using polyelectrolyte multilayers, within which the antibacterial substances can be trapped, or replacing one of the charged layers with the antibacterial compounds [4].

Despite all the advancements, the primary focus of current research is on creating antibacterial coatings that have the ability to destroy germs without inherent limits while also having limited cytotoxicity [5]. Here, instead of concentrating on traditional techniques, we want to highlight the most cutting-edge, modern ways. Additionally, a discussion on the antibacterial nanocoatings' many application areas will be held, with a focus on the most innovative products and solutions. We also draw attention to antibacterial coatings' toxicological impacts and safety concerns [6]. Strains of *Staphylococcus aureus, Pseudomonas aeruginosa, Escherichia coli,* coagulase-negative *Staphylococcus,* and *Enterococcus* species are the primary targets of antibacterial coatings [5].

Due to the introduction of novel strains of bacteria, nanocoatings that are solely operating by an individual lethal mechanism are not adequately efficient. The initial stage toward creating more auspicious antibacterial nanocoatings that exhibit superior performance in diverse applications involved combining numerous antibacterial ingredients to attain mutually reinforcing effects. The antibacterial substances that have been administered may function in this way in a similar way or in a different way [5].

When silver (Ag) particles and quaternary ammonium substances are combined, the contact-killing ability of the tertiary ammonium molecule can succeed the severe biocide effect of the Ag release, thus preventing the deterioration of the Ag reservoirs. Hydrogel comprising Ag/AgCl nanocrystals and oxide of zinc (ZnO) nanorods might benefit through the photosensitivity of Zn oxides generating radicals with exposure to light and the long-term release of Ag1 and Zn21 [7].

By mixing these photocatalysts with cationic polymers, the generated ROS's low penetration into Gram-negative bacteria with complicated cell walls can be impeded. Additional characteristics including mechanical strength, biocompatibility, corrosion and wear resistance, and durability may also be necessary, depending on the antibacterial coatings' intended use [8].

Thus efforts to create stable, multifunctional nanocoatings have recently been made. Compared to traditional coating techniques, the plasma deposition method offers both mechanical and chemical stabilities, making it a successful way for immobilizing antibacterial chemicals. After the development of stimuli-responsive substances, antibacterial nanocoatings that are sensitive to the bactericidal environment with triggered-antibacterial activity have received tremendous interest. The most innovative and successful methods of bacterial killing tactics are self-defensive and clever "kill-and-release" antibacterial coatings [9].

Self-defeating nanocoatings that exhibit active biocidal characteristics solely when bacteria are present were created to overcome the drawbacks of conventional antibacterial coatings, which included reduced efficacy in killing, elevated toxicity towards healthy cells, exhaustion of biocides, and the development of resistant bacteria [5].

One of the earliest self-responsive coatings created in reaction to the pH drop that occurs during bacterial infection of the surface was pH-responsive self-defensive antibacterial nanocoatings. The LBL assembly approach could be used to create these nanocoatings [7]. Antibiotics that kill bacteria are released from such polyelectrolyte multilayers when the electrostatic equilibrium of the ionic pairing between the cationic antibiotics and anionic components is disrupted. Apart from the LBL multifaceted paints, a multipurpose fused polyacrylic acid (PAA) packed using the cationic antibiotic tobramycin (TOB)/chitosan bilayer structure was created. This bilayer's hydration and swelling under acidic conditions led to the release of TOB and the killing of bacteria while the reduction of the electrostatic interaction between PAA and TOB also played a role [10].

Another method for causing bacteria-induced localized acidification to trigger the antibiotic-release behavior was to coat the surface with hydrogel [9]. pH-responsive antibacterial nanocoatings were also created by introducing acid-sensitive linkages, such as Schiff-base, which rapidly hydrolyzed in a weakly acidic environment, or metal coordination bonds, which, in reaction to bacteria-induced acidification, emit loaded antibacterial compounds through the dissociation of coordination bonds [4].

An additional form of self-defense antibacterial nanocoatings relies on an enzyme-triggered biocide-release technique. This involves the progressive degradation of hydrogel coatings or multilayered films with enzyme-sensitive components by hydrolytic enzymes, releasing preloaded biocides that kill bacteria [9]. A few fruitful studies have also concentrated on the combined impact of pH/enzyme dual-responsive systems, which show promise in terms of antibacterial efficacy and a decreased likelihood of bacterial resistance [11].

Additionally, methods for delivering bacterial strain-specific drugs were created to target particular strains of bacteria and reduce their resistance [6]. A recent strategy for creating antibacterial nanocoatings relies on agent-free bactericidal techniques like phototherapy, microwave therapy, radiofrequency therapy, and ultrasound therapy, which cause localized hyperthermia by converting light energy into thermal energy, which lyses the target bacteria [4].

The phototherapy technologies were coupled with biocide agents, such as metal NPs and antibiotics, to prevent the negative effects of these systems, such as harm to normal biological concerns. In these settings, the localized hyperthermia broke down the structure of the bacterial membranes, making it easier for biocide chemicals to enter the bacteria [12].

Another intriguing recent discovery demonstrated that the membrane of bacteria can be destroyed by localized hyperthermia, which facilitates the release of Ag ions when the configuration of the NPs of Ag (AgNPs) is changed to triangular or small plates, which adsorb light with a specific wavelength and convert the light energy to heat [13].

One of the major issues with antibacterial nanocoating systems is recontamination of surfaces due to constant accumulation of dead bacteria that deactivate the bacterial functional groups and provide nutrition for subsequent adhesion of bacteria producing inflammation. As a result, an ingenious "kill-and-release" technique was created to release the deceased bacteria in response to an appropriate signal [7]. The initial attempts consisted of combining the benefits of biocide efficacy of cationic polymers as well as the antibacterial characteristic of zwitterionic polymers via a convertible manner, as their chemical structure can be evolved from cationic form via bacteria-killing capability to zwitterionic form alongside bacteria-repelling behavior. Usually, when the pH transformed, these systems responded [12].

Biofouling is a significant area in which nanopaints can be applied. One of the most important uses of nanocoatings is to stop or reduce the growth of microbes on various surfaces, a process referred to as biofouling or biocontamination. The discovery of antibiotics was first considered the greatest scientific achievement of the twentieth century; however, resistance of bacteria to antibiotics has since been acknowledged as a major challenge [6].

Antibacterial nanocoatings are seen as a significant development that will help overcome the antibiotics' limitations. Any substrate in a maritime environment will eventually become coated in organisms, a process known as biofouling [11].

Unwanted growth on the submerged surfaces of micro-(bacteria and protists) and macro-(invertebrates and algae) organisms is known as biofouling. Microorganisms have the ability to build biofilms by colonizing organic particles and molecules on any submerged substrate which can happen very quickly [14].

Numerous types of bacteria, microalgae, and diatoms can be found in marine biofilms. As a result, biofilms can either increase or decrease the settling of invertebrate and algae spores and larvae [7].

The biofouling of marine installations causes billions of US dollars' worth of losses to the shipping sector and naval forces worldwide. On exposed surfaces, industries employ antifouling coatings to prevent biofouling. Antimicrobial activity is a property of antifouling coatings, which typically inhibit micro- and macrofouling. Most antifouling coatings include organic (isothiazolone) and inorganic (copper) biocides, which are poisonous and seep out of the coating to kill biofouling organisms. Toxic biocides, however, also harm marine animals that are not their intended targets and build up in marine ecosystems [15].

The market's nontoxic antifouling techniques are frequently more expensive and less efficient than conventional biocidal solutions. Thus it is vital to develop solutions that are environmentally friendly therefore nanotechnology-based technologies can be of considerable assistance to successfully construct harmless or low-toxic antifouling coatings [14].

Nanotechnology is the process of modifying materials at the nanoscale scale to enhance and create new material qualities. Atoms can self-assemble to create nanostructures with a wide range of tunable and controlled features. These nanostructures might be one or two dimensional (nanowires), or three-dimensional (arrays, hierarchical structures), or zero dimensions (NPs) [11].

Materials with distinct size-dependent characteristics that are different from their bulk exist at the nanoscale and can be used in a variety of fields, including water treatment, solar cells, sensors, food, medicine, electronics, and fuel. Biofouling prevention has been achieved with nanocoatings. High degrees of nanoroughness in C_{18} coatings were proven to prevent Ulva spore settlement and to have an antifouling effect that was superior to that of a fouling release coating that was sold commercially. Significant micro- and macrofouling agents were kept from settling by superhydrophobic coverings with nanoscale roughness [16].

Conversely, engineered nanomaterials have been widely used in antifouling coatings, electronics, medicines, and renewable energy, among other fields. Combining the qualities of various materials to create hybridized nanocomposites of organic and inorganic materials allows for the prevention of biofouling [14].

Antifouling capabilities are exhibited by metal and oxides of metal NPs, including zinc oxide (ZnO), titanium dioxide (TiO_2), and Ag. Furthermore, coatings containing carbon nanotubes (CNTs) impede the settling and adherence of larvae, preventing macrofouling. Because photocatalysis produces redox reactions, nanostructured metal oxides, such as ZnO and TiO_2, can absorb visible and ultraviolet (UV) light and hinder the growth of microorganisms. Microorganisms cannot grow on surfaces because of the reactive oxygen species (ROS) produced during photocatalysis, such as superoxides, peroxides, and hydroxyl radicals. Because ROS exhibit localized surface toxicity and are short-lived, metal oxide nanostructures may be more advantageous for antifouling applications than other forms of nanocoatings [17]. The main strategies and active components used in antifouling coatings are shown in the schematic diagram of Fig. 7.1.

Recent research has examined the antifouling and antialgal properties of coatings that incorporate nanorods made of ZnO and TiO_2 NPs. The primary goal of this chapter is to give a summary of current developments in antibacterial and antifouling nanopaints [11].

7.2 Antibacterial coatings: applications

Simple doorknobs, flooring made of laminate, cooking areas and sanitary areas, air conditioning and ventilation equipment, preparation of food and packaging materials, the more advanced fields of healthcare and medical-related tasks, army clothes, and marine transport are just a few of the many uses for antibacterial nanocoatings. In this segment, we endeavor to emphasize the most current uses of antibacterial nanocoatings together with their current state of the market. The usage of Ag antimicrobial paints in all applications is expanding extremely quickly due to a growing need for use towards hospital-acquired diseases in the healthcare business. This is because Ag antimicrobial coatings are active against a wide range of bacteria [18].

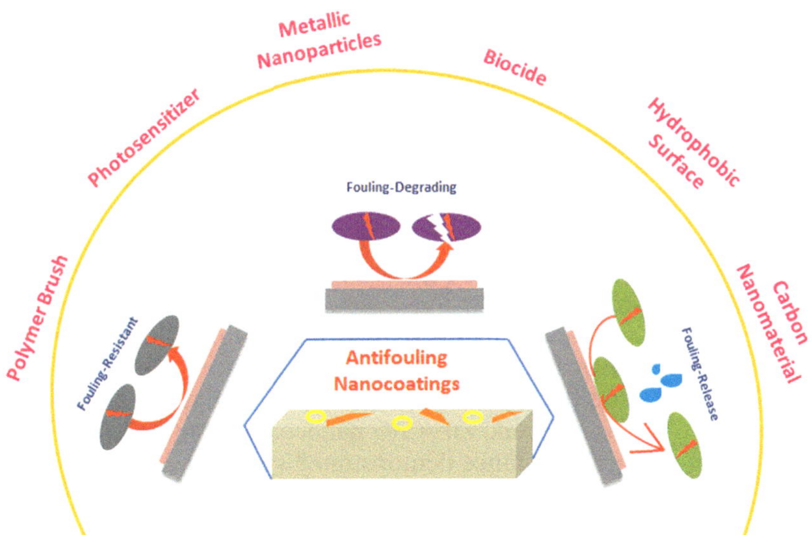

Figure 7.1 Diagrammatic representation of the main strategies and active components used in antifouling coatings.

7.2.1 Medicines and health

Scientists began working on biocompatible implant surfaces with antibacterial qualities and the ability to resist bacterial adhesion concurrently with the development of biomedical devices due to infections caused by these devices. The majority of materials used in implants are polymeric polymers, stainless steel, cobalt-chromium, titanium and its alloys, and other metals. These materials are very prone to biofilm infections from Gram-positive and Gram-negative bacteria [14].

Despite the effective functioning of dentistry, orthopedic, and pulmonary implants, the potential for infection necessitates antibiotic medication, implant replacement, and reoperation. The three infections that infect implants most frequently are *P. aeruginosa*, *E. coli*, and *Staphylococci*. The implanted surfaces' ability to withstand bacterial adherence is significantly influenced by hydrophilicity, charge, chemistry, and surface topographical characteristics [9].

While some researchers have suggested that a rougher nanosurface reduces bacterial adherence, other studies have found the converse to be true. The varying bacterial strains, their disparate cell properties, membrane stiffness, physical architecture, and variances in the nanoscale surface are the causes of the contentious results [19].

Implanted devices can generally be coated with polymers, like poly(ethylene glycol), to give them antifouling qualities that prevent microbes from adhering to their surface. NPs, like TiO_2, are typically applied to surfaces to confer antibacterial properties. Furthermore, bactericidal organic and inorganic materials could be employed to create active antibacterial surfaces. This includes nonmetals like graphene or selenium as well as inorganic metals and metal oxides, as well as biocidal and biocide-releasing polymers, polymerized biocides, and biocidal polymers [12].

Modifying the plasma surface to immobilize antibacterial coatings and change the surface's chemistry and roughness to confer antibacterial capabilities is an intriguing alternative. In a fairly recent work, Zn–CuO nanocoated contact lenses with sonochemical assistance demonstrated strong and stable antibacterial activity while maintaining the fundamental characteristics of a silicon hydrogel contact lens [4].

7.2.2 Construction

The traditional building sector (materials including coatings, glass, concrete, metals, wood, and plastic) has seen substantial changes with the emergence of nanocoatings, offering improved qualities and additional capabilities. Glass, pipes, and interior and exterior building surfaces have all been proven to benefit from nanocoatings [1]. They function as antibacterial and waterproofing layers, enhancing durability and esthetics, in addition to providing protection against UV rays, scratches, and rust. Whereas most research investigations cantered on the utilization of nanocoatings for better mechanical strength, abrasion strength, and self-cleaning qualities, there are some newer approaches toward the efficacy of nanocoatings on material construction against biological attacks. Successful examples include "GreenBuilding" paints that contain Ag and copper NPs with potential antibacterial efficiency or ZnO and copper oxide NPs that increase the resilience of black pine timber against fungi and mold. Numerous studies have concentrated on the application of semiconductors made of metal oxide because the majority of construction goods are in illuminated environments [18].

Applying a titania coating to limestone and marble in an outdoor environment effectively halted the growth of the fungus *Aspergillus niger*. Utilizing titania and Ag together produced a multifunctional surface with antibacterial and antifungal qualities in a synergistic way. For example, Ag–TiO_2 thin-film ceramic tiles, produced via liquid phase deposition, have good antibacterial activity against *S. aureus* and *E. coli*. Under light irradiation, TiO_2/SiO_2 nanocomposites that contained hydroxyl-terminated polydimethylsiloxane (PDMS) were similarly effective in killing microorganisms on stone surfaces [8].

Additionally, researchers reported that covering building structures with CNTs might eradicate 80% of *E. coli* germs. Nevertheless, the use of CNTs is prohibited because of their harmful effects on ecosystems. There exists a significant disparity between research findings and the practicalities of the building sector regarding the ability of the nanocoatings and nanotechnologies to mitigate the negative environmental effects and energy consumption of structures, while also improving functionalities, safety, and lowering costs of civil infrastructure. Consequently, there remains a need to identify low-cost, long-lasting, and sustainable solutions [19].

7.2.3 Food and packaging

A significant issue facing the food business that has an impact on human health, safety, and quality is bacterial contamination. A good way to get around the environmental issues associated with plastic packaging is to use nanocoating. Edible nanocoatings are one type of coating that can be applied directly to food surfaces and consumed as a component of the food itself. Raw meat, cheese, fresh produce, prepared meals, and baked goods all have thin edible nanocoatings applied to their exteriors [5].

Antibacterial nanocoatings extend the shelf life of food by halting the growth of harmful germs and preventing food degradation. Antibacterial compounds could be added to polymeric matrix to offer antibacterial qualities, in addition to using biopolymers having built-in antimicrobial activity, such chitosan [12].

Hydrocolloids, which include proteins (gelatin, wheat gluten, maize gluten, soybean protein, casein, and whey proteins), carbohydrate (cellulose derivatives, starch, pectin, and gums), and lipids (waxes, fatty acids, fatty acids, and resins) make up the majority of the structural matrix. To increase mechanical strength and flexibility, additional chemicals such plasticizers and food-grade surfactants are required [10].

In general, the coatings ought to possess promising antibacterial properties together with water vapor obstruction, structural stability, cohesiveness, avoidance of humidity loss during preservation, lipid oxidation rate decrease, and reduced aroma transfer. Among the most common materials used to create edible antibacterial coatings are extracts from plants, organic acids, and bactericides [4].

Additionally, some examples of how CNTs, nanochitosan, metallic and metal oxide-nanoparticles (MO-NPs), nonmetallic substances (selenium and silicates), alkaline earth metals (calcium and magnesium), nanocellulose, ceramic nanoclays, and starch particles are used to improve the characteristics of polymeric films and create antibacterial nanocoatings. By physically interacting with the food product it contains or the headspace, active packaging can limit or inhibit microbial development, thereby extending its shelf life. Instead of using these lists, research should be done on the possibility that NPs will migrate from the coatings to the food items after employing them in food packaging applications. This is especially important when high temperature, pH, and temporal conditions are present [13].

Commercial use of NPs in food packaging requires a thorough grasp of the safety concerns as well as the implications for the environment and public health. It is still difficult to improve food stability, safety, and quality while lessening the impact on consumers and the environment. The most affordable active packaging solutions made of natural, renewable materials that are accepted by consumers must be introduced [11].

Additionally, intelligent packaging solutions for food have been created that have the capacity to enclose an active ingredient, make it easier to work with, shield it from the outside world, and allow it to be released when an external stimulus—such as a change in temperature, magnetic attraction, pressure, or light—is applied. For example, a successful research describes the coating of commercial packaging films (polylactic acid and polyethylene) with photoresponsive tiny capsules containing essential oils that have antibacterial properties [18].

Utilization of hydrogels to manage the dampness of food packing has been also involved, inhibiting the development of yeast, mold, and spoiling bacteria. It has also been recorded that the hydrogel has potential uses in food packaging methods when combined with various antibacterial substances, including NPs or bioactive substances [20].

AgNP-infused hydrogels containing chitosan and gelatin shown antibacterial activity against *S. aureus*. The low mechanical and water-resistant properties of hydrogels can also be mitigated by adding NPs into the gel [17].

7.2.4 Textile

Textiles have historically been known to be substrates for the growth of microbes, which can have a negative impact on the textiles themselves as well as the users. These include the potential harm to human health as well as an unpleasant smell, diminished mechanical strength, staining, and discoloration [4].

Antibacterial textiles have many uses, including home, military apparel, space suits, healthcare textiles, sportswear, hygiene, and architectural and automotive structures. They are also referred to as technical textiles. Wearers' psychological comfort in relation to odors in casual clothing is just one of these applications. Apart from the integration of NPs into the polymetric framework in the synthetic fiber manufacturing process, textile substrates have also been surface-coated to get antibacterial characteristics [10].

All the usual nanocoating processes comprising vapor deposition, plasma-assisted/ion-beam-assisted procedures, pulse laser deposition, magnetic sputtering, self-assemblage, LBL coatings, dip coatings, sol–gel coatings, electrochemical deposition, and sonochemical could be employed for antibacterial coating. Additionally, clever nanocoatings that can detect illnesses and release antibacterial agents when bacteria are present have been created. While a variety of natural, synthetic, natural, and artificial antibacterial agents have been used for antibacterial nanocoating of textiles and clothing, AgNPs have been used in most research investigations instead of triclosan, metal oxides, polyhexamethylene biguanide, quaternary ammonium compounds, or N-halamine [7].

Antimicrobial fabrics are utilized in applications for wound healing as well as healthcare settings to stop the spread of infections caused by pathogens between patients and physicians. The usage of encapsulated components in the medical field, where fabric textiles can be treated with micro/nanocapsules possessing antibacterial qualities, is one of the most intriguing themes of late [6].

Natural biopolymers are employed as the wall material in these healing wound systems, while natural substances with antiinflammatory and antioxidant properties, like essential oils, function as the core. The human body is compatible with essential oils as well. The antimicrobial chemical diffused into the fabric surface and reached the bacteria as a result of the slow disruption of the nanocapsule wall [9].

The most significant uses of antibacterial fabrics in the military are in protecting soldiers from infections caused by fungi, skin irritations, decreased deterioration of tents, soldiers' uniforms, and boots, as well as in providing comfort to improve the soldier's performance in the battle zone. Sport shoes, insoles, and antibacterial clothing were created to shield athletes from odors and germs while shielding their muscles from deterioration [16].

Additionally, antimicrobial nanocoatings were applied to stadium floor coverings, improving the longevity and hygienic conditions of the flooring without compromising athlete performance or safety. Antimicrobial fabrics are becoming increasingly important for long-duration space flight, as they reduce waste and mass in spaceships, in part because of the limits connected with the usage of washing equipment in spacecraft and space stations [11].

Antimicrobial fabrics are necessary for sweat handling in space since there is no convection of perspiration due to the low gravitational pull of space, which increases the risk of bacteria development and warming of the body. Despite the high need, antimicrobial textile uses in space are a relatively young field of study with little research to date. Apart from the practical attributes of transportation interior textiles, such water resistance, decreased combustibility, stain and abrasion resistance, odor, and hygiene are also among the difficult issues, necessitating the usage of antibacterial fabrics in automobile applications [14].

To provide antiodor and antibacterial qualities to vehicle seat textiles, chitosan-based coatings have been successfully developed as appropriate scent carriers. The fastest-growing trend will be the use of synthetic organic molecules since they are less expensive than metal, metallic salts, and bio-based agents. With the prohibition of some substances like triclosan and polyhexamethylene biguanides, quaternary ammonium compounds will be emphasized more in this context. Among the commonly used textile substrates are cotton, polyester, and polyamide; polyethylene, polypropylene, acrylic, and cellulose acetate follow. The most popular technique for creating commercial antimicrobial fabrics is the pad-dry-cure procedure [7].

In contrast to the previously suggested antimicrobial textile market trend, the necessary characteristics for antibacterial textiles are durability, interoperability with common textile industries, and nontoxicity to humans as well as the environment. It is necessary to develop greener antibacterial agents and methods that are less hazardous to the environment and human health. These should be based on natural components [8].

Ag-based nanocoatings rank highest among antimicrobial coatings, followed by titanium, zinc, copper, chitosan, and quaternary ammonium compounds. However, titanium and chitosan were typically employed as transmitters for other release-based antibacterial agents [21]. Antimicrobial drugs with toxicity, including metal NPs, have been linked to cell absorption and passage into lysosomes, which damage DNA by producing ROS.

Furthermore, antimicrobial resistance could be brought on by the gradual infusion of active substances. The coating architecture affects the ejection dynamics of antibacterial agents like Ag [14].

Controlling the expulsion of Ag particles from nanocoatings to give them greater durability has been the subject of several attempts. These include the application of polymer layers of different thicknesses and pore sizes between the coating surface and the Ag, the utilization of additives for the firm adherence or covalent liaison of small particles with the protective coatings the placement of Ag in conjunction with reducing agents, like dopamine, to lessen the expulsion of oxidative Ag, as well as the encapsulation of metal NPs in core–shell structures [15].

Although there is some research available on the ejection rates of Ag NPs through antimicrobial coatings, comparability of the data is limited since different compounds are utilized in different applications, and the compounds also vary in dose. We are unable to assess how well the present antimicrobial nanocoatings balance risk and benefit due to a lack of sufficient evidence [6].

Before hailing antibacterial nanocoatings as humanity's rescuer against microbes and a guarantee for human health and hygiene, we think it is necessary to get unequivocal evidence of the much greater benefits of antimicrobial usage over the associated hazards. If not, the majority of the items on the market today will be banned due to harmful impacts on human and environmental health, even at the risk of irreversible harm [14].

7.3 Antifouling solutions based on nanotechnology

To address diverse forms of biofoulings, a variety of antifouling techniques, including fouling-resistant, fouling-release, as well as fouling-degrading, have been investigated in the creation of antifouling coatings [7].

Foulants like proteins, algae, or bacteria are prevented from adhering by coatings that are resistant to fouling and have a highly hydrated surface. These coatings operate as a physical and free-energy barrier. The fouling-release coatings, on the other hand, permit a weak adherence of foulant, which can subsequently be eliminated by an external force as water flushing. With these two coating techniques, the surface is endowed with hydrophilic and hydrogen bonding properties, wherein the firmly attached water molecules create a barrier layer that inhibits adsorption [11].

The third kind of coatings, known as fouling-degrading coatings, uses antimicrobial moieties to oxidize or perform other bactericidal functions against the seated bacteria or microorganisms [9].

7.3.1 Biofouling prevention through inorganic ingredient-based coatings

It has been claimed that antifouling marine paints using NPs give superhydrophobicity, exceptionally durable, water repellent, antisticking, microbiological obstruction, elastomeric, and anticorrosive qualities. These are innovative solutions for the maritime industries' sustainable growth. Antifouling coating compositions can benefit from the efficient usage of NPs, which leads to better qualities [4].

Antifouling agents, such as graphene, CNTs, and AgNPs are frequently employed. Furthermore, a number of metal oxide semiconductors, such as oxide of zinc (ZnO) NPs, and TiO_2, have been employed in the creation of nanocomposite materials exhibiting antifouling properties [6]. The synergistic protection against biofoulings provided by phenylmethylsilicone oil, PDMS, and polystyrene coating is shown in Fig. 7.2.

According to a review of the literature, metal oxide nanostructures like ZnO and TiO_2 were the subject of the majority of antifouling investigations. Carbon nanostructures like CNT are not given as much attention. The antifouling properties of composite nanocoatings, carbon nanostructures, nanostructured metal oxides, and nanosilver are discussed below [8].

7.3.1.1 Silver nanoparticles

It is commonly recognized that surfaces coated with Ag or coated with Ag have bacteriostatic/bactericidal, fungistatic/fungicidal, and algistatic/algicidal qualities. It is well known that Ag inhibits the growth of certain bacteria [22]. As a result, it finds use in consumer and medical goods as well as environmental applications. Micron-sized metallic Ag, that is, NPs of Ag, have found growing applicability as an antibacterial agent due to its large surface to volume ratios and improved surface reactivity. Numerous investigations about the antibacterial properties of NPs of Ag and their mode of action have been documented [12].

Ag formulations comprising ions of Ag or NPs of Ag are usually regarded as biologically innocuous as well as less hazardous than any other hefty metal-based treatments.

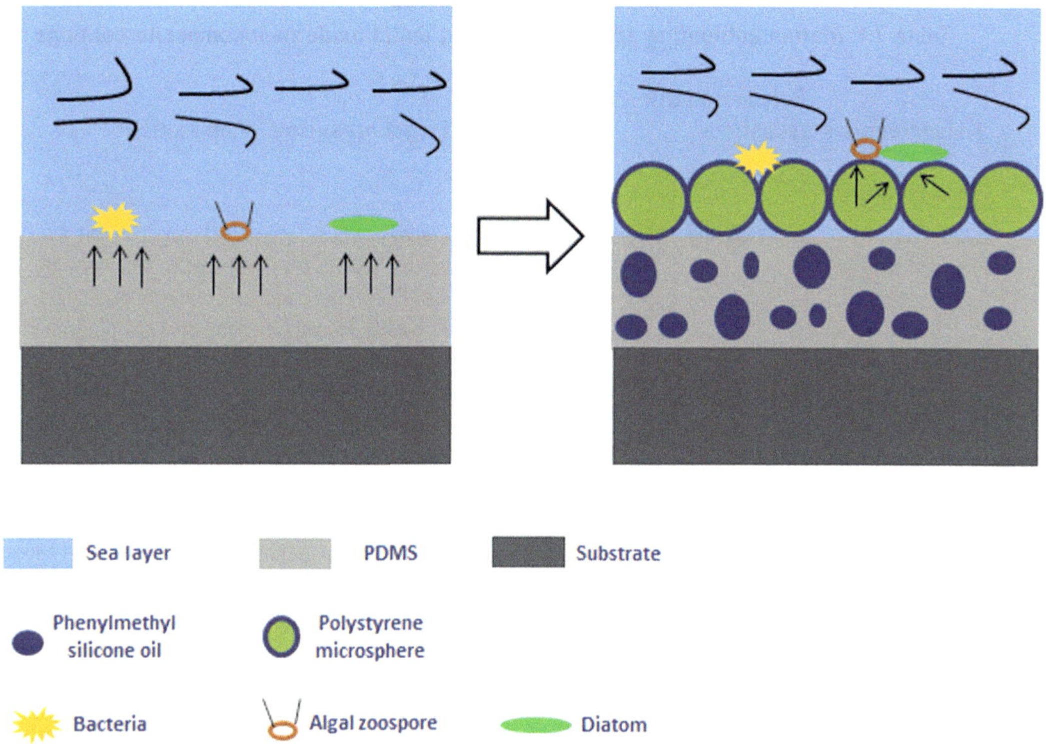

Figure 7.2 Synergistic protection against biofoulings provided by phenylmethylsilicone oil, polydimethylsiloxane, and polystyrene coating.

In applications aimed at preventing microfouling, nanosilver's bacteriostatic and antibacterial qualities may also prove beneficial [23]. The impact of nanosilver on eukaryotic or macrofouling organisms was the subject of very few investigations. For instance, it has been demonstrated that freshwater and marine algae cannot adhere to a coating made of AgNPs. In a different study, nanosilver altered the framework of biofilms to prevent mussels from settling. It is still unclear and up for debate what mechanism gives AgNPs their antifouling characteristics [6]. The marine antifouling applications of metal oxide nanocomposite coatings are given in Table 7.1.

Ag^+ ions' capacity to adhere to cell walls is the foundation of some of the most widely recognized processes. It has been demonstrated that Ag^+ ions enter bacterial cells and bind with the thiol groups of the majority of essential enzymes, deactivating the enzyme and ultimately stopping bacterial growth and cell death. According to the eukaryotic study, the primary

Table 7.1 Marine antifouling applications using metal oxide nanocomposite coatings.

Nanostructures	Polymer matrix /substrate	Properties	Target organisms	Applications
Antimicrobial				
Cu_2O nanoparticles	Epoxy coating	Antimicrobial	*Staphylococcus aureus, Escherichia coli, Bacillus sp., and Klebsiella pneumonia* (bacteria)	Cu_2O provided protection against surface damage (such as scratches and abrasions) and a decrease in the production of biofilms
ZnO nanoparticles (polyaniline/zinc oxide nanocomposites)	Polyurethane film	Antimicrobial	*Vibrio harveyi* (bacterium) *Bacillus licheniformis* (bacteria)	Nanocomposites as a biofouling prevention method
ZnO nanoparticles	Chitosan	Antimicrobial	*Pseudoalteromonas nigrifaciens* (bacterium) *Navicula* sp. (diatom)	Hybrid coatings to manage marine creatures that cause fouling
ZnO nanoparticles	PDMS	Antimicrobial	*Bacillus flexus* (bacterium) *Navicula* sp. (diatom)	Coating with antimicrobial fouling release
ZnO nanorods	Chitosan	Antifouling	*Bacillus subtilis, Escherichia coli* (bacteria)	Biofilm-preventing coating
Ag nanoparticles	PDMS	Antimicrobial	Marine bacteria	Coating with antimicrobial fouling release
TiO_2 nanoparticles	Chitosan	Antimicrobial	Marine bacteria	Coating with antimicrobial fouling release
TiO_2 nanoparticles	Chitosan	Antimicrobial	*Dunaliella salina* (microalga)	Hybrid coatings for handling marine life that cause fouling
TiO_2 nanoparticles	Fluorinated acrylic paint	Antimicrobial	Bacteria and diatom	Antifouling paint for marine environments
Al_2O_3 (alumina) and zeolite nanoparticles	Polysulfone	Antimicrobial	*Escherichia coli* (bacterium) *Pseudomonas aeruginosa* (bacteria)	Antimicrobial, antiadhesion coatings
Prevention of micro- and macrofouling				
ZnO nanoparticles	Alkyd resin	Antimacrofouling	Marine organisms	Proven antifouling characteristics in situ
ZnO nanorods	Epoxy/polyaniline	Antimicrobial and antifouling	Marine organisms and microbes	Proven antifouling characteristics in situ

(*Continued*)

Table 7.1 (Continued)

Nanostructures	Polymer matrix /substrate	Properties	Target organisms	Applications
ZnO nanorods	Silicone	Antimicrobial and antifouling	Marine organisms and microbes	Highly hydrophobic coatings for fouling release
CuO nanoparticles	Polymethyl methacrylate	Antimicrobial and antifouling	Marine organisms and microbes	Antifouling paint for marine environments
CuO nanoparticles	Polydimethylsiloxane	Antimicrobial and antifouling	Microalgal community	Antifouling to prevent larval settling and algal fouling
ZnO–SnO$_x$ core–shell nanoparticles	Chitosan	Antimicrobial and antifouling	Marine organisms	Antifouling coating resistant to chemicals

Ag, Silver; *PDMS*, polydimethylsiloxane; *TiO$_2$*, titanium dioxide; *ZnO*, zinc oxide.

antifouling mechanism is Ag^+ ions' ability to destroy cells by contact [24]. Another theory is that AgNPs cause free radicals to develop. Studies using electron spin resonance spectroscopy revealed that when AgNP-produced free radicals come into contact with bacteria, they cause damage to cell membranes, which leads to rupture of the membrane and ultimately cell death [4].

AgNPs' antibacterial activity is contingent upon a number of elements, including physico-chemical properties including size, shape, and crystallinity, as well as the coatings on the surfaces of NPs and environmental microbiological species [17].

7.3.1.2 Carbon nanotubes and graphene

An allotrope comprising only one layer of atoms of carbon organized in a hexagonal lattice with sp2 bonds is called graphene. They are found in nature as the components of graphite. Graphene materials are interesting for a variety of environmental applications, including photocatalysis and energy production and storage, due to their significant surface area, electron conductivity, and thermal stability [12].

The hollowed out concentric cylindrical structure known as CNTs usually has a measurement of a few nanometers and an overall length that can range anywhere from a few nanometers up to a few microns (100 μm) to a few millimeters (4 mm) [5,25].

Single-walled nanotubes of carbon (SWCNTs), double-walled carbon nanotube, and multiwalled carbon nanotubes (MWCNTs) are the three main types of nanotubes. The creation of nanocomposite components with biocidal qualities and fouling releasing activities is the primary way that CNTs are applied to the environment [2].

7.3.1.3 Nanotubes of carbon

When it comes to bacteria, both Gram-positive and Gram-negative, as well as bacterial spores, CNTs and graphene show strong antibacterial action. The antibacterial activity of CNTs, particularly SWCNTs, was demonstrated to be substantially higher than that of MWNTs. This is likely because SWCNTs' smaller size allows for membrane disruption [20].

Furthermore, CNTs have the ability to influence the influx of macrofouling organisms. Nanosized carbon black more efficiently prevented settling among the barnacle Amphibalanus amphitrite larvae comparison to single-layer graphene oxide (GO). The nature of biofilms was altered by a mere 0.5 wt.% of CNT, which led to a decrease in Bacteroidetes and an increase in Proteobacteria, which in turn prevented *Mytilus coruscus* from settling [9].

In 2008, one of the earliest CNT-based coatings for antifouling was created. Its foundation was the addition of artificial MWCNTs to silicone coatings [25]. These days, the construction, gas, and oil sectors actively use CNT-based coatings as marine antifouling and anticorrosion coatings. When CNTs are used as protective coatings, they are often incorporated into a polymeric matrix. The majority of applications for epoxy resins have involved the reinforcement of various CNT types (single-walled, double-walled, and multiwalled). The enormous area of the surface, π-bond electron on the surface, which provide more chemically active regions on the nanotubes, and the hollow and layered architectures of CNTs make them an effective adsorbent [4].

CNTs possess a high tensile strength as well as Young's modulus; they are hundreds of times stronger and have a greater Young's modulus than steel [9].

Coatings and paints have better mechanical qualities when the paint matrix is reinforced with CNTs. Self-polishing antifouling resin-based paints are discovered to have a higher mechanical strength when the amount of MWCNT in the paint is increased. This is because CNTs have an effect on the mechanical strength of the paint [25]. The impact resistance of antifouling paints containing 0.54% (w/w) and 0.75% (w/w) of MWCNTs was significantly increased. Another study that assessed the impact of MWCNT and GO reinforcement on the mechanical characteristics of PDMS coatings for marine use provided additional support for this one [5].

PDMS marine coatings' tensile strength and percentage of elongation both rose when MWCNTs were added to the composites [24]. According to these investigations, CNT and GO can enhance the mechanical characteristics of nanocomposite coatings in addition to their antifouling qualities. The precise mechanism behind the antifouling function of GO and CNTs remains unclear despite a number of hypothesized theories. According to a study, many antifouling mechanisms might be at play. Firstly, bacteria and the larvae of macrofouling species are poisonous to CNTs and GO. It is well known that CNTs are poisonous [2].

CNTs have the ability to raise oxidative stress, inflammation, and cell death. Second, by merely interfering with the cyprid larvae's ability to attach and adhere, CNTs and GO can prevent larval settling in a nontoxic manner. The antifouling action of CNTs is dependent on several factors, including concentration, treatment duration, diameter, surface area, length, and other physical characteristics. The impact of the dimension of SWCNTs' on antibacterial responses to *Salmonella* cells employing three distinct lengths of SWCNTs [12].

Shorter SWCNTs were shown to cluster between each other without incorporating many bacterial cells, whereas longer SWCNTs were claimed to demonstrate higher antibacterial efficacy by clustering bacterial cells more effectively. Another crucial factor in the deactivation of bacteria is the tube diameter. Reduced diameters Although CNTs with large diameters (15–30 nm) mostly engage bacteria through their sidewalls, they can also cause harm to cell membranes through their engagement with the cell surface [5].

7.4 Metal oxides and hybrid metal/metal oxide nanostructure

Because of their strong photoactivity, metal oxide NPs, such as ZnO, TiO_2, and SnO_2 are employed in antibacterial, self-cleaning, self-healing, anticorrosion, and antibiofouling applications. Researchers looked on the antifouling properties of hybrid nanostructures, wires, rods, and metal oxide NPs in general [14].

The most widely employed systems for the suppression of biofouling were ZnO, TiO_2, and the hybrid metal–metal oxide nanostructures, in addition to hybrid carbon–metal oxide nanostructures. Relative to other metal oxides, ZnO NPs have garnered greater attention since they are readily available and reasonably priced. ZnO has been listed as a safe material for a variety of uses by the Food and Drug Administration, United States [26]. Because of its low toxicity, biocompatibility, and antibacterial qualities, ZnO NPs find application in both the processing of food and packaging industry as well as in agriculture. Both ZnO and TiO_2 semiconductor oxides have been used to photocatalyze the removal of many harmful organic dyes as well as inorganic contaminants from industrial effluent under light irradiation. TiO_2 is a common ingredient in self-cleaning paints and coatings, but because they need UV light to activate, they can only be used outside. Gram-negative (*E. coli* and *P. aeruginosa*) and Gram-positive (*Bacillus subtilis* and *Micrococcus*) pathogens are the most often used microorganisms in testing [4].

Conversely, the most often utilized eukaryotes in research on metal oxide nanostructures are environmental microalgae species and invertebrate larvae. Most commonly utilized as an antibacterial, AgNPs may be produced in an environmentally benign manner and are not harmful to animals. Because the oxide of titanium nanotubes has a superior particular area of surface on the inside as well as the outside of the tubular structure, they have been employed for enhancing the functional characteristics of AgNPs [9].

A novel two-step hydrothermal synthesis method was used to create a composite material consisting of AgNPs evenly distributed throughout the nanomaterial, exhibiting nanotubular TiO_2 structures. Examining the composite material's ability to suppress the production of biofilm, it was shown that Ag/trynitro toluene (TNT) significantly reduced the formation of biofilm in the ocean bacterium *Halomonas pacifica* around 99% when compared to titania nanotubes alone [12].

Furthermore, Ag/TNT inhibited the growth of *Isochrysis* sp. and *Dunaliella tertiolecta*, two marine microalgae. GO coupled with alumina tiny rods (GO-Al_2O_3 NRs) and PDMS were combined to create a ternary composite (PDMS/GO-Al_2O_3 NRs) that is superhydrophobic. This was achieved by the solution casting process. The findings demonstrated that these well-dispersed GO-γ-Al_2O_3 NRs hybrid sheets improved micro-nanoroughness on the surface, reduced surface-free energy, and raised the contact angle (151 degrees). As a result, the newly created nanocomposites may have exciting antifouling uses in the maritime sector [17].

A recent study developed a pair novel superhydrophobic nanocomposites based on silicon and graphene. The results showed that the PDMS/GO-γ-AlOOH nanorod composite possessed better antibacterial activity against various bacterial strains than the PDMS/RGO nanocomposite. This could be attributed to the large surface area and sustaining impacts provided by the GO-γ-AlOOH hybrid nanofillers. The GO-γ-AlOOH (3 wt.%) nanostructured coating exhibited strong superhydrophobic antifouling characteristics because of its high WCA of 151 degrees, rough surface, and uniformity. One of the most durable inorganic materials is aluminum oxide, which is also reasonably priced and nontoxic [5].

Boehmite is a particle made of oxide of aluminum hydroxide (γ-AlOOH) with hydroxyl groups on the surface [27]. The surface of boehmite NPs is covered in OH groups, giving them an orthorhombic structure. Because boehmite NPs have more hydroxyl groups on their outermost layer, they can increase the membrane's hydrophilicity and surface characteristics and lessen membrane fouling. Boehmite has the highest hydrophilicity and hydrated surface of all the alumina compounds. Certain biological species are more susceptible than others to the toxicity and antifouling qualities of metal oxide NPs [9].

Additionally, the physico-chemical characteristics of nanomaterials, such as surface groups, charges on the surface, and particle size and shape, influence their toxicity. Because of their smaller size, NPs are easily able to pass through living things' cellular membranes and other obstacles. As particle size increases, there is a decrease in the rate of uptake of nanomaterials [6].

The underlying concept of metal oxide-driven photocatalytic disinfection, nevertheless, involves the formation of extremely reactive intermediate species, and such species contribute to the photocatalytic annihilation of adherent species (organic matter, microorganisms, and macroorganisms) [24]. Variable surface groups and charges on nanostructures show varied levels of surface oxidation reactivity, which leads to the production of ROS at variable intensities. Because of their large bandgaps and high rates of charge recombination from photogenerated energy, TiO_2 and ZnO can only be used in UV light irradiation [15].

Thus efforts have been directed at utilizing these materials to achieve high photocatalytic efficiency, particularly in the presence of visible light. Various techniques have been developed to make semiconductor materials like ZnO, TiO_2, SnO_2, and others appropriate for visible light photocatalysis [28]. These techniques include modifying the surface of semiconductor materials, oxygen vacancy creation, injecting metal/nonmetal atomic particles, placing noble metals, as well as linking to additional electronic devices or carbon material [29]. Visible light is absorbed by ZnO and TiO_2 when they are doped with metal or nonmetal [25]. Doping semiconductors with metallic or nonmetal atoms can produce levels of impurity in the bandgap that change the photoelectric characteristics and cover the spectrum absorption in the visible range [27]. The expulsion of metal ions, such as Ag^+ ions and Zn^{2+} ions, from the corresponding NPs as well as the generation of ROS, are the basis of the antifouling processes associated with nanostructured metal oxides. When UV or visible light is applied to metal oxides, electrons are transferred from the valence band to the conduction band, creating electron–hole pairs. If the exciton formed by photosynthesis is separated, it may efficiently decrease and/or oxidize a chemical that has been adsorbed on its surface by producing O_2—anions or OH radicals [4].

Pollutants can react with both radicals and anions, converting them into less hazardous by-products or degrading them. In the same way, ROS and metal anions inhibit macro- and microorganism fouling [12].

Generation of ROS at very low levels results in severe oxidative stress, which damages genetic material like DNA and RNA, oxidizes lipids, modifies proteins, inhibits enzymes, etc. Increased amounts of ROS cause both eukaryotic and prokaryotic cell death. Only those who are alive in close proximity to the created ROS are impacted by them because of their brief half-life. The ease of controlled production techniques of metal oxide nanorod and nanowire-based photocatalytic coatings has garnered attention recently [20].

In laboratory and field investigations, ZnO nanorod-based photocatalytic films have proven efficient against both macro- and microfouling on glass substrates and fishing net substrates, respectively. Compared to ZnO NPs, ZnO nanorods exhibit greater stability, less NP loss, and the capacity to be formed onto a variety of support materials. Additionally, studies using bacteria, microalgae, invertebrate larvae, and adult mussels have demonstrated that ZnO nanorods are less hazardous than ZnO NPs [8]. The antifouling characteristics of few common metal oxide NPs are given in Table 7.2.

Table 7.2 Common metal oxide nanostructures' antifouling characteristics.

Nanostructure	Organisms	Mechanisms	Applications
Prokaryotes			
TiO_2 nanoparticles	*Stenotrophomonas maltophilia*	Photocatalytic effect	Submerged archeological sites in the sea
Ag–TiO_2 hybrid	*Halomonas pacifica*	Ninty-eight percent less biofilm adhesion for	Effect of antimicrofouling
Ag@TA–SiO_2 nanoparticles	*Escherichia coli*, and *Staphylococcus aureus*	*Staphylococcus aureus* and *Escherichia coli*, coating reduced antibacterial efficacy by 82.7% and 99.1%, respectively, and inhibited 98.6% of protein adsorption.	Antimicrofouling effect
ZnO nanoparticles	*Staphylococcus aureus* (bacterium)	Effects of photocatalysis and toxicity	Mitigation of microfouling on lenses
ZnO nanowires	*Bacillus subtilis* and *Escherichia coli*	Changes to cell membranes and the generation of ROS	Antimicrobial agents and photocatalysis
ZnO nanorods	*Escherichia coli*	The generation of ROS causes the lipids in membranes to oxidize, damaging the membrane and triggering cell lysis	Antibiofouling abilities
ZnO nanostructures	*Pseudomonas aeruginosa*	Increased oxygen vacancies lead to an increase in ROS generation and improved antibacterial qualities	Enhanced resistance against biofouling
MgO nanoparticles	*Escherichia coli* and *Salmonella enterica*	Damage to the cell membrane that causes the contents of the cell to seep out	Food surface antifouling
Al_2O_3 nanoparticles	*Escherichia coli* and *Pseudomonas aeruginosa*	High surface energy and small particle size promote rupture of the cell wall	Antifouling coating with the capacity to resist adhesion
PDMS/GO-Al_2O_3 NR	*Micrococcus* sp., *Pseudomonas putida*, and *Aspergillus niger*	Reduction of surface-free energy and the creation of micronano roughness	Superhydrophobic antifouling self-cleaning surface
CeO_2nanoparticles	*Pseudomonas aeruginosa* and *Staphylococcus epidermidis*	Stress on cells results in cell death	Applications against infections
Graphene oxide (GO) nanoparticles	*Bacillus* sp. and *Pseudomonas* sp.	Notable decrease in biomass thickness and biofilm	Properties that prevent biofouling and corrosion
Eukaryotes			
TiO_2 solution	*Chlorella mirabilis* and *Chroococcidiopsis fissurarum*	Degradation of photocatalysis when exposed to UV light	Clay bricks

(*Continued*)

Table 7.2 (Continued)

Nanostructure	Organisms	Mechanisms	Applications
TiO_2 nanoparticles	Marine biofouling	Hydrophobic wettability and the structure of the riblet surface	Antibiofouling surface
ZnO nanorods	*Dunaliella salina* (microalga)	Generation of ROS	Prevention from membrane fouling
ZnO nanorods	*Bugula neritina* (bryozoan) *Tetraselmis* sp. (microalga)	Generation of ROS	Prevention from fouling on glass substrata
$ZnO-SnO_x$ core–shell nanoparticles	*Bacillariophyceae* spp. and *Navicula* spp. (diatoms)	Generation of ROS	Prevention from fouling on glass substrata
TiO_2 and Nb_2O_5 nanoparticles	Barnacles, hydroids, and oysters	Generation of ROS	Prevention from fouling on glass substrata
$Ag-TiO_2$ hybrid	*Dunaliella tertiolecta* and *Isochrysis* sp.	Antifouling action	Fouling microalgae

Ag, Silver; *PDMS*, polydimethylsiloxane; ROS, reactive oxygen species; *TiO_2*, titanium dioxide; *ZnO*, zinc oxide.

7.5 Prevention of marine biofouling with inorganic polymer nanocomposites coatings

In addition to pure polymeric and organic molecule mixed coatings, the fouling-release and fouling-degrading properties of organic/inorganic hybrid composites, which are made by combining nanosized inorganic fillers in a polymer matrix, have been studied. Among various coatings, nanomaterial-based coatings for antifouling have shown to be one of the most effective options for preventing biofouling. Because of their solubility, metal oxide nanostructures have limited efficacy in avoiding both micro- and macrofouling [18].

Furthermore, NPs have the potential to disappear and harm the ecosystem. Metal oxide NPs are typically bonded to support substrates or integrated into a robust coating matrix to improve stability, lessen environmental impact, and increase the antifouling effect. For instance, under artificial sunshine, ZnO nanorods affixed to glass substrates showed efficacy against the marine microalga *Tetraselmis* sp., the bryozoan *Bugula neritina*, and the marine bacterial *Acinetobacter species* sp. The most often utilized NPs are ZnO and TiO_2; however, nanocomposite coatings can also contain Cu_2O, CuO, and Al_2O_3. As a coating, PDMS offers a few benefits. It is utilized for fouling discharge antifouling applications because of its hydrophobic surface [2].

Nanocomposite antifouling coatings that are environmentally benign have been created using biopolymers like chitosan. It has been observed that chitosan and ZnO or TiO_2-based nanocomposite coatings work well against a variety of marine microorganisms, such as the diatom *Navicula* sp., *Pseudoalteromonas nigrifaciens*, and the biofilm-forming bacterium *Vibrio fischeri*. The inclusion of chitosan improves the antifouling characteristics of metal oxides in dark settings, whereas the metal oxides' photocatalytic properties inhibit biofouling under light conditions [30]. The majority of research involving ZnO nanorods or NPs in nanocomposite antifouling coatings was carried out in lab settings. Testing coatings under field conditions is crucial to demonstrate their antifouling performance because biofouling organisms come in a variety of species and are subjected to a range of physical and biological stimuli [8].

7.6 Concluding remarks and perspectives

In this chapter, we attempted to draw attention to the potential toxicological hazards associated with these developing technologies while simultaneously highlighting the many application areas of antimicrobial nanotechnology coatings and the advantages they offer.

Antibacterial products are not suitable for use if the risk ratio associated with using antimicrobial nanocoating greatly exceeds the benefits involved, both during the product's design process and during its useful life. The significance of further study on safe-by-design, environmentally friendly methods for creating antibacterial nanocoatings is highlighted in this chapter. In addition, biofouling poses a persistent threat to marine installations and is a significant issue for the maritime sectors.

Conventional antifouling techniques mostly include the use of biocidal agents, which appear to be unfriendly to the environment because they eliminate or deter nearby species. The alternatives to toxic antifouling paints that exhibit marine antifouling effects by fouling resistance, releasing, or degrading mechanisms were outlined in this chapter for polymeric and polymer–inorganic composites. The antifouling performance of hydrogel and polymer antifouling coatings is often determined by the coating process and surface chemistry. Fouling degradation in inorganic–polymer nanocomposite is caused by intrinsic properties of inorganic NPs. Micro- and macroorganisms can be killed or rendered inactive by ROS produced by photocatalytic nanostructures in the presence of light using photocatalytic nanostructure-based antifouling applications. The produced ROS are short-lived and only have an impact on species that are in proximity to the coatings, in contrast to harmful biocides.

In comparison to conventional antifouling coatings, photocatalytic nanocoatings are therefore less harmful to aquatic life. Coatings based on metal oxide nanostructures are one of the most promising environmentally friendly antifouling technology approaches that can be used in maritime applications. Conversely, there aren't many studies that have documented the suppression of macrofouling in field conditions.

This may be because nanocomposite antifouling coatings are less effective in the field or because scaling up these coatings is proving to be challenging. An antifouling coating should be substrate independent, economical, easy to apply, dependable, long-lasting, and environmentally beneficial. Realizing all the desired needs at once is still exceedingly difficult. There is currently no one antifouling technique that works for all situations. Rather, combining multiple antifouling techniques into one multifunctional coating with synergistic capabilities is likely the best approach to achieve a good antifouling coating. Zwitterionic polymers are gaining a lot of interest in the production of subsequent-generation antifouling materials due to their ease of synthesis, ease of application, abundance of raw materials, and wide range of functional groups. This is due in part to the substantial quantity of high-quality and advanced research that is being conducted in this area.

In addition, inorganic NPs are quite advantageous when it comes to antifouling preparation and use. Consequently, given their affordability and efficacy, inorganic–polymer composite coatings offer yet another potentially marketable option. To increase antifouling performance, new polymers or composites must still be developed, with a focus on long-term endurance in dynamic as well as static situations.

To effectively migrate these coatings from laboratory experiments to large-scale applications, coating processes must be developed in addition to efficiency enhancement. Therefore to handle the physisorption processes for coating, which are often LBL or spray-painting, the frequently employed chemical grafting approaches must be made simpler. Crosslinkers or catch-bonds to improve mechanical stability and basic coating techniques to increase antifouling durability could lead to the development of technologically advanced antifouling coatings that can be applied widely on a variety of materials, including pipes, fishing nets, and vessels, in real maritime environments. Lastly, a more thorough investigation of the effects of such coatings' toxicity to the environment must be conducted.

Acknowledgments

Authors are thankful to Director, Amity Institute of Applied Sciences (AIAS), and Head, Department of Chemistry for providing the facilities. One of the authors, Tejendra K. Gupta is thankful to the Council of Science & Technology, Lucknow, Uttar Pradesh, India for the financial support (Vide No. CST/CHEM/D-1431).

References

[1] M. Al-Fori, S. Dobretsov, M.T.Z. Myint, J. Dutta, Antifouling properties of zinc oxide nanorod coatings, Biofouling 30 (7) (2014) 871–882. Available from: https://doi.org/10.1080/08927014.2014.942297.
[2] L.A. Alshehri, N.F. Attia, Reinforcement and antibacterial properties of hand embroidery threads based on green nanocoatings, Coatings 13 (4) (2023). Available from: https://doi.org/10.3390/coatings13040747, https://www.mdpi.com/journal/coatings.
[3] J. Amalric, P.H. Mutin, G. Guerrero, A. Ponche, A. Sotto, J.P. Lavigne, Phosphonate monolayers functionalized by silver thiolate species as antibacterial nanocoatings on titanium and stainless steel, Journal of Materials Chemistry 19 (1) (2009) 141–149. Available from: https://doi.org/10.1039/b813344a.
[4] M. Montazer, T. Harifi, Antibacterial nanocoatings, Nanotoxicity, Elsevier, 2020, pp. 399–413.
[5] Z. Liu, X. Zheng, H. Zhang, W. Li, R. Jiang, X. Zhou, Review on formation of biofouling in the marine environment and functionalization of new marine antifouling coatings, Journal of Materials Science 57 (39) (2022) 18221–18242. Available from: https://doi.org/10.1007/s10853-022-07791-8.
[6] J.W. Molling, J.W. Seezink, B.E.J. Teunissen, I. Muijrers-Chen, P.J.A. Borm, Comparative performance of a panel of commercially available antimicrobial nanocoatings in Europe, Nanotechnology, Science and Applications 7 (2014) 97–104. Available from: https://doi.org/10.2147/NSA.S70782, http://www.dovepress.com/getfile.php?fileID = 22332.
[7] H.Y. Lee, H.K. Park, Y.M. Lee, K. Kim, S.B. Park, A practical procedure for producing silver nanocoated fabric and its antibacterial evaluation for biomedical applications, Chemical Communications (28)(2007) 2959–2961. Available from: https://doi.org/10.1039/b703034g, http://pubs.rsc.org/en/journals/journal/cc.

[8] S. Archana, B. Sundaramoorthy, Review on biofouling prevention using nanotechnology, Journal of Entomology and Zoology Studies 7 (2019) 640–648.
[9] S. Kumar, F. Ye, S. Dobretsov, J. Dutta, Nanocoating is a new way for biofouling prevention, Frontiers in Nanotechnology 3 (2021). Available from: https://doi.org/10.3389/fnano.2021.771098.
[10] P.C. Balaure, A.M. Grumezescu, Recent advances in surface nanoengineering for biofilm prevention and control. part I: molecular basis of biofilm recalcitrance. Passive anti-biofouling nanocoatings, Nanomaterials 10 (6) (2020) 1230. Available from: https://doi.org/10.3390/nano10061230.
[11] C.I. Idumah, C.M. Obele, E.O. Emmanuel, A. Hassan, Recently emerging nanotechnological advancements in polymer nanocomposite coatings for anti-corrosion, anti-fouling and self-healing, Surfaces and Interfaces 21 (2020) 100734. Available from: https://doi.org/10.1016/j.surfin.2020.100734.
[12] A.A. Ionin, A.K. Ivanova, R.A. Khmel'nitskii, Y.V. Klevkov, S.I. Kudryashov, A.O. Levchenko, et al., Antibacterial effect of the laser-generated Se nanocoatings on *Staphylococcus aureus* and *Pseudomonas aeruginosa* biofilms, Laser Physics Letters 15 (1) (2018) 015604. Available from: https://doi.org/10.1088/1612-202x/aa897f.
[13] M. Bulwan, K. Wójcik, S. Zapotoczny, M. Nowakowska, Chitosan-based ultrathin films as antifouling, anticoagulant and antibacterial protective coatings, Journal of Biomaterials Science, Polymer Edition 23 (15) (2012) 1963–1980. Available from: https://doi.org/10.1163/092050611x601711.
[14] S.M. Hammad, N.A. El-Wassefy, M.S. Shamaa, A. Fathy, Evaluation of zinc-oxide nanocoating on the characteristics and antibacterial behavior of nickel-titanium alloy, Dental Press Journal of Orthodontics 25 (4) (2020) 51–58. Available from: https://doi.org/10.1590/2177-6709.25.4.051-058.oar, http://www.scielo.br/pdf/dpjo/v25n4/2176-9451-dpjo-25-04-51.pdf.
[15] J. Butler, R.D. Handy, M. Upton, A. Besinis, Review of antimicrobial nanocoatings in medicine and dentistry: mechanisms of action, biocompatibility performance, safety, and benefits compared to antibiotics, ACS Nano 17 (8) (2023) 7064–7092. Available from: https://doi.org/10.1021/acsnano.2c12488.
[16] M.L. Chen, Y.Y. Qu, L. Yang, H. Gao, Structures and antifouling properties of low surface energy non-toxic antifouling coatings modified by nano-SiO_2 powder, Science in China Series B: Chemistry 51 (9) (2008) 848–852. Available from: https://doi.org/10.1007/s11426-008-0069-5.
[17] L. Graziani, E. Quagliarini, M. D'Orazio, The role of roughness and porosity on the self-cleaning and anti-biofouling efficiency of TiO_2-Cu and TiO_2-Ag nanocoatings applied on fired bricks, Construction and Building Materials 129 (2016) 116–124. Available from: https://doi.org/10.1016/j.conbuildmat.2016.10.111, https://www.journals.elsevier.com/construction-and-building-materials.
[18] H.J. Cox, I. Cooper, H.F. Kaspar, M.A. Packer, J.P.S. Badyal, Anti-biofouling functional surfaces for marine aquaculture, Colloids and Surfaces A: Physicochemical and Engineering Aspects 639 (2022) 128313. Available from: https://doi.org/10.1016/j.colsurfa.2022.128313.
[19] A. Dwivedi, A. Mazumder, N. Nasongkla, Layer-by-layer nanocoating of antibacterial niosome on orthopedic implant, International Journal of Pharmaceutics 547 (1-2) (2018) 235–243. Available from: https://doi.org/10.1016/j.ijpharm.2018.05.075, http://www.elsevier.com/locate/ijpharm.

[20] S. Hwangbo, H. Jeong, J. Heo, X. Lin, Y. Kim, M. Chang, et al., Antibacterial nanofilm coatings based on organosilicate and nanoparticles, Reactive and Functional Polymers 102 (2016) 27–32. Available from: https://doi.org/10.1016/j.reactfunctpolym.2016.03.004.
[21] S. Sinha, R. Kumar, J. Anand, R. Gupta, A. Gupta, K. Pant, et al., Nanotechnology-based solutions for antibiofouling applications: an overview, ACS Applied Nano Materials 6 (14) (2023) 12828–12848. Available from: https://doi.org/10.1021/acsanm.3c01539.
[22] C.X. Shuai, Y. He, P. Su, Q. Huang, D. Pan, Q. Xu, et al., Integration of PEGylated polyaniline nanocoatings with multiple plastic substrates generates comparable antifouling performance, Langmuir 36 (31) (2020) 9114–9123. Available from: https://doi.org/10.1021/acs.langmuir.0c01223, http://pubs.acs.org/journal/langd5.
[23] H. Shan, J. Liu, X. Li, Y. Li, F.H. Tezel, B. Li, et al., Nanocoated amphiphobic membrane for flux enhancement and comprehensive anti-fouling performance in direct contact membrane distillation, Journal of Membrane Science 567 (2018) 166–180. Available from: https://doi.org/10.1016/j.memsci.2018.09.038.
[24] M.S. Selim, M.A. Shenashen, S.A. El-Safty, S.A. Higazy, M.M. Selim, H. Isago, et al., Recent progress in marine foul-release polymeric nanocomposite coatings, Progress in Materials Science 87 (2017) 1–32. Available from: https://doi.org/10.1016/j.pmatsci.2017.02.001.
[25] M.S. Selim, S.A. El-Safty, M.A. Shenashen, S.A. Higazy, A. Elmarakbi, Progress in biomimetic leverages for marine antifouling using nanocomposite coatings, Journal of Materials Chemistry B 8 (17) (2020) 3701–3732. Available from: https://doi.org/10.1039/c9tb02119a, http://pubs.rsc.org/en/journals/journal/tb.
[26] D. Nazarov, L. Kozlova, E. Rogacheva, L. Kraeva, M. Maximov, Atomic layer deposition of antibacterial nanocoatings: a review, Antibiotics 12 (12) (2023) 1656. Available from: https://doi.org/10.3390/antibiotics12121656.
[27] P. Prabhakar, R. Kumar Sen, M. Patel, Shruti, N. Dwivedi, S. Singh, et al., Development of copper impregnated bio-inspired hydrophobic antibacterial nanocoatings for textiles, Colloids and Surfaces B: Biointerfaces 220 (2022) 112913. Available from: https://doi.org/10.1016/j.colsurfb.2022.112913.
[28] C.Y.N. Nicoliche, A.M. Pascon, Í.R.S. Bezerra, A.C.H. de Castro, G.R. Martos, J. Bettini, et al., In situ nanocoating on porous pyrolyzed paper enables antibiofouling and sensitive electrochemical analyses in biological fluids, ACS Applied Materials & Interfaces 14 (2) (2022) 2522–2533. Available from: https://doi.org/10.1021/acsami.1c18778.
[29] I.C. Nwuzor, C.I. Idumah, S.C. Nwanonenyi, O.E. Ezeani, Emerging trends in self-polishing anti-fouling coatings for marine environment, Safety in Extreme Environments 3 (1) (2021) 9–25. Available from: https://doi.org/10.1007/s42797-021-00031-3.
[30] P. Sathe, K. Laxman, M.T.Z. Myint, S. Dobretsov, J. Richter, J. Dutta, Bioinspired nanocoatings for biofouling prevention by photocatalytic redox reactions, Scientific Reports 7 (1) (2017). Available from: https://doi.org/10.1038/s41598-017-03636-6.

8

Tribology and antifriction nanopaints

Prateek Mittal[1], Jimmy Mehta[1], Pallav Gupta[2], Dinesh Chawla[1] and Sushant Sharma[1]

[1]*Department of Mechanical Engineering, School of Engineering and Technology, Manav Rachna International Institute of Research and Studies, Faridabad, Haryana, India* [2]*Department of Mechanical Engineering, Madan Mohan Malaviya University of Technology, Gorakhpur, Uttar Pradesh, India*

8.1 Introduction

Coatings are application of a layer of liquid or solid substance on a solid surface (substrate) in order to fulfill a desired purpose like protection from corrosion, protection from wear and tear, protection from environmental hazards, protection from bacterial and fungal actions etc [1]. Under the vast umbrella of coatings lie the paints which not only protect the substrate but also provide esthetic looks to the object like shine and smoothness [2]. Paint is a coating which is usually applied in liquid form and hardens post application to impart a specific feature to the substrate apart from giving lusture and smoothness. Coatings are generally not nano in structure but the constituents used in synthesis of the coatings are nano in structure. Nano-paints consist of a suspension of microscopic particles known as nanotubes which change their behavior with changing external conditions [3]. The layer of nano-paints is generally applied in liquid form and is ultrathin, in some cases this layer may be just one molecule thick. The nano-coating is usually optically colorless and because of its atomic size may not be detected by touch of hand [1]. The purpose of the nano-paint may vary according to the application, some commonly known applications are imparting scratch resistance, hardness, bacterial resistance, UV resistance to the substrate along with lusture and smoothening.

This chapter throws light on the role of nano-paints in friction and wear reduction including materials for nano-paints, characterization methods, mechanism of wear reduction,

Nanopaints. DOI: https://doi.org/10.1016/B978-0-443-26512-9.00008-2

applications in diverse fields and challenges. Nano refers to the size of the offering consummate reduction in friction and wear resistance [4]. Synthesized at the nanoscale, these innovative coatings leverage nanomaterials to impart superior lubrication properties, thereby enhancing the durability and performance of various surfaces [5–7]. Furthermore, their environment friendly formulations and low volatile organic compound content make them sustainable choices for surface protection. As the demand for advanced surface protection solutions continues to grow, the development of next-generation antifriction nano-paints holds immense promise. However, challenges such as cost-effectiveness, scalability, and regulatory compliance must be addressed to accelerate their widespread adoption.

Antifriction nano-paints represent a paradigm shift in surface protection technology, offering extraordinary friction reduction and wear resistance. By utilizing the unique properties of nanomaterials, these innovative coatings provide multifaceted benefits across a wide range of industrial applications. As research and development in nanotechnology continue to advance, antifriction nano-paints are expected to contribute significantly to the field of surface engineering and tribology.

8.2 Materials for nano-paints

Antifriction nano-paints have gained popularity in recent years due to their ability to enhance surface performance and longevity. At the core of these innovative coatings lie advanced nanomaterials, which are synthesized for their lubricating properties, mechanical strength, and compatibility with paint formulations. This chapter presents some basic nanomaterials used in antifriction nano-paints, shedding light on the role of various nanomaterials in friction reduction and surface protection. Nano composites have also become popular as coating materials due to their high hardness and ability to withstand high temperatures [8–12]. Especially, hybrid nano composites are being used as wear-resistant materials due to their excellent tribological behavior under sliding friction conditions [13–16]. Graphene, a two-dimensional sheet of carbon atoms, is a versatile nanomaterial widely employed in antifriction nano-paints. Due to its exceptional strength, flexibility, and lubricating properties, graphene acts as an effective solid lubricant, reducing friction and wear between the surfaces. Graphene-based nanomaterials are renowned for their ability to form robust layers on the boundary, providing durable protection against mechanical abrasion and surface damage. Molybdenum disulfide (MoS_2) is another prominent material for antifriction nano-paints, known for its lamellar structure and low shear strength. When dispersed within paint formulations, MoS_2 nanoparticles serve as effective lubricants, minimizing frictional losses and preventing surface-to-surface contact. MoS_2-based nanomaterials exhibit excellent chemical stability and adhesion to substrate surfaces, making them excellent choice for enhancing surface protection. Tungsten disulfide is renowned for its outstanding tribological properties, including low friction coefficient (cof) and high load-bearing capacity. In antifriction nano-paints, WS_2 nanoparticles act as solid lubricants, forming long-lasting lubricating films that reduce friction and wear. WS_2-based nanomaterials excel in harsh operating environments, offering exceptional resistance to temperature, pressure, and chemical degradation. Carbon nanotubes (CNTs) exhibit extraordinary mechanical, tribological and thermal properties, making them suitable for antifriction nano-paints. With their high aspect ratio and intrinsic lubrication capabilities, CNTs facilitate the formation of protective layers, thereby reducing friction and enhancing surface durability. Also, CNT- based nanomaterials can improve the electrical and thermal properties of painted surfaces. Hybrid nanomaterials, consisting of more than two components such as Graphene-MoS_2 or WS_2-CNTs, impart combined properties for antifriction nano-paints. By leveraging the complementary properties of different nanomaterials, hybrid formulations can achieve enhanced lubrication, mechanical strength, and surface adhesion. Hybrid nanomaterials pave the way for multifunctional antifriction paints capable of addressing diverse surface challenges.

The selection of appropriate nanomaterials is critical to the performance and efficacy of antifriction nano-paints. Graphene, MoS_2, tungsten disulfide, CNTs, and hybrid nanomaterials each contribute unique properties to these coatings, enabling superior friction reduction and surface protection. Fig. 8.1 indicates the materials used for synthesizing wear-resistant nano-coatings and paints.

8.2.1 Graphene-based nano-paints

Graphene-based nano-paints represent a good solution for enhancing surface protection. Leveraging the remarkable properties of graphene, these innovative coatings offer superior lubrication, wear resistance, and durability. Graphene, a two-dimensional sheet of carbon atoms, has gained considerable

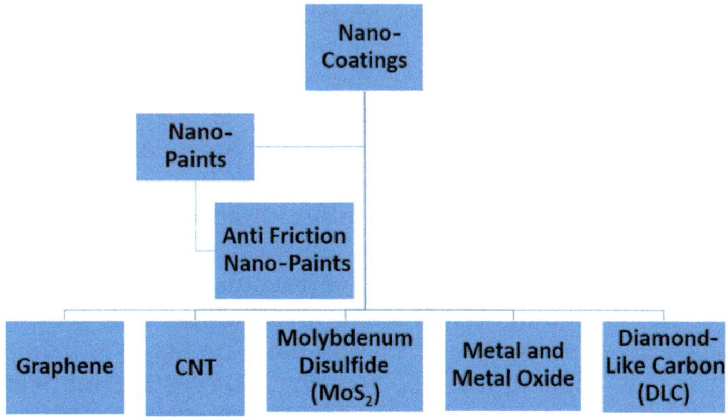

Figure 8.1 Materials for antifriction nano-paints.

popularity in recent years due to its exceptional mechanical, electrical, and thermal properties. Graphene-based nano-paints capitalize on these unique characteristics to provide robust surface protection and friction reduction. Graphene-based nano-paints exhibit a plethora of desirable characteristics, including high mechanical strength, chemical inertness, and superior lubricating properties. The incorporation of graphene into paint formulations enhances surface durability, reduces frictional losses, and provides long-lasting protection against wear and corrosion. Graphene's high aspect ratio and two- dimensional structure contribute to the formation of robust lubricating layers, ensuring smooth sliding and minimized surface damage [17]. The friction-reducing mechanisms employed by graphene-based nano-paints involve the formation of a graphene-based lubricating film on the surface. Graphene sheets possess inherent ability of solid lubrication, enabling them to reduce friction and wear by facilitating sliding between contacting surfaces. Graphene can be prepared by several methods but simplest of them is mechanical exfoliation. Graphene has been well known for its high strength and extraordinary wear resistance properties. Correlation between the number of graphene layers and corresponding friction behavior has been studies by various researchers in the past. It was found that friction and wear reduced with the increase in the no. of graphene layers [18–20]. The change in the tribological behavior of graphene surface could not be attributed to the change in structure, difference in contact stiffness or potential of the contact. This could be attributed to the force of adhesion between graphene and the substrate [19]. This was particularly noted when SiC was the substrate, which was also seen in other substrates like SiO_2 and Si.

In loosely attached sheets of graphene, a puckering effect was observed for thinner layers of graphene, usually exhibiting deformation in the direction normally outwards to the plane due to low flexural stiffness of graphene. This thickness dependence ceases to exist when graphene is strongly bound to the substrate as in case of mica. Apart from the number of layers, the method of deposition and nature of substrate material are also seen to have significant influence on the coefficient of friction and wear behavior. Graphene has not only been found useful as a coating material to reduce wear for smooth and flat surfaces but has also been found to be of considerable effectiveness in case of metals. Graphene-based nano-paints find diverse applications across various industries, including automotive, aerospace, marine, and manufacturing. In the automotive sector, these paints are utilized to reduce engine friction, enhance fuel efficiency, and for increasing component lifespan. In aerospace applications, graphene-based nano-paints protect critical components from wear and corrosion, ensuring reliable operation in demanding environments. Moreover, their versatility makes them suitable for use in coatings for consumer electronics, sporting goods, and infrastructure. The adoption of graphene-based nano-paints offers several notable advantages, including enhanced surface protection, reduced maintenance costs, and improved operational efficiency.

8.2.2 Carbon nanotube coatings and paints

CNTs are sheets of graphene which are rolled in the form of cylinders having diameter in the range of 0.5–2.0 nm. Graphene and thus CNTs can be said to be one of the allotropes of carbon. Graphene is made up of graphite which is pure carbon and is cut in sheets having planar hexagonal structure. The hexagonal arrangement of atoms is in such a way that it forms a web or a network. CNT coatings and paints have emerged as promising solutions for enhancing surface durability and reducing friction across various industrial applications. Leveraging the exceptional properties of CNTs, these coatings offer superior mechanical strength, electrical conductivity, and corrosion resistance [21]. CNTs are nanostructures known for their remarkable mechanical, electrical, and thermal properties. CNT coatings and paints utilize these unique attributes to provide robust surface protection and minimize friction [22–28]. CNT coatings and paints possess desirable characteristics such as high tensile strength, electrical conductivity, and chemical inertness. Incorporating CNTs into coating formulations enhances surface durability, reduces friction, and ensures strong adhesion to substrate surfaces. CNTs may be classified as single walled carbon nano tubes (SWCNTs) and multi walled carbon nano tubes (MWCNTs). SWCNTs have single layer of graphene sheet while MWCNTs have multiple layers of graphene sheet. It is difficult to synthesize SWCNT due to the difficulty in controlling its growth and atmospheric conditions, also the purity of SWCNT is low as compared to MWCNTs. MWCNTs have high purity but it has a very complex structure which is a little difficult to characterize. SWCNTs and MWCNTs both are quite useful as antifriction coatings and paints.

Polyimide (Pi) coating of 6–7 μm on Si substrate was applied with SWCNT as an additive [22]. The wear test was conducted using ball-on-disc apparatus with a constant pressure of 370 MPa. It was found that cof increased a little bit but it was less as compared to the cof for bare Si substrate. However, the increase in wear life was observed to be exceptional which reached a high of upto 7.2×10^3 cycles with cof below 0.3. This enhanced wear life could be attributed to increase in hardness and Young's modulus by 60%–70% after addition of CNTs to the Pi coating. Also, changes in structure of coating upon being subjected to wear test resulting in re-alignment of CNTs also contributed to increased wear life. MWCNTs were also tested and were found to be quite effective as coating material. MWCNTs are found to increase the toughness and wear resistance of the coating through inhibition to crack propagation [28]. Also, self-lubricating and load-bearing abilities of the MWCNTs make them suitable material for wear-resistant coatings and paints [26]. Wear behavior of MWCNT reinforced alumina coatings was tested on a tungsten-carbide ball-on-disc test rig. The test was conducted at room temperature and at elevated temperatures of 573 K and 873 K. The only-alumina coating was found to exhibit reduced wear resistance with increase in temperature. However, MWCNT reinforced alumina coating was found to exhibit enhanced wear resistance at elevated temperatures which could be attributed to coverage of wear tracks by large protective film formed by MWCNTs and bridging effect of MWCNTs on between splats.

Furthermore, the nanoscale dimensions of CNTs contribute to the formation of durable tribofilms, enhancing tribological performance. CNT coatings and paints find diverse applications in automotive, aerospace, electronics, and biomedical industries. In automotive applications, these coatings reduce friction in engines, improve fuel efficiency, and extend component lifespan. Aerospace applications benefit from CNT coatings by providing protection against wear, erosion, and thermal degradation. Additionally, the electrical conductivity of CNTs makes them suitable for use in electronics, sensors, and energy storage devices. Adopting CNT coatings and paints offers significant advantages including enhanced surface protection, reduced frictional losses, and improved functionality. These coatings contribute to extended service life, lower maintenance costs, and enhanced performance. Furthermore, the electrical and thermal conductivity of CNTs enables multifunctional coatings suitable for various electronic devices and advanced technologies.

8.2.3 Molybdenum disulfide coatings and nano-paints

Solid lubricant coatings are becoming increasingly popular due to their capability to reduce the coefficient of friction and thus the wear in case of metal-metal contact conditions, where liquid or mist lubrication systems have been found to be inappropriate and incapable [29]. Amongst solid lubricants sulfides, selenides or tellurides of tungsten, niobium and molybdenum are quite popular. MoS_2 has a tendency to interact with oxygen present in atmospheric air. MoS_2, a layered transition metal dichalcogenide, has attracted considerable interest due to its inherent lubricating properties and structural stability. MoS_2 coatings and nano-paints exploit these characteristics to provide effective surface protection and friction reduction. Also, the hardness and strength of MoS_2 are low for a good wear-resistant coating. In order to improve the effectiveness of the coating in ambient air environment to which most often than not these coatings are subjected to and to impart sufficient strength and hardness to these coatings, several additives may be added to the coating. These additives may include Nickel, Chromium, Titanium, Tungsten, Silver etc [29]. The combination of metal with MoS_2 as a coating is gaining popularity to prevent wear losses in metal-metal contact applications. Several techniques of depositing the coating material on the substrate surface are available, physical vapor deposition (PVD) comes out as the simplest and the most convenient amongst them. leveraging the unique properties of MoS_2, these coatings offer exceptional lubrication, wear resistance, and corrosion protection.

MoS_2 coatings and nano-paints exhibit unique characteristics such as high lubricity, chemical inertness, and thermal stability [29,30]. The incorporation of MoS_2 nanoparticles into coating formulations enhances surface durability, reduces friction, and provides excellent adhesion to substrate surfaces. Moreover, MoS_2-based coatings offer superior resistance to wear, corrosion, and environmental degradation. MoS_2 coatings and nano-paints reduce friction and wear through mechanisms such as solid-state lubrication and boundary film formation. The lamellar structure of MoS_2 facilitates sliding between contacting surfaces, minimizing frictional losses and surface damage. Additionally, the presence of MoS_2 nanoparticles promotes the formation of robust tribofilms, further enhancing tribological performance. MoS_2 coatings and nano-paints find diverse applications in automotive, aerospace, manufacturing, and energy sectors. In automotive applications, these coatings are utilized to reduce friction in engine components, improve fuel efficiency, and extend service life. Aerospace industries benefit from MoS_2 coatings due to their ability to provide protection against wear, corrosion, and high-temperature environments. Moreover, MoS_2 nano-paints offer opportunities for enhancing surface properties in manufacturing processes and energy applications.

Furthermore, the lubricating properties of MoS_2 enable smooth operation and reliability in various mechanical systems and industrial processes. Despite their potential, MoS_2 coatings and nano-paints face challenges such as scalability, cost-effectiveness, and compatibility with existing coating technologies. Future research efforts should focus on optimizing coating formulations, exploring novel synthesis methods, and assessing environmental impacts. Collaboration between researchers, industry partners, and regulatory bodies is essential for advancing the commercialization and widespread adoption of MoS_2 coatings and nano-paints. MoS_2 coatings and nano-paints represent a promising avenue for surface protection and friction reduction across diverse industrial applications. Leveraging the unique properties of MoS_2, these advanced coatings offer enhanced durability, lubrication, and corrosion resistance. Continued research and innovation in MoS_2 coatings and nano-paints hold the potential to drive efficiency, reliability, and sustainability in surface engineering and tribological applications.

8.2.4 Metal and metal oxide coatings and nano-paints

Metal nano particles (MNPs) are quite popularly used as wear-resistant coating materials due to their low melting point, low shear strength and high surface area which imparts self-repairing capability to MNPs to resist wear for longer durations. MNPs are also used as additives in various lubricants. Metal and metal oxide nanoparticles (MoNPs) have gained significant attention in the field of surface protection and functional coatings [31]. These nanoparticles, with sizes ranging from a few to hundreds of nanometers, exhibit distinctive physical and chemical properties compared to their bulk counterparts. Incorporating metal and MoNPs into coatings and paints allows for tailoring surface properties such as hardness, adhesion, and corrosion resistance, thus expanding the scope of applications across diverse industries [31]. Metal and MoNPs exhibit a wide range of properties depending on their composition, size, and surface morphology. Metal nanoparticles such as silver, gold, and copper offer excellent antimicrobial properties and catalytic activity. On the other hand, MoNPs like titanium dioxide and zinc oxide possess photocatalytic properties, UV resistance, and self-cleaning capabilities. These nanoparticles also provide optical transparency, making them suitable for applications in transparent coatings and paints. The incorporation of metal and MoNPs into coatings and paints enhances surface properties through various mechanisms. Metal nanoparticles act as antimicrobial agents, inhibiting the growth of bacteria and fungi on coated surfaces. Additionally, MoNPs provide UV protection by absorbing and scattering harmful UV radiation, thereby preventing degradation of underlying materials.

Typically used MNPs are Cu, Bi, Zn, Ni, Al, Sn, Pd, and Co. Three-fold wear mechanism is generally observed in case of MNPs being used as a nano-paint or coating to resist wear and friction. First, the high-temperature and high pressure conditions created during surface-surface sliding action lead to reduction in the gap of wear track thereby promoting the MNP to get sintered imparting them the self-repairing attribute. Second, friction and wear between the two sliding surfaces tends to reduce as a result of rolling action of MNPs between the two interacting surfaces. Third, the formation of friction film or mixed layer between the two sliding surfaces acts as a shield and prevents the material at the interface from getting worn-out upon being subjected to wear. Fig. 8.2 indicates the scanning electron microscopy (SEM) micrograph of Fe nano particle coated surface.

Metal oxide layers made up of MoNPs also referred to as glaze layers which offer reduced cof and enhanced wear resistance for metal-metal sliding contact [33–36]. The aforementioned wear mechanism for MNPs are quite applicable to

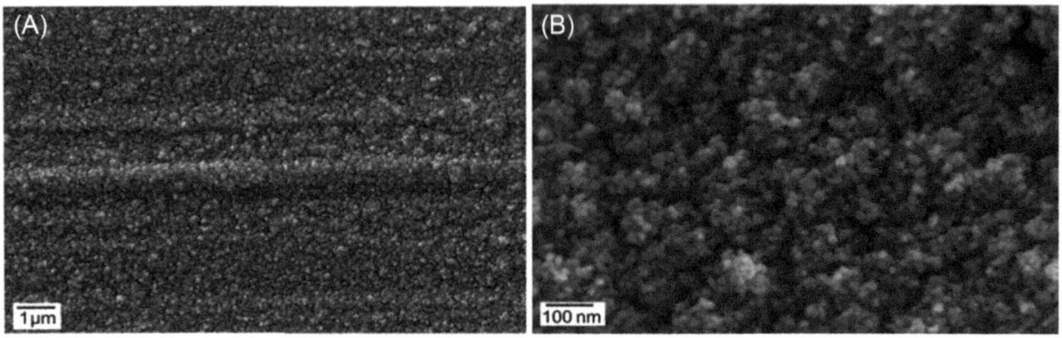

Figure 8.2 Scanning electron microscopy images of iron nano particle coated surfaces (A) At lower magnification, (B) At higher magnification [32].

MoNPs as well. The formation of wear-resistant oxide layers is enabled by formation of oxide debris and agglomeration of nano particles. These enablers generally get triggered during the initial tribo/friction interaction between the two sliding surfaces [37]. Post removal of the material through sliding friction and wear, the nano oxide layers take their place separating the two surfaces and forming an extremely hard wear-resistant interlayer formed as a result of compaction and sintering of the oxide debris [38,39]. The frictional interaction between two metal surfaces are shielded by glaze layer which generally provide better wear resistance as compared to compacted debris of MoNPs [40].

The incorporation of metal and MoNPs offers several advantages for coatings and paints, including enhanced antimicrobial properties, UV resistance, and corrosion protection. These nanoparticles contribute to improved durability, longevity, and esthetics of coated surfaces. Additionally, the catalytic activity of metal nanoparticles enables advanced functionalities such as self- cleaning and pollutant degradation, further enhancing the value proposition of nanoparticle-enhanced coatings and paints. While metal and MoNPs hold immense promise for coatings and paints, several challenges must be addressed to realize their full potential. These include scalability of nanoparticle synthesis, cost- effectiveness, and environmental considerations. Future research efforts should focus on developing sustainable synthesis methods, optimizing nanoparticle dispersion and stability in coating formulations, and assessing long-term performance and environmental impacts. Collaboration between academia, industry, and regulatory agencies is essential for addressing these challenges and accelerating the commercialization of nanoparticle-enhanced coatings and paints. Metal and MoNPs represent a promising avenue for enhancing the performance and functionality of coatings and paints. Leveraging their unique properties, these nanoparticles enable tailored surface properties such as antimicrobial activity, UV resistance, and self-cleaning capabilities. Continued research and innovation in nanoparticle synthesis, dispersion, and application techniques are essential for unlocking the full potential of nanoparticle-enhanced coatings and paints across diverse industries.

8.2.5 Diamond like coatings

Diamond-like carbon (DLC) coatings and paints have gained significant attention in recent years due to their exceptional mechanical properties, high hardness, and chemical inertness. These coatings offer superior wear resistance, low friction, and excellent biocompatibility, making them ideal for various industrial and biomedical applications. DLC coatings are thin films composed of amorphous carbon with properties resembling those of diamond. These coatings exhibit remarkable hardness, low friction, and excellent resistance to wear and corrosion. DLC coatings and paints have gained widespread interest across industries for their ability to enhance surface durability and functionality. Diamond-like carbon is SP^3 hybridized metastable amorphous carbon with good mechanical optical chemical and tribological properties. DLC film has carbon bonded atoms of SP^2 and SP^3 hybridization with sufficient quantity of hydrogen. In dry conditions DLC provide extraordinary low coefficient of friction that is of the order of less than 0.01. The major we're in DLC occurs in the run-in period which generally precedes the state where cof below 0.01 is achieved [41]. DLC is where in hydrogen was absent one not seen to perform well when subjected to wear load for longer periods. The use of hydrogen gas during wear test contributes towards decrease in number of carbon atoms involved in frictional contact at the interface of two rubbing regions thereby reducing addition and friction. Under ordinary conditions DLC provide lowest coefficient of friction as compared to the coating materials of varying nature including steel, titanium carbonitride (TiCN), titanium aluminide (TiAl), TiN. DLC coatings possess the following qualities like impermeability, chemical inertness, biocompatibility, scratch resistance, wear resistance and antisticking property which make them suitable for nearly

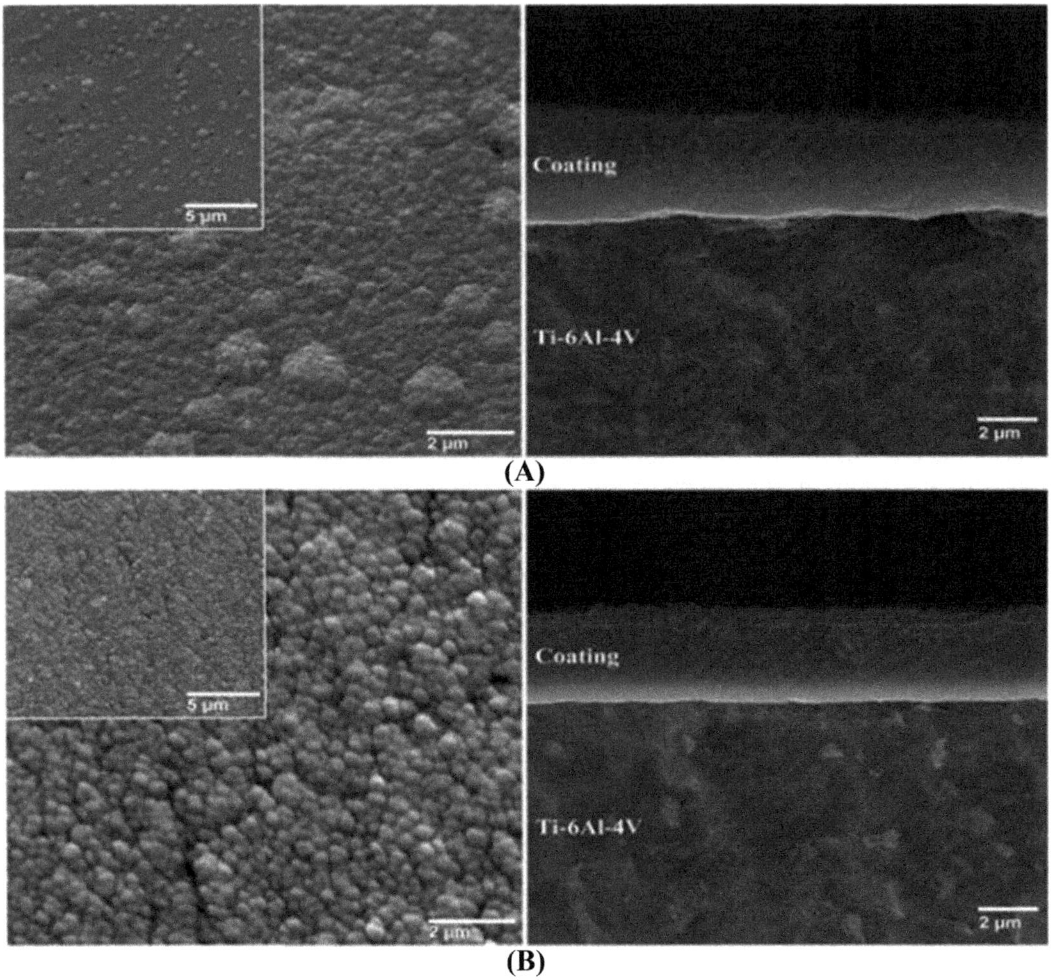

Figure 8.3 Scanning electron microscopy micrograph depicting cross-sectional view of Diamond-like carbon coated surface obtained through (A) plasma immersion ion deposition (PIID) and (B) PIID + plasma enhanced magnetron sputtering [42].

all applications. Fig. 8.3 indicates the SEM micrographs of DLC-coated surface.

DLC coatings find diverse applications across numerous industries, including automotive, aerospace, medical devices, and cutting tools [43]. In automotive and aerospace applications, DLC coatings are utilized to reduce friction in engine components, improve fuel efficiency, and enhance wear resistance in bearings and gears. In the medical field, DLC-coated implants offer biocompatibility and reduced wear, promoting tissue integration and longevity. The adoption of DLC coatings and paints offers several significant advantages, including enhanced surface protection, reduced frictional losses, and improved wear resistance. These coatings contribute to extended service life, reduced maintenance costs, and enhanced performance in various applications. Furthermore, the biocompatibility of DLC coatings makes them suitable for medical implants and devices, offering superior wear resistance and reduced inflammatory response. DLC used

8.3 Fabrication techniques

8.3.1 Chemical vapor deposition

Chemical vapor deposition (CVD) involves dissociation or chemical reactions or combination of both to form a stable solid product which may be in powder form or film or coating in the presence of heat, light and plasma or any one or two of them. The chemical reaction which produces the solid product is either completely gas phase reaction or it may originate from a heterogeneous chemical interaction near a heated surface. CVD traces its origin back to historic times then soot (unburnt carbon particles) produced due to burning of wood used to get deposited on other surfaces. The process has seen much development in the last 50 years. The process got popularity when electronic circuits were coated with a protective coating using CVD process. Post this application lot of other industrial application used this process to impart wear resistance, corrosion resistance, prevention from oxidation, prevention from exposure to sudden temperature variations etc. The prominent areas of applications for coating purposes include machine tools, blades of rotating machine, solar cells, coating of composite materials etc. The process now is widely used in surface modifications, electronic materials, ceramic fibers and composites. The potential applications of CVD are being worked upon rigorously in the recent research works. Tiny structures whose dimensions are of the order of 10^1-10^2 nm may be referred to as nano structures. This structures may have different forms such as tubes, platelets, particles etc. Amongst several nano structures CNTs are quite prominent. CNTs have been discussed in Section 8.2.2. CVD is used to synthesize CNTs. Also, directionally oriented CNTs embedded on the substrate to attain special thermos-physical properties may be

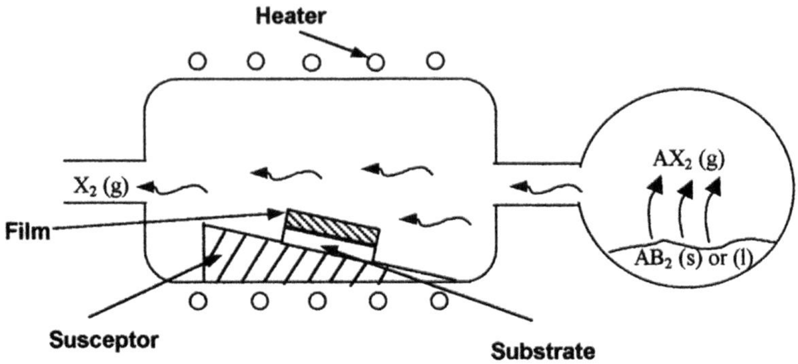

Figure 8.4 Schematic diagram of chemical vapor deposition process [44].

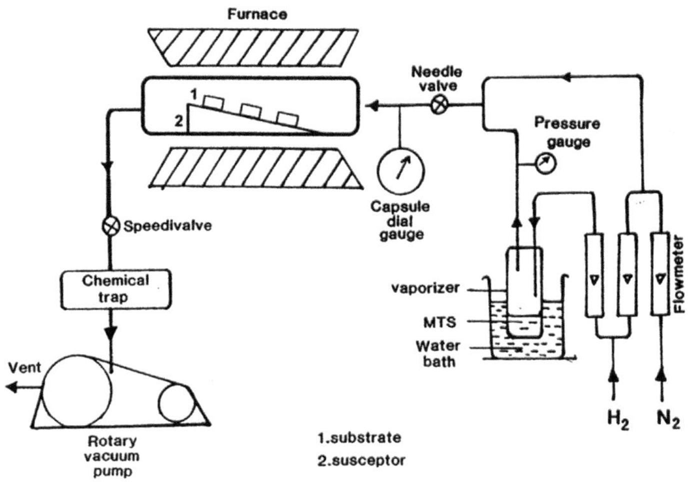

Figure 8.5 A laboratory setup of chemical vapor deposition process for deposition of SiC coating [44].

obtained through CVD process. Figs. 8.4 and 8.5 indicate the schematic of CVD process and setup respectively.

8.3.2 Physical vapor deposition

PVD is a thin-film deposition process where a target material is atomized or vaporized, enabling atom-by-atom coating growth on a substrate. This modifies both the surface and the interface between the substrate and coating, with substrate properties also influencing the final film. PVD can occur in various environments (vacuum, gas, plasma, electrolytic), with vacuum chambers minimizing contamination. Recent decades have seen significant advancements in PVD techniques, targeting enhanced coating properties and faster deposition rates while emphasizing surface cleanliness. This progress is particularly notable in carbide and nanocomposite transition metal nitride substrates. While industry prioritizes improving deposition efficiency, research continues to refine coating characteristics. Evaporation and sputtering are the dominant gaseous-state PVD methods. In evaporation atoms are evaporated thermally and are allowed to condense on the surface of the substrate while in sputtering the material gets deposited with the help of high energy ion flow or gas particles. Evaporation involves high-vacuum environments, typically resulting in lower adhesion and less pure coatings but with faster deposition. Sputtering offers better adhesion but may be slower. PVD is environment friendly and it significantly improves wear and corrosion resistance. It finds widespread use in functional applications like tools, decorative pieces, optics, molds, dies, and blades [45–47]. PVD offers numerous advantages, enhancing product durability and value. This technique plays a crucial role in machining, where tools require exceptional hardness, elevated temperature resistance, abrasion resistance, chemical stability, toughness, and stiffness [48]. Furthermore, PVD produces coatings with excellent adhesion, homogeneity, customizable structures, graduated properties, controllable morphology, and vast material diversity.

PVD processes enable the deposition of mono-layered, multilayered, and multigraduated coating systems with unique alloys and structures. The ability to continuously vary coating characteristics throughout the film is a key advantage [49].

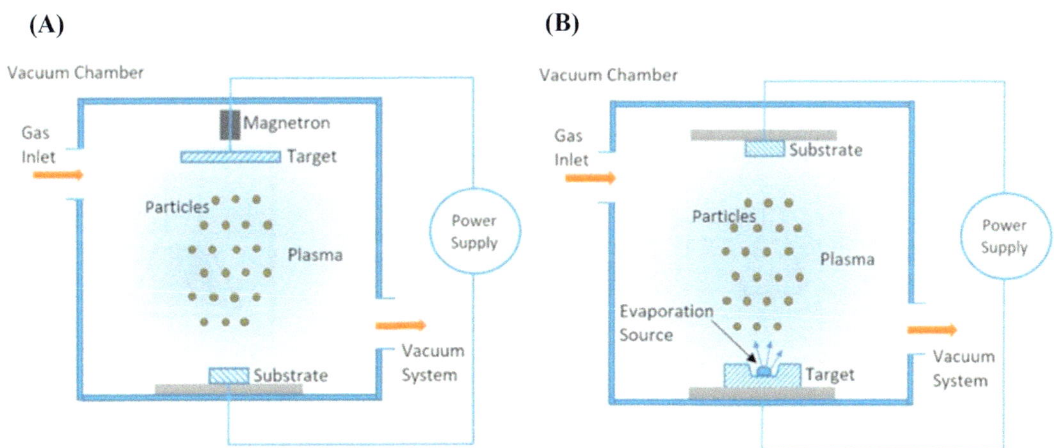

Figure 8.6 Schematic diagram of (A) physical vapor deposition sputtering and (B) physical vapor deposition evaporation processes [50].

Market adaptability has spurred the development of numerous PVD process variants. Fig. 8.6 indicates the schematic diagram of PVD sputtering and evaporation process respectively.

8.3.3 Sol-gel technique

The sol-gel process stands as a flexible and widely employed technique for the fabrication of nano-coatings and paints. The sol-gel synthesis route entails the transformation of a precursor solution into a solid-state material via hydrolysis and condensation reactions. Metal salts are usually solubilized to form a 'sol.' This sol undergoes gelation, establishing a three-dimensional network. Further processing of the resultant gel by drying and heat treatment yields nano-coatings or paints. Within the sol-gel process, precursor molecules experience hydrolysis and condensation reactions, generating nanoparticles or nanoclusters suspended in a liquid medium. Deposition techniques such as dip, spin, or spray coating serve to deposit these nanoparticles onto target substrates, forming thin films or coatings. Ensuing heat treatments promote the densification of nanoparticles into a homogenous coating. Nano-coatings and paints derived via sol-gel find extensive use across automotive, aerospace, electronics, and biomedical sectors. These coatings impart benefits like corrosion resistance, scratch resistance, optical clarity, and biocompatibility, catering to protective, functional, and decorative demands. The ease of coating synthesis is facilitated by this technique through moderate processing temperatures, substrate compatibility, adjustable coating composition and properties, and industrial scalability. Sol-gel coatings also demonstrate exceptional substrate adhesion and adaptability to application-

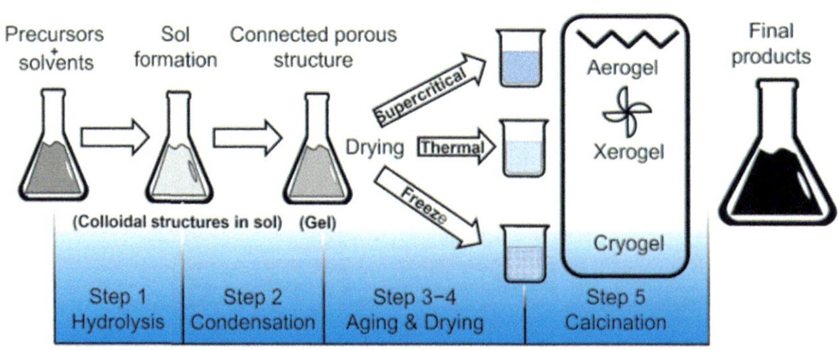

Figure 8.7 Schematic diagram of sol-gel technique [51].

specific needs. However, the sol-gel method presents challenges of controlling coating thickness and porosity, optimizing deposition parameters, and ensuring reproducibility. Fig. 8.7 indicates the schematic diagram of sol-gel technique.

8.3.4 Atomic layer deposition

Atomic layer deposition (ALD) has emerged as a versatile technology that provides excellent thickness, composition and compatibility control for the synthesis of nano-coatings and paints. ALD is based on self-adjusting surface shielding between alternative precursor gases and the substrate surface. Continuous pulses of precursor gases enter the reaction chamber, where they react with the substrate surface to form a layer of material. The process is repeated in a cycle to ensure controlled growth of the film with atomic precision. During the ALD process, precursor molecules undergo surface reactions with functional groups on the surface of the substrate, leading to the deposition of atomic layers of the product. The advantages include level control of layer thickness, conformal deposition on high aspect ratio substrates, excellent uniformity, and scalability for industrial production. Additionally, ALD facilitates the deposition of complex coatings consisting of different materials with customized properties and functionalities. The self-limiting nature of ALD process enables it to deposit uniform, coatings on difficult to coat substrates providing room for precise thickness control and composition maneuverability. Reactivity of precursor, compatibility of the substrate and rate of deposition are some of the limitations of ALD despite having several advantages. Future research directions include the development of advanced reactors, advanced reactor design, and in situ monitoring systems to overcome these challenges and commercialization of ALD for different applications. Fig. 8.8 indicates the schematic diagram of ALD process.

8.3.5 Spray coating

Spray coating is an effective technique for synthesizing nano-coatings and paints. It offers benefits including scalability, uniformity, and procedural simplicity. Spray coating offers distinctive advantages for nano-coating and paint synthesis, allowing for the deposition of thin films with controlled thickness and morphology. The spray coating process involves atomizing liquid precursor solutions into fine droplets through techniques like air-brushing, ultrasonic spraying, or pneumatic spraying.

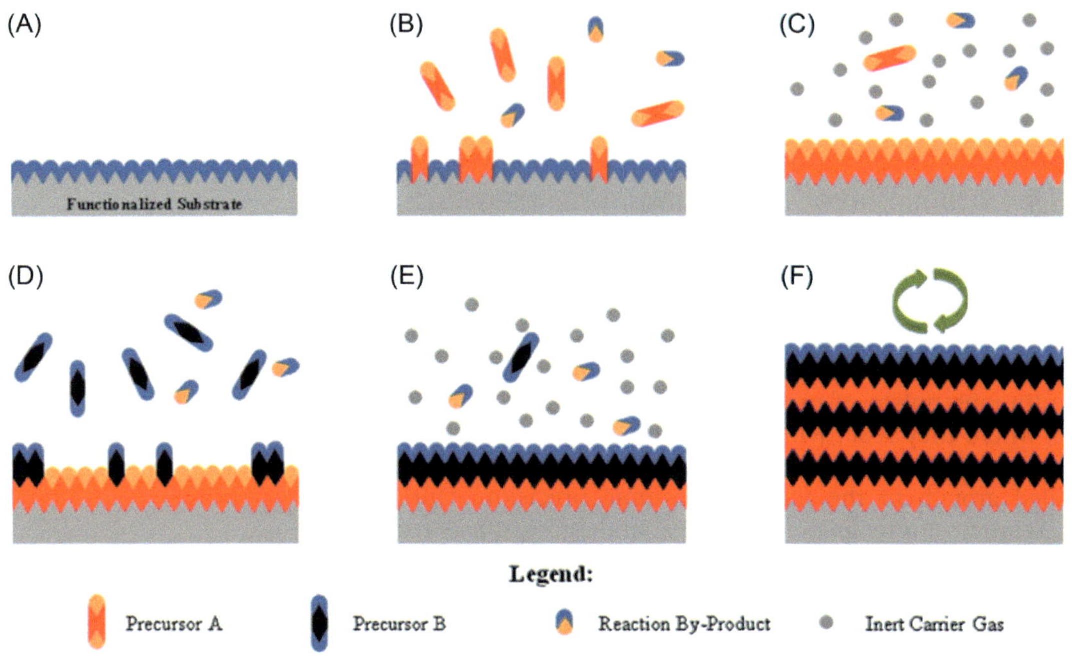

Figure 8.8 Schematic diagram of atomic layer deposition process. (A) Substrate surface; (B) reaction of precursor A with surface; (C) inert gas purged as a carrier for excess reaction by-products; (D) reaction of precursor B with the surface; (E) reaction by-products are purged with inert gas, (F) repetition of steps until the attainment of desired thickness.

These droplets are projected onto a substrate surface, coalescing into a uniform coating layer. Spray coating facilitates decent control over thickness and coating area, making it suitable for applications where thin and uniform coatings are required. During spray coating, atomized droplets strike the surface of the material, undergoing spreading, coalescence, and drying. Solvent evaporation leaves behind a thin-film of the deposited material. The process of depositing the coating material can be refined by modifying parameters such as spray pressure, nozzle geometry, and substrate temperature, leading to coatings with desired characteristics. Spray-coated nano-coatings and paints have applications in industrial sectors namely, defense, automobile, aviation, medical, and energy storage. These coatings impart benefits like corrosion resistance, wear resistance, and optical transparency thereby imparting suitability for protective, functional, and esthetic purposes. Spray coating provides multiple advantages for nano-coating and paint synthesis. These advantages include commercialization for mass production, uniform coating deposition across complex geometries, and compatibility with a large number of substrates and materials.

Moreover, spray coating facilitates control over coating thickness and coverage, enabling the creation of coatings customized for specific applications. Despite its numerous strengths, spray coating-based synthesis of nano-coatings and paints have challenges, such as overspray, substrate contamination, and possibility of porosity. Development of advanced nozzle designs, optimizing spray parameters, and utilizing characterization techniques to elevate coating quality and performance will further enhance the usability of spray coating process. Spray coating still remains an effective, simple and widely used technique for synthesizing nano-coatings and paints with precise control over their properties and morphology.

8.4 Characterization

Characterizing the surface, mechanical, tribological, and chemical properties of nano-coatings and paints is essential for understanding their performance and suitability for various applications. This article provides a comprehensive overview of characterization methods employed to analyze nanostructured coatings and paints, covering techniques for surface morphology, mechanical behavior, tribological properties, and chemical composition. By exploring the principles, capabilities, and applications of these characterization methods, this article aims to facilitate a deeper understanding of nano-coatings and paints for optimizing their performance in diverse industrial and technological applications. Nano-coatings and paints exhibit unique properties at the nanoscale, necessitating specialized characterization techniques for their comprehensive analysis. Understanding the surface morphology, mechanical behavior, tribological properties, and chemical composition of these materials is crucial for tailoring their properties to specific applications [8,9].

Surface characterization techniques such as SEM and atomic force microscopy (AFM) provide insights into the surface morphology and topography of nano-coatings and paints. SEM offers high-resolution imaging of surface features, while AFM enables nanoscale surface profiling and roughness analysis. These methods aid in understanding surface structure and quality.

Mechanical characterization methods, including nano-indentation and scratch testing, assess the mechanical properties and adhesion strength of nano-coatings and paints. Nano-indentation measures hardness and elastic modulus at the nanoscale, while scratch testing evaluates coating durability and resistance to wear. These techniques help in optimizing coating performance and durability. Tribological characterization techniques such as tribometer testing and wear analysis evaluate the friction, wear, and lubrication properties of nano-coatings and paints. Tribometer testing measures coefficient of friction and wear rate under different conditions, while wear analysis assesses wear mechanisms and surface damage. These methods are essential for understanding coating performance in tribological applications.

Chemical characterization methods, including X-ray photoelectron spectroscopy (XPS) and Fourier-transform infrared spectroscopy (FTIR), provide information about the chemical composition and molecular structure of nano-coatings and paints. XPS detects elemental composition and chemical states, while FTIR identifies functional groups and chemical bonds. These techniques aid in identifying coating components and assessing chemical changes.

Characterization methods for nano-coatings and paints play a vital role in assessing their surface, mechanical, tribological, and chemical properties, enabling researchers and engineers to optimize their design and functionality for specific applications. By employing a combination of surface imaging, mechanical testing, tribological analysis, and chemical characterization techniques, comprehensive insights into the structure, composition, and behavior of nanostructured coatings can be obtained, facilitating the development of advanced materials for diverse industrial and technological applications.

8.4.1 Surface morphology analysis methods

Understanding the surface morphology of nano-coatings and paints is crucial for optimizing their performance in various applications. This article provides a comprehensive overview of surface morphology analysis methods, including SEM, AFM, transmission electron microscopy (TEM), and profilometry. Each technique's principles, capabilities, and applications in characterizing nanostructured coatings and paints are discussed. By employing these methods, researchers and engineers can gain insights into surface features, topography, and roughness, facilitating the optimization of coating processes and properties. Nano-coatings and paints possess intricate surface structures at the nanoscale, necessitating precise characterization techniques for analysis. Surface morphology influences properties like adhesion, friction, and optical behavior, making its analysis critical for optimizing performance. Various methods play a crucial role in elucidating surface features, offering valuable insights into coating quality and suitability for specific applications.

SEM is a versatile technique for imaging surface morphology at high-resolution. By scanning a focused electron beam across the sample, SEM generates detailed images of surface features such as particles, cracks, and pores. It provides qualitative and quantitative data for optimizing coating processes and assessing coating quality. TEM offers unparalleled resolution for studying surface morphology at the atomic level. By transmitting electrons through a thin sample, TEM produces high- resolution images of surface structures and interfaces. It is invaluable for analyzing nanoscale features and defects, providing insights into coating microstructure and composition. AFM is renowned for its capability to provide high-resolution surface topography images. It employs a sharp tip to scan the sample surface, detecting forces to generate topographic images. AFM enables visualization of surface roughness, grain morphology, and defects, aiding in process optimization and quality control. Profilometry quantitatively measures surface roughness and texture. Using a stylus or optical probe, it traces surface profiles and measures height variations. Profilometry parameters like average roughness (Ra) and peak-to-valley height provide quantitative measures for quality control and process optimization. Surface morphology analysis methods such as SEM, TEM, AFM and profilometry are indispensable for characterizing nano-coatings and paints. By leveraging these techniques, researchers can gain valuable insights into surface features and topography, enabling optimization of coating processes and properties for diverse applications.

8.4.2 Chemical analysis methods

Chemical analysis methods play a crucial role in characterizing the composition and molecular structure of nano-coatings and paints, providing insights into their chemical properties and behavior. This article provides an extensive overview of chemical analysis techniques employed for analyzing nanostructured coatings and paints, including XPS, FTIR, and Raman spectroscopy. Each method's principles, capabilities, and applications in chemical analysis of nano-coatings and paints are discussed, highlighting their importance in understanding coating composition, chemical bonding, and functional groups. By employing these techniques, researchers and engineers can gain valuable insights into the chemical properties of nanocoatings and paints, facilitating their optimization for specific industrial and technological applications. Nano-coatings and paints exhibit complex chemical compositions at the nanoscale, necessitating advanced techniques for chemical analysis. Understanding the chemical properties and behavior of these materials is essential for tailoring their performance and functionality for various applications. Chemical analysis methods enable researchers to identify coating components, characterize chemical bonding, and assess molecular structure, providing valuable insights into coating composition and behavior. XPS is a powerful technique for chemical analysis of nano-coatings and paints. By bombarding the sample with X-rays, XPS generates photoelectrons whose energy distribution provides information about the elemental composition and chemical states of the sample's surface. XPS is capable of detecting elements from hydrogen to uranium and provides valuable insights into surface chemistry, oxidation states, and chemical bonding in nano-coatings and paints.

FTIR is widely used for chemical analysis of nano-coatings and paints. FTIR measures the absorption of infrared radiation by the sample, revealing information about functional groups, molecular vibrations, and chemical bonds present in the material. By analyzing FTIR spectra, researchers can identify coating components, assess chemical changes, and elucidate molecular structures, facilitating the characterization and optimization of nano-coatings and paints for specific applications.

Raman spectroscopy is a nondestructive technique for chemical analysis of nano-coatings and paints. Raman spectroscopy measures the scattering of monochromatic light by the sample, providing information about molecular vibrations and crystal structures. It is particularly useful for identifying carbon-based materials, detecting impurities, and characterizing surface functional groups in nanostructured coatings and paints. Raman spectroscopy offers high sensitivity and spatial resolution, making it suitable for analyzing small sample areas and detecting subtle chemical changes.

Chemical analysis methods such as XPS, FTIR, and Raman spectroscopy play a vital role in characterizing the composition and molecular structure of nano-coatings and paints. By employing these techniques, researchers and engineers can gain valuable insights into coating composition, chemical bonding, and functional groups, facilitating the optimization of coating properties for specific industrial and technological applications.

8.4.3 Mechanical testing methods

Mechanical testing methods play a crucial role in evaluating the mechanical properties and performance of nano-coatings and paints, providing valuable insights into their durability, adhesion, and resistance to wear. This article provides a comprehensive overview of mechanical testing techniques employed for analyzing nanostructured coatings and paints, including nano-indentation and scratch testing. Each method's principles, capabilities, and applications in mechanical characterization of nano-coatings and paints are discussed, highlighting their importance in assessing coating hardness, elasticity, and adhesion strength. By employing these techniques, researchers and engineers can gain valuable insights into the mechanical behavior of nano-coatings and paints, facilitating their optimization for specific industrial and technological applications. Nano-coatings and paints often face mechanical challenges in various applications, such as abrasion, scratching, and wear. Understanding their mechanical properties is crucial for ensuring their performance and longevity. Mechanical testing methods provide quantitative measures of coating hardness, elasticity, adhesion, and resistance to wear, enabling researchers to optimize coating formulations and processes for enhanced mechanical performance. Nano-indentation is a widely used technique for assessing the mechanical properties of nano-coatings and paints. It involves applying a controlled load to the sample surface using a sharp indenter, typically a diamond tip, and measuring the resulting indentation depth. By analyzing the load-displacement curve, parameters such as hardness, elastic modulus, and indentation resistance can be determined. Nano-indentation offers high spatial resolution and can characterize mechanical properties at the nanoscale, making it suitable for evaluating thin films and nanostructured coatings. Scratch testing is employed to evaluate the adhesion strength and resistance to wear of nano-coatings and paints. In scratch testing, a sharp stylus or indenter is drawn across the coating surface under controlled conditions, such as load and speed. The critical load at which coating failure occurs, as well as the nature and extent of damage, provide insights into adhesion strength and coating durability. Scratch testing is valuable for assessing coating performance in real-world conditions and optimizing coating formulations for enhanced mechanical properties. Mechanical testing methods such as nano-indentation and scratch testing are essential for evaluating the mechanical properties and performance of nano-coatings and paints. By employing these techniques, researchers and engineers can gain valuable insights into coating hardness, elasticity, adhesion strength, and resistance to wear, facilitating the optimization of coating formulations and processes for specific industrial and technological applications.

8.4.4 Tribological testing methods

Tribological testing methods are essential for evaluating the friction, wear, and lubrication properties of nano-coatings and paints, providing crucial insights into their performance in real-world applications. This article presents a comprehensive overview of tribological testing techniques employed for analyzing nanostructured coatings and paints, encompassing tribometer testing and wear analysis. Each method's principles, capabilities, and applications in tribological characterization of nano-coatings and paints are discussed, emphasizing their significance in assessing coating friction, wear resistance, and lubrication effectiveness. By employing these techniques, researchers and engineers can gain valuable insights into the tribological behavior of nano-coatings and paints, facilitating their optimization for specific industrial and technological applications. Nano-coatings and paints often encounter tribological challenges in various applications, necessitating a thorough understanding of their tribological properties. Tribological testing methods offer quantitative measures of coating friction, wear resistance, and lubrication effectiveness, crucial for ensuring coating reliability and durability. By employing these techniques, researchers can tailor coating formulations and processes to enhance tribological performance in diverse applications. Tribometer testing serves as a fundamental approach to evaluating the friction and wear behavior of nano-coatings and paints. This method involves subjecting the coating surface to controlled sliding or rotational motions against a counterpart material under specific operating conditions, including load, speed, and lubrication. Tribometer testing provides crucial data such as coefficient of friction, wear rate, and surface damage, offering insights into coating performance under simulated operational environments. It plays a pivotal role in optimizing coating formulations and surface treatments to improve tribological properties. Wear analysis entails quantifying wear mechanisms and surface damage in nano-coatings and paints. Through techniques such as SEM and profilometry, worn surfaces are examined to identify wear mechanisms such as abrasive, adhesive, and fatigue wear. Wear analysis offers valuable insights into coating wear resistance, durability, and failure modes, enabling the development of wear-resistant coatings for demanding applications. Tribological testing methods, including tribometer testing and wear analysis, are indispensable for evaluating the friction, wear, and lubrication properties of nano-coatings and paints. By leveraging these techniques, researchers and engineers can gain critical insights into coating performance under realistic operating conditions, facilitating the optimization of coating formulations and processes for specific industrial and technological applications.

Changes in applied load have a significant impact on the tribological behavior of antifriction nano-paints. Higher loads can increase contact pressures between sliding surfaces, leading to elevated friction coefficients and wear rates. Conversely, lower loads may result in reduced friction and wear, as the contact area and pressure decrease. Understanding these load-dependent effects is crucial for optimizing the performance of nano-paints under different loading conditions.

Sliding speed also plays a critical role in determining the tribological behavior of antifriction nano-paints . Higher sliding speeds can generate more significant frictional heat and wear debris, potentially leading to increased friction coefficients and wear rates. Lower sliding speeds, on the other hand, may promote boundary lubrication and reduce friction and wear. Analyzing the effects of sliding speed on tribological behavior is essential for assessing the suitability of nano-paints for specific operating conditions.

8.5 Tribological behavior

Understanding the tribological behavior of antifriction nano-paints is essential for optimizing their performance in various industrial applications. This article provides a comprehensive overview of the tribological characteristics of antifriction nano-paints, including their frictional properties, wear resistance, and lubrication effectiveness. Tribological behavior is a critical aspect of their performance, influencing their effectiveness in mitigating frictional losses and enhancing mechanical efficiency.

8.5.1 Frictional properties

The frictional properties of antifriction nano-paints are influenced by various factors, including coating composition, surface roughness, and lubricating additives. These coatings typically exhibit low coefficients of friction due to their smooth surface morphology and lubricating properties. The presence of nanoparticles in the paint matrix reduces interfacial friction and adhesion, leading to improved tribological performance. Understanding the frictional behavior of antifriction nano-paints is crucial for predicting their performance in real-world applications and optimizing their formulation for specific tribological requirements.

8.5.2 Wear resistance

Antifriction nano-paints demonstrate enhanced wear resistance compared to conventional coatings due to their unique nanostructure and composition. Nanoparticles dispersed within the paint matrix act as reinforcing agents, strengthening the coating and reducing wear-induced damage. Additionally, the presence of lubricating additives or self-lubricating nanoparticles further improves wear resistance by reducing frictional contact and surface damage. Evaluating the wear resistance of antifriction nano-paints is essential for assessing their durability and longevity in demanding tribological environments. Antifriction nano-paints employ various mechanisms to reduce friction coefficients, including surface smoothening, boundary lubrication, and tribofilm formation. Nanoparticles in the paint matrix act as solid lubricants, filling surface asperities and reducing direct contact between mating surfaces. Additionally, nanoparticles can form tribofilms under sliding conditions, providing a protective layer that further reduces friction and wear. Surface roughness plays a crucial role in the friction reduction capabilities of antifriction nano-paints. By filling in surface irregularities and creating a smoother surface, nano-paints mitigate the effects of roughness-induced friction. Moreover, nanoparticles can adhere to surface defects and protrusions, minimizing contact stresses and reducing friction coefficients.

8.5.3 Lubrication effectiveness

The lubrication effectiveness of antifriction nano-paints plays a crucial role in minimizing friction and preventing wear in mechanical systems. These coatings form a lubricating film on the contact surfaces, reducing frictional forces and protecting against surface damage. Nanoparticles or additives with lubricating properties enhance the effectiveness of antifriction nano-paints, ensuring smooth and efficient operation of mechanical components. Investigating the lubrication effectiveness of these coatings is essential for optimizing their performance and extending the service life of machinery and equipment. The lubrication properties of antifriction nano-paints significantly influence their friction reduction capabilities. Nanoparticles with intrinsic lubricating properties, such as graphene or MoS_2, provide effective boundary lubrication, reducing friction coefficients under boundary and mixed lubrication regimes. Additionally, surface-modified nanoparticles can enhance the lubricating properties of the paint matrix, further reducing friction. The dispersion of nanoparticles within the paint matrix is critical for achieving uniform lubrication and friction reduction. Proper dispersion prevents agglomeration of nanoparticles, ensuring homogeneous coverage and effective lubrication. Moreover, the incorporation of synergistic additives, such as surfactants or dispersants, enhances nanoparticle dispersion and improves the overall tribological performance of antifriction nano-paints.

Antifriction nano-paints hold immense potential for enhancing wear resistance and extending the service life of coated surfaces in various applications. Through the utilization of nanoparticle reinforcement, surface modification techniques, and nanomaterial additives, these paints effectively mitigate wear-induced damage and improve the durability of coated surfaces. By understanding the mechanisms underlying wear resistance enhancement, researchers can develop advanced antifriction nano-paints with superior wear protection capabilities for diverse industrial and technological applications.

8.6 Effectiveness of antifriction nano-paints at different temperatures and environmental conditions

The effectiveness of antifriction nano-paints in mitigating friction and wear is crucially influenced by temperature variations and environmental conditions. This article investigates the performance of antifriction nano-paints across a range of temperatures and environments, exploring the mechanisms underlying their tribological behavior under diverse conditions. Through a comprehensive review of recent studies, we analyze the impact of temperature fluctuations, humidity, and chemical exposure on the tribological properties of nano-paints. Furthermore, we examine strategies to enhance the stability and efficacy of antifriction nano-paints in challenging operating environments. Understanding the effectiveness of antifriction nano-paints in different temperatures and environments is essential for their successful application in various industrial and technological settings. Antifriction nano-paints have emerged as promising solutions for reducing friction and wear in diverse applications. However, their performance can vary significantly depending on temperature fluctuations and environmental factors.

8.6.1 Temperature variations

Temperature fluctuations can significantly affect the tribological properties of antifriction nano-paints. At elevated temperatures, the viscosity of the paint matrix may decrease, leading to reduced lubrication effectiveness and increased friction coefficients. Conversely, at low temperatures, the paint matrix may become more viscous, affecting its ability to flow and form a protective lubricating film. Understanding these temperature-dependent effects is crucial for optimizing the performance of antifriction nano-paints across different temperature ranges.

8.6.2 Environmental conditions

Environmental conditions such as humidity and chemical exposure can also impact the effectiveness of antifriction nano-paints. High humidity levels can promote moisture absorption and swelling of the paint matrix, affecting its mechanical properties and tribological performance. Chemical exposure to corrosive substances may degrade the paint matrix or alter the surface chemistry, compromising its friction-reducing capabilities. Assessing the tribological behavior of nano-paints in different environments is essential for ensuring their reliability and performance in real-world applications. To enhance the stability and efficacy of antifriction nano-paints in challenging environments, several strategies can be employed. These may include incorporating additives or surface modifiers to improve resistance to temperature and humidity variations.

8.6.3 Stability under varying environmental conditions

To enhance the stability and efficacy of antifriction nano-paints in challenging environments, several strategies can be employed. These may include incorporating additives or surface modifiers to improve resistance to temperature and humidity variations. Additionally, selecting nanoparticles with inherent stability and corrosion resistance can enhance the durability of nano-paints in harsh environments. By implementing these strategies, researchers can develop robust antifriction nano-paints capable of maintaining their tribological performance across a wide range of temperatures and environments.

Additionally, selecting nanoparticles with inherent stability and corrosion resistance can enhance the durability of nano-paints in harsh environments. By implementing these strategies, researchers can develop robust antifriction nano-paints capable of maintaining their tribological performance across a wide range of temperatures and environments. The effectiveness of antifriction nano-paints is influenced by temperature variations and environmental conditions, which can significantly impact their tribological behavior. By understanding the mechanisms underlying their performance under different temperatures and environments, researchers can optimize the formulation and design of antifriction nano-paints for enhanced stability and efficacy. This knowledge is essential for their successful application in diverse industrial and technological settings, where reliable friction reduction and wear protection are paramount.

8.7 Applications

Antifriction nano-paints have gained significant attention across a wide range of industries due to their exceptional tribological properties and versatility. This article explores the diverse applications of antifriction nano-paints in the automotive industry, aerospace industry, manufacturing machinery, biomedical applications, and renewable energy systems. Through an extensive review of recent studies and technological advancements, we highlight the role of antifriction nano-paints in enhancing performance, reducing friction and wear, and extending the service life of components in these key sectors. By understanding the specific applications and benefits of antifriction nano-paints, researchers and engineers can leverage their potential to address challenges and drive innovation in various industrial and technological domains. Antifriction nano-paints represent a cutting-edge solution for improving the performance and durability of components across multiple industries. Their unique properties, including low friction, high wear resistance, and corrosion protection, make them highly desirable for a wide range of applications. This article provides insights into the diverse applications of antifriction nano-paints in key sectors, highlighting their transformative impact on automotive, aerospace, manufacturing, biomedical, and renewable energy technologies.

8.7.1 Automotive industry

In the automotive industry, antifriction nano-paints find applications in engine components, transmission systems, and chassis parts. By reducing friction and wear, these paints improve fuel efficiency, extend component lifespan, and enhance overall vehicle performance. Additionally, antifriction nano-paints can be applied to bearings, gears, and brake systems to minimize frictional losses and ensure smooth operation.

8.7.2 Aerospace industry

In the aerospace industry, where reliability and performance are paramount, antifriction nano-paints offer significant benefits. These paints can be used on aircraft engines, landing gear components, and airframe surfaces to reduce friction, prevent wear, and enhance fuel efficiency. By improving the tribological properties of critical aerospace components, antifriction nano-paints contribute to safer and more efficient flight operations.

8.7.3 Manufacturing machinery

Antifriction nano-paints play a crucial role in enhancing the performance and longevity of manufacturing machinery. They can be applied to cutting tools, molds, and machine components to reduce friction, minimize tool wear, and improve machining accuracy. By incorporating antifriction nano-paints into manufacturing processes, companies can achieve higher productivity, lower maintenance costs, and improved product quality.

8.7.4 Biomedical application

In the biomedical field, antifriction nano-paints are utilized in medical implants, prosthetics, and surgical instruments. These paints offer biocompatibility, corrosion resistance, and reduced friction, making them ideal for medical devices and implants. By applying antifriction nano-paints to biomedical surfaces, healthcare providers can enhance patient comfort, minimize tissue damage, and improve the longevity of medical implants.

8.7.5 Renewable energy

In renewable energy systems such as wind turbines and solar panels, antifriction nano-paints play a critical role in improving energy efficiency and reliability. These paints can be applied to turbine blades, bearings, and tracking systems to reduce frictional losses, increase power generation, and prolong equipment lifespan. By utilizing the benefits of antifriction nano-paints, renewable energy systems can achieve higher performance and lower maintenance requirements.

Antifriction nano-paints offer versatile solutions for improving performance, reducing friction and wear, and extending the service life of components across various industries. Their applications in the automotive, aerospace, manufacturing, biomedical, and renewable energy sectors demonstrate their transformative potential in addressing critical challenges and driving innovation. By leveraging the unique properties of antifriction nano-paints, industries can enhance efficiency, reliability, and sustainability in diverse industrial and technological applications.

8.8 Challenges and future perspectives of antifriction nano-paints: durability, scalability, environmental impact, and emerging trends

Antifriction nano-paints have emerged as promising solutions for reducing friction and wear in various applications, but they also face several challenges that must be addressed to realize their full potential. This article delves into the durability issues, scalability of fabrication techniques, environmental impact, and emerging trends in nano-paintings research. Through a comprehensive review of recent studies and technological developments, we highlight the key challenges facing antifriction nano-paints and discuss potential strategies for overcoming them. Additionally, we explore emerging trends in nano-painting research, including advances in materials science, fabrication techniques, and environmental sustainability. By addressing these challenges and embracing emerging trends, researchers and engineers can propel the field of antifriction nano-paints forward, unlocking new opportunities for innovation and application.

Antifriction nano-paints hold immense promise for enhancing the performance and durability of surfaces across various industries. However, several challenges persist, ranging from durability issues and scalability concerns to environmental impact. This article discusses the challenges and future perspectives of antifriction nano-paints, aiming to provide insights into potential solutions and emerging trends in the field.

8.8.1 Durability issues

One of the primary challenges facing antifriction nano-paints is ensuring long-term durability under harsh operating conditions. Over time, factors such as mechanical stress, temperature variations, and environmental exposure can degrade the performance of nano-paints, leading to reduced friction reduction and wear resistance. Addressing durability issues requires optimizing paint formulations, selecting robust nanoparticles, and developing effective surface bonding mechanisms to enhance adhesion and cohesion.

8.8.2 Scalability of fabrication techniques

Another challenge is scaling up fabrication techniques to meet industrial demands while maintaining consistent quality and performance. Many existing fabrication methods for antifriction nano-paints are suitable for small-scale production but may face challenges in scalability. Developing scalable fabrication techniques, such as continuous manufacturing processes or high-throughput synthesis methods, is essential for widespread adoption of antifriction nano-paints in industrial applications.

8.8.3 Environmental impact

The environmental impact of antifriction nano-paints is also a significant concern, particularly regarding nanoparticle toxicity and waste disposal. While nano-paints offer environmental benefits such as reduced energy consumption and extended equipment lifespan, their production and disposal processes may pose risks to human health and the environment. Addressing these concerns requires adopting green synthesis methods, minimizing nanoparticle release during application, and implementing responsible waste management practices. Fig. 8.9 indicates the impact of sustainable tribology on local and global level.

8.8.4 Emerging trends in nano-paints research

Despite these challenges, several emerging trends in nano-paint research offer opportunities for advancement. These include the development of multifunctional nano-paints with integrated properties such as self-healing, anticorrosion, and antimicrobial capabilities. Additionally, advancements in nanomaterials

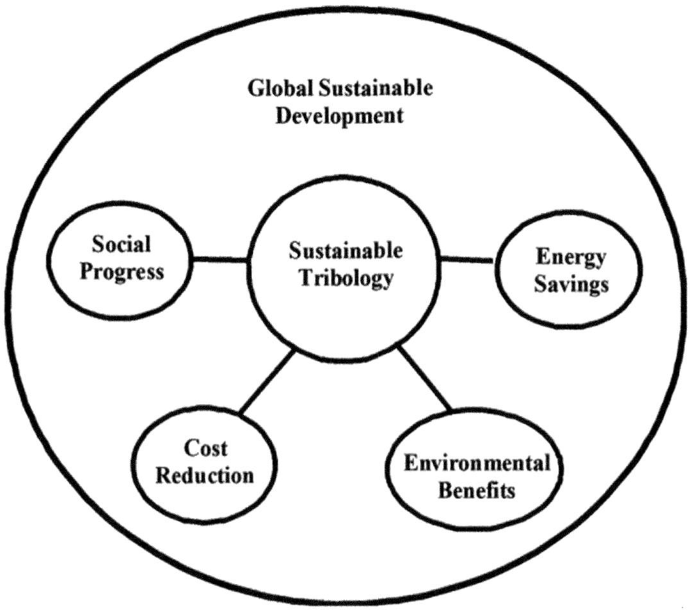

Figure 8.9 Sustainable tribology and its impact [53].

synthesis, characterization techniques, and computational modeling are driving innovation in nano-paint research, enabling the design of tailored coatings for specific applications.

8.9 Conclusion

In conclusion, the field of tribology and the development of antifriction nano-paints represent a significant advancement in surface protection technology. Through the exploration of various nanomaterials and fabrication techniques, researchers have made considerable progress in enhancing frictional properties, wear resistance, and lubrication effectiveness. Graphene-based nano-paints, CNT coatings, MoS_2 coatings, metal and metal oxide coatings, and diamond-like coatings have all demonstrated promising tribological behavior across a range of testing methods and environmental conditions.

The effectiveness of antifriction nano-paints has been observed in diverse industrial applications, including the automotive, aerospace, manufacturing machinery, biomedical, and renewable energy sectors. These coatings have the potential to improve efficiency, reliability, and sustainability in critical components and systems.

However, challenges remain in ensuring the durability, scalability, and environmental impact of antifriction nano-paints. Addressing issues related to durability, scalability of fabrication techniques, and environmental impact is crucial for the widespread adoption of these coatings. Future research efforts should focus on optimizing fabrication techniques, enhancing durability, mitigating environmental concerns, and exploring emerging trends to realize the full potential of antifriction nano-paints in advancing surface protection technology. With continued innovation and collaboration, antifriction nano-paints hold promise for revolutionizing tribological applications and driving progress in various industrial sectors.

References

[1] E.R. Sadiku, O. Agboola, I.D. Ibrahim, P.A. Olubambi, B. Avabaram, B. Manjula, et al., Nanotechnology in paints and coatings, Advanced Coating Materials (2018) 175–233. Available from: https://doi.org/10.1002/9781119407652.ch7.

[2] S. Mohan, A. Mohan, Wear, friction and prevention of tribo-surfaces by coatings/nanocoatings, in: Anti-Abrasive Nanocoatings, Woodhead Publishing, Elsevier, 2015, pp. 3–22.

[3] A. Raina, M.I.U. Haq, P.I. Victoria, S.R.J. Mohan, A. Anand (Eds.), Nanomaterials for Sustainable Tribology, Routledge, United Kingdom, 2023.

[4] R. Solano, D. Patino-Ruiz, A. Herrera, Preparation of modified paints with nano-structured additives and its potential applications, Nanomaterials and Nanotechnology 10 (2020). 1847980420909188.

[5] Z.N. Farhat, Wear resistant composite coatings, Materials Characterization 60 (4) (2009) 337–345.

[6] D.J. Branagan, W.D. Swank, D.C. Haggard, J.R. Fincke, Wear-resistant amorphous and nanocomposite steel coatings, Metallurgical and Materials Transactions A 32 (2001) 2615–2621.

[7] Y. Bautista, J. Gonzalez, J. Gilabert, M.J. Ibáñez, V. Sanz, Correlation between the wear resistance, and the scratch resistance, for nanocomposite coatings, Progress in Organic Coatings 70 (4) (2011) 178–185.

[8] P. Mittal, V. Acharya, S. Kumar, M. Paswan, K.K. Sadasivuni, P. Gupta, Synthesis and characterization of stir casted Cu-ZrO_2-graphite hybrid metal matrix composites for thermal management structures, Proceedings of the Institution of Mechanical Engineers, Part C: Journal of Mechanical Engineering Science 236 (14) (2022) 7927–7941.

[9] P. Mittal, M.K. Paswan, K.K. Sadasivuni, P. Gupta, Structural, wear and thermal behaviour of Cu–Al_2O_3–graphite hybrid metal matrix composites, Proceedings of the Institution of Mechanical Engineers, Part L: Journal of Materials: Design and Applications 234 (8) (2020) 1154–1164.

[10] A. Jamwal, P. Mittal, R. Agrawal, S. Gupta, D. Kumar, K.K. Sadasivuni, et al., Towards sustainable copper matrix composites: manufacturing routes with structural, mechanical, electrical and corrosion behavior, Journal of Composite Materials (2020). 0021998319900655.

[11] P. Mittal, M. Paswan, P. Gupta, Synthesis and characterization of CNT-TiO_2 reinforced copper matrix composites, Materials Today: Proceedings 67 (2022) 736–741.
[12] G. Kumar, A. Sharma, B. Sharma, P. Mittal, CNT-reinforced metal matrix composites: A review, in: Biennial International Conference on Future Learning Aspects of Mechanical Engineering, Springer Nature Singapore, Singapore, 2022, August, pp. 357–368.
[13] J. Mehta, P. Mittal, S. Mehta, Impact of coating blends and coating techniques on metal matrix composites, in: Metal Matrix Composites, CRC Press, United Kingdom, 2023, pp. 117–134.
[14] P. Mittal, J. Mehta, S. Mahto, S. Mehta, Copper matrix composites: synthesis and applications, in: Metal Matrix Composites, CRC Press, United Kingdom, 2023, pp. 47–62.
[15] P. Mittal, V. Raghav, D. Chawla, J. Mehta, M.K. Paswan, P. Gupta, Structural, wear and thermal behavior of copper metal matrix composites: a review, in: B. P. Sharma, G.S. Rao, S. Gupta, P. Gupta, A. Prasad (Eds.), Advances in Engineering Materials. Lecture Notes in Mechanical Engineering, Springer, Singapore, 2021. Available from: https://doi.org/10.1007/978-981-33-6029-7_30.
[16] P. Mittal, et al., Challenges and opportunities in synthesis of hybrid Cu-Al_2O_3-C and Cu-ZrO_2-C composites through stir casting route, in: B.P. Sharma, G.S. Rao, S. Gupta, P. Gupta, A. Prasad (Eds.), Advances in Engineering Materials. Lecture Notes in Mechanical Engineering, Springer, Singapore, 2021. Available from: https://doi.org/10.1007/978-981-33-6029-7_1.
[17] O. Penkov, H.J. Kim, H.J. Kim, D.E. Kim, Tribology of graphene: a review, International Journal of Precision Engineering and Manufacturing 15 (2014) 577–585.
[18] T. Filleter, R. Bennewitz, Structural and frictional properties of graphene films on SiC (0001) studied by atomic force microscopy, Physical Review B 81 (15) (2010) 155412.
[19] Q. Li, C. Lee, R.W. Carpick, J. Hone, Substrate effect on thickness-dependent friction on graphene, Physica Status Solidi (B) 247 (11-12) (2010) 2909–2914.
[20] C. Lee, Q. Li, W. Kalb, X.Z. Liu, H. Berger, R.W. Carpick, et al., Frictional characteristics of atomically thin sheets, Science 328 (5974) (2010) 76–80.
[21] G. Hatipoglu, M. Kartal, M. Uysal, T. Cetinkaya, H. Akbulut, The effect of sliding speed on the wear behavior of pulse electro Co-deposited Ni/MWCNT nanocomposite coatings, Tribology International 98 (2016) 59–73.
[22] N. Satyanarayana, K.S. Rajan, S.K. Sinha, L. Shen, Carbon nanotube reinforced polyimide thin-film for high wear durability, Tribology Letters 27 (2007) 181–188.
[23] B.H. Ryu, A.J. Barthel, H.J. Kim, H.D. Lee, O.V. Penkov, S.H. Kim, et al., Tribological properties of carbon nanotube–polyethylene oxide composite coatings, Composites Science and Technology 101 (2014) 102–109.
[24] H.D. Lee, O.V. Penkov, D.E. Kim, Tribological behavior of dual-layer electroless-plated Ag–carbon nanotube coatings, Thin Solid Films 534 (2013) 410–416.
[25] S. Arai, T. Kanazawa, Electroless deposition and evaluation of Cu/multiwalled carbon nanotube composite films on acrylonitrile butadiene styrene resin, Surface and Coatings Technology 254 (2014) 224–229.
[26] A.K. Keshri, V. Singh, J. Huang, S. Seal, W. Choi, A. Agarwal, Intermediate temperature tribological behavior of carbon nanotube reinforced plasma sprayed aluminum oxide coating, Surface and Coatings Technology 204 (11) (2010) 1847–1855.

[27] M.A. Samad, S.K. Sinha, Mechanical, thermal and tribological characterization of a UHMWPE film reinforced with carbon nanotubes coated on steel, Tribology International 44 (12) (2011) 1932–1941.
[28] X. Li, Y. Zhou, J. Sun, H. Zhang, W. Wu, Tribological behavior of the electroless Ni-P-CNTs-SiC (nanometer) composite coating, Xiyou Jinshu Cailiao yu Gongcheng (Rare Metal Materials and Engineering) 36 (2007) 712–714.
[29] S. Banday, M.F. Wani, Nanoscratch resistance and nanotribological performance of Ti/MoS_2 coating on Al-Si alloy deposited by pulse laser deposition technique, Journal of Tribology 141 (2) (2019) 022003.
[30] O. Elianov, S. Garusi, R. Rosentsveig, S.R. Cohen, Y. Feldman, I. Pinkas, et al., Deposition of metal coatings containing fullerene-like MoS_2 nanoparticles with reduced friction and wear, Surface and Coatings Technology 353 (2018) 116–125.
[31] Y.J.J. Jason, H.G. How, Y.H. Teoh, H.G. Chuah, A study on the tribological performance of nanolubricants, Processes 8 (11) (2020) 1372.
[32] J.C. Walker, S.R. Saranu, A.H. Kean, R.J.K. Wood, Fe nano-particle coatings for high temperature wear resistance, Wear 271 (9-10) (2011) 2067–2079.
[33] E. Rabinowicz, Lubrication of metal surfaces by oxide films, ASLE Transactions 10 (4) (1967) 400–407.
[34] I.A. Inman, P.K. Datta, H.L. Du, J.S. Burnell-Gray, S. Pierzgalski, Q. Luo, Studies of high temperature sliding wear of metallic dissimilar interfaces, Tribology International 38 (9) (2005) 812–823.
[35] F.H. Stott, J. Glascott, G.C. Wood, Factors affecting the progressive development of wear-protective oxides on iron-base alloys during sliding at elevated temperatures, Wear 97 (1) (1984) 93–106.
[36] F.H. Stott, D.S. Lin, G.C. Wood, The structure and mechanism of formation of the 'glaze'oxide layers produced on nickel-based alloys during wear at high temperatures, Corrosion Science 13 (6) (1973) 449–469.
[37] F.H. Stott, G.C. Wood, The influence of oxides on the friction and wear of alloys, Tribology International 11 (4) (1978) 211–218.
[38] J. Jiang, F.H. Stott, M.M. Stack, The role of triboparticulates in dry sliding wear, Tribology International 31 (5) (1998) 245–256.
[39] F.H. Stott, High-temperature sliding wear of metals, Tribology International 35 (8) (2002) 489–495.
[40] J. Jiang, F.H. Stott, M.M. Stack, A generic model for dry sliding wear of metals at elevated temperatures, Wear 256 (9-10) (2004) 973–985.
[41] A.A. Ala'A, O. Eryilmaz, A. Erdemir, S.H. Kim, Nano-texture for a wear-resistant and near-frictionless diamond-like carbon, Carbon 73 (2014) 403–412.
[42] A. Hatem, J. Lin, R. Wei, R.D. Torres, C. Laurindo, P. Soares, Tribocorrosion behavior of DLC-coated Ti-6Al-4V alloy deposited by PIID and PEMS + PIID techniques for biomedical applications, Surface and Coatings Technology 332 (2017) 223–232.
[43] J. Wang, J. Ma, W. Huang, L. Wang, H. He, C. Liu, The investigation of the structures and tribological properties of F-DLC coatings deposited on Ti-6Al-4V alloys, Surface and Coatings Technology 316 (2017) 22–29.
[44] K.L. Choy, Chemical vapour deposition of coatings, Progress in Materials Science 48 (2) (2003) 57–170.
[45] G. Fox-Rabinovich, J.M. Paiva, I. Gershman, M. Aramesh, D. Cavelli, K. Yamamoto, et al., Control of self-organized criticality through adaptive behavior of nano-structured thin film coatings, Entropy 18 (8) (2016) 290.
[46] H. Hoche, S. Groß, M. Oechsner, Development of new PVD coatings for magnesium alloys with improved corrosion properties, Surface and Coatings Technology 259 (2014) 102–108.

[47] H. Korhonen, A. Syväluoto, J.T. Leskinen, R. Lappalainen, Optically transparent and durable Al_2O_3 coatings for harsh environments by ultra short pulsed laser deposition, Optics & Laser Technology 98 (2018) 373–384.
[48] A. Inspektor, P.A. Salvador, Architecture of PVD coatings for metalcutting applications: a review, Surface and Coatings Technology 257 (2014) 138–153.
[49] G. Skordaris, K.D. Bouzakis, T. Kotsanis, P. Charalampous, E. Bouzakis, O. Lemmer, et al., Film thickness effect on mechanical properties and milling performance of nano-structured multilayer PVD coated tools, Surface and Coatings Technology 307 (2016) 452–460.
[50] A. Baptista, F.J.G. Silva, J. Porteiro, J.L. Míguez, G. Pinto, L. Fernandes, On the physical vapour deposition (PVD): evolution of magnetron sputtering processes for industrial applications, Procedia Manufacturing 17 (2018) 746–757.
[51] D. Navas, S. Fuentes, A. Castro-Alvarez, E. Chavez-Angel, Review on sol-gel synthesis of perovskite and oxide nanomaterials, Gels 7 (4) (2021) 275.
[52] R.W. Johnson, A. Hultqvist, S.F. Bent, A brief review of atomic layer deposition: from fundamentals to applications, Materials Today 17 (5) (2014) 236–246.
[53] I. Tzanakis, M. Hadfield, B. Thomas, S.M. Noya, I. Henshaw, S. Austen, Future perspectives on sustainable tribology, Renewable and Sustainable Energy Reviews 16 (6) (2012) 4126–4140.

9

Conducting and thermal barrier nanopaints

Priti[1] **and Radha Sachan**[2]

[1]*Department of Paint Technology, Mahamaya Polytechnic of Information Technology, Chail, Kaushambi, Uttar Pradesh, India* [2]*Department of Paint Technology, School of Chemical Technology, Harcourt Butler Technical University, Kanpur, Uttar Pradesh, India*

9.1 Introduction

A paint is defined as a coating that can be applied to the substrate in liquid or solid solution, which adheres and solidifies as it spreads across the substrate, forming a protective, esthetic film or adds a special feature to the substrate on which it is applied [1]. To understand nano applications in coatings, it is important to note that the coating material is the nano, while the coating is not. This is widely used. A nanoparticle is defined as a particle with a diameter between 1 and 100 nanometers (nm). Like conventional paints, nanopaint is a coating that allows users to customize a substrate property according to their needs. Nanopaint represents a breakthrough in materials science, using the unique properties of nanotechnology to enhance conventional paints. The addition of nanoparticle paints provides advanced functionalities and applications beyond paint a commonly used on. Nanopaint demonstrates how nanotechnology can transform traditional materials, providing numerous advantages that enhance the overall properties and performance of the coating. As research and development advances, the applications and effectiveness of nanopaint will expand, making it an important component in a variety of industries. Nanopaint, which utilizes the potential of nanoparticles, is at the leading edge of new materials invented in coating technology [2].

Among many important characteristics of nano paints here we are discussing mainly thermal barrier and thermal conductivity properties of nanopaints.

Nanopaints. DOI: https://doi.org/10.1016/B978-0-443-26512-9.00009-4

Thermal barrier coatings (TBCs) are designed to reduce damage due to high temperatures. By incorporating nanoscale materials, nanopaints can significantly improve the thermal insulation properties, which is important for applications in a variety of high-temperature environments. The use of thermal coatings provides increased heat transfer between the surface of coating and the substrate thus increasing the range of working temperature and performance [3].

Therefore a TBC is required to increase the operating temperature of the power generation. The thermal coating is usually a steel coating usually a superalloy with an intermediate metallic bond coat used to reduce the thermal expansion between the substrate. The topcoat is a ceramic coating applied plasma spraying technique or electron beam physical vapor deposition (EBPVD) is used. The deposition method effects the properties and performances of the TBCs. Nanostructured TBCs have shown promising ability to enhance thermal insulation and prolong thermal cycle life compared to conventional TBCs causing TBCs to spall or de-bond upon heat transfer under high cycle conditions. Thermal shock behavior is one of important performance indicator of TBCs [4,5].

Nanopaint reinforced with nanoparticles can exhibit unique thermal conductivity tailored for specific applications. While many nanopaints have been designed to provide thermal insulation, there are formulations aimed at improving heat transfer in areas where efficient heat dissipation is necessary. However, the thermal importance of polymers is often overlooked in the design of thermal conductivity nanopaints because they tend to be very low compared to fillers. Nanofiller type, loading rate, nanofiller size, and nanofiller shape have significant effects on the thermal conductivity of nanopaint. In addition, the spatial arrangement and orientation of nanofillers are also important for heat transfer. Many types of fillers have nonspherical (plate-like or needle-like) shapes. They can be oriented during processing, resulting in composites with anisotropic thermal conductivity (e.g., if plate-like fillers are tightly aligned, the "in-plane" thermal conductivity will be higher than the normal or "through-plane" thermal conductivity). The interaction between nanofiller and polymer matrix is also very important for thermal conductivity. A poor interface between them could result in high thermal resistance at the interface and reduced thermal conductivity. Enhancing the interfacial interactions between polymer and filler can boost nanopaint thermal conductivity [6].

The objective of this chapter is to present an overview regarding properties and performance of thermal conductive and TBCs.

9.2 Thermal barrier coatings

TBC can be manufactured by reducing the heat transfer to the coating, increasing the temperature of the coating, resulting an increase in the overall system efficiency. The materials used to design TBCs degrade slowly at higher temperature exposure. Porosity is commonly used in TBCs applications to reduce heat transfer and enhance the thermal insulation. Porosity is important in the control of thermal processes, as it depends on density. The size and location of the porosity can vary depending on the method parameters. Two types of pores were observed: micropores (1–10 μm) and nanopores (10–30 nm). Young's modulus decreases with increasing porosity of the coating. Thus controlled porosity is always necessary to enhance the mechanical properties of the coating. When the porosity exceeds a certain value, the mechanical properties deteriorate due to the decrease in cohesion [7]. Like size and location, the orientation and size of pores and cracks influence the thermomechanical properties of these coatings. Increased cracks which are mainly formed horizontally and perforations can improve thermal resistance and reduce heat transfer. For example, EBPVD created top coats with more cracking, whereas atmospheric plasma spraying (APS) created coatings which contain microcracks and horizontally moving pores. Both were found to be highly effective in reducing overall thermal conductivity [8]. Nano-submicron pores affected the thermal properties more than large pores, and increasing the number of nanopores is a right way to enhance coating thermal barrier properties.

9.2.1 Design of thermal barrier coatings

Fig. 9.1 shows the basic TBC design contains several layers. The first layer is a bond coat, the second layer is a TGO layer, and the final ceramic topcoat layer. All three layers are deposited on a Ni-based superalloy substrate.

Let us add further, in TBCs, there is primer coat knowns as the bond coat always act as an interfacial layer between the substrate and the subsequent coats resulted increase in the bond of complete system of TBCs on the substrate. The topcoat

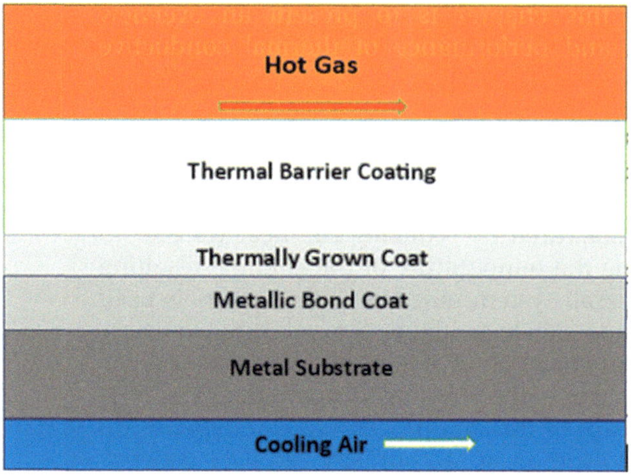

Figure 9.1 Basic design of thermal barrier coating system.

of the TBC system is made of ceramic materials because they act as a thermal barrier to the substrate [9].

9.2.1.1 Topcoat thermal barrier materials

There are many properties of thermal barrier materials such as thermal resistance, thermal expansion during heat cycling, phase-stability and, strong adhesive adhesion to the substrate, and protection of the substrate from oxidation and corrosion [7].

Since the barrier coating has excellent sintering properties these coatings withstand prolonged use at temperatures above 1200°C, leading to thermal conductivity degradation [8].

9.2.1.2 Bond Coat Materials

Depending on the application method, two kinds of bond coats are mainly used in basic TBC system applications: diffusion and overlay. Diffusion coatings consist a sandwich layer of a diffuse intermetallic materials between the substrate and the bond coat, which acts as a boundary for diffusion of oxygen [6]. Overlay coatings, such as MCrAlX bond coats, in which "M" stands for the base material, which is usually blended with Ni, Fe or Co which provides a bond coat system of higher melting point than typical diffusion coatings [10] and "X" indicates Y or Zr with a high reactivity to oxygen, resulting in an oxide layer. This ceramic coating with grown aluminum oxide improves thermal resistance and thermal compatibility.

However, Al in this coating mainly acts as a protective oxide against oxidation, and the addition of Cr enhances the effective chemical effect of Al. Similarly, the M is intended to improve compatibility with the substrate material [11].

9.2.1.3 Dual ceramic layer coatings

The introduction of dual ceramic layer coating is one way to enhance TBC performance by introducing a layer of zirconate to conventional Yttria-stabilized zirconia (YSZ) topcoats.

The incorporation of these materials into ceramic coatings has been proposed to improve thermal stability as well as oxidation resistance. The overall thermal conductivity decreases and the increased phase-stability and found mismatch in thermal expansion can reduce the rate of oxidation and prevent layer-breaking in TBCs [12].

The thermal barrier and mechanical properties of TBC system can be improved by using new fabrication techniques that allow to control the pore size.

9.2.1.4 Functionally graded thermal barrier coating materials

The use of functionally graded (FGMs) is one of the most interesting TBC innovations [10]. As shown in Fig. 9.2 a functionally graded TBC which contains a gradient in interface

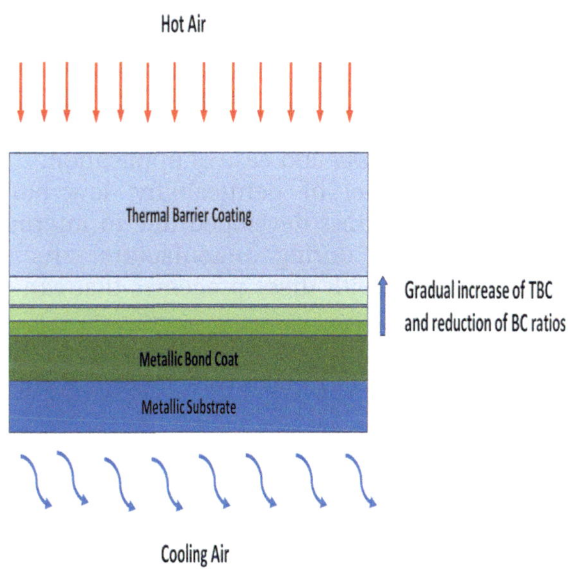

Figure 9.2 Diagram of FGC system. *TBC*, Thermal barrier coating.

configuration between topcoat and bond coat. In heat-resistant applications, the main objective of FGM is to reduce developed stresses due to temperature as well as cracks at the interfaces of metal and ceramic, However, they also demonstrate tremendous potential in a variety of industries [12]. Functionally graded coatings (FGC) is an intermetallic layer of bond coat that separates the metal substrate layer and the topcoat layer of ceramic materials of the TBC.

Through graded layers, heat transfer and oxygen diffusion can be spatially controlled. A method for reducing heat transfer through a bond coat and substrates by improving thermal resistance, reducing residual-thermal stress, and slowing the growth rate of oxide layer. Moreover, the residual stress in conventional TBC was found to be compressive and constant, while that in standard TBC it is fluctuated with time and temperature [13].

Furthermore, TBCs experience various harms in the harsh environment, which need to be investigated. Tailoring the bond coat material and FGC porosity, as well as controlling the desired microstructure, can result in thermally compatible topcoat layers for various metallic substrates.

9.2.2 Thermal barrier coating fabrication techniques

Several coating methods have been used, including Ni-dispersion coating, diffusion coating, electric arc wire spray coating, thermal spray, magnetron sputtering systems and physical vapor deposition (PVD) [14]. Plasma spray techniques have grown in popularity for ceramic coating production due to their versatility and ease of application.

High level of permeability, low bond strength with the substrate, stress developed due to internal stress and oxidation occurrence during manufacture are some of the issues associated with these processes that can shorten the lifetime of TBC [15].

Methods a commonly used for TBC manufacturing are Electron Beam PVD as well as plasma spray deposition [16].

9.2.2.1 Electron-beam physical vapor deposition

EBPVD bombards the target anode with a high-energy electron beam filled with tungsten wires in a elevated-temperature atmosphere [17]. A beam of electrons deflects several atoms from their desired surface in the gas phase. These high-energy atoms eventually produce a thin, dense layer of anodic material, filling the entire vacancy. It is also called as an "Electron Evaporation System" because the incoming electron beam plays a significant role in evaporating the coating and depositing it on the substrates. Coating thickness can vary from 1 nm to a few microns, and can provide thermal and optical properties which has improved substrate production [11].

9.2.2.2 Plasma spray deposition

Plasma sprays are arranged in controlled areas to minimize the adverse effects of volatile flying particles in contact with the environment, such as oxidation and unwanted contamination [14]. They are further divided into three main categories: low-pressure/vacuum plasma spray, suspension/dissolution precursor APS, and suspension/dissolution precursor plasma spray. Recently on TBC topcoat deposition new thermal spray techniques have been developed so like suspension plasma spraying, plasma spray-PVD is extensively studied [18].

High-quality thermal sprays are made with low and very low pressures. The pressures used vary, with low-pressure plasma spraying typically ranging from 4 thousand to 40 thousand Pa and very low-pressure plasma spraying down to 100 Pa. When applying pressure is even less than 100 Pa then this process is known as "Vacuum Plasma Spray". Although a popular deposition method, plasma spraying produces a ceramic coating with microporous pores, which is desirable to reduce heat transfer [15].

9.2.2.3 Electrophoretic deposition technique

Recently, These processes are used to fabricate coatings for metal and ceramic substrate, and metal oxides and oxides are used to coat cathodes that control aqueous metal salts for electrical storage [19]. Electrolytic deposition deposits an oxide layer using the target metal's soluble salts, whereas electrolytic deposition deposits solid metal oxide particles suspended in a solution of electrolyte [20]. Electric mold covers are a more modern method of coating. The electrophoresis method is a process for producing effective coatings at the low-temperature which have enhanced sintering ability and uniform microstructure [21].

9.2.3 Properties and performances

Thermal barrier coatings have various performances as mentioned below:

9.2.3.1 Physical properties

9.2.3.1.1 Density and porosity

Pores are gaps between particles that are deposited on the substrate at high speeds during the TBC spraying process. Porosity is an indicator of coating compactness. Porosity is a critical parameter in TBCs, influencing both thermal conductivity and strain tolerance. Closed porosity, like open porosity, cannot be calculated directly; it needs to be calculated using a theoretical density value [22].

Archimedes method The Archimedes method is used to weigh TBC samples and estimate coating volume. The weight of the TBC coated sample (W_{TBC}) was determined. Using the Archimedes technique, the volume of the coated sample was measured by the volume of the displaced water (V_{TBC}). However, during the test the TBC coating needs to be sealed. Coating porosity can be determined as written below in (9.2) [23]:

$$\text{Density of the coated sample: } \rho_{TBC} = W_{TBC}/V_{TBC} \tag{9.1}$$

$$\%\text{Porosity} = \left(1 - \rho_{TBC}/\rho_0\right) \times 100 \tag{9.2}$$

where $\rho_0 = 5.73\ \text{g.cm}^{-3}$

Microscopic method The thickness of the TBC coating was estimated by taking the mean of the thickness at 10 different equidistant places on a photomicrograph. The desired volume of coating (V) was calculated with the help of measured the average thickness as well as surface area. The mass of the coating (W) was calculated as the increase in mass of the sample (substrate and bond coat) following TBC deposition [24]. The density was calculated as:

$$\rho = W/V \tag{9.3}$$

The percent porosity (P) is calculated as below:

$$\%P = \left(1 - \rho/\rho_0\right) \times 100 \tag{9.4}$$

where ρ_0 is the theoretical density: 5.73 gms cm^{-3}.

Point count method The porosity is measured using optical micrographs. A 6×6 transparent glass grid is kept over the photomicrograph, and a pore is defined as a juncture of a grid location and a void. The porosity is determined by the fraction of grid intersections that overlap with voids. The final porosity was calculated as an average value of five determinations [23].

Eddy current method By this method only thickness of TBC is measured and rest porosity is calculated as for microscopic method or capacitance techniques. It is possible to combine either of these techniques with eddy current in a single instrument to assess TBC thickness and porosity uniformity [25].

Water immersion method Dry coated specimens are weighed (suppose it is W_d) and next immersed for 15 minutes in boiling water. To cool down the samples they are left in the normal water at the ambient temperature. The samples are then taken out from the water, the surface of the samples is quickly cleaned off and the samples are instantly weighed (suppose it W_w) [23]. The mass water which is absorbed is calculated as the difference of weight of the "wet sample" and "dry sample" that is $W_w - W_d$.

The volume water which was absorbed is calculated with the mass of the water (W_w–W_d) and its density data (assumed it is 1.00 g cm^{-3}).

$$V_w = (W_w - W_d)/1 = (W_w - W_d) \quad (9.5)$$

The volume of dry coating was calculated from the mass of the dry coating (W_d) and the theoretical density (ρ_o) as below:

$$V = W_d/\rho_o \quad (9.6)$$

Porosity is then calculated as

$$\%P = \left(V_w/V_w + V\right) \times 100 \quad (9.7)$$

9.2.3.1.2 Phase and lattice structure analysis

Thermally grown oxide (TGO) of a specific thickness form on the substrate after being exposed to high temperatures. X-ray diffraction (XRD) and Raman analysis revealed that the main compound in TGO is Al_2O_3. The characteristic peak of alumina appears in the XRD spectrum. TGO always shows a dense structure and the grains confirm the uniform size distribution [26].

9.2.3.2 Mechanical properties

TGO which is sandwich layer of ceramic coat and bond coat. Characterizing the mechanical properties of the TGO layer is critical to simulate stress and predicting TBC failures. Nanoindentation and tensile methods are used to determine the hardness, elastic modulus, and fracture strength of TGO. A good amount of TGO was deposited on the substrate surface at high-temperature conditions and then released from residual stress and annealed under vacuum at a temperature as high as about 1000° C for as long as 9–10 hours beforehand high-temperature oxidation is making it a more uniform alloy. Before oxidation, samples must be polished and cleaned manually.

9.2.3.2.1 Elastic moduli and hardness

Nanoindentation test Because this test is delicate to roughness of the surface, all the oxidized samples must be carefully polished with an alumina suspension. The hardness and elastic modulus are measured using a displacement-based nanoindentation apparatus [19,20]. Prior to the test, the system must be calibrated against a standard sample of fused silica. Following calibration, the prepared TGO sample should be tested with a series of loads while bulk alumina is tested with a list of loads. To prevent interference, the distance between each indentation point should be at least 10 μm. To ensure accuracy, every load was run a minimum of five times (Fig. 9.3).

This test is carried out using the digital image correlation (DIC) system, as shown in the figure. The sample is clamped, and a central Composite Design (CCD) camera is positioned in opposite of its surface to capture the deformation process in real time. The displacement and fatigue areas of the TGO surface beneath tensile stress can be obtained using the DIC method by observing the change in position of speckle patterns on the analyzed surface, and the same can then be used to expect the commencement and proliferation of the crack [21,22]. The speckle patterns are designed by inorganic glue and

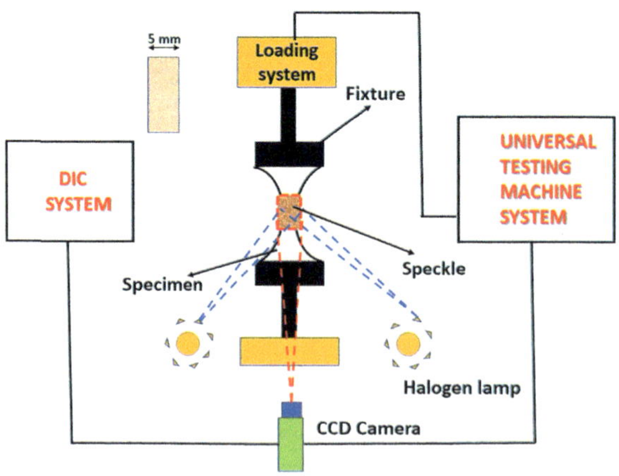

Figure 9.3 Diagram of tensile test.

cobalt oxide (CoO) powder, which are used to apply on the TGO surface using an airbrush gun.

The sample is then kept in a high-temperature furnace like muffle furnace to get dried for 1 hour at 40°C. During the tensile test, the load is applied at a specific rate, and the CCD camera captures an image in every second.

As Kick model says [23], it is well-known in the indentation experiment, the applied load (P) is related with proportional to the square of the indentation depth (h^2), as written below:

$$P \alpha h^2 \tag{9.8}$$

We can say that the indentation depth increases with increase in load; however, the hardness and elastic modulus values reduce with increase in indentation depth. Gao and Nix [16] reported that the indentation hardness is altered with indentation depth which can be calculated using the following formula:

$$H/H_0 = \sqrt{1 + \frac{h^*}{h}} \tag{9.9}$$

where, H denotes the hardness at a specific indentation depth, H_0 represents the hardness at infinite indentation depth, and h^* is a typical length parameter associated with properties of the material. The experiment yields the values of both H and h whereas H_0 and h^* can be obtained by curve fitting with $(H/H_0)^2$ on the y axis and 1/h on the x axis. Curve fitting the results of the indentation experiment with Formula (9.2) yields a linear relationship. The slope and length of intercept of the line can be calculated directly from the linear curve.

When the indentation method is employed for determining the elastic modulus of materials, the results are inaccurate; however, the experimental instrument must be considered.

We used a linear fitting approach similar to that used to analyze hardness. Furthermore, elastic modulus and indentation depth are related as mentioned by following Formula (9.3).

$$\left(\frac{E}{E_0}\right)^4 = 1 + \frac{h^*}{h} \tag{9.10}$$

where,

E: The elastic modulus at a particular indentation depth,

E_0: The elastic modulus at infinite indentation depth

Upon curve fitting between $\left(\frac{E}{E_0}\right)^4$ and $\frac{1}{h}$ a linear relationship is obtained from which with the help of slope and interception the value of intrinsic modulus of elasticity (E_0) can be calculated.

9.2.3.2.2 Bond strength

Although TBCs are manufactured without any unwanted defects, measuring their bond strength values is critical to validating the microstructural results. The coating and respective substrate are bonded to loading fixtures using a good quality of glue-like epoxy based to prevent impregnation of the ceramic coating and influence on the adhesion result, the full assembly is then subjected to a tensile load that is perpendicular to the plane of the coating. The epoxy glue to assemble the testing set-up. A testing machine is used with certain dynamic load and run at a constant speed [27].

9.2.3.2.3 Erosion test

Before performing the erosion tests, all the test samples are weighted with a sensitive weighing machine to a minimum count of 0.1 mg. The erosion tests are carried out using an air-jet erosion tester on TBC samples at room temperature. The erodent material, such as alumina sand, impacts the samples at two angles 90° and 15°. Fig. 9.4 shows a schematic diagram of the erosion rig. The erodent is fed at a certain rate in g/min with a specified erosion velocity in min/s for a set period, typically seconds. The test parameters are chosen to allow comparisons between the various TBCs. The eroded specimens are ultrasonically cleaned with compressed air to remove loose erosion debris that could affect weight loss. Each 90° impacted sample undergo five independent erosion tests in different areas under identical conditions, yielding five indents. Each sample impacted at 15° underwent two erosion tests, yielding two indents. Weight loss is determined after each independent

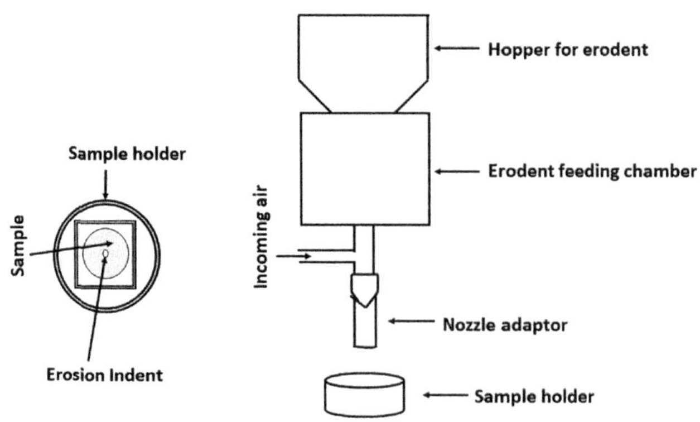

Figure 9.4 Basic diagram of the erosion test.

erosion test. All erosion tests and weight loss measurements are performed by Siemens Energy [28].

9.2.3.2.4 Residual stress test

During high-temperature oxidation, alumina forms on the TBC-treated substrate. The variation in thermomechanical properties amongst TGO and substrate generates residual stress in TBC. The residual stress has a substantial impact on the tensile fracture strength of TBC. The residual stress that occurs in the coating is determined by the procedure used to form it. The residual pattern shifts from compressive to tensile as the coating process and temperature change. Additionally, to this change, residual stress can vary with thickness [29].

In general, the residual stress near the substrate is higher than near the surface of the coating.

9.2.3.3 Thermal properties

9.2.3.3.1 Sintering and thermal aging analysis

The mismatch found in calculated in-plane stress at the TGO/TBC interface layer is always related to the TBC modulus, as shown in Formula (9.4). Because of sintering, the TBC modulus parameter increases with an increase in in-plane stress. At equilibrium, the in-plane and out-of-plane pressure gradients are balanced at the interface. Therefore the radial strain at the interface of TGO/TBC is also likely to growth with increasing TBC.

$$\sigma_{\Delta} = \Delta T \times \Delta CTE \times \frac{E_{TBC}}{(1 - \nu_{TBC})} \tag{9.11}$$

To achieve the desired coating thickness, TBC requires multiple spray passes. Thus there is porosity between the consecutive TBCs, which can lead to poor binding. Furthermore, many micro-cracks are discovered within individual layer of TBC [13]. The first kind are referred to as intersplat defects, and the latter as intra-splat defects. Due to these flaws, fresh TBC [14] has a significantly lower modulus than dense zirconia.

When TBC is exposed to preeminent temperatures for a sufficiently long time, the intersplat defects are eliminated, thereby reducing TBC porosity. This reduction in porosity degrades the protection of the coating when Young the coating modulus rises, the stress tolerance decreases, and subsequent spallation of the coating can occur.

The TBC in the substrate is subjected to high temperatures, causing the rate of sintering or parameter growth to vary in thickness. The TBC modulus is affected by the stress state due to the thermal expansion difference between the TBC and the substrate [30].

9.2.3.3.2 Thermal cycling and thermal shock test

Thermal cycling tests, are carried out using two different methods.

Furnace test During this test, samples are heated in a tube furnace to a maximum temperature of thousands of degrees Celsius. After a few minutes in the furnace at that temperature, they are cooled with compressed air. The cooling time for air quenching is always shorter than that for heating [31].

Gas burner test Burner test facilities that use propane and oxygen. To provide a temperature gradient, the substrates are cooled from the back using compressed air. The surface temperature is measured using a pyrometer. A thermocouple mounted in a hole drilled from one side into the substrate's center measures its temperature. The surface temperature is adjusted between two temperatures, while the substrate temperature is always kept below the surface temperature. In the test facilities, gas burners with a broad flame are used to ensure a uniform temperature distribution in the center of the samples. After a few seconds of heating, the maximum temperature is nearly reached [32].

After that, the burner is automatically removed from the surface for a set period, and compressed air is used to cool the surface from the front. The cycles are repeated for a few minutes of heating followed by a few minutes of cooling, until failure occurs. Cycling is stopped when a visible rupture of the coating occurs.

9.2.3.3.3 Hot salt corrosion analysis

The hot salt corrosion test for TBCs simulates and analyzes the degradation behavior of these coatings when exposed to high temperatures and salt contaminants. This test is especially important for components in gas turbines, jet engines, and other high-temperature environments where salt particles (such as sodium sulfate) can accumulate on the surface, causing accelerated corrosion and degradation of the TBC.

The hot salt corrosion test is designed to determine how well TBCs withstand corrosive environments at high temperatures, which aids in assessing their durability, performance, and durability in real-world applications. During the tests, TBC test samples are prepared. To ensure uniform thickness, a specific amount of salt, typically sodium sulfate, is sprayed, dipped, or applied to the surface of TBC test samples. The coated samples are then exposed to high temperatures in a furnace. The temperatures typically range from 800°C–1100°C, simulating the operating conditions of turbines and engines. The samples are kept at these high temperatures for a predetermined time, which can range from a few hours to several hundred hours, depending on the test requirements. Thermal cycling is used in some tests to simulate the real thermal cycles encountered in service by repeatedly heating and cooling the samples.

Following the test, the samples are examined for corrosion damage. Visual inspection, microscopic examination, chemical analysis, and mechanical testing are all options for determining the extent of corrosion, flake off, and other forms of degradation [33].

9.2.3.3.4 Calcium- magnesium- aluminum- silicon analysis

Understanding the degradation mechanisms of TBCs. TBCs are commonly used in high-temperature applications, such as gas turbines and jet engines, to shield underlying metal components from extreme temperatures and thermal cycling. Calcium- magnesium- aluminum- silicon (CMAS) deposits consist mainly of calcium, magnesium, aluminum, and silicon oxides. These deposits are formed from a variety of sources, including volcanic ash, sand, and dust, which can be ingested by engines. When these deposits melt at high temperatures, they can penetrate the porous structure of TBCs, causing significant degradation. Scanning electron microscopy (SEM), energy dispersive x-ray spectroscopy (EDS), XRD, thermogravimetric analysis (TGA), and Fourier transform infrared spectroscopy (FTIR) can all be used to investigate the effect of CMAS on TBCs [34].

9.2.3.3.5 Water vapor corrosion performance

Where the combustion process generates large amounts of water vapor. The presence of water vapor at high temperatures can speed up the degradation of TBCs, reducing their protective capabilities. Various analytical methods can be used to evaluate the performance of TBCs under water vapor conditions, including TGA, XRD, SEM, EDS, and FTIR. Water vapor corrosion resistance is an important aspect of TBC durability in high-temperature environments [34].

9.3 Thermal conductive coatings

These are specialized materials applied to surfaces to enhance their ability to transfer heat. These coatings are designed to facilitate efficient heat dissipation and distribution, which is crucial in various applications where effective thermal management is required.

The primary purpose of thermal conductive coatings (TCCs) is to improve the thermal conductivity of the surfaces they cover, thereby enhancing heat transfer. Implement appropriate curing processes to harden the coating and enhance its properties. This may involve thermal curing, UV curing, or other methods. These coatings often incorporate materials with high thermal conductivity such as metals (e.g., aluminum, copper), ceramics (e.g., AlN, boron nitride), and carbon-based materials (e.g., graphite, graphene). Commonly used in electronics, automotive, aerospace, and industrial sectors to manage heat in components like heat sinks, printed circuit boards, engine parts, and heat exchangers. Improved thermal management leads to better performance, increased reliability, and extended lifespan of the coated components [6]. TCCs are applied on Electronics, Automotive, Aerospace and Industrial Equipment etc.

9.3.1 Design of thermal conductive coating

Designing TCCs involves several key steps to ensure that the coating effectively enhances heat transfer while maintaining durability and compatibility with the substrate [35]. Here is a detailed guide on the design process:

Materials to be used for TCCs are always choose those materials which are well-known for their excellent thermal conductivity.

9.3.1.1 Metallic fillers

Adding metallic particles to a polymer can improve both thermal and electrical conductivity in composite materials. However, when significant metal loadings are added to the polymer matrix, the density increases, limiting applications that require lightweighting. Powdered aluminum, silver, copper, and nickel are examples of metallic particles used to enhance heat transfer. Polymers containing metallic particles include polyethylene [36], polypropylene [37], polyamide [38], polyvinyl chloride [39], and epoxy [40].

9.3.1.2 Ceramic fillers

Ceramic powder reinforced polymer materials have been widely used as electronic materials. Ceramic materials like silicon carbide, beryllium oxide, aluminum nitride (AlN), and boron nitride (BN) exhibit high electrical conductivity due to their metallic particles. Thermally conductive fillers have gained popularity due to their high conductivity and electrical resistivity [41]. Composites with ceramic filler have different thermal conductivities depending on factors such as filler density [42], particle size and distribution [43], surface treatment [44], and mixing methods.

9.3.1.3 Carbon-based fillers

Carbon-based fillers appear to be the best promising fillers, coupling high thermal conductivity and lightweight. Graphite, carbon fiber and carbon black are well-known traditional carbon-based fillers.

Graphite is usually recognized as the best conductive filler because of its good thermal conductivity, low cost and fair dispersibility in polymer matrix citations.

Single graphene sheets constituting graphite show intrinsically high thermal conductivity citations, this determining the high thermal conductivity of graphite. Expanded graphite (EG), which is an exfoliated form of graphite with layers of 20–100 nm thickness, has also been used in polymer composites citation, the thermal conductivity depends on the degree of exfoliation [45], its dispersion in matrix [46] and the aspect ratio of the EG [47].

Carbon nanotubes (CNTs) are cylindrical nanostructures with exceptional thermal and electrical properties. They have gained significant interest for use in TCCs due to their high heat transfer properties, mechanical strength, and unique structural properties. CNT enhance the mechanical strength and wear resistance of the coating as well as provides high thermal performance without adding significant weight, crucial for aerospace and automotive applications.

CNTs based conductive coating significantly enhances heat dissipation in electronic components, batteries, and heat sinks.

9.3.1.4 Composite materials

Often, a combination of materials is used to achieve the desired properties. For instance, a composite of ceramic and metal particles can balance thermal conductivity and electrical insulation [48].

9.3.1.5 *Binders and additives*

Select appropriate binders to hold the conductive particles together and additives to enhance adhesion, flexibility, or other properties [49].

9.3.2 Thermal conductive coating fabrication techniques

9.3.2.1 *Spray coating deposition*

Spray coating is a versatile fabrication technique used in various industries, including thermal management applications where TCCs are required. Here's a rundown of how spray coating can be utilized for applying TCCs. This technique is suitable for large or irregular surfaces, ensuring even coverage [50].

9.3.2.1.1 Substrate preparation

Before applying the TCC, the substrate surface needs to be cleaned thoroughly to remove any contaminants such as dust, grease, or oxides. This ensures proper adhesion of the coating.

9.3.2.1.2 Selection of coating material

Various materials can be used as TCCs, including metallic powders (such as aluminum, copper, or silver), ceramic powders (such as BN or aluminum oxide), or polymer-based materials with added thermally conductive fillers. The choice of material depends on factors like thermal conductivity requirements, substrate material, and application conditions [51].

9.3.2.2 *Dip coating deposition*

This process contains following steps:

9.3.2.2.1 Immersion

The substrate is immersed into the coating solution, ensuring complete coverage of the surface by the liquid.

9.3.2.2.2 Withdrawal

The substrate is then withdrawn from the coating solution at a controlled rate. The withdrawal speed can influence the thickness and uniformity of the coating.

9.3.2.2.3 Drainage

As the substrate is withdrawn, excess coating solution drips off the surface due to gravity, leaving a uniform coating adhered to the substrate.

Dip coating offers advantages such as simplicity, uniformity of coating thickness, and the ability to coat complex-shaped substrates. However, it may not be suitable for all substrate geometries or coating materials. Manufacturers should consider factors such as viscosity, surface tension, and drying/curing properties when optimizing the dip coating process for TCCs [52].

9.3.2.3 Brush coating deposition

This application contains following steps:

9.3.2.3.1 Coating preparation

The coating material is typically mixed or prepared according to the manufacturer's instructions. It may need to be thinned with a compatible solvent to achieve the desired consistency for brush application.

9.3.2.3.2 Brush application

Using a suitable brush, the coating material is applied evenly onto the substrate surface. Care should be taken to ensure uniform coverage and consistent coating thickness. The brush strokes should be applied in a methodical manner to minimize streaks or unevenness in the coating.

9.3.2.3.3 Multiple coats

Depending on the desired thickness and properties of the TCC, multiple coats may be applied. Each coat should be allowed to dry or cure according to the manufacturer's recommendations before applying the next coat.

Brush coating offers advantages such as simplicity, control over coating thickness, and suitability for small-scale or irregularly shaped substrates. This application allows for precision application in specific areas. However, it may require more labor and time compared to automated coating techniques. Manufacturers should consider factors such as brush selection, coating material compatibility, and application consistency when using brush coating for TCCs [53].

9.3.2.4 Vacuum deposition (physical vapor deposition, chemical vapor deposition)

Vacuum deposition techniques, such as PVD and chemical vapor deposition (CVD), are advanced methods for applying thin films of TCCs onto substrates. These techniques offer precise control over coating thickness, composition, and properties, making them highly suitable for various applications requiring thermal management. Here's how PVD and CVD can be utilized for TCCs [54].

9.3.2.4.1 Physical vapor deposition

PVD involves the deposition of a thin film of coating material onto a substrate surface through physical processes, such as evaporation or sputtering, under vacuum conditions [55]. TCCs can be applied using various PVD methods, including:

Evaporation In this process, the coating material is heated in a vacuum chamber until it vaporizes. The vaporized atoms or molecules then condense onto the substrate surface, forming a thin film. Evaporation techniques include resistance heating, electron beam heating, or thermal evaporation.

Sputtering Sputtering involves bombarding a target material (the coating material) with energetic ions in a vacuum chamber. This dislodges atoms or molecules from the target, which then deposit onto the substrate surface. Sputtering can be performed using techniques such as magnetron sputtering or ion beam sputtering.

PVD offers advantages such as high purity coatings, precise control over film thickness, uniformity, and excellent adhesion to the substrate. It is suitable for depositing metallic, ceramic, or composite TCCs with tailored properties.

9.3.2.4.2 Chemical vapor deposition

CVD is a process where a thin film of coating material is deposited onto a substrate surface through chemical reactions of precursor gases in a vacuum chamber. TCCs can be deposited using CVD by controlling the composition and deposition conditions to achieve the desired properties [56]. The steps involved in CVD typically include:

Precursor delivery Gaseous precursor materials containing the elements of the desired coating material are introduced into the vacuum chamber.

Chemical reaction The precursor gases undergo chemical reactions on the substrate surface, resulting in the deposition of the coating material as a thin film.

Film growth The coating material accumulates on the substrate surface, forming a thin film with controlled thickness and composition.

CVD offers advantages such as excellent conformal coating of complex geometries, precise control over film composition, and the ability to deposit high-quality coatings at relatively low temperatures. It is suitable for depositing ceramic, metallic, or carbon-based TCCs with tailored microstructures and properties.

The choice between PVD and CVD depends on factors such as coating material, substrate material, coating thickness, and desired properties of the final coating.

9.3.3 Properties and performance

TCCs have various performances as mentioned below:

TCCs play a crucial role in enhancing heat transfer and thermal management in various applications. The properties and performance of these coatings are influenced by factors such as the coating material, application method, substrate material, operating conditions, and environmental factors. Here are some key properties and performance characteristics of TCCs:

9.3.3.1 Environmental properties

Under these testing the coating's performance is evaluated under expected operational conditions, including temperature extremes, humidity, and exposure to chemicals etc. Resistance to corrosion, oxidation, chemical attack, and environmental degradation is critical for maintaining coating integrity and performance over time [57].

9.3.3.2 Mechanical properties

Under these testing of TCC the coating's adhesion, hardness, and resistance to wear and abrasion are accessed [58].

9.3.3.2.1 Adhesion strength

Good adhesion between the coating and the substrate is essential for ensuring long-term performance and reliability. Strong adhesion prevents delamination, cracking, or flaking of the coating, especially under thermal cycling or mechanical stress conditions.

9.3.3.2.2 Thickness and uniformity

The thickness and uniformity of the coating layer influence its thermal conductivity, adhesion, and overall performance. Uniform coating thickness ensures consistent heat dissipation across the substrate surface and minimizes thermal resistance at the coating-substrate interface.

9.3.3.2.3 Surface finish

The surface finish of the coating affects its contact resistance with adjacent surfaces and the efficiency of heat transfer. Smooth, flat surfaces promote better thermal contact and reduce thermal resistance, enhancing overall heat transfer performance.

9.3.3.2.4 Mechanical durability

TCCs should withstand mechanical stresses, vibrations, and handling during assembly, operation, and maintenance. Mechanical durability prevents coating damage, delamination, or wear that could compromise thermal performance.

9.3.3.2.5 Compatibility

TCCs should be compatible with the substrate material and other components in the system. Compatibility ensures proper adhesion, thermal contact, and long-term reliability without causing adverse effects on substrate materials or nearby components.

9.3.3.3 Thermal properties

Use techniques like laser flash analysis, guarded hot plate method, or thermocouples to measure thermal conductivity [59].

9.3.3.3.1 Thermal conductivity

The primary function of TCCs is to facilitate the efficient transfer of heat from the substrate to the surrounding environment or other heat sink components. High thermal conductivity is a desirable property for these coatings, as it allows for effective heat dissipation and temperature regulation.

Testing methods of thermal properties of TCCs are divided in two parts one is steady-state methods and another is transient methods. Steady-state methods are applied when the system has achieved stability, while the transient methods are used during the process of heating up or cooling down a material.

Steady-state methods Steady state condition means that temperature varies across the sample but is generally time independent. Heat transfer is based on measurements of the heat flux and heat rise. The key steady-state methods are mentioned above.

The guarded hot plate The guarded hot plate method is used for testing materials with low thermal conductivity. The guarded hot plate device is made up of a hot plate, one or two cold plates, thermal insulation, and a guard heating system. The technique's concept is to create a temperature gradient by applying a known heat flux to one side of the plate-like sample and allowing the heat to pass through to a cold plate on the opposite side of the sample. When the system reaches steady-state citation, the temperatures on both sides of the samples are measured. Thermal conductivity can be determined via heat flux, the sample thickness, the surface area, and variation in temperature [60] (Fig. 9.5).

Formula used to calculate thermal conductivity of TCC, as below

$$\lambda = \frac{\dot{Q} \times d}{A \times \Delta T} \tag{9.12}$$

Where,

λ: Thermal conductivity
$\dot{Q}$: Heat flow (Equivalent to the supplied electrical power)
d: Thickness of the sample
A: Area of the heating plate ($2A$ for two specimen setup)
ΔT: Temperature difference over the sample

The axial flow method The axial flow method is the most widely used technique to determine thermal conductivity at lower temperatures. Horizontal heating is accomplished in this method, as well as in the Guarded Hot Plate method, by

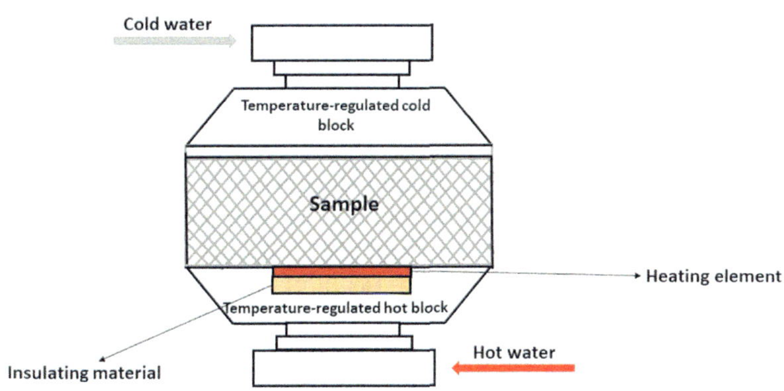

Figure 9.5 Diagram of the guarded hot plate.

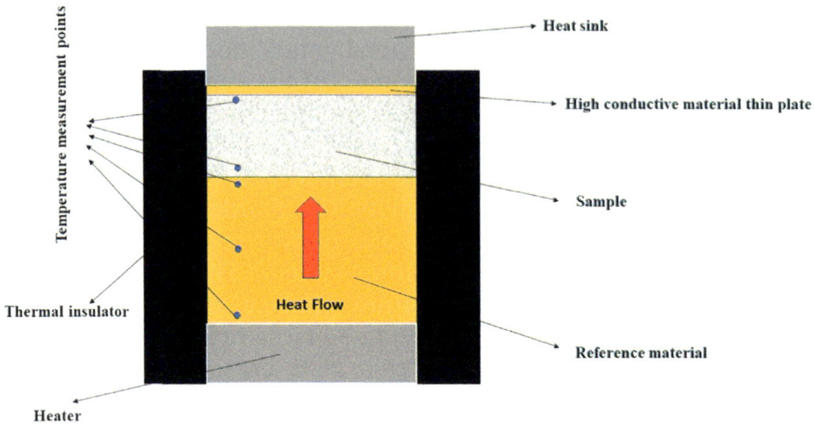

Figure 9.6 Diagram of the axial flow method.

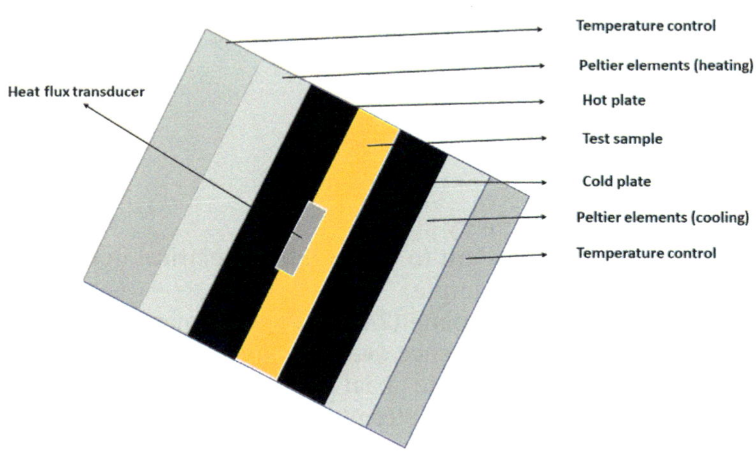

Figure 9.7 Diagram of the heat flow meter.

sandwiching the test sample between two predefined temperature samples to form a specimen column containing temperature and temperature regulators at both ends [61] (Fig. 9.6).

The heat flow meter method This method is like" The guarded hot plate apparatus," excepting a heat flux sensor replaces the main heater. Temperature differences were measured across a thermal resistor containing a series of thermocouples (Fig. 9.7).

The pipe method The pipe method is also known as radial heat flow form. The device usually contains a core heater located on the central axis of the cylinder-shaped specimen and a heat sink surrounding the outer specimen Thermocouples are placed on the specimen to record the temperature gradient. To

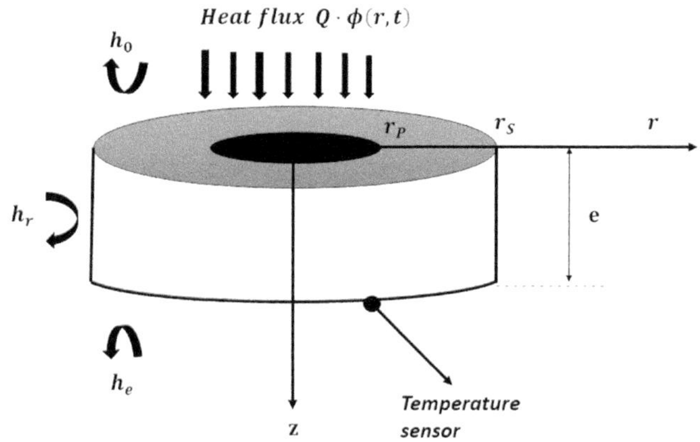

Figure 9.8 Diagram of the pipe method.

reduce heat loss at the top and bottom, the length of the specimen to thickness ratio should be high [62] (Fig. 9.8).

The transient-state methods Usually, delayed methods are used to measure thermal diffusion by recording temperature as the time preceding instantaneous or periodical heating of the sample surface. The thermal conductivity can be calculated through the following equation.

$$\boldsymbol{k} = \alpha \times \boldsymbol{C_p} \times \rho \tag{9.13}$$

where,

k is the thermal conductivity,
α is the thermal diffusivity,
ρ is the density,
Cp is the specific heat.

The transient methods are significantly faster than the steady-state methods. A steady-state measurement takes an hour, but A transient method requires only couple of minutes or sub second periods of time. However, these methods typically have lesser accuracy and necessitate additional complex data study.

The flash method The flash method is the most popular method to measure solid thermal transport properties due to its broad thermal conductivity. This method involves heating a small disk specimen in a furnace with a short laser pulse on its front surface and monitoring the temperature growth on the reverse surface with an infrared detector [63] (Fig. 9.9).

The transient hot wire method This method is a modified version of the steady-state pipe method with radial heat flux. This

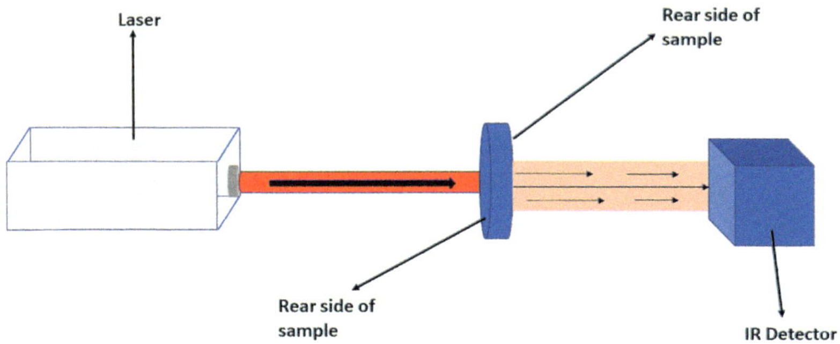

Figure 9.9 Diagram of the flash method.

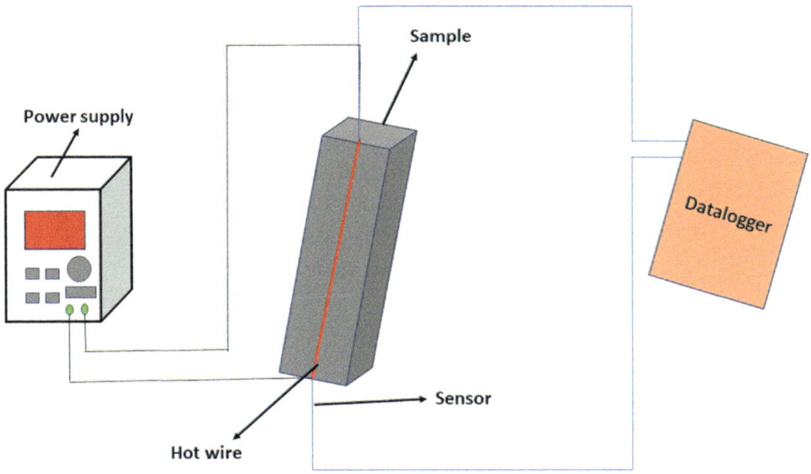

Figure 9.10 Diagram of the transient hot wire method.

method is based on mechanism of a linear heat source (hot wire) which is embedded in the test specimen. This hot wire works as a temperature sensor and heater. A constant current supply is applied to the embedded wire. Thermal conductivity can be measured based on the rate of temperature rise and the voltage generated by the wire [64] (Fig. 9.10).

The transient plane source method Thermal conductivity and diffusivity can be measured using this transient pulse heating method, which is also referred to as the hot disk method. With this method, a thin, spiral-shaped resistive element that is electrically insulated serves as a source of heat and a temperature gauge. The heating element is sandwiched between two identical test samples. Measure the resistance increase over

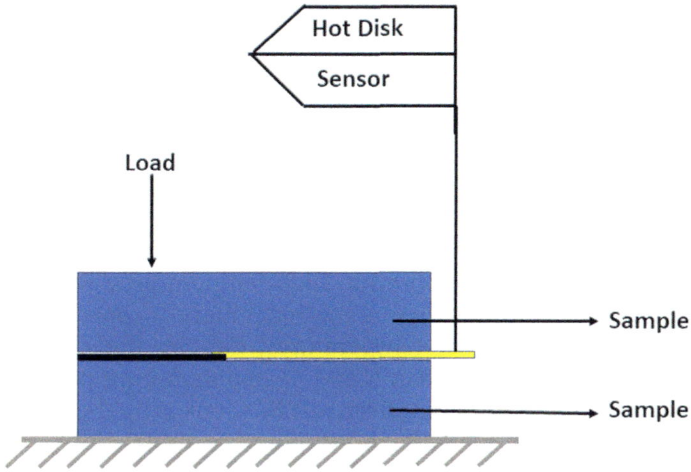

Figure 9.11 Diagram of the transient plane source method.

time when heating with an electrical current pulse to find the thermal conductivity [65] (Fig. 9.11).

Each of these techniques has its own advantages and limitations, and is suitable for only a limited range of materials, depending on the thermal properties, sample configuration, and measurement temperature.

9.4 Thermal stability

TCCs should be stable and maintain their thermal properties over a wide range of temperatures encountered during operation. Thermal stability ensures consistent heat transfer performance and prevents degradation or changes in coating properties under thermal cycling or high-temperature conditions [66,67].

9.5 Electrical insulation

In many applications, TCCs are required to provide electrical insulation between components or substrate surfaces. Electrical insulation properties prevent short circuits or electrical interference while maintaining high thermal conductivity [68].

Overall, the properties and performance of TCCs are optimized to meet specific application requirements, such as thermal management in electronic devices, heat exchangers, automotive components, aerospace systems, and industrial processes. By understanding these key properties, manufacturers can select and tailor TCCs to achieve optimal heat transfer performance and reliability in their applications.

9.6 Conclusions

Both thermal barrier and TCCs are crucial for managing temperature in high-performance applications.

TBCs are essential for enhancing the performance and longevity of components exposed to high temperatures. By providing thermal insulation, they allow machinery to operate more efficiently and at higher temperatures. Despite the challenges they face, continuous research and development are paving the way for more resilient and effective TBCs, ensuring their critical role in modern and future high-temperature applications.

TCCs are vital for efficient heat dissipation in numerous high-performance applications. They enhance the thermal management of components, ensuring optimal performance and longevity by preventing overheating. Despite existing challenges, continuous innovation in material science and coating technologies is paving the way for more effective and versatile TCCs, meeting the evolving thermal management demands of various industries.

In contrast, TCCs facilitate efficient heat dissipation, essential for preventing overheating and ensuring the reliability of heat-sensitive components. Advancements in materials science and coating technologies continue to enhance the effectiveness, durability, and application range of both types of coatings, addressing the evolving needs of various industries.

AI disclosure

During the preparation of this work the author(s) used Quillbot for rephrasing. After using this tool/service, the author (s) reviewed and edited the content as needed and took full responsibility for the content of the publication.

References

[1] I.J. Primo, Coating & Painting Inspection, ASTM International, Estados Unidos, 2020, p. 478.

[2] L. Liu, S. Wang, B. Zhang, et al., Present status and prospects of nanostructured thermal barrier coatings and their performance improvement strategies: a review, Journal of Manufacturing Processes 97 (2023) 12–34. Available from: https://doi.org/10.1016/j.jmapro.2023.04.052.

[3] H. Xu, H. Guo, S. Gong, Thermal Barrier Coatings, Elsevier BV, Tucson, Arizona, 2008, pp. 476–491. Available from: https://doi.org/10.1533/9781845694258.2.476.
[4] D. Zhang, Thermal Barrier Coatings Prepared by Electron Beam Physical Vapor Deposition (EB–PVD), Elsevier BV, Sawston, Cambridge, UK, 2011, pp. 3–24. Available from: https://doi.org/10.1533/9780857090829.1.3.
[5] D.E. Crawmer, Plasma Spray Coatings, Elsevier BV, Ronse, Belgium, 2001, pp. 7035–7040. Available from: https://doi.org/10.1016/b0-08-043152-6/01248-1.
[6] J. Li, J. Liang, Y. Liu, High-thermal conductive coating used on metal heat exchanger, Chinese Journal of Chemical Engineering 22 (5) (2014) 596–601. Available from: https://doi.org/10.1016/s1004-9541(14)60068-9.
[7] A.K. Saini, D. Das, M.K. Pathak, Thermal barrier coatings -applications, stability and longevity aspects, Procedia Engineering 38 (2012) 3173–3179. Available from: https://doi.org/10.1016/j.proeng.2012.06.368.
[8] R. Hashaikeh, J.A. Szpunar, Electrolytic processing of MgO coatings, Journal of Physics: Conference Series (2009) 165. Available from: https://doi.org/10.1088/1742-6596/165/1/012008.
[9] K. Mondal, L. Nuñez, C.M. Downey, I.J. Van Rooyen, Thermal barrier coatings overview: design, manufacturing, and applications in high-temperature industries, Industrial and Engineering Chemistry Research 60 (17) (2021) 6061–6077. Available from: https://doi.org/10.1021/acs.iecr.1c00788.
[10] W.J. Nowak, T. Kubaszek, M. Góral, B. Wierzba, Durability of underaluminized thermal barrier coatings during exposure at high temperature, Surface and Coatings Technology (2020) 382. Available from: https://doi.org/10.1016/j.surfcoat.2019.125236.
[11] P. Zhang, Performance of MCrAlX Coatings: Oxidation, Hot Corrosion and Interdiffusion (2019).
[12] X. Guo, Z. Lu, H.-Y. Park, et al., Thermal properties of $La_2Zr_2O_7$ double-layer thermal barrier coatings, Advances in Applied Ceramics 118 (3) (2018) 91–97. Available from: https://doi.org/10.1080/17436753.2018.1510820.
[13] X. Zhao, P. Xiao, Residual stresses in thermal barrier coatings measured by photoluminescence piezospectroscopy and indentation technique, Surface and Coatings Technology 201 (3–4) (2006) 1124–1131. Available from: https://doi.org/10.1016/j.surfcoat.2006.01.035.
[14] D.T.M.R. Rad, A.M.T. Hussain, Beyond Traditional Coatings: A Review on Thermal-Sprayed Functional and Smart Coatings, Springer, Berlin, Heidelberg, 2019.
[15] D. Naumenko, V. Shemet, L. Singheiser, W.J. Quadakkers, Failure mechanisms of thermal barrier coatings on MCrAlY-type bondcoats associated with the formation of the thermally grown oxide, Journal of Materials Science 44 (7) (2009) 1687–1703. Available from: https://doi.org/10.1007/s10853-009-3284-3.
[16] R.A. Miller, Current status of thermal barrier coatings - an overview, Surface and Coatings Technology 30 (1) (1987) 1–11. Available from: https://doi.org/10.1016/0257-8972(87)90003-X.
[17] S. Gong, Q. Wu, Processing, Microstructures and Properties of Thermal Barrier Coatings by Electron Beam Physical Vapor Deposition (EB–PVD), Elsevier BV, Sawston, Cambridge, UK, 2011, pp. 115–131. Available from: https://doi.org/10.1533/9780857090829.2.115.
[18] W. He, Deposition mechanisms of thermal barrier coatings (TBCs) manufactured by plasma spray-physical vapor deposition, PS-PVD. Forschungszentrum Jülich GmbH, Germany ,398 (2017).

[19] E. Bolster, E. Amherst, R.M. Murphy, P.E. Skapars, United States Patent (19), 1982.
[20] A. Allanore, Features and challenges of molten oxide electrolytes for metal extraction, Journal of the Electrochemical Society 162 (1) (2014) E13–E22. Available from: https://doi.org/10.1149/2.0451501jes.
[21] A.R. Boccaccini, C. Kaya, K.K. Chawla, Use of electrophoretic deposition in the processing of fibre reinforced ceramic and glass matrix composites: a review, Composites Part A: Applied Science and Manufacturing 32 (8) (2001) 997–1006. Available from: https://doi.org/10.1016/s1359-835x(00)00168-8.
[22] S. Lokachari, K. Leng, R.A. Rincon, et al., Processing–microstructure–properties of columns in thermal barrier coatings: a study of thermo-chemico-mechanical durability, ACS Applied Materials & Interfaces 16 (8) (2024) 10646–10660. Available from: https://doi.org/10.1021/acsami.3c16681.
[23] M. Van Roode, B. Beardsley, Porosity determination of thermal barrier coatings, in: Proceedings of the ASME Turbo Expo, 1988, 5, doi:10.1115/88-GT-278.
[24] A. Portinha, V. Teixeira, J. Carneiro, et al., Characterization of thermal barrier coatings with a gradient in porosity, Surface and Coatings Technology 195 (2-3) (2005) 245–251. Available from: https://doi.org/10.1016/j.surfcoat.2004.07.094.
[25] B. Rogé, A. Fahr, J.S.R. Giguère, K.I. McRae, Nondestructive measurement of porosity in thermal barrier coatings, Journal of Thermal Spray Technology 12 (4) (2003) 530–535. Available from: https://doi.org/10.1361/105996303772082279.
[26] J.L. Smialek, A. Garg, B.J. Harder, M.D. Cuy, Crystallographic and TEM features of a TBC/Ti_2AlC MAX phase interface after 1300°C burner rig oxidation, Crystals 13 (4) (2023). Available from: https://doi.org/10.3390/cryst13040691.
[27] M. Gell, J. Eric, V. Krishnakumar, et al., Bond strength, bond stress and spallation mechanisms of thermal barrier coatings, Surface and Coatings Technology 120-121 (1999) 53–60. Available from: https://doi.org/10.1016/s0257-8972(99)00338-2.
[28] M. Kirschner, T. Wobst, B. Rittmeister, C. Mundt, Erosion Testing of thermal barrier coatings in a high enthalpy wind tunnel, Journal of Engineering for Gas Turbines and Power 137 (3) (2015). Available from: https://doi.org/10.1115/1.4028469.
[29] B. Ekici, Effects of residual stress patterns on TBC systems, Surface Engineering 20 (1) (2004) 37–42. Available from: https://doi.org/10.1179/026708404225010694.
[30] X. Zhao, X. Wang, P. Xiao, Sintering and failure behaviour of EB-PVD thermal barrier coating after isothermal treatment, Surface and Coatings Technology 200 (20-21) (2006) 5946–5955. Available from: https://doi.org/10.1016/j.surfcoat.2005.09.006.
[31] A. Bolcavage, A. Feuerstein, J. Foster, P. Moore, Thermal shock testing of thermal barrier coating/bondcoat systems, Journal of Materials Engineering and Performance 13 (4) (2004) 389–397. Available from: https://doi.org/10.1361/10599490419883.
[32] J. Schloesser, M. Bäker, R. Joachim, Laser cycling and thermal cycling exposure of thermal barrier coatings on copper substrates, Surface and Coatings Technology 206 (7) (2011) 1605–1608. Available from: https://doi.org/10.1016/j.surfcoat.2011.07.084.
[33] C. Holländer, S. Kiliani, W. Stamm, et al., Hot corrosion of TBC-coated components upon combustion of low-sulfur fuels, Materials and Corrosion

72 (10) (2021) 1643–1655. Available from: https://doi.org/10.1002/maco.202112371.
[34] CMAS (Calcium- Magnesium- Aluminium- Silicon) Analysis.Pdf.
[35] A.A. Hamid, D. Periathamby, S. Azizan, C.E. Tan, S.K.S. Nadarajan, Design and characterization of thermal conductive wafer coating in thin small outline package for automotive product application, Proceedings of the IEEE/CPMT International Electronics Manufacturing Technology (IEMT) Symposium, 2016, 2016. Available from: https://doi.org/10.1109/IEMT.2016.7761955.
[36] P. Sobolčiak, A. Abdulgader, M. Mrlik, et al., Thermally conductive polyethylene/expanded graphite composites as heat transfer surface: mechanical, thermo-physical and surface behavior, Polymers 12 (12) (2020). Available from: https://doi.org/10.3390/polym12122863.
[37] N. Song, D. Cao, X. Luo, Q. Wang, P. Ding, L. Shi, Highly thermally conductive polypropylene/graphene composites for thermal management, Composites Part A: Applied Science and Manufacturing (2020) 135. Available from: https://doi.org/10.1016/j.compositesa.2020.105912.
[38] T. Xu, S. Zhou, F. Jiang, N. Song, L. Shi, P. Ding, Polyamide composites with improved thermal conductivity for effective thermal management: The three-dimensional vertically aligned carbon network, Composites Part B: Engineering (2021) 224. Available from: https://doi.org/10.1016/j.compositesb.2021.109205.
[39] H. Zhang, Y. Jia, J. Yang, Z. Wang, J. Zhang, Enhanced thermal conductivity of plasticized polyvinyl chloride composites based on 2D structural regulation of flaky graphite, Diamond and Related Materials (2022) 129. Available from: https://doi.org/10.1016/j.diamond.2022.109360.
[40] L. Chen, H. Liu, Z. Liu, Q. Song, Thermal conductivity and anti-corrosion of epoxy resin based composite coatings doped with graphene and graphene oxide, Composites Part C: Open Access (2021) 5. Available from: https://doi.org/10.1016/j.jcomc.2021.100124.
[41] H. Tahir, M. Saad, N. Shafi, F. Muslim, Composite, in: Interface Science and Technology, vol. 33, Elsevier B.V., Pakistan, 2021, 211–278. Available from: https://doi.org/10.1016/B978-0-12-818805-7.00004-7.
[42] H. He, R. Fu, Y. Han, Y. Shen, X. Song, Thermal conductivity of ceramic particle filled polymer composites and theoretical predictions, Journal of Materials Science 42 (16) (2007) 6749–6754. Available from: https://doi.org/10.1007/s10853-006-1480-y.
[43] A. Devpura, P.E. Phelan, R.S. Prasher, Size effects on the thermal conductivity of polymers laden with highly conductive filler particles, Microscale Thermophysical Engineering 5 (3) (2001) 177–189. Available from: https://doi.org/10.1080/108939501753222869.
[44] M. Derradji, J. Wang, W. Liu, Ceramic-based phthalonitrile micro- and nanocomposites, Phthalonitrile Resins and Composites 35 (2018) 295–375. Available from: https://doi.org/10.1016/b978-0-12-812966-1.00006-8. Elsevier BV.
[45] G. Mariz, D.O. Barra, A.D.A. Lucas, Expanded Graphite as a Multifunctional Filler for Polymer Nanocomposites, Elsevier Inc, Norwich, NY, 2015.
[46] G. Czel, A. Sycheva, D. Janovszky, Effect of different fillers on thermal conductivity, tribological properties of Polyamide 6, Scientific Reports 13 (1) (2023). Available from: https://doi.org/10.1038/s41598-023-27740-y.
[47] D. Japić, S. Kulovec, M. Kalin, et al., Effect of expanded graphite on mechanical and tribological properties of polyamide 6/glass fibre composites, Advances in Polymer Technology 2022 (2022) 1–8. Available from: https://doi.org/10.1155/2022/9974889.

[48] N. Mehra, L. Mu, T. Ji, J. Zhu, Thermal Conduction in Polymer Composites, Elsevier BV, Amsterdam, The Netherlands, 2019, pp. 77–110. Available from: https://doi.org/10.1016/b978-0-12-815067-2.00003-2.

[49] N.A. Salleh, S. Kheawhom, A. Ashrina, N. Hamid, W. Rahiman, A.A. Mohamad, Electrode polymer binders for supercapacitor applications: a review, Journal of Materials Research and Technology 23 (2023) 3470–3491. Available from: https://doi.org/10.1016/j.jmrt.2023.02.013.

[50] L. Gao, H. Guo, L. Wei, C. Li, H. Xu, Microstructure, thermal conductivity and thermal cycling behavior of thermal barrier coatings prepared by plasma spray physical vapor deposition, Surface and Coatings Technology 276 (2015) 424–430. Available from: https://doi.org/10.1016/j.surfcoat.2015.06.033.

[51] X. Hu, M. Huang, N. Kong, F. Han, R. Tan, Q. Huang, Enhancing the electrical insulation of highly thermally conductive carbon fiber powders by SiC ceramic coating for efficient thermal interface materials, Composites Part B: Engineering (2021) 227. Available from: https://doi.org/10.1016/j.compositesb.2021.109398.

[52] X. Tang, X. Yan, Dip-coating for fibrous materials: mechanism, methods and applications, Journal of Sol-Gel Science and Technology 81 (2) (2017) 378–404. Available from: https://doi.org/10.1007/s10971-016-4197-7.

[53] H. Jiang, L. Cooke, K. Srivilliputhur, et al., Copper–carbon nanotube composites enabled by brush coating for advanced conductors, ACS Applied Nano Materials 7 (10) (2024) 11176–11183. Available from: https://doi.org/10.1021/acsanm.4c00679.

[54] D.M. Mattox, The Foundations of Vacuum Coating Technology, 1st ed., Elsevier, Amsterdam, Netherlands, 2003.

[55] S.M. Shang, W. Zeng, Conductive Nanofibres and Nanocoatings for Smart Textiles, Woodhead Publishing Limited, Elsevier BV, Sawston, Cambridge, UK, 2013, pp. 92–128. Available from: https://doi.org/10.1533/9780857093530.1.92.

[56] M. Tortello, I. Pasternak, K. Zeranska-Chudek, W. Strupinski, R.S. Gonnelli, A. Fina, Chemical-vapor-deposited graphene as a thermally conducting coating, ACS Applied Nano Materials 2 (5) (2019) 2621–2633. Available from: https://doi.org/10.1021/acsanm.8b02243.

[57] G. Liebig, U. Kirstein, S. Geißendörfer, O. Zahid, F. Schuldt, C. Agert, The impact of environmental factors on the thermal characteristic of a lithium–ion battery, Batteries 6 (1) (2020). Available from: https://doi.org/10.3390/batteries6010003.

[58] S.I. Hussein, A.M. Abd-Elnaiem, T.B. Asafa, H.I. Jaafar, Effect of incorporation of conductive fillers on mechanical properties and thermal conductivity of epoxy resin composite, Applied Physics A 124 (7) (2018). Available from: https://doi.org/10.1007/s00339-018-1890-0.

[59] Y. Lin, Y. Jia, G. Alva, G. Fang, Review on thermal conductivity enhancement, thermal properties and applications of phase change materials in thermal energy storage, Renewable and Sustainable Energy Reviews 82 (2018) 2730–2742. Available from: https://doi.org/10.1016/j.rser.2017.10.002.

[60] Version D., Description of the Guarded Hot Plate Method for Thermal Conductivity Measurement with the EP500, 2019.

[61] E. Gariboldi, L.P.M. Colombo, D. Fagiani, Z. Li, Methods to characterize effective thermal conductivity, diffusivity and thermal response in different classes of composite phase change materials, Materials 12 (16) (2019). Available from: https://doi.org/10.3390/ma12162552.

[62] M. Koller, J. Unterkofler, G. Glanz, D. Lager, A. Bergmann, H. Popp, Radial Thermal conductivity measurements of cylindrical lithium-ion batteries—an uncertainty study of the pipe method, Batteries 8 (2) (2022). Available from: https://doi.org/10.3390/batteries8020016.
[63] W.J. Parker, R.J. Jenkins, C.P. Butler, G.L. Abbott, Flash method of determining thermal diffusivity, heat capacity, and thermal conductivity, Journal of Applied Physics 32 (9) (1961) 1679–1684. Available from: https://doi.org/10.1063/1.1728417.
[64] J.J. Healy, J.J. de Groot, J. Kestin, The theory of the transient hot-wire method for measuring thermal conductivity, Physica B + C 82 (2) (1976) 392–408. Available from: https://doi.org/10.1016/0378-4363(76)90203-5.
[65] P. Johansson, B. Adl-Zarrabi, C.E. Hagentoft, Using transient plane source sensor for determination of thermal properties of vacuum insulation panels, Frontiers of Architectural Research 1 (4) (2012) 334–340. Available from: https://doi.org/10.1016/j.foar.2012.09.004.
[66] T.S. Jang, S.W. Myoung, H.S. Kim, et al., Effect of top coat thickness on thermal stability in thermal barrier coatings, Applied Mechanics and Materials 261-262 (2013) 460–465. Available from: https://doi.org/10.4028/www.scientific.net/AMM.260-261.460.
[67] M. Matsumoto, K. Aoyama, H. Matsubara, et al., Thermal conductivity and phase stability of plasma sprayed $ZrO_2-Y_2O_3-La_2O_3$ coatings, Surface and Coatings Technology 194 (1) (2005) 31–35. Available from: https://doi.org/10.1016/j.surfcoat.2004.04.078.
[68] S.-W. Xiong, P. Zhang, Q. Zou, Y. Xia, M.-ying Jiang, J.-G. Gai, High thermal conductivity and electrical insulation of liquid alloy/ceramic/epoxy composites through the construction of mutually independent structures, Composites Science and Technology (2021) 202. Available from: https://doi.org/10.1016/j.compscitech.2020.108596.

10

Self-healing and water-repellent nanopaint

Chinnala Kalyani[1], Kanak[1], Shweta Gomey[1], Rohit Verma[2] and Manoj Raula[1]

[1]Department of Chemistry, Amity Institute of Applied Sciences, Amity University, Noida, Uttar Pradesh, India [2]Department of Physics, Amity Institute of Applied Sciences, Amity University Uttar Pradesh, Noida, Uttar Pradesh, India

10.1 Introduction

The formulation of liquid-repellent coatings is difficult, but they offer great potential for usage in industry due to their fast self-healing and powerful substrate adhesion. Basically, protective organic paints comprise polymeric composites with a great deal of chemical heterogeneity and complexity in their macro- and microscale structures, surface/interfacial characteristics, and structure. Materials with the ability to mend themselves automatically in the event of damage or wear and tear are known as self-healing materials. These materials have drawn a lot of interest lately because of their prospective uses in a number of industries, such as construction, electronics, automotive, aerospace and nanopaints. These materials are capable of naturally recovering from injury, much like biological systems do. For example, human skin has the capacity to organically repair itself, become more durable, and adjust to environmental changes. The needs have led to the study and development of smart materials, of which self-healing materials are in high demand because of their exceptional ability to automatically recover and adapt to environmentally friendly changes in a dynamic way, setting them apart from antiquated static and tough materials. Conversely, the industrial community and basic scientists have expressed interest in superhydrophobic surfaces because of their special qualities, which are achieved by combining surface roughness with decreased surface energy. A vast array of uses for superhydrophobic surfaces, including self-cleaning windows or solar panels and anticorrosion coating,

Nanopaints. DOI: https://doi.org/10.1016/B978-0-443-26512-9.00010-0

have been developed. These uses are inspired by nature, such as lotus leaves, insect legs, and moth wings. When hydrophobic molecules are chemically grafted onto a smooth surface, the surface energy is altered, creating a hydrophobic surface with a contact angle of up to 90 degrees. An apparent contact angle of greater than 90 degrees is significantly improved by creating a rough surface, which also has a major impact on wetting. An extremely hydrophobic coating having a contact angle greater than 150 degrees results from increasing the roughness.

In this review, the benefits and drawbacks of popular autonomous and nonautonomous healing methods in protective organic coatings are reviewed in depth. This study also examines the agents and materials that are currently available for self-healing, emphasizing their uses, critiques, and outcomes in a number of research articles. Nevertheless, encapsulating chemical healing agents exhibits more successful results due to improved reactivity among unhydrated particles of cement and healing nanopaints. This analysis suggests that the geometry and size of fractures play an essential role in the selection of self-healing procedures.

Additionally, will go over the several techniques for creating a stable, controlled-roughness water-repellent coating. Additionally, recent advancements in research on a number of functional coating categories, with a focus on their applications are emphasized. Water-repellent and self-healing coatings were among them.

10.2 Different forms of self-healing materials

Typical forms of self-healing materials include:

Polymer-based materials: these substances are made of polymers that have different processes for self-repair. One method includes adding microcapsules filled with adhesives or solvents healing agents to the polymer matrix. The capsules burst when the material is broken, releasing the healing chemicals, which subsequently react and fill in the holes or cracks [1].

Materials made of metals: a few metals and alloys have self-healing capabilities. For instance, when heated, shape memory alloys can recall and return to their previous shape, effectively repairing any distortion. By forming shielding oxide layers that stop further corrosion, other metal-based materials can repair themselves [2].

Materials with self-healing properties based on coatings: coatings with self-healing properties are applied to the surface of materials to offer protection and repair capabilities. These coatings may include enclosed healing agents or substances that can react to outside stimuli, such as heat or light, to begin the healing process [3].

Materials with self-healing properties based on concrete: concrete is the most used building material, and self-healing mechanisms can increase its durability. For instance, when exposed to moisture, bacteria in concrete can create calcium carbonate, which can be used to fill gaps [4].

10.3 Mechanisms of self-healing

10.3.1 Autonomous healing mechanisms

Without any outside physical assistance, autonomous healing processes can restore their functional or structural integrity [5]. Incorporating extrinsic polymerizable healing agents into the coating is one of the simplest ways to achieve autonomous healing. These healing substances are frequently contained in microcapsules. The capsules burst in reaction to mechanical impacts when the coating is damaged, releasing the healing substances, which then transform to form a film of protection, restoring the coating's barrier property.

The ability of self-healing materials to fix damage on their own without outside assistance is one of their fundamental ideas. This is accomplished by including different healing agents or dynamic linkage networks that can be turned on in response to damage. Corrosion inhibitors are used as the healing agents in a different types of autonomous self-healing protective coatings. The coating flaw could be affected by the corrosion inhibitor that leaches into it and occurrences of electrochemical processes at the exposed metal substrate. The healing of the inhibitor-based coating is also thought to be an autonomous process because such triggering factors are easily accessible in corrosion events at the coating damage [6].

10.3.1.1 Based on the effect of filling defects

In this method, a mechanical damage causes coating flaws that subsequently cause the components that can be polymerized to cure themselves to spontaneously release. These substances can polymerize into a film with a specific strength and thickness to repair the coating defect and restore the barrier qualities of the coating by reacting with either catalysts coexisting in the coating or with moisture or oxygen in the environment [7]. For these components to remain reactive for a longer period of time and to contact the bulk coating materials as little as possible, the healant must be well encapsulated. In this study, dicyclopentadiene monomers were stored in polyurea-formaldehyde (PUF) microcapsules with a dimension of 220 μm. The Grubbs catalyst that can polymerize the monomers was not mixed with these microcapsules when they were disseminated in the coating matrix. Approximately 60% of the initial fracture stress was healed efficiently.

This group then generated a number of other groundbreaking instances utilizing similar concepts [8–10]. The same idea was also used more recently to create a self-healing anticorrosion covering [11]. The distinct encapsulation of the catalysts and healing agents (such as prepolymers and monomers) maintains the components' reactivity when the outer layer is intact but improves their interaction at the injury site. The healing agents that emerge from the shattered capsules should ideally have a chemical composition and mechanical properties that will allow them to re-bond the fracture interface. Because they are widely used as coating resins and have exceptional adherence and strength, epoxies can be regarded as great choices in this regard [10,12–14]. Recently, Yi et al. prepared microcapsules containing either a tetraethylenpentamine (TEPA) or epoxy curing agent using a Pickering emulsion, and then they embedded the microcapsules in a coating of the same composition [12]. The epoxy and the amine curing compound overflowed and came into contact to create a layer of protection covering the exposed steel substrate whenever the coating was damaged. The inconsistent distribution of microcapsules across the coating is one drawback of utilizing two-part healants; this makes it challenging to combine and cure the two portions according to the required ratios. Furthermore, self-healing facilitated by catalyst-free, mono reactive healants is desirable given the high cost of catalysts [15–22]. Liquid isocyanates and silanes are two attractive options for one-part polymerizable healants [23,24]. Both species have special benefits for applications involving corrosion protection because of how they react with water. As one-part healing agents for coatings, Wang et al. created microcapsules containing isophorone diisocyanate, hexamethylene diisocyanate (HDI) [16], and a less toxic HDI trimer [20]. Water seeped into the coating/scratch interface, causing the exposed microcapsules to enlarge and split. This also led the isocyanates to start polymerizing, filling the scratch at the same time. Nevertheless, it was discovered that it was impossible to cure scratches larger than 30 μm, presumably due to the insufficiency of the healing agents exposed and their disappearance before the healing process was finished [17]. The short-lived self-healing capacity of water responsive healants is a serious problem. Moisture infiltrates the undamaged coating in service conditions and reacts with its healing agents, which lessens the coating's capacity to repair future damage. To overcome this issue, Yang et al. used a double-wall microcapsule made of an outside wall made of PUF and an inner wall made of TEPA reacting with isocyanates to enclose the 4,4′-methylenebis (cyclohexyl isocyanate) healing agent. The double-wall microcapsule demonstrated enhanced long-term anticorrosion efficacy and higher water resistance, despite having a total wall thickness of only 1.5 μm. Traditional coating materials, such as tung and linseed oils, are drying oils that can be oxidized by air to generate cross-linked films. Additionally, these oils have been employed as inexpensive, eco-friendly healing agents for protective layers put on steel substrates [25–28]. PUF microcapsules containing tung oil, either by combining them with primer directly or by encasing the microcapsules in two layers of primer [22]. It was discovered that the latter technique produced coatings with greater adherence to the metal substrate and improved self-healing capabilities. Linseed oil was recently the subject of an intriguing study that described the use of self-assembled graphene oxide microcapsules to encapsulate the oil and provide a self-healing effect in aqueous polyurethane coatings (Fig. 10.1) [29]. By adding just a little cobalt ions as drying agents, the solidification duration of oils produced from ruptured capsules can be shortened from days to hours. The use of drying oils may improve the coated surface's lubricating qualities in addition to its capacity for self-healing [28]. A review of the current coatings that use the defect-filling process to heal reveals that a fine balance between the mechanical and chemical characteristics, size, and number of the microcapsules is crucial for achieving a desirable level of self-healing capability. Since the majority of the reactive healing agents in these

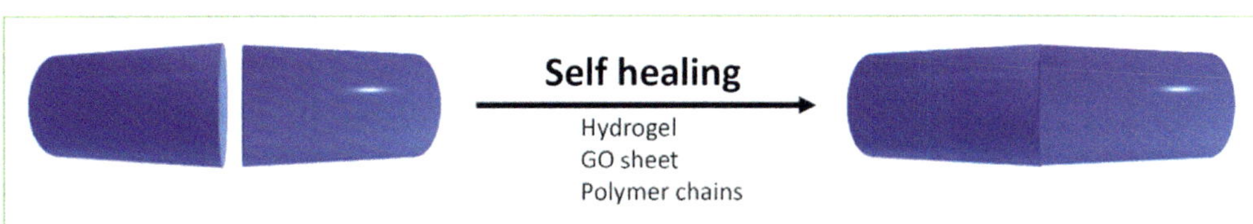

Figure 10.1 Hydrogels made of various kind of composite and their suggested self-healing mechanism [32,33].

coatings are liquid, the walls of an ideal microcapsule, must be sufficiently stiff to maintain both the coating strength and the integrity of the capsule [30,31]. The capsules must, however, also be sufficiently brittle to shatter upon deformation. Minimizing the preferential penetration of water through the coating interactions also requires a strong adhesion between the microcapsule outer surface and the coating matrix. Good resistance and stability to water, oxygen, and other chemicals in the corrosive environment are also important qualities for the capsule materials to extend the shelf-life of the self-healing capacity. The microcapsules should preferably be combined with the primer and positioned close to the metal substrate to seal the defect efficiently. It has been advised to add the microcapsules in a different layer on top of the primer because their inclusion may lessen the primer's adhesion to the metals [24].

10.3.1.2 Corrosion inhibitor-based self-healing

Adding corrosion inhibitors to coatings is another way to achieve autonomous self-healing. The organic corrosion inhibitors used in self-healing coatings include benzotriazole (BTA), mercaptobenzothiazole (MBT), imidazoline, 8-hydroxyquinoline (8-HQ), and aliphatic amines [34]. The most extensively studied inorganic corrosion inhibitors included in these coatings are phosphates, vanadates nitrite, molybdates, tungstates, borates, and rare earth salts [35]. In this process, the inhibitors leached from the coating matrix can promptly reduce anodic breakdown and cathodic chemical reactions at the exposed substrate in the coating defect. While inhibitors that cause the oxide and hydroxide to precipitate at the cathodic sites frequently slow down cathodic processes, they can also be used to limit anodic dissolution by increasing the passivity of the oxide film on exposed metals. Additionally, there exist mixed-type inhibitors, which consist of a wide range of chemical compounds [36,37]. These inhibitors may impede the pace of anodic and cathodic reactions of corrosion on metal substrates through complexation, chemical interactions, or physical adsorption. In fact, owing to their straightforward conceptual designs and immediate impact on the corrosion process, inhibitor-based coatings make up the bulk of documented self-healing coatings for corrosion protection. To encourage a self-healing effect, coating materials might be directly combined with corrosion inhibitors. For instance, in the early research, organic coatings were doped with cerium nitrates and phosphates to enable self-healing on surfaces made of zinc, galvanized steel, and aluminum alloys

[36,38,39]. Lithium salts, including lithium carbonate, have more recently been added straight into polyurethane coatings. Leached lithium salts have the ability to produce an alkaline coating scratch environment, which can lead to the development of a three-layer protective layer of aluminum oxide/hydroxide made up of columnar, porous, and dense areas [40]. The total ability of the coatings to mend themselves is significantly influenced by the compatibility of the inhibiting agent and the coating substance. To maintain the inhibitory efficiency over time, it is necessary to reduce undesired interactions among the inhibitors and coatings. Additionally, inhibitor aggregation is to be avoided since it creates pathways for quicker water penetration. Given these factors, the field has concentrated on further encapsulating inhibitors in the form of micro- or nanocontainers to protect them and release them gradually under regulated conditions. The inhibitors can be released from these containers without breaking [41].

Consequently, in comparison to those containing polymerizable healants, their sizes can be significantly smaller. Self-healing coatings have employed a broad range of nanoparticles, including solid nanoparticles, nanotubes, and porous nanostructures, as carriers of corrosion inhibitors [42–44]. These nanocontainers' modest sizes enable their application in thin organic coatings and sol–gel pretreatments, and they may also enhance the homogeneity of their dispersion within the coating matrix. Because of the mechanical and environmental instability of polymeric nanoparticles (such as heat and solvent), inorganic nanoparticles have been utilized in most investigations, and their chemistries vary. Additionally, these nanoparticles can frequently be surface-grafted or coated to release feed-back active inhibitors in response to certain environmental stimuli [45]. In addition, as the nanoparticles dissolve, ions, such as cerium and molybdate which are also inhibitors are simultaneously released. Cerium molybdate was also synthesized into amorphous nanofibers, which are more soluble than their crystalline counterparts, to attain a quicker release. Organic corrosion inhibitors, such as 8-HQ, MBT, and 1*H*-benzotriazole-4-sulfonic acid, were transported by hollow nanoparticles made of crystalline cerium molybdate [46–48]. The dealloying of the active copper rich the S phase (Al_2CuMg) intermetallics can lead to the release of organic inhibitors that, upon dissolution of the nanoparticles, can form impermeable chelates on aluminum alloys and block pitting corrosion. Positively charged metal hydroxides are stacked layers known as layered double hydroxides (LDHs), in which solvent and anionic molecules are intercalated [49,50]. LDHs are promising receptacles for both releasing anionic corrosion inhibitors and trapping harmful anions due to their capacity to interchange anions. The fact that LDHs have high aspect ratio 2D geometries may improve the coatings' barrier qualities, which is another important advantage of using them as coating fillers. LDH nanoparticles were added to cerium molybdate nanoparticles as a further container for the MBT in a 2 μm thick model epoxy primer that was applied to galvanized steels. On the other hand, dissolving cerium molybdate nanoparticles takes longer and releases the corrosion inhibitors that are made up of MBT, cerium ions, and molybdates [50–52]. This kind of synergy is advantageous for getting a longer-lasting and better self-healing mechanism against corrosion that happens at various stages. A self-healing epoxy covering has also shown similar synergistic effects. Recent research has revealed that LDHs can grow in situ on metal oxide layers to provide a similar form of active corrosion protection, in addition to being employed as fillers for organic coatings. Corrosion inhibitors can be released slowly from containers with a large loading capacity to facilitate long-term self-healing [53]. By using nanotube with a large interior area of surface (or volume) and small apertures at the ends, such properties can be achieved. Halloysite is a cheap and readily available option for the encapsulation and controlled release of various corrosion inhibitors because of its two-layered aluminosilicate nanotubular structure. The diameter of the hollow lumen of halloysite nanotubes is only about 15 nm, despite their length being several microns. This greatly reduces their ability to load inhibitors. The lumen can be made larger and the BTA loading can reach 40 wt.% by selectively etching the interior alumina coating with sulfuric acid. This is similar to the 40–60 wt.% BTA loading found in polymer capsules. Nevertheless, the mechanical strength of halloysite nanotubes treated in this way is slightly reduced [54,55]. By creating a metal–inhibitor combination, the ends of the halloysite nanotubes can be additionally capped by Cu^{2+}, Fe^{2+}, Fe^{3+}, and Co^{2+} ions, giving the nanotubes a more prolonged and pH-sensitive releasing behaviur. Graphene oxide, mesoporous silicon dioxide, mesoporous ZrO_2, montmorillonite nanoparticles, zeolites nanoparticles, titanium dioxide, and other naturally occurring particles are among the various nanostructures that have been used as inhibitor carriers in addition to halloysite nanotubes [56,57]. Because of its extremely enormous specific surface area ($>1000\,m^2\,g^{-1}$), mesoporous SiO_2 has the largest known inhibitor loading among these materials [58,59]. In addition to micro- and nanoparticles, corrosion-preventing agents for self-healing coatings have also been stored in fibrous or particulated porous scaffolds. The low compatibility between the inert nanoparticle surface and the organic coating resin may be one restriction on the direct application of nanoparticles that are inorganic in self-healing coatings. The barrier and adhesion qualities of the coating are negatively impacted by particle aggregation and flaws at the particle/coating interface. This can be a particular issue if the particles are positioned in close proximity to metal substrates. Fig. 10.2 describes the comparison of the physical and chemical processes of self-healing.

By adding organic molecules to the outside of the nanoparticle, these problems can be fixed or at least mitigated. For instance, to improve the compatibility with polyester and epoxy coatings, mesoporous silica nanoparticles were treated with organosilanes [49].

It is possible to create containers that can release inhibitors more intelligently in response to changes in the surroundings at the coating flaws by using organic compounds with greater reactivity to components. Because the anodic and cathodic reactions that occur at the metal interfaces induce a pH fluctuation, this environmental trigger is most frequently used. The organic molecules may be protonated (or deprotonated) for greater swelling, or their hydrolytic decomposition may be

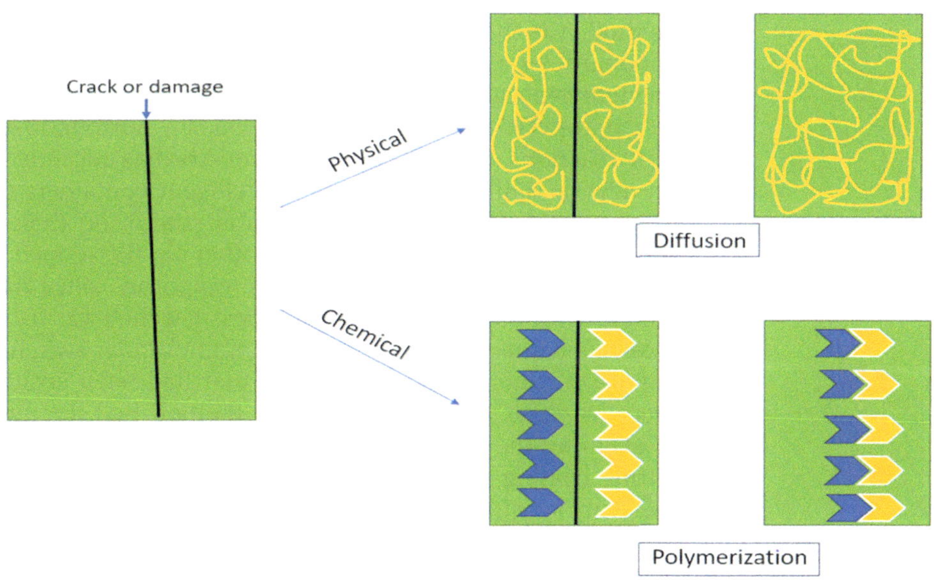

Figure 10.2 Physical versus chemical process of self-healing.

accelerated, both of which could result in an increase in inhibitor release, depending on how the local pH changes [60,61]. The composite nanocontainers, which combine nanoparticle cores with polyelectrolyte multilayer shells, are a prime illustration of this tactic. Easily applied to the nanoparticle surface, layer-by-layer procedures allow for the gradual accumulation of polyelectrolytes with opposing charges to create the multilayers [45,46]. Corrosion inhibitors might be included in polyelectrolyte multilayers or be contained inside nanoparticles. In the latter instance, the loading capacity is rather modest and is contingent upon the multilayers' deposition cycles. The polyelectrolyte conformation may be influenced and the shells may open to let out the corrosion inhibitors within the coating damage site due to the shift in local pH values brought on by the anodic and cathodic corrosion reactions. The pH restoration shuts the shells to protect the inhibitors when the reactions that cause corrosion are inhibited. Apart from their intelligent release characteristic, the polyelectrolytes could also function as buffers, mitigating the pH gradient that is formed among the anodic and cathodic areas within the coating damage region. As a result, it is anticipated that the addition of these smart containers might reduce coating disbonding brought on by cathode alkalization. Li et al. described a more generalized but comparable method in which they used surface-graft precipitation polymerization to coat silica nanotubes with poly(methacrylic acid) to achieve pH-responsive release of the inhibitor BTA. The same method, which is fundamentally applicable to other kinds of nanoparticles as well, was also utilized to graft redox-responsive poly(ethylene glycol) methacrylate and thermoresponsive poly (*N*-isopropylacrylamide) chains [47,59,62]. Surrounding corrosion reactions, supramolecular chemistry has recently opened up new possibilities for building intelligent containers that can react to more complex environmental stimuli on their own. A sol–gel layer was put on top of another Ce (IV)-doped sol–gel layer, containing redox-responsive mesoporous silica nanocontainers with supramolecular switches. Leached Ce (IV) at the coating's damage site activated a supramolecular switch, releasing the *p*-coumaric acids inhibitor that had been accumulated there by redox processes [63]. This method had the advantage of avoiding the problem of early pH-induced inhibitor leakage during coating preparation because the supramolecular switch was only sensitive to redox reactions and not pH. The number of inhibitors present at the coating damage location determines how well inhibitor-based coatings self-heal, much like coatings that use monomers or drying oils as healing agents [64]. It is challenging to satisfactorily cure a significant region of coating loss when the liberated inhibitor is not enough to cover the whole exposed metal substrate. This process was therefore primarily discovered in primers and prior treatment of the layers wherein the inhibitors are in close proximity to the metal substrates [65]. Nevertheless, to maintain the coating's overall barrier and adhesion qualities, an excess of the inhibitor needs to be avoided. To reduce the number of containers, it is preferable to choose inhibitors with strong anticorrosion performance and high inhibitor loading capacities. Nevertheless, this level needs to be kept constant so that the coated surface's capacity for self-healing is present consistently [66]. For a while, the corrosion inhibitor's activity inhibits corrosion reactions, but it does not totally restore the coating's physical barrier. As a result, the creation of self-healing coatings depends on the proper inhibitor selection. The performance of the inhibitor is substrate-dependent. To achieve long-term healing, two main steps would need to be taken: (1) a thorough screening of the inhibitors of corrosion taking into account the local electrolytes chemistry at the site of damage and the reactions of corrosion at the metallic substrate; and (2) an optimized container design taking into account the stability in the environment and adaptability of the container as well as the compatibility of the inhibitors with the coating [34,67]. Furthermore, long-term corrosion testing in simulated and, more importantly, actual service conditions should be conducted before the healing performance is verified. More research is required to understand how inhibitors are released from containers that are partially or completely covered in the coating matrix and transferred to the defect site, despite the fact that numerous studies have shown the stimuli-responsive discharge of inhibitors from free containers [68,69].

10.3.2 Nonautonomous mechanisms of healing

By employing external stimuli unrelated to corrosion processes, coatings' ability to prevent corrosion can also be restored, or at least substantially enhanced, nonautonomously. Since heat and light can be applied easily in service contexts, they are the most typical stimuli for this sort of coating [5,6]. The majority of cases in this category are also examples of inherent healing systems. These coatings are repaired by the restoration of the inherent chemical linkages and physical arrangements of the polymer matrices inside the coating matrix, as opposed to those that are self-healing through extrinsic healing agents. In contrast to inhibitor-based systems, their healing capacity is therefore essentially independent of the metal substrate [70–72]. Because it supplies the activation energy needed for bond breaking or reconstruction, the external stimulus is essential. The flow characteristics of the polymers are enhanced by heat stimulation, which can promote the reactivity of the broken bonds by drawing them closer together. Heat can be produced artificially or naturally (by sunshine, abrasion, etc.) to restore coatings. Sunlight, UV light, and near-infrared light are common light stimuli that are utilized to trigger self-healing effects. A complimentary corrosion sensing component that detects minute pH variations in early stages of corrosion may also aid in the stimuli-responsive healing actions of these coatings [72]. This makes it easier to identify coating deterioration and allows for quicker, more effective repair. These intrinsically repaired polymers in bulk have been evaluated more regularly and have far larger applications in the healthcare, aerospace, construction, and electronics industries than the methods based on extrinsic healing agents. While a thorough examination of intrinsic repairing polymer compounds is outside the purview of this piece, the part that follows discusses a few systems that are more directly connected to protective coatings. There are two key reasons why a coating is typically harder to cure than its bulk counterpart: initially, given the coating's thinness, the polymer networks below the stimulus need to be sufficiently mobile to seal the damage. Secondly, the adhesion of the coating to the substrate may limit the mobility of the polymer matrices throughout the healing process [6,73–75].

10.3.2.1 Self-healing through dynamic connections

Materials with self-healing properties have frequently been created via ionic interactions, hydrogen bonds, and dynamic covalent bonds [76]. For instance, at a specific temperature, thermally reversible bonds can break down, enabling the polymer chains to flow to the defect and re-crosslink to fix it. According to theory, the process of repair can be repeated an endless number of times without the need for second phase healing agents, which is a major benefit of this kind of self-healing technology [77]. A thorough analysis of the typical chemistries of reversible connections that can be used for self-healing polymers can be found elsewhere. The most common example involves the Diels–Alder (DA) reaction, that is, a [4 + 2] cycloaddition between a dienophile and a conjugated diene, like the well-known combination of maleimide and furan [78,79]. For instance, by synthesizing furfuryl glycidylether and *N,N*-diglycidyl furfurylamine, Wouters et al. [78] and Zhang's [79] study illustrated the feasibility of adding reversible Furan and maleimide linkages to traditional epoxy coatings. In recent years, self-healing coatings for corrosion protection on cold rolled steel were created using a trifunctional maleimide, a trifunctional furan, and a tetra-functional furan-capped anilinetrimer. The scratched coating's corrosion protection efficiency, which was determined by measuring the density of corrosion currents in the polarization curve, completely recovered after heating it for 1 hour at 140°C to initiate the retro-DA reaction and for another 24 hours at 80°C to finish the re-crosslinking process through the DA reaction. The examinations also revealed that the trifunctional furan's flexible structure may have had a role in the coating's good self-healing properties. The majority of these examinations show that the healing effects made possible by a retro-DA decyclization frequently call for hours of heating above 100°C, which is unfeasible for the majority of corrosion protection coatings. For DA-based self-healing materials, the inclusion of plasticizers might potentially be useful in achieving softer healing conditions if 10% benzyl alcohol plasticizer is applied to a polymer resin that contains bismaleimides and trifunctional and difunctional furans [80,81]. After heating at 120°C for a period of 5 minutes, the resultant polymer was able to perfectly close a 100 μm surface scratch, while the plasticizer-free control demonstrated nearly no ability to close the damage under similar circumstances [82]. The homo polymerization of maleimides, which reduces the self-healing efficacy, and the side reactions between the maleimides and amines are additional possible disadvantages of the DA-based self-healing process. Reducing the amount of reactive compounds in the polymer network and reducing the length of time exposed to extremely high healing temperatures may help to resolve this problem. Another typical stimulant that might cause self-healing effects is light [83]. These methods have a number of significant advantages over thermally induced self-healing, including the ability to initiate healing instantly, remotely, and as needed. A light stimulus can be carefully administered to only the injury site, unlike heating. The possibility of adverse effects and deterioration in the intact coating throughout the healing process is decreased, which makes this property intriguing in theory for self-healing coatings. Light might be an easy way to start the self-healing process for coatings used outside [84,85]. However, the majority of the chemistry and characteristics of the coating need to be stable against prolonged exposure to sunlight, aside from the light-sensitive linkages. Additionally, this use of light stimulation has been combined with additional self-healing processes, such as regulating the release of inhibitors of corrosion from containers and causing the polymerization of monomers streaming into the coating damaged locations [86].

10.4 Process/techniques adopted for self-healing process

10.4.1 Categorization of self-healing techniques

According to the length and width of the fractures, several forms of self-healing treatments are typically applied in the target environment (as shown in Fig. 10.1).

10.4.1.1 Autogenous/natural healing

In conventional/traditional concrete, about 20%–30% of the cement is left unhydrated. Concrete begins to fracture as water seeps through the unhydrated cement particles. The resulting reaction produces hydration products to seal the gaps and restarts the hydration process. This inherited self-healing process is known as autogenous healing. Fig. 10.3 depicts various autogenous self-healing methods. The following scenarios can result in autogenous healing [87]:

1. Hydration of cement particles that are not hydrated.
2. Crack blocking with $CaCO_3$ or Ca $(OH)_2$.
3. Impurities in the water obstructing cracks.
4. Calcium-silicate hydrates (CSH) gel swelling or expansion.

All of the above-mentioned steps may take place at once. Furthermore, some of the previously listed methods can only partially seal the gaps, leaving the rest unfilled. The most efficient natural healing mechanism is the production of $CaCO_3$ and $Ca(OH)_2$. Numerous research was conducted to find out what influences autogenous healing and how effective it is. It has been discovered that autogenous healing processes work well for small fractures; nevertheless, larger cracks (less than 200 μm) can occasionally be healed in the presence of water in addition to cracks larger than 300 μm [92].

Key elements influencing autogenous healing are dependent upon: (1) the age and makeup of the concrete, (2) the presence of water, and (3) size and form of the crack.

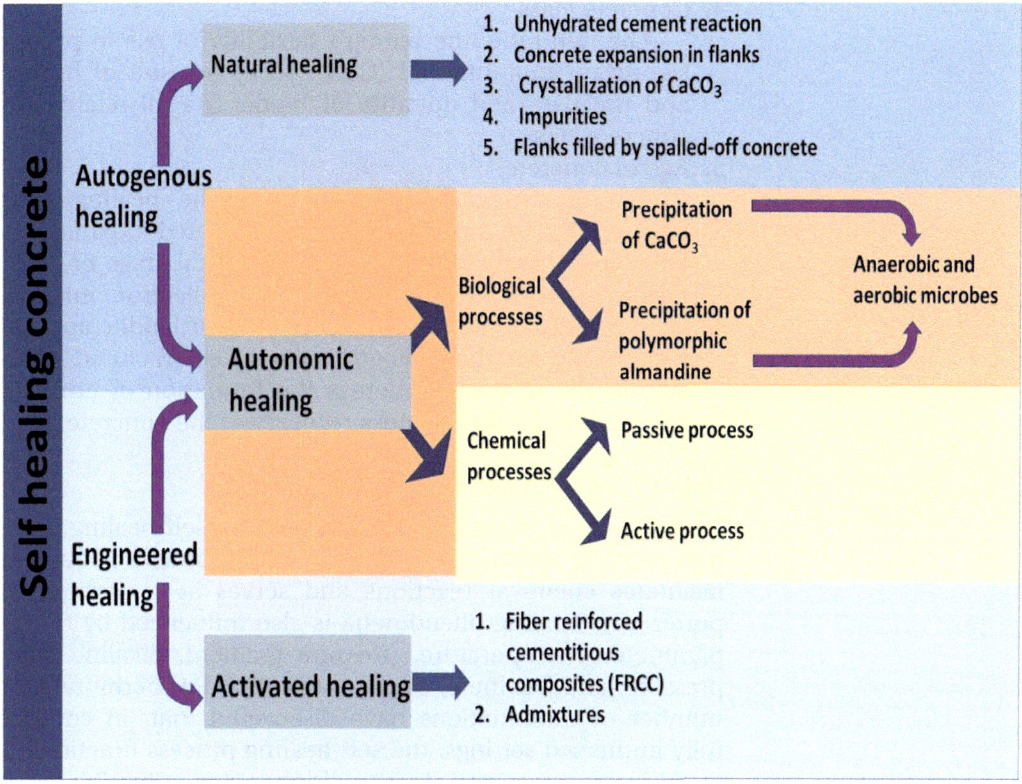

Figure 10.3 Autogenous self-healing methods [88–91].

10.4.1.1.1 The age and makeup of the concrete

The following criteria are included in the concrete composition that determines the autogenous healing potential:

1. Cement's clinker content

   The kind of cement used has little impact on the composition of concrete. But the amount of clinker in cement ultimately shows how the matrix can precipitate calcium carbonate, which controls the supply of calcium ions.

2. Addition of silicates

   By affecting pozzolanic responses the length of the healing mechanism, and the consumption of $Ca(OH)_2$, such addition to the concrete mix influences the kind of mix prepared.

3. Type of aggregate

   This element influences the cracking pattern and the subsequent healing process.

4. **Concrete class**
   The w/c ratio, the binder's flexibility of use in producing significant quantities of (CSH) gel as a result of hydration, and the type and quantity of binder are all related to the concrete class.
5. **Age of concrete**
   Age of concrete is essential for the healing process. Because it contains extremely unhydrated bonding molecules to generate new CSH gel, the initial stage of concrete becomes more profound. A scanning electron microscope analysis showed that the formation of portlandite and CSH is responsible for the strength recovery of fractured concrete submerged in water, whereas the formation of ettringite is responsible for the stiffness recovery of the concrete [93].

10.4.1.1.2 Water content

Exposure to water is said to be ideal for self-healing. Water is necessary for the autogenous healing phenomena to occur, as it facilitates chemical reactions and serves as a particle transporter. The healing phenomena is also influenced by the many parameters (temperature, pressure gradient, alkaline content, pressure, and hardness) of the water [94]. Furthermore, a small number of examinations have discovered that, in contrast to fully immersed settings, the self-healing process functions flawlessly in both wet and dry conditions. This is feasible because the abundance of carbon dioxide in the air makes $CaCO_3$ easy to produce.

10.4.1.1.3 Size and form of cracks

In terms of geometry, cracks indicate the degree of autogenous healing by designating the width, length, and height of the cracks as well as their pattern (branched or aggregated).

By restricting the diameter of the fracture and strengthening the healing mechanism, narrower cracks can also be repaired swiftly and simply. To improve hydration (and healing), controlling the creation and growth of cracks can help store and move water throughout the matrix [95]. Furthermore, to improve autogenous healing, fracture shape and age of the initial crack are also significant factors.

10.4.1.2 Engineered self-healing

Depending on the crack diameter, several engineered self-healing techniques for concrete cracks have been investigated thus far. A practical solution is still lacking for many of the self-healing concrete techniques that have been investigated outside of engineered concrete. For instance, while the bacterial approach is an effective method, most bacteria are killed during the mixing and placement of concrete, making it extremely difficult for them to maintain their existence [94]. Alternative techniques for concrete healing, including the vascular technique, can induce prematuration of healing materials before to the onset of cracks. Even though the previously described ways to engineered self-healing concrete yield inferior outcomes, these technologies make use of fiber reinforcement or engineered cement-based composites. Because these techniques are less expensive, effective at restoring durability, and do not require any specific components, they produce better results in the real world.

10.4.2 Practical methods for self-healing

10.4.2.1 Encapsulation

The most popular and efficient way to introduce a healing agent into material composites, like concrete, is by encapsulation. Depending on desired efficiency, a range of materials can be used to manufacture healing agents and capsules [96]. Encapsulation is only a successful process if the internal forces of the concrete mixing are balanced against the mechanical resistance of the capsules. Capsules viscosity, shell thickness, diameter, and surface all affect encapsulation healing procedures. This method can release the pattern of chemical and biological healing ingredients in concrete and increase its active duration [97]. Numerous studies have indicated that this procedure is adaptable and successful due to its capacity to restore the mechanical and durable qualities of concrete. Furthermore, the repeatability of encapsulation techniques is being challenged by the repeated damage to concrete buildings and the limited number of healing chemicals that capsules can supply in a concrete fracture, making long-term usability unfeasible. Currently, however, research is concentrated on the active release of healing agents, proposing the production of capsules that will enable several healing cycles [98]. To conduct this research, a healing material in a tube was used; however, this decreased the recovery of durable properties each time the healing process began. Encapsulation techniques are divided into two categories based on where the capsules are dispersed: tabular glass, which distributes the capsules at projected failure points, and microcapsules, which distribute the capsules uniformly throughout the matrix [99,100]. Chemical agents and bacterial precipitation are two general categories into which healing produced by encapsulation self-healing can be divided.

10.4.2.1.1 Chemical agents

The development of the self-healing phenomena in cement matrix involves the employment of various encapsulating materials, such as polymers, lightweight particles, glass and ceramic tubes, etc. Polymeric capsules, one of the encapsulating materials previously described, are widely employed after being prepared using the polymeric material's oil-in-water dispersion phenomena. Most often, NH_2CONH_2 and CH_2O combine in the liquid phase to create powerful formaldehyde (CH_2O) and urea (NH_2CONH_2) capsule shells. To give capsules a slightly rough texture, the prepolymer created by the aforementioned reaction in the liquid phase can be removed [33,101,102]. The bonding with the cement matrix can be improved by using the method described. Other materials used in the production and processing of capsules include glass and acrylic tubes, the polyurethane and melamine-based encapsulated materials [32]. MMA monomer, sodium silicate solution, epoxy, isocyanate prepolymer, cyanoacrylate, and other healing agents have been added to the aforementioned capsules [103]. The geometry and design of the capsules used to encapsulate healing substances are crucial to their proper operation. This method is advantageous because it can effectively restore the mechanical characteristics of concrete in cases of multiple cracking. However, this approach is not environmentally friendly, weakens concrete, and does not allow for the recycling of concrete that has been mixed with encapsulated chemical agents.

10.4.2.1.2 Bacterial precipitation

Generally, several types of materials, including melamine-based and diatomaceous earth, are utilized to manufacture capsules to encapsulate microbial organisms [104]. Furthermore, the sorts of microbial species employed in the aforementioned capsules vary based on a number of variables, including temperature, concrete type, and crack breadth. *Pseudomonas aeruginosa*, *Bacillus sphaericus*, and *Bacillus pseudofirmus* are typical examples of these microorganisms. The aforementioned bacteria have been effectively encapsulated to support concrete's self-healing characteristics, and their impacts have varied on the material's qualities. Fig. 10.4 shows the several steps which concrete undergoes during self-healing. It includes bacterial participation also. The primary drawbacks of this

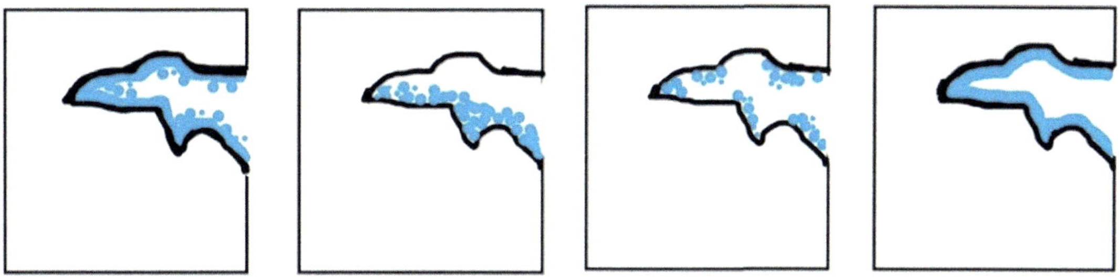

Figure 10.4 Concrete undergoes several steps that include encapsulated bacterial precipitation.

technology are that it is costly, repeatable, and moisture-dependent; on the other hand, its main advantages are that it is environmentally friendly as well as compatible with cement matrix. Figure illustrates the steps that must be taken to incorporate encapsulated bacterium precipitation in concrete.

10.5 Water-repellent nanopaint: synthesis

Because of their special qualities, superhydrophobic surfaces which are produced by combining surface roughness with decreased surface energy have caught the attention of both basic scientists and companies in the manufacturing industry. Numerous uses for superhydrophobic surfaces have been developed, including self-cleaning windows or solar panels and anticorrosion coatings [105]. These uses are inspired by nature, such as lotus leaves, insect legs, and insect wings. The development of superhydrophobic coatings has involved the application of a number of techniques, including electrospinning, laser ablation, spray coating, sol–gel, and electroless deposition [106–108]. Fig. 10.5 diagrammatically shows the film production process. When hydrophobic molecules are chemically grafted onto a smooth surface, the surface energy is altered, creating a hydrophobic surface with a contact angle of up to 90 degrees. Raising the surface roughness has a major impact on wetting and increases the perceived contact angle (>90 degrees) considerably. Roughness increases result in a surface that is superhydrophobic having a contact angle greater than 150 degrees [109]. It has been observed that hierarchical structures with at least one dual-scale roughness are an effective nanostructure for achieving minimal contact angle hysteresis with a large contact angle [110,111]. The solid–liquid

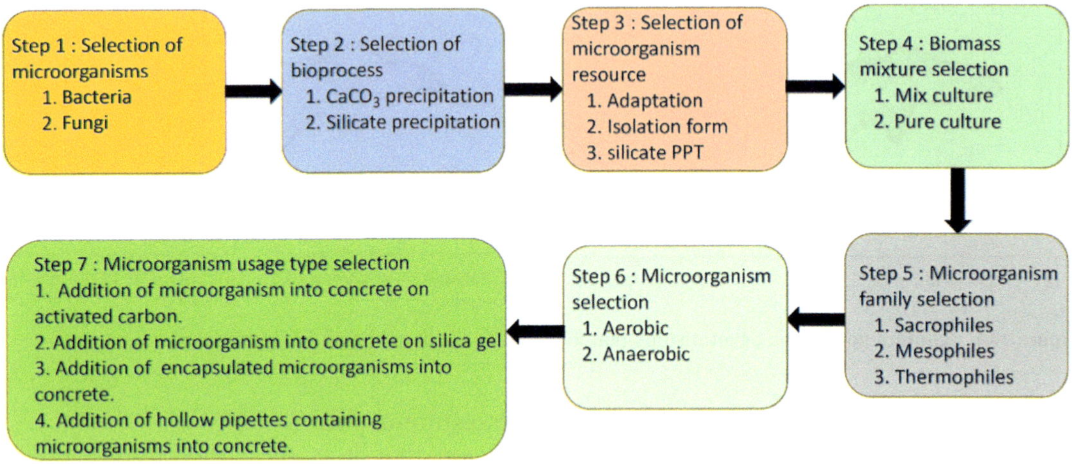

Figure 10.5 Diagrammatic depiction of the film's production process.

interaction region is decreased by this composite surface that has pockets of air in the valleys within the asperities. One of the many living examples that demonstrates the effectiveness of mixing such disparate scales is the surface of the Lotus leaf, where nanosized tubules and micro-sized papillose epidermal cells coexist. Controllable-size silica nanoparticles produced by Stöber soft chemistry are recommended as a means of regulating surface roughness at the nano and microscales [112]. According to recent research, nanoparticle aggregation can be used to produce nanoscale and microscale roughness that result in a hierarchical structure. By raising the liquid droplets' breakthrough pressure, hierarchical roughness is in fact necessary for the stability of the system and the resilience of the liquid-repellent surfaces [113]. While the results demonstrate a synergistic impact, the literature currently in publication suggests that the moderately dense surface coverage and randomly distributed monoscale nanoparticles allow for the achievement of superhydrophobic characteristics. As the surface is covered more thoroughly, the surface roughness diminishes [114].

10.5.1 Synthesis

10.5.1.1 Design of coatings

Two layers made up the water-repellent coating, which was designed to regulate the roughness. The initial layer, which was in contact with the substrate, was developed by employing hybrid silica nanoparticles of precisely regulated size. By adjusting the silica nanoparticle size, the roughness could be controlled by varying the spacing between the particles in the topmost layer (where the size is fixed at 100 nm). Each top layer nanoparticle is triangulated, or uniformly arranged, by a minimum of three other nanoparticles. The nanoparticles are arranged in a random triangulation, with the triangulation vertices formed by the cores of the particles. As a result, changing the size of the nanoparticles in the first layer will change both the spacing and arrangement of the nanoparticles in the top layer (triangle or squared forms). Hence, altering the nanoparticles' sizes, spacing between them (also known as distance), and organization (also known as periodicity) are crucial factors in determining whether or not water permeates the surface texture. The surface covering density and, thus the spacing between neighboring nanoparticles may be effectively controlled according to this model's building design. Consequently, the water contact angle will be changed by adjusting the roughness. Similar claims have been made by Wang et al. on the design of an intricate textured surface that resembles mushrooms [115]. The breakthrough pressure, or point at which the droplet penetrates or not, is controlled by adjusting the size of pillars that resemble mushrooms (Fig. 10.6) [116].

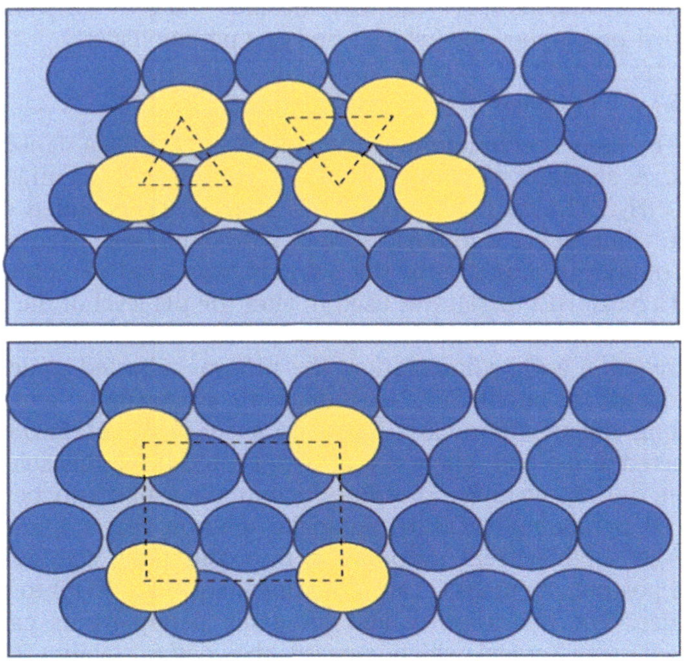

Figure 10.6 Hierarchical perspective on self-healing methods [8,117–122].

10.5.1.2 Content materials

These are the reagents that were used: Aldrich offered tetraethylorthosilicate (TEOS, 98%), vinyltriethoxysilane (VTES, 98%), sodium dodecylbenzene sulfonate (SDBS, $M = 348.48\ \text{g mol}^{-1}$ Aldrich), ethanol absolute (≥99.8%, Aldrich), 2-hydroxy-2-methyl-propiophenone (97%, Aldrich), hydrochloric acid (HCl, 37%), and 1*H*,1*H*,2*H*,2*H*-perfluorooctyltriethoxysilane (98%, Aldrich). Without any additional purification, all of the reagents were used just as they were obtained from the manufacturers [116].

10.5.1.3 Synthesis of hybrid silica nanoparticles

The process of creating hybrid silica nanoparticles involved using a microemulsion. In brief, the vinyl group-functionalized V-silica nanoparticles were created at 50°C [123]. Under normal circumstances, an emulsion was created by vigorously stirring a mixture of VTES (3.8 g) and SDBS (1.3×10^{-3} g [~150 nm]; 2.6×10^{-3} g [~100 nm]; 3.9×10^{-3} g [~70 nm]; 6.5×10^{-3} g [~40 nm]) in 30 mL of water. Addition of $NH_3 \cdot H_2O$ (0.5 mL) dropwise resulted in the emulsion pH being fixed at 11.5. A temperature of 50°C was maintained for 48 hours on the reaction mixture. A temperature of 50°C was maintained for 48 hours on the reaction mixture. The result was a uniform colloidal dispersion. Before coating, 4 mL of each solution was sonicated for 30 minutes in 25 mL of 100% ethanol (pH <2) that included 1.0 g of 2-hydroxy-2-methyl-propiophenone (which acted as a photoinitiater).

10.5.1.4 Synthesis of silica nanoparticles

The Stöber technique was used to create the silica nanoparticles. A solution of water (14 mL), pure ethanol (15 mL), and $NH_3 \cdot H_2O$ (1.5 mL) was mixed with 1 mL of TEOS added dropwise while being vigorously stirred. Distillation was used to get rid of the ammonia after the solution was aged for 1 hour at 40°C. Next, HCl (37%), was used to alter the pH level of the solution containing colloidal particles to less than 2. To prevent aggregation and sedimentation, the pH was adjusted under the isoelectric value of SiO_2 to induce an electrostatic repulsion among the nanoparticles. To create a sol, 1 mL of TEOS was injected at the end. The top layer was coated using this solution. After the dip-coating procedure, a thin silica film can be produced by adding TEOS to the acidic solution that is known as xerogel. Both solvent evaporation throughout the coating step and postheating at 70°C cause the precursor (i.e., TEOS) to polymerize. The hybrid silica and inorganic nanoparticles can be linked using Xerogel, which increases the coating stability [116].

10.5.1.5 Method of coating

The polymeric substrate, also known as the polycarbonate substrate, was first thoroughly cleaned with a detergent, then rinsed with deionized water and immersed in ethanol under ultrasonication over a duration of 15 minutes. An enlarged Plasma Cleaner (Harrick Scientific model PDC-001) was used to plasma-treat the substrate for 60 minutes at a flow of oxygen (0.5 mL min^{-1}). The initial exposure duration was adjusted to minimize contact angle and prevent surface etching. The primary objective of the plasma treatment was to oxidize the substrate surface to produce hydroxyl groups in addition to eliminating impurities from the surface. The primary objective of the plasma treatment was to oxidize the substrate surface to produce hydroxyl groups in addition to eliminating impurities from the surface. The substrate's plasma-activation enhances the V-Silica nanoparticles' adherence. The substrate was dipped in a solution of colloidal particles at a velocity of 20 mm min^{-1} to create the first layer. To further elaborate on the latter, 1.0 g of 2-hydroxy-2-methyl-propiophenone (as a photon initiator) and 4.5 mL of the mother solution containing V-silica nanoparticles are dispersed in 25 mL of pure ethanol. A sufficient amount of HCl (37%) was added to the solution to bring its pH down to 2. 10 dipping cycles were conducted to create a uniform layer. After the samples were dried for 30 minutes at 70°C, they were exposed to UV light to cause a photo-grafting reaction among the groups of vinyl on the surface of the polymer and the silica nanoparticles. Then, to get rid of the nonbonded nanoparticles, the treated surface was thoroughly cleaned with distilled water. The second layer was created by dip-coating the substrate in a solution comprising TEOS precursor and SiO2 ($d = 100$ nm) nanoparticles at the same speed. To create the Xerogel and evaporate the solvent, the ultimate coating was dried for 30 minutes at 70°C. The coated substrate was submerged for 30 minutes in an anhydrous solution of 1*H*,1*H*,2*H*,2*H*-perfluorodecyltrichlorosilane in heptane (5 g L^{-1}) to create a hydrophobic surface. As a result, ethanol and water were used to thoroughly wash the surfaces [116].

10.6 Recent advancements

10.6.1 UV-curable self-healing nanopaints

Because of its affordability, energy efficiency, and environmental friendliness, UV-curable coating has found widespread application as protective and decorative surface coating. When exposed to UV light, the coating can be fully cured in a matter of seconds [124]. Nevertheless, because UV-curable coating is thin, it can be damaged by external factors like bumping and scraping, which can lead to the formation of microscopic fissures that are hard to see with the unaided eye.

10.6.1.1 Exchange of disulfide bonds

The energy of disulfide bonds is lower than that of C–C bonds and C–H bonds because they are often produced by two linking two mercaptan groups. Functional materials can have self-healing properties when disulfide bonds are added to them due to the reversible nature of these bonds [125]. At some point chain exchange reactions or temperature in the nucleophile system can happen among disulfide-bonded cross-linked structures, starting the self-healing process. A novel molecular structure was prepared by Hideyuki et al. [126] utilizing a 400 W high-pressure reactor using alkyl disulfide as the substrate and found that disulfide bonds might form in a mercury lamp between several molecules to produce a novel chemical, suggesting that disulfide bonds take involvement in both heat and light exchange. To discover how the presence of disulfide bonds affects the UV-curable coating's self-healing rate was designed by Zhao et al. [127] a range of UV-curable, self-healing polyurethane acrylic resins containing varying amounts of disulfide bonds with the addition of 2,2-dithio diethylene glycol, polytetramethylene ether glycol, and isocyanate into the molecular structure of the resin. As the amount of disulfide bonds in the UV-curable coatings increased, so did their self-healing efficacy. In addition to being compatible with commercial acrylic resin, the coating continued to show positive self-healing properties when a UV light source was employed in place of the high-pressure mercury lamp to reduce the heat effect of the lamp. Li et al. [128] used sulfiydryl-ene photopolymerization to create UV-curable based on water polyurethane self-healing coatings. The primary component of these coatings is polysulfide prepolymer. The produced coating displayed outstanding anticorrosion efficacy and self-healing capabilities due to the interchange of disulfide bonds, suggesting a new avenue for advancement of multifunctional anticorrosion coatings [129].

10.6.1.2 Hydrogen bonds

A hydrogen bond is substantially weaker than an ionic or covalent bond and slightly stronger than an intermolecular force.

Whenever an acceptor of protons and donors approach one another within a specific distance, a hydrogen bond can develop. Strong interaction hydrogen bonds, common and weak interaction hydrogen bonds, and unconventional hydrogen bonds are the three categories of hydrogen bonds based on their bond energies [130,131]. To get elevated self-healing efficacy, experts frequently integrate hydrogen bonds into the molecular structure blueprint of self-repairing materials [132]. To create polymers with high conductivity, Han et al. [133] synthesized a variety of acrylic monomers with imidazole groups and exposed them to UV radiation. The addition of the imidazole ring raised the polymer's glass transition temperature, and the hydrogen bond between it and the hydroxyl group in the polymer gave the material the ability to mend itself. Tetra-hydrogen bond-containing polymers have higher rates of self-healing and more favorable mechanical characteristics than pure supramolecular polymers. Liu et al. [134,135] created quadruple hydrogen bonds to examine the impact of hydrogen bonds on the self-healing effectiveness of coatings, the resultant oligomers were highly adhesive, glossy, and rigid with good self-healing properties. Active diluents can be used to modify the self-healing coating's additional characteristics. The fluidity of the coating's polymer chain can be increased by raising the temperature to break down the hydrogen bonds present in the damaged coating. The hydrogen bonds reform to finish the self-healing process when the temperature drops. Using a rigid hexachlorocyclophosphazene group with a polyphenyl ring as its core, Liu et al. [136] developed and synthesized a UV-curable polyurethane covering with numerous soft arm structures made of polyurethane produced with polycarbonate glycol. The coating's outstanding hardness and remarkable self-healing capabilities show that a rigid core can successfully increase a coating's hardness. An efficient way to increase the hardness of a self-healing coating is to incorporate soft/hard phases into its polymer structure [129].

10.7 Conclusion

To prolong the lifespan of organic coatings for corrosion protection, a thorough examination of both autonomous and non-autonomous self-healing mechanisms is presented in this paper. A thorough taxonomy covering every approach that might be employed to create self-healing concrete with various healing agents and efficaciousness is also provided. Concrete made by encasing chemical healing agents is a preferable choice since the reaction that occurs between the unhydrated particles of cement and the healing agents is well-received. This article also includes how to construct and manage the roughness of a nanopaint that repels water. Inorganic and hybrid silica nanoparticles are used in the creation of this extremely hydrophobic nanopaint. Additionally highlighted new developments in the study of certain functional coating categories, emphasizing their applications. Among these were self-healing and water-repellent nanopaints.

References

[1] K. Urdl, A. Kandelbauer, W. Kern, U. Müller, M. Thebault, E. Zikulnig-Rusch, Self-healing of densely crosslinked thermoset polymers—a critical review, Progress in Organic Coatings 104 (2017) 232–249. Available from: https://doi.org/10.1016/j.porgcoat.2016.11.010.

[2] V. Srivastava, M. Gupta, Parametric assessments of self-healing characteristics in AA2014–NiTi-based metallic composites through destructive and nondestructive evaluation, Russian Journal of Nondestructive Testing 56 (12) (2020) 1064–1082. Available from: https://doi.org/10.1134/S1061830920120104.

[3] Y.-K. Song, H.-W. Kim, C.-M. Chung, Repeatable self-healing of a protective coating based on vegetable-oil-loaded microcapsules, Polymers 14 (10) (2022) 2013. Available from: https://doi.org/10.3390/polym14102013.

[4] M. Sabzi, S. Mersagh Dezfuli, A study on the effect of compositing silver oxide nanoparticles by carbon on the electrochemical behavior and electronic properties of zinc-silver oxide batteries, International Journal of Applied Ceramic Technology 15 (6) (2018) 1446–1458. Available from: https://doi.org/10.1111/ijac.13047.

[5] Y. Yang, M.W. Urban, Self-healing polymeric materials, Chemical Society Reviews 42 (17) (2013) 7446. Available from: https://doi.org/10.1039/c3cs60109a.

[6] M.L. Zheludkevich, A.E. Hughes, Delivery Systems for Self Healing Protective Coatings (2016) 157–199. http://doi.org/10.1007/978-94-017-7540-3_8.

[7] S.R. White, et al., Autonomic healing of polymer composites, Nature 409 (6822) (2001) 794–797. Available from: https://doi.org/10.1038/35057232.

[8] K.S. Toohey, N.R. Sottos, J.A. Lewis, J.S. Moore, S.R. White, Self-healing materials with microvascular networks, Nature Materials 6 (8) (2007) 581–585. Available from: https://doi.org/10.1038/nmat1934.

[9] J. Yang, M.W. Keller, J.S. Moore, S.R. White, N.R. Sottos, Microencapsulation of isocyanates for self-healing polymers, Macromolecules 41 (24) (2008) 9650–9655. Available from: https://doi.org/10.1021/ma801718v.

[10] S.H. Cho, S.R. White, P.V. Braun, Self-healing polymer coatings, Advanced Materials 21 (6) (2009) 645–649. Available from: https://doi.org/10.1002/adma.200802008.

[11] Z.P. Zhang, M.Z. Rong, M.Q. Zhang, C. Yuan, Alkoxyamine with reduced homolysis temperature and its application in repeated autonomous self-healing of stiff polymers, Polymer Chemistry 4 (17) (2013) 4648. Available from: https://doi.org/10.1039/c3py00679d.

[12] H. Yi, Y. Deng, C. Wang, Pickering emulsion-based fabrication of epoxy and amine microcapsules for dual core self-healing coating, Composites Science and Technology 133 (2016) 51–59. Available from: https://doi.org/10.1016/j.compscitech.2016.07.022.
[13] P. Vijayan P., M.A.S.A. Al-Maadeed, TiO_2 nanotubes and mesoporous silica as containers in self-healing epoxy coatings, Scientific Reports 6 (1) (2016) 38812. Available from: https://doi.org/10.1038/srep38812.
[14] X. Cai, D. Fu, A. Qu, Optimization of preparation conditions of epoxy-containing nanocapsules, Science and Engineering of Composite Materials 24 (1) (2017) 155–161. Available from: https://doi.org/10.1515/secm-2014-0469.
[15] M. Huang, J. Yang, Facile microencapsulation of HDI for self-healing anticorrosion coatings, Journal of Materials Chemistry 21 (30) (2011) 11123. Available from: https://doi.org/10.1039/c1jm10794a.
[16] W. Wang, L. Xu, F. Liu, X. Li, L. Xing, Synthesis of isocyanate microcapsules and micromechanical behavior improvement of microcapsule shells by oxygen plasma treated carbon nanotubes, Journal of Materials Chemistry A 1 (3) (2013) 776–782. Available from: https://doi.org/10.1039/C2TA00612J.
[17] W. Wang, L. Xu, X. Li, Z. Lin, Y. Yang, E. An, Self-healing mechanisms of water triggered smart coating in seawater, Journal of Materials Chemistry A 2 (6) (2014) 1914–1921. Available from: https://doi.org/10.1039/C3TA13389C.
[18] G. Wu, J. An, D. Sun, X. Tang, Y. Xiang, J. Yang, Robust microcapsules with polyurea/silica hybrid shell for one-part self-healing anticorrosion coatings, Journal of Materials Chemistry A 2 (30) (2014) 11614–11620. Available from: https://doi.org/10.1039/C4TA01312C.
[19] M. Huang, J. Yang, Salt spray and EIS studies on HDI microcapsule-based self-healing anticorrosive coatings, Progress in Organic Coatings 77 (1) (2014) 168–175. Available from: https://doi.org/10.1016/j.porgcoat.2013.09.002.
[20] W. Wang, L. Xu, H. Sun, X. Li, S. Zhao, W. Zhang, Spatial resolution comparison of AC-SECM with SECM and their characterization of self-healing performance of hexamethylene diisocyanate trimer microcapsule coatings, Journal of Materials Chemistry A 3 (10) (2015) 5599–5607. Available from: https://doi.org/10.1039/C5TA00529A.
[21] N.W. Khun, H. Zhang, D.W. Sun, J.L. Yang, Tribological behaviors of binary and ternary epoxy composites functionalized with different microcapsules and reinforced by short carbon fibers, Wear 350–351 (2016) 89–98. Available from: https://doi.org/10.1016/j.wear.2016.01.007.
[22] D. Sun, H. Zhang, X.-Z. Tang, J. Yang, Water resistant reactive microcapsules for self-healing coatings in harsh environments, Polymer 91 (2016) 33–40. Available from: https://doi.org/10.1016/j.polymer.2016.03.044.
[23] S.J. García, et al., Self-healing anticorrosive organic coating based on an encapsulated water reactive silyl ester: synthesis and proof of concept, Progress in Organic Coatings 70 (2–3) (2011) 142–149. Available from: https://doi.org/10.1016/j.porgcoat.2010.11.021.
[24] F. Cotting, I.V. Aoki, Smart protection provided by epoxy clear coating doped with polystyrene microcapsules containing silanol and Ce(III) ions as corrosion inhibitors, Surface and Coatings Technology 303 (2016) 310–318. Available from: https://doi.org/10.1016/j.surfcoat.2015.11.035.
[25] A. Kumar, L.D. Stephenson, J.N. Murray, Self-healing coatings for steel, Progress in Organic Coatings 55 (3) (2006) 244–253. Available from: https://doi.org/10.1016/j.porgcoat.2005.11.010.

[26] M. Samadzadeh, S.H. Boura, M. Peikari, A. Ashrafi, M. Kasiriha, Tung oil: an autonomous repairing agent for self-healing epoxy coatings, Progress in Organic Coatings 70 (4) (2011) 383–387. Available from: https://doi.org/10.1016/j.porgcoat.2010.08.017.

[27] P.D. Tatiya, P.P. Mahulikar, V.V. Gite, Designing of polyamidoamine-based polyurea microcapsules containing tung oil for anticorrosive coating applications, Journal of Coatings Technology and Research 13 (4) (2016) 715–726. Available from: https://doi.org/10.1007/s11998-015-9780-2.

[28] H. Li, et al., Preparation of high thermal stability polysulfone microcapsules containing lubricant oil and its tribological properties of epoxy composites, Journal of Microencapsulation 33 (3) (2016) 286–291. Available from: https://doi.org/10.3109/02652048.2016.1171917.

[29] J. Li, et al., Self-assembled graphene oxide microcapsules in Pickering emulsions for self-healing waterborne polyurethane coatings, Composites Science and Technology 151 (2017) 282–290. Available from: https://doi.org/10.1016/j.compscitech.2017.07.031.

[30] D.A. Leal, I.C. Riegel-Vidotti, M.G.S. Ferreira, C.E.B. Marino, Smart coating based on double stimuli-responsive microcapsules containing linseed oil and benzotriazole for active corrosion protection, Corrosion Science 130 (2018) 56–63. Available from: https://doi.org/10.1016/j.corsci.2017.10.009.

[31] M. Mahmoudian, E. Nozad, M.G. Kochameshki, M. Enayati, Preparation and investigation of hybrid self-healing coatings containing linseed oil loaded nanocapsules, potassium ethyl xanthate and benzotriazole on copper surface, Progress in Organic Coatings 120 (2018) 167–178. Available from: https://doi.org/10.1016/j.porgcoat.2018.03.014.

[32] Q. Liu, et al., Preparation and characterization of self-healing microcapsules embedding waterborne epoxy resin and curing agent for asphalt materials, Construction and Building Materials 183 (2018) 384–394. Available from: https://doi.org/10.1016/j.conbuildmat.2018.06.185.

[33] M.R. Kessler, N.R. Sottos, S.R. White, Self-healing structural composite materials, Composites Part A: Applied Science and Manufacturing 34 (8) (2003) 743–753. Available from: https://doi.org/10.1016/S1359-835X(03)00138-6.

[34] M. Behzadnasab, S.M. Mirabedini, M. Esfandeh, R.R. Farnood, Evaluation of corrosion performance of a self-healing epoxy-based coating containing linseed oil-filled microcapsules via electrochemical impedance spectroscopy, Progress in Organic Coatings 105 (2017) 212–224. Available from: https://doi.org/10.1016/j.porgcoat.2017.01.006.

[35] K. Aramaki, Self-healing mechanism of an organosiloxane polymer film containing sodium silicate and cerium(III) nitrate for corrosion of scratched zinc surface in 0.5 M NaCl, Corrosion Science 44 (7) (2002) 1621–1632. Available from: https://doi.org/10.1016/S0010-938X(01)00171-8.

[36] W. Trabelsi, P. Cecilio, M.G.S. Ferreira, M.F. Montemor, Electrochemical assessment of the self-healing properties of Ce-doped silane solutions for the pre-treatment of galvanised steel substrates, Progress in Organic Coatings 54 (4) (2005) 276–284. Available from: https://doi.org/10.1016/j.porgcoat.2005.07.006.

[37] M.F. Montemor, R. Pinto, M.G.S. Ferreira, Chemical composition and corrosion protection of silane films modified with CeO_2 nanoparticles, Electrochimica Acta 54 (22) (2009) 5179–5189. Available from: https://doi.org/10.1016/j.electacta.2009.01.053.

[38] J. Mardel, et al., The characterisation and performance of $Ce(dbp)_3$-inhibited epoxy coatings, Progress in Organic Coatings 70 (2–3) (2011) 91–101. Available from: https://doi.org/10.1016/j.porgcoat.2010.10.009.

[39] J. Carneiro, et al., Chitosan-based self-healing protective coatings doped with cerium nitrate for corrosion protection of aluminum alloy 2024, Progress in Organic Coatings 75 (1–2) (2012) 8–13. Available from: https://doi.org/10.1016/j.porgcoat.2012.02.012.
[40] Z. Mirzakhanzadeh, A. Kosari, M.H. Moayed, R. Naderi, P. Taheri, J.M.C. Mol, Enhanced corrosion protection of mild steel by the synergetic effect of zinc aluminum polyphosphate and 2-mercaptobenzimidazole inhibitors incorporated in epoxy-polyamide coatings, Corrosion Science 138 (2018) 372–379. Available from: https://doi.org/10.1016/j.corsci.2018.04.040.
[41] Y. Liu, et al., Protective film formation on AA2024-T3 aluminum alloy by leaching of lithium carbonate from an organic coating, Journal of the Electrochemical Society 163 (3) (2016) C45–C53. Available from: https://doi.org/10.1149/2.0021603jes.
[42] C.D. Dieleman, P.J. Denissen, S.J. Garcia, Long-term active corrosion protection of damaged coated-AA2024-T3 by embedded electrospun inhibiting nanonetworks, Advanced Materials Interfaces 5 (12) (2018). Available from: https://doi.org/10.1002/admi.201800176.
[43] H. Choi, K.Y. Kim, J.M. Park, Encapsulation of aliphatic amines into nanoparticles for self-healing corrosion protection of steel sheets, Progress in Organic Coatings 76 (10) (2013) 1316–1324. Available from: https://doi.org/10.1016/j.porgcoat.2013.04.005.
[44] D. Shchukin, H. Möhwald, A coat of many functions, Science 341 (6153) (2013) 1458–1459. Available from: https://doi.org/10.1126/science.1242895.
[45] J. Fu, et al., Acid and alkaline dual stimuli-responsive mechanized hollow mesoporous silica nanoparticles as smart nanocontainers for intelligent anticorrosion coatings, ACS Nano 7 (12) (2013) 11397–11408. Available from: https://doi.org/10.1021/nn4053233.
[46] D.G. Shchukin, M. Zheludkevich, K. Yasakau, S. Lamaka, M.G.S. Ferreira, H. Möhwald, Layer-by-layer assembled nanocontainers for self-healing corrosion protection, Advanced Materials 18 (13) (2006) 1672–1678. Available from: https://doi.org/10.1002/adma.200502053.
[47] N. Pirhady Tavandashti, M. Ghorbani, A. Shojaei, Y. Gonzalez-Garcia, H. Terryn, J.M.C. Mol, pH responsive Ce(III) loaded polyaniline nanofibers for self-healing corrosion protection of AA2024-T3, Progress in Organic Coatings 99 (2016) 197–209. Available from: https://doi.org/10.1016/j.porgcoat.2016.04.046.
[48] P. Ren, D. Zhang, C. Dong, X. Li, Preparation and evaluation of intelligent corrosion inhibitor based on photo-crosslinked pH-sensitive hydrogels, Materials Letters 160 (2015) 480–483. Available from: https://doi.org/10.1016/j.matlet.2015.08.005.
[49] E. Alibakhshi, E. Ghasemi, M. Mahdavian, B. Ramezanzadeh, A comparative study on corrosion inhibitive effect of nitrate and phosphate intercalated Zn-Al-layered double hydroxides (LDHs) nanocontainers incorporated into a hybrid silane layer and their effect on cathodic delamination of epoxy topcoat, Corrosion Science 115 (2017) 159–174. Available from: https://doi.org/10.1016/j.corsci.2016.12.001.
[50] M. Mohedano, et al., Active protective PEO coatings on AA2024: role of voltage on in-situ LDH growth, Materials and Design 120 (2017) 36–46. Available from: https://doi.org/10.1016/j.matdes.2017.01.097.
[51] K.A. Yasakau, J. Tedim, M.F. Montemor, A.N. Salak, M.L. Zheludkevich, M.G. S. Ferreira, Mechanisms of localized corrosion inhibition of AA2024 by cerium molybdate nanowires, The Journal of Physical Chemistry C 117 (11) (2013) 5811–5823. Available from: https://doi.org/10.1021/jp3124633.

[52] D. Snihirova, M. Taryba, S.V. Lamaka, M.F. Montemor, Corrosion inhibition synergies on a model Al–Cu–Mg sample studied by localized scanning electrochemical techniques, Corrosion Science 112 (2016) 408–417. Available from: https://doi.org/10.1016/j.corsci.2016.08.008.
[53] I.A. Kartsonakis, G. Kordas, Synthesis and characterization of cerium molybdate nanocontainers and their inhibitor complexes, Journal of the American Ceramic Society 93 (1) (2010) 65–73. Available from: https://doi.org/10.1111/j.1551-2916.2009.03310.x.
[54] X. Liu, et al., Comprehensive understandings into complete reconstruction of precatalysts: synthesis, applications, and characterizations, Advanced Materials 33 (32) (2021). Available from: https://doi.org/10.1002/adma.202007344.
[55] J. Tedim, et al., Zn–Al layered double hydroxides as chloride nanotraps in active protective coatings, Corrosion Science 55 (2012) 1–4. Available from: https://doi.org/10.1016/j.corsci.2011.10.003.
[56] E. Abdullayev, A. Joshi, W. Wei, Y. Zhao, Y. Lvov, Enlargement of halloysite clay nanotube lumen by selective etching of aluminum oxide, ACS Nano 6 (8) (2012) 7216–7226. Available from: https://doi.org/10.1021/nn302328x.
[57] A. Chenan, S. Ramya, R.P. George, U. Kamachi Mudali, Hollow mesoporous zirconia nanocontainers for storing and controlled releasing of corrosion inhibitors, Ceramics International 40 (7) (2014) 10457–10463. Available from: https://doi.org/10.1016/j.ceramint.2014.03.016.
[58] X. Ma, L. Xu, W. Wang, Z. Lin, X. Li, Synthesis and characterisation of composite nanoparticles of mesoporous silica loaded with inhibitor for corrosion protection of Cu–Zn alloy, Corrosion Science 120 (2017) 139–147. Available from: https://doi.org/10.1016/j.corsci.2017.02.004.
[59] E. Shchukina, D. Grigoriev, T. Sviridova, D. Shchukin, Comparative study of the effect of halloysite nanocontainers on autonomic corrosion protection of polyepoxy coatings on steel by salt-spray tests, Progress in Organic Coatings 108 (2017) 84–89. Available from: https://doi.org/10.1016/j.porgcoat.2017.03.018.
[60] J. Cleverly, et al., Dynamics of component carbon fluxes in a semi-arid *Acacia* woodland, central Australia, Journal of Geophysical Research: Biogeosciences 118 (3) (2013) 1168–1185. Available from: https://doi.org/10.1002/jgrg.20101.
[61] Y. Guo, J. Wang, D. Zhang, T. Qi, G.L. Li, pH-responsive self-healing anticorrosion coatings based on benzotriazole-containing zeolitic imidazole framework, Colloids and Surfaces. A, Physicochemical and Engineering Aspects 561 (2019) 1–8. Available from: https://doi.org/10.1016/j.colsurfa.2018.10.044.
[62] M.J. Hollamby, D. Fix, I. Dönch, D. Borisova, H. Möhwald, D. Shchukin, Hybrid polyester coating incorporating functionalized mesoporous carriers for the holistic protection of steel surfaces, Advanced Materials 23 (11) (2011) 1361–1365. Available from: https://doi.org/10.1002/adma.201003035.
[63] T. Yang, et al., Multicomponent intermetallic nanoparticles and superb mechanical behaviors of complex alloys, Science 362 (6417) (2018) 933–937. Available from: https://doi.org/10.1126/science.aas8815.
[64] K. Zhang, L. Wang, G. Liu, Copper(II) 8-hydroxyquinolinate 3D network film with corrosion inhibitor embedded for self-healing corrosion protection, Corrosion Science 75 (2013) 38–46. Available from: https://doi.org/10.1016/j.corsci.2013.05.014.

[65] P.J. Denissen, S.J. Garcia, Cerium-loaded algae exoskeletons for active corrosion protection of coated AA2024-T3, Corrosion Science 128 (2017) 164–175. Available from: https://doi.org/10.1016/j.corsci.2017.09.019.
[66] J.M. Falcón, T. Sawczen, I.V. Aoki, Dodecylamine-loaded halloysite nanocontainers for active anticorrosion coatings, Frontiers in Materials 2 (2015). Available from: https://doi.org/10.3389/fmats.2015.00069.
[67] M. Serdechnova, et al., PEO coatings with active protection based on in-situ formed LDH-nanocontainers, Journal of the Electrochemical Society 164 (2) (2017) C36–C45. Available from: https://doi.org/10.1149/2.0301702jes.
[68] I.A. Kartsonakis, et al., Multifunctional epoxy coatings combining a mixture of traps and inhibitor loaded nanocontainers for corrosion protection of AA2024-T3, Corrosion Science 85 (2014) 147–159. Available from: https://doi.org/10.1016/j.corsci.2014.04.009.
[69] I.A. Kartsonakis, A.C. Balaskas, E.P. Koumoulos, C.A. Charitidis, G.C. Kordas, Incorporation of ceramic nanocontainers into epoxy coatings for the corrosion protection of hot dip galvanized steel, Corrosion Science 57 (2012) 30–41. Available from: https://doi.org/10.1016/j.corsci.2011.12.037.
[70] D. Mata, N. Scharnagl, S.V. Lamaka, E. Malheiro, F. Maia, M.L. Zheludkevich, Validating the early corrosion sensing functionality in poly (ether imide) coatings for enhanced protection of magnesium alloy AZ31, Corrosion Science 140 (2018) 307–320. Available from: https://doi.org/10.1016/j.corsci.2018.05.034.
[71] T.L.P. Galvão, et al., Improving the functionality and performance of AA2024 corrosion sensing coatings with nanocontainers, Chemical Engineering Journal 341 (2018) 526–538. Available from: https://doi.org/10.1016/j.cej.2018.02.061.
[72] N. Dararatana, F. Seidi, D. Crespy, pH-sensitive polymer conjugates for anticorrosion and corrosion sensing, ACS Applied Materials & Interfaces 10 (24) (2018) 20876–20883. Available from: https://doi.org/10.1021/acsami.8b05775.
[73] Z. Wei, et al., Self-healing gels based on constitutional dynamic chemistry and their potential applications, Chemical Society Reviews 43 (23) (2014) 8114–8131. Available from: https://doi.org/10.1039/C4CS00219A.
[74] P. Michael, D. Döhler, W.H. Binder, Improving autonomous self healing via combined chemical/physical principles, Polymer 69 (2015) 216–227. Available from: https://doi.org/10.1016/j.polymer.2015.01.041.
[75] X.K.D. Hillewaere, F.E. Du Prez, Fifteen chemistries for autonomous external self-healing polymers and composites, Progress in Polymer Science 49–50 (2015) 121–153. Available from: https://doi.org/10.1016/j.progpolymsci.2015.04.004.
[76] N. Roy, B. Bruchmann, J.-M. Lehn, DYNAMERS: dynamic polymers as self-healing materials, Chemical Society Reviews 44 (11) (2015) 3786–3807. Available from: https://doi.org/10.1039/C5CS00194C.
[77] R.J. Wojtecki, M.A. Meador, S.J. Rowan, Using the dynamic bond to access macroscopically responsive structurally dynamic polymers, Nature Materials 10 (1) (2011) 14–27. Available from: https://doi.org/10.1038/nmat2891.
[78] M. Wouters, E. Craenmehr, K. Tempelaars, H. Fischer, N. Stroeks, J. van Zanten, Preparation and properties of a novel remendable coating concept, Progress in Organic Coatings 64 (2–3) (2009) 156–162. Available from: https://doi.org/10.1016/j.porgcoat.2008.09.023.
[79] Q. Tian, M.Z. Rong, M.Q. Zhang, Y.C. Yuan, Synthesis and characterization of epoxy with improved thermal remendability based on Diels–Alder

reaction, Polymer International 59 (10) (2010) 1339–1345. Available from: https://doi.org/10.1002/pi.2872.
[80] K.K. Oehlenschlaeger, et al., Adaptable hetero Diels–Alder networks for fast self-healing under mild conditions, Advanced Materials 26 (21) (2014) 3561–3566. Available from: https://doi.org/10.1002/adma.201306258.
[81] K.K. Oehlenschlaeger, et al., Fast and catalyst-free hetero-Diels–Alder chemistry for on demand cyclable bonding/debonding materials, Polymer Chemistry 4 (16) (2013) 4348. Available from: https://doi.org/10.1039/c3py00476g.
[82] G. Postiglione, S. Turri, M. Levi, Effect of the plasticizer on the self-healing properties of a polymer coating based on the thermoreversible Diels–Alder reaction, Progress in Organic Coatings 78 (2015) 526–531. Available from: https://doi.org/10.1016/j.porgcoat.2014.05.022.
[83] D.H. Turkenburg, H.R. Fischer, Diels-Alder based, thermo-reversible cross-linked epoxies for use in self-healing composites, Polymer 79 (2015) 187–194. Available from: https://doi.org/10.1016/j.polymer.2015.10.031.
[84] D. Habault, H. Zhang, Y. Zhao, Light-triggered self-healing and shape-memory polymers, Chemical Society Reviews 42 (17) (2013) 7244. Available from: https://doi.org/10.1039/c3cs35489j.
[85] G.L. Fiore, S.J. Rowan, C. Weder, Optically healable polymers, Chemical Society Reviews 42 (17) (2013) 7278. Available from: https://doi.org/10.1039/c3cs35471g.
[86] Y. Meng, J.-C. Yang, C.L. Lewis, J. Jiang, M. Anthamatten, Photoinscription of chain anisotropy into polymer networks, Macromolecules 49 (23) (2016) 9100–9107. Available from: https://doi.org/10.1021/acs.macromol.6b01990.
[87] M. de Rooij, K. Van Tittelboom, N. De Belie, E. Schlangen (Eds.), Self-Healing Phenomena in Cement-Based Materials, 11, Springer Netherlands, Dordrecht, 2013. Available from: 10.1007/978-94-007-6624-2.
[88] Sybrand van der Zwaag, Self-healing materials: an alternative approach to 20 centuries of materials science, Chemistry International—Newsmagazine for IUPAC 30 (6) (2008). Available from: https://doi.org/10.1515/ci.2008.30.6.20.
[89] M. Wu, B. Johannesson, M. Geiker, A review: self-healing in cementitious materials and engineered cementitious composite as a self-healing material, Construction and Building Materials 28 (1) (2012) 571–583. Available from: https://doi.org/10.1016/j.conbuildmat.2011.08.086.
[90] P. Byoungsun, C.C. Young, Investigating a new method to assess the self-healing performance of hardened cement pastes containing supplementary cementitious materials and crystalline admixtures, Journal of Materials Research and Technology 8 (6) (2019) 6058–6073. Available from: https://doi.org/10.1016/j.jmrt.2019.09.080.
[91] A. Talaiekhozani, A. Keyvanfar, A. Shafaghat, R. Andalib, A Review of Self-healing Concrete Research Development (2014). Available: http://www.jett.dormaj.com (online).
[92] C.-M. Aldea, W.-J. Song, J.S. Popovics, S.P. Shah, Extent of healing of cracked normal strength concrete, Journal of Materials in Civil Engineering 12 (1) (2000) 92–96. Available from: https://doi.org/10.1061/(ASCE)0899-1561(2000)12:1(92).
[93] H. Huang, G. Ye, D. Damidot, Effect of blast furnace slag on self-healing of microcracks in cementitious materials, Cement and Concrete Research 60 (2014) 68–82. Available from: https://doi.org/10.1016/j.cemconres.2014.03.010.
[94] H. Ma, S. Qian, Z. Zhang, Effect of self-healing on water permeability and mechanical property of medium-early-strength engineered cementitious

composites, Construction and Building Materials 68 (2014) 92–101. Available from: https://doi.org/10.1016/j.conbuildmat.2014.05.065.
[95] J.A.O. Barros, L. Ferrara, E. Martinelli (Eds.), Recent Advances on Green Concrete for Structural Purposes, Springer International Publishing, Cham, 2017. Available from: 10.1007/978-3-319-56797-6.
[96] H.M. Jonkers, A. Thijssen, G. Muyzer, O. Copuroglu, E. Schlangen, Application of bacteria as self-healing agent for the development of sustainable concrete, Ecological Engineering 36 (2) (2010) 230–235. Available from: https://doi.org/10.1016/j.ecoleng.2008.12.036.
[97] K. Van Tittelboom, N. De Belie, W. De Muynck, W. Verstraete, Use of bacteria to repair cracks in concrete, Cement and Concrete Research 40 (1) (2010) 157–166. Available from: https://doi.org/10.1016/j.cemconres.2009.08.025.
[98] N.-X. Han, F. Xing, A comprehensive review of the study and development of microcapsule based self-resilience systems for concrete structures at Shenzhen University, Materials 10 (1) (2016) 2. Available from: https://doi.org/10.3390/ma10010002.
[99] C. Joseph, A.D. Jefferson, B. Isaacs, R. Lark, D. Gardner, Experimental investigation of adhesive-based self-healing of cementitious materials, Magazine of Concrete Research 62 (11) (2010) 831–843. Available from: https://doi.org/10.1680/macr.2010.62.11.831.
[100] B. Dong, Y. Wang, G. Fang, N. Han, F. Xing, Y. Lu, Smart releasing behavior of a chemical self-healing microcapsule in the stimulated concrete pore solution, Cement and Concrete Composites 56 (2015) 46–50. Available from: https://doi.org/10.1016/j.cemconcomp.2014.10.006.
[101] K. Van Tittelboom, N. De Belie, D. Van Loo, P. Jacobs, Self-healing efficiency of cementitious materials containing tubular capsules filled with healing agent, Cement and Concrete Composites 33 (4) (2011) 497–505. Available from: https://doi.org/10.1016/j.cemconcomp.2011.01.004.
[102] E. Brown, S. White, N. Sottos, Retardation and repair of fatigue cracks in a microcapsule toughened epoxy composite—part II: in situ self-healing, Composites Science and Technology 65 (15–16) (2005) 2474–2480. Available from: https://doi.org/10.1016/j.compscitech.2005.04.053.
[103] R.A. Chowdhury, et al., Self-healing epoxy composites: preparation, characterization and healing performance, Journal of Materials Research and Technology 4 (1) (2015) 33–43. Available from: https://doi.org/10.1016/j.jmrt.2014.10.016.
[104] J.Y. Wang, N. De Belie, W. Verstraete, Diatomaceous earth as a protective vehicle for bacteria applied for self-healing concrete, Journal of Industrial Microbiology and Biotechnology 39 (4) (2012) 567–577. Available from: https://doi.org/10.1007/s10295-011-1037-1.
[105] P.M. Barkhudarov, P.B. Shah, E.B. Watkins, D.A. Doshi, C.J. Brinker, J. Majewski, Corrosion inhibition using superhydrophobic films, Corrosion Science 50 (3) (2008) 897–902. Available from: https://doi.org/10.1016/j.corsci.2007.10.005.
[106] N. Karthik, M.G. Sethuraman, Fabrication of micro-nanocomposite coatings with lotus leaf like texture by combining electroless and candle soot depositions, New Journal of Chemistry 39 (5) (2015) 3337–3340. Available from: https://doi.org/10.1039/C4NJ02380C.
[107] S.G. Lee, D.S. Ham, D.Y. Lee, H. Bong, K. Cho, Transparent superhydrophobic/translucent superamphiphobic coatings based on silica–fluoropolymer hybrid nanoparticles, Langmuir 29 (48) (2013) 15051–15057. Available from: https://doi.org/10.1021/la404005b.

[108] X. Men, Z. Zhang, J. Yang, X. Zhu, K. Wang, W. Jiang, Spray-coated superhydrophobic coatings with regenerability, New Journal of Chemistry 35 (4) (2011) 881. Available from: https://doi.org/10.1039/c0nj00954g.
[109] I. Sas, R.E. Gorga, J.A. Joines, K.A. Thoney, Literature review on superhydrophobic self-cleaning surfaces produced by electrospinning, Journal of Polymer Science, Part B: Polymer Physics 50 (12) (2012) 824–845. Available from: https://doi.org/10.1002/polb.23070.
[110] L. Gao, T.J. McCarthy, How Wenzel and Cassie were wrong, Langmuir 23 (7) (2007) 3762–3765. Available from: https://doi.org/10.1021/la062634a.
[111] D.C. Pease, The significance of the contact angle in relation to the solid surface, The Journal of Physical Chemistry 49 (2) (1945) 107–110. Available from: https://doi.org/10.1021/j150440a007.
[112] W. Stöber, A. Fink, E. Bohn, Controlled growth of monodisperse silica spheres in the micron size range, Journal of Colloid and Interface Science 26 (1) (1968) 62–69. Available from: https://doi.org/10.1016/0021-9797(68)90272-5.
[113] H. Vahabi, W. Wang, S. Movafaghi, A.K. Kota, Free-standing, flexible, superomniphobic films, ACS Applied Materials & Interfaces 8 (34) (2016) 21962–21967. Available from: https://doi.org/10.1021/acsami.6b06333.
[114] X.Y. Ling, I.Y. Phang, G.J. Vancso, J. Huskens, D.N. Reinhoudt, Stable and transparent superhydrophobic nanoparticle films, Langmuir 25 (5) (2009) 3260–3263. Available from: https://doi.org/10.1021/la8040715.
[115] J. De Coninck, F. Dunlop, T. Huillet, Is superhydrophobicity robust with respect to disorder? The European Physical Journal E 36 (9) (2013) 104. Available from: https://doi.org/10.1140/epje/i2013-13104-2.
[116] J. Conti, J. De Coninck, M.N. Ghazzal, Design of water-repellant coating using dual scale size of hybrid silica nanoparticles on polymer surface, Applied Surface Science 436 (2018) 234–241. Available from: https://doi.org/10.1016/j.apsusc.2017.12.017.
[117] S. Al-Thawadi, S.M. Al-Thawadi, A. Info, Ureolytic Bacteria and Calcium Carbonate Formation as a Mechanism of Strength Enhancement of Sand (2011). Available: https://www.researchgate.net/publication/230603500 (online).
[118] C. Navneet, R. Anita, S. Rafat, Calcium carbonate precipitation by different bacterial strains, African Journal of Biotechnology 10 (42) (2011) 8359–8372. Available from: https://doi.org/10.5897/AJB11.345.
[119] W. De Muynck, K. Cox, N. De Belie, W. Verstraete, Bacterial carbonate precipitation as an alternative surface treatment for concrete, Construction and Building Materials 22 (5) (2008) 875–885. Available from: https://doi.org/10.1016/j.conbuildmat.2006.12.011.
[120] U.K. Gollapudi, C.L. Knutson, S.S. Bang, M.R. Islam, A new method for controlling leaching through permeable channels, Chemosphere 30 (4) (1995) 695–705. Available from: https://doi.org/10.1016/0045-6535(94)00435-W.
[121] H.M. Jonkers, Self Healing Concrete: A Biological Approach, in: S. van der Zwaag (Ed.), Self Healing Materials, in: Springer Series in Materials Science, 100, Springer, Dordrecht, 2007. Available from: https://doi.org/10.1007/978-1-4020-6250-6_9.
[122] R. Siddique, N.K. Chahal, Effect of ureolytic bacteria on concrete properties, Construction and Building Materials 25 (10) (2011) 3791–3801. Available from: https://doi.org/10.1016/j.conbuildmat.2011.04.010.
[123] W. Wang, J. Salazar, H. Vahabi, A. Joshi-Imre, W.E. Voit, A.K. Kota, Metamorphic superomniphobic surfaces, Advanced Materials 29 (27) (2017). Available from: https://doi.org/10.1002/adma.201700295.

[124] T. Liu, et al., Synthesis and characterization of UV-curable waterborne polyurethane acrylate possessing perfluorooctanoate side-chains, Journal of Polymer Research 19 (2) (2012) 9741. Available from: https://doi.org/10.1007/s10965-011-9741-0.
[125] J. Rong, et al., Study on waterborne self-healing polyurethane with dual dynamic units of quadruple hydrogen bonding and disulfide bonds, Polymer 221 (2021) 123625. Available from: https://doi.org/10.1016/j.polymer.2021.123625.
[126] H. Otsuka, S. Nagano, Y. Kobashi, T. Maeda, A. Takahara, A dynamic covalent polymer driven by disulfidemetathesis under photoirradiation, Chemical Communications 46 (7) (2010) 1150–1152. Available from: https://doi.org/10.1039/B916128G.
[127] D. Zhao, S. Liu, Y. Wu, T. Guan, N. Sun, B. Ren, Self-healing UV light-curable resins containing disulfide group: synthesis and application in UV coatings, Progress in Organic Coatings 133 (2019) 289–298. Available from: https://doi.org/10.1016/j.porgcoat.2019.04.060.
[128] T. Li, Z.P. Zhang, M.Z. Rong, M.Q. Zhang, Self-healable and thiol–ene UV-curable waterborne polyurethane for anticorrosion coating, Journal of Applied Polymer Science 136 (26) (2019). Available from: https://doi.org/10.1002/app.47700.
[129] Z. Guo, Research advances in UV-curable self-healing coatings, RSC Advances 12 (50) (2022) 32429–32439. Available from: https://doi.org/10.1039/D2RA06089B.
[130] M. Wei, et al., Novel poly(tetramethylene ether)glycol and poly(ε-caprolactone) based dynamic network via quadruple hydrogen bonding with triple-shape effect and self-healing capacity, ACS Applied Materials & Interfaces 7 (4) (2015) 2585–2596. Available from: https://doi.org/10.1021/am507575z.
[131] D. Zhu, Q. Ye, X. Lu, Q. Lu, Self-healing polymers with PEG oligomer side chains based on multiple H-bonding and adhesion properties, Polymer Chemistry 6 (28) (2015) 5086–5092. Available from: https://doi.org/10.1039/C5PY00621J.
[132] J. Han, H. Gong, S. Jiang, Y. Gao, J. Nie, F. Sun, Effect of imidazolium monomer structure on properties of imidazolium-functionalized self-healing UV-cured polymers for flexible electronic devices, Macromolecular Chemistry and Physics 220 (23) (2019). Available from: https://doi.org/10.1002/macp.201900362.
[133] M.V. Biyani, E.J. Foster, C. Weder, Light-healable supramolecular nanocomposites based on modified cellulose nanocrystals, ACS Macro Letters 2 (3) (2013) 236–240. Available from: https://doi.org/10.1021/mz400059w.
[134] R. Liu, X. Yang, Y. Yuan, J. Liu, X. Liu, Synthesis and properties of UV-curable self-healing oligomer, Progress in Organic Coatings 101 (2016) 122–129. Available from: https://doi.org/10.1016/j.porgcoat.2016.08.006.
[135] F. Gao, et al., Properties of UV-cured self-healing coatings prepared with PCDL-based polyurethane containing multiple H-bonds, Progress in Organic Coatings 113 (2017) 160–167. Available from: https://doi.org/10.1016/j.porgcoat.2017.09.011.
[136] J. Liu, J. Cao, Z. Zhou, R. Liu, Y. Yuan, X. Liu, Stiff self-healing coating based on UV-curable polyurethane with a 'Hard Core, Flexible Arm' structure, ACS Omega 3 (9) (2018) 11128–11135. Available from: https://doi.org/10.1021/acsomega.8b00925.

11

Superhydrophobic nanopaints

Radha Sachan[1] and Priti[2]

[1]Department of Paint Technology, School of Chemical Technology, Harcourt Butler Technical University, Kanpur, Uttar Pradesh, India [2]Department of Paint Technology, Mahamaya Polytechnic of Information Technology, Chail, Kaushambi, Uttar Pradesh, India

11.1 Introduction

Superhydrophobic nanopaints represent an innovative fusion of nanotechnology and superhydrophobic surface principles. These paints incorporate nanoscale materials, which play a crucial role in defining the unique properties of the coating. Whether through the addition of nanomaterials that significantly influence the coating's characteristics or through the nanoscale morphology inherent in certain components, superhydrophobic nanopaints exemplify the intersection of nanotechnology and surface engineering. The inspiration for superhydrophobic nanopaints stems from observations of natural phenomena, particularly the remarkable water repellency demonstrated by lotus leaves. This phenomenon, often referred to as the "lotus effect," serves as a model for achieving superhydrophobicity in artificial surfaces. The intricate structure of lotus leaves provides valuable insights into the requirements for developing superhydrophobic coatings, emphasizing the significance of both low surface energy and surface roughness in achieving extreme water repellency [1].

By harnessing nanotechnology, superhydrophobic nanopaints aim to replicate and enhance the properties observed in natural systems like the lotus leaf. Nanomaterials incorporated into these paints enable precise control over surface characteristics, such as roughness and energy, leading to superior water-repellent properties. Moreover, the nanoscale features inherent in these coatings contribute to their enhanced performance and durability, furthering their applicability in various industries and applications [2].

In essence, superhydrophobic nanopaints represent a groundbreaking advancement in surface engineering, capitalizing on nanotechnology to achieve unparalleled water repellency inspired

Nanopaints. DOI: https://doi.org/10.1016/B978-0-443-26512-9.00011-2

by nature's own designs [3]. As research and development efforts continue to refine these coatings, they hold immense potential for revolutionizing fields ranging from consumer products to industrial applications, offering solutions that are both innovative and sustainable [4].

The term superhydrophobicity is not very old. In 1996, it was proposed by Onda et al. when they published two articles based on the wettability of fractal surfaces [5,6]. Since then, researchers are continuously working on the fabrication of superhydrophobic surfaces. At present, the superhydrophobic nanopaints are in their fourth phase. In the first phase, engineered, ordered/disordered, nanotextured surfaces are treated with a hydrophobic material to get a high water contact angle. These surfaces are fabricated using microelectronics-based photolithography, differential etching phase separated borosilicate glass, self-assembled-monolayer on the cones' surfaces, and by treating diatomaceous earth (DE) with hydrophobic silane to make superhydrophobic DE. This first phase of superhydrophobic nanopaints has various limitations, such as high cost, poor nanostructure stability, low durability, no water vapor repellency or wetting in condensation conditions, impingement issues, and wetting problems with surfactant or oil. This shortcoming leads to the second phase of superhydrophobic nanopaints that is known as "NeverWet" which is comprised of functionalized nanoparticles that are loosely bonded with a top-coat. The limitation with "NeverWet" is that if the topcoat gets damaged then superhydrophobicity will be lost. The third phase is known as volumetric superhydrophobic paints which is superhydrophobic throughout the entire volume of paint and there is no problem of durability. The fourth phase of superhydrophobic paints are commercially viable optically thin films that are durable and provide optical clarity and superhydrophobicity as well. Superhydrophobic nanopaints gained the high interest of researchers and industrialists due to the wide range of application areas. This includes self-cleaning surfaces, anticorrosion coating, antiicing surfaces, antifouling coatings, heat transfer enhancement, electronics devices, viscous drag reduction, desalination, textiles and fabrics, and oil and water separation [7–10].

11.2 Wetting theory

11.2.1 Contact angle

To understand the superhydrophobicity, the basics of the wetting phenomenon are needed. The wetting property is due to the

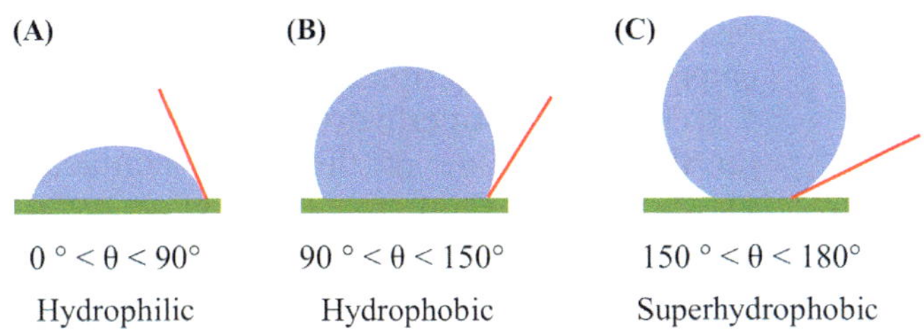

Figure 11.1 Static contact angles and wetting phenomenon for (A) hydrophilic, (B) Hydrophobic, and (C) Superhydrophobic.

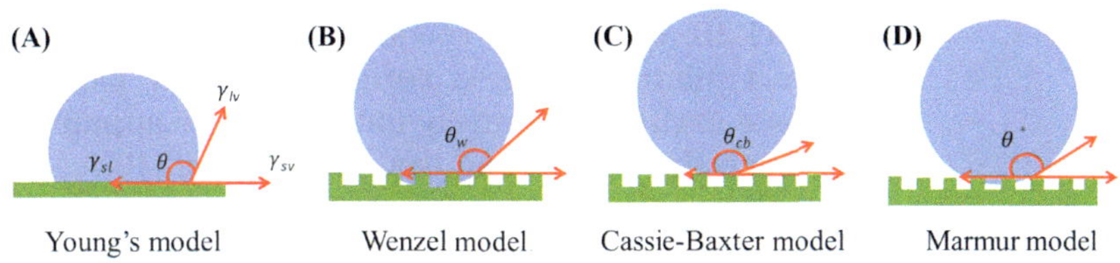

Figure 11.2 Various states of droplets on a solid surface. (A) Young's model, (B) Wenzel model, (C) Cassie-Baxter model, and (D) Marmur model.

interaction between liquid molecules and substrate. The interaction forces between the substrate surface and liquid play a vital role in wetting and spreading phenomena. A strong contact between liquid molecules and solid surfaces results in surface wetting and liquid spreading throughout the surface. In the case of low interaction forces between liquid molecules and substrate, the water droplets attracted to each other and stayed away from the surface which resulted in the hydrophobic phenomenon.

The angle that exists between a droplet edge and a solid surface is called the contact angle as shown in Fig. 11.1. The contact angle helps to determine the surface free energy (SFE) and is used to represent the degree of wettability.

In 1612, Galileo published his paper "Bodies That Stay Atop Water, or Move in it," which contained the results of the first study on wetting phenomena [11]. Wetting theories, which have made great progress in the recent few decades, have been used to define wetting models. Thomas Young first presented the fundamental law of wetting with a water droplet on a smooth, level surface in 1805, as shown in Fig. 11.2 [6] and presented in the form of the below equation given as:

$$\text{Cos}\theta = \frac{\gamma_{sv} - \gamma_{sl}}{\gamma_{lv}} \tag{11.1}$$

where the solid–vapor, solid–liquid, and liquid–vapor interfacial tensions are represented by the values of γ_{sv}, γ_{sl}, and γ_{lv}, respectively, and θ is the static contact angle. The surfaces are categorized as hydrophilic ($0° < \theta \leq 90°$), hydrophobic ($90° < \theta \leq 150°$), and superhydrophobic ($150° < \theta \leq 180°$) depending on the magnitude of the static contact angle as shown in Fig. 11.1.

In 1936, Wenzel modified the Young model, incorporating the roughness factor (r) to account for the influence of surface micro-structure on wettability. This adjustment was necessary because Young's equation was originally formulated for smooth and homogeneous surfaces, failing to accurately represent the behavior of textured surfaces. The Wenzel model, illustrated in Fig. 11.2B, describes a scenario where a droplet makes contact with a micro-structured surface and fully penetrates into the surface cavities, effectively increasing the apparent contact area between the droplet and the solid surface [10].

$$\mathrm{Cos}\theta = r\left(\frac{\gamma_{sv} - \gamma_{sl}}{\gamma_{lv}}\right) = r\ \mathrm{Cos}\theta \qquad (11.2)$$

r = roughness = actual surface/planar surface.

where θ_w is the static contact angle of the Wenzel model. When considering rough surfaces, the roughness factor typically exceeds 1 because such surfaces have a greater surface area compared to flat ones. Consequently, if the contact angle (θ) is less than 90 degrees, an increase in roughness (r) will result in a reduction of the apparent contact angle (θ_w). On the other hand, an increase in roughness causes an increase in θ_w if θ is larger than 90 degrees. The Wenzel equation is applicable only in scenarios where wetting is homogeneous. However, on rough surfaces, liquid drops may not fully penetrate the surface cavities in all cases, resulting in air being trapped within these cavities and creating a composite interface [12].

In 1945, Cassie-Baxter introduced the concept of a composite interface wetting state for droplets on rough surfaces, as depicted in Fig. 11.2C. The Cassie-Baxter model considers the apparent contact angle as influenced by the contributions of two distinct phases. This model accounts for the interactions between the liquid droplet and both the solid surface and the air trapped within the surface features. The equation describing this composite interface wetting state is as follows:

$$\mathrm{Cos}\theta = f_{sl}\mathrm{Cos}\theta_{sl} + f_{lv}\mathrm{Cos}\theta_{lv} \qquad (11.3)$$

where the θ_{cb} is the static contact angle under the Cassie-Baxter state, f_{sl}, and f_{lv} stands for surface fractions of the solid–liquid and liquid–vapor phases, respectively. Additionally, θ_{sl} and θ_{lv}

correspond to the contact angles at the solid–liquid and liquid–vapor interfaces, respectively. Here $Cos\theta_{sl} = Cos\theta_w$ as it corresponds to solid–liquid interface and the second fraction $f_{lv} = 1 - f_{sl}$ and $\theta_{lv} = 180°$ so the Eq. (11.3) becomes:

$$Cos\theta_{cb} = f_{sl}\ (1 + Cos\theta_w) - 1 \tag{11.4}$$

A mixed model of Wenzel and Cassie-Baxter is defined by Marmur as shown in Fig. 11.2D [13]. He derived the intermediate state of droplets which partially wet the surface and partially stay on air pockets. So, In this instance, the roughness factor (r) and the solid–liquid surface percentage determine the contact angle, and are presented as Eq. (11.5).

$$Cos\theta^* = r_f\ f_{sl}\ (1 + Cos\theta_w) - 1 \tag{11.5}$$

Here, r_f is the roughness of the portion of the solid surface that touches the droplet.

11.2.1.1 Static contact angle

The angle at which a liquid droplet touches a solid surface is termed as static contact angle on a flat surface.

11.2.1.2 Dynamic contact angle

When a water droplet is placed on a slanted surface, its volume influences the contact angles observed during both advancing (θ_a) and receding stages (θ_r), as illustrated in Fig. 11.3. Contact angle hysteresis (CAH) describes the difference between the advancing and receding contact angles [14].

11.2.1.3 Roll-off angle

The roll-off angle (θ_R), as depicted in Fig. 11.4, signifies the angle at which a water droplet rolls off a flat slanted surface. Superhydrophobic surfaces are known for their exceptional water-repelling properties, often characterized by roll-off angles of less than 10 degrees. In fact, superhydrophobic surfaces of superior quality may attain roll-off angles of less than 1 degree, demonstrating their remarkable ability to efficiently shed water droplets.

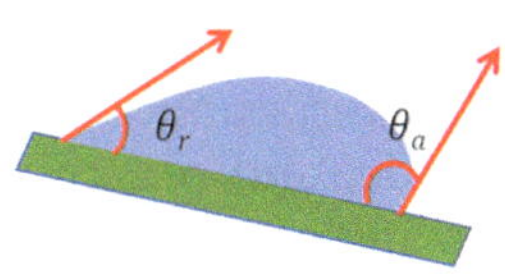

Figure 11.3 Receding and advancing dynamic contact angles.

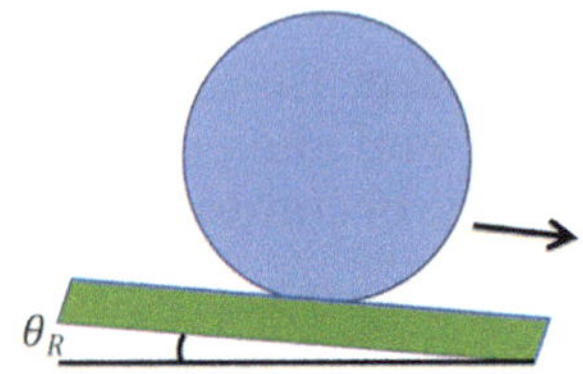

Figure 11.4 Roll-off angle of water droplet.

11.2.2 Surface free energy

The SFE is defined as the energy required to enlarge an interface between two different phases without causing a volume change. Because atoms on the surface come into touch with fewer molecules than those in the bulk phase, the potential energy of surface molecules is higher than that of molecules inside the material. The higher energy zone is intended to be minimized by solid surfaces. The liquid droplet seems to have a spherical form when the surface area/volume ratio is at its lowest [15]. The SFE and surface tension are physically equivalent. The unit for SFE is millijoule per square meter (mJ m^{-2}) whereas the unit for surface tension is millinewton per meter (mN m^{-1}). Every system aspires to the lowest feasible level of free energy. Because of the surface tension (ST), liquids adopt the lowest surface area for a given volume and form spherical droplets while in weightlessness. Solids are unable to diminish their surface via deformation so that they form an interface with a liquid, to lower free energy, and can be wetted. For this reason, a solid's wettability and SFE are strongly connected [16]. Liquids with defined surface tensions are used to calculate the SFE derived from the liquid/solid contact angle specified. The liquid being utilized might have an impact on the SFE. The SFE of polar and nonpolar liquids is calculated using many methods, including the Fowkes method, Owens–Wendt method, Van-Oss-Chaudry-Good method, Zisman method, Neumann method, Young's equation, Dupre equation, and Berthelot hypothesis as well. Polar liquids are subject to concentration on certain molecular interactions within the surface energy components while nonpolar surfaces are not the subject of Table 11.1 concentration. Van der Waals forces and hydrophobic contacts are examples of nonpolar interactions, whereas OH, C = O, amide, and nitrate units are the focus of polar interactions [17]. Table 11.1 illustrates the SFE of various substrates.

Table 11.1 Surface free energy of various liquid and solid materials [18–21].

Material	Surface free energy ($mJ\ m^{-2}$) or surafce tension ($mN\ m^{-1}$)
H_2O	72.8
IPA	21.3
Decane	23.4
Dodecane	24.9
Hexadecane	27.1
Formamide	58.0
Diiodomethane	50.8
Glycerol	64.0
Nitromethane	34.9
Ehtylene glycol	47.3
DMSO	42.9
Perfluorooctane	14.0
Acrylonitrile butadiene styrene	42
Polyethylene	33–35
Polypropylene	30
Polystyrene	25
Polymethylmethacrylate	40
Polytetrafluoroethylene	20–22
Aluminum foil	41–42
Polyvinylchloride	40–42
Alumina	39.7
Calcite	57
Cellulose nitrate	45.1
Dolomite	42.5
Rutile	47.1
Hematite	53.4

Note: *IPA*, Isopropyl alcohol; *DMSO*, Dimethyl sulfoxide.

11.3 Natural superhydrophobic surfaces

Superhydrophobic surfaces have existed naturally since the inception of our flora and fauna world. Researchers got inspired by the "Lotus effect" and tried to develop a similar kind of surface with the help of nanotechnology to make it more durable. It is well known that the lotus plant is an Asian-origin aquatic plant and its leaves do not get wet. Because of this, when rainfall comes into touch with the leaf's surface, it forms spherical droplets that glide freely, collecting all debris and keeping the leaf dry, clean, and free of bacterial colonies even though it lives in contaminated waters. The chemical composition and morphology of lotus leaves were studied by biologists to understand their superhydrophobic characteristics and self-cleaning behavior. They discovered that the leaves exhibited two levels of structure that accounted for this behavior: a nanostructure level created by tiny hairs, and a micro-structure level that has been made up of 20–40 μm epicuticular wax coating lumps, which collectively enable the lotus leaf surface to repel water and establish it as superhydrophobic surface (Fig. 11.5A–C). Lotus leaves have a very stable superhydrophobic surface when compared to many other plants that have a water contact angle of 160 degrees. The lotus effect is a perfect example of superhydrophobic self-cleaning based on traits found in lotus plants. Other significant factors that contributed to the high performance of lotus superhydrophobic surfaces were the formation of microscopic epicuticular wax crystals along leave surfaces, which are responsible for low SFE, the surface's epidermal structures designated with nanoroughness, and the stability of the superhydrophobic layer under conditions of moisture condensation. Dettre and Johnson examined this specific phenomenon in 1964 [22].

Living things in the natural world have developed complex surface chemistries and morphologies that provide them with better qualities needed for survival. The surface of lotus leaves is not the only plant material with water-nonwettability properties; other plant leaves, like rice leaves, and insect wings, like butterfly wings, also exhibit these properties. Rice leaves exhibited superhydrophobicity due to entrapped air holes on the micro/nanostructures' surface grooves. Axial texturing, minimal CAH of $<10^\circ$, microroughness along the leaf, and different micro/nanosurface pins were observed in rice leaves (Fig. 11.5H and I). The top ends of rice leaves feature eggbeater structures with four cohered hairs, whereas the waxy nanobumps with longitudinal grooves and a transverse sinusoidal pattern make up the leaf's surface core. These ridge terminals improve their capacity to hold air under submerged sources, resulting in superhydrophobicity. Taro leaves have micro/nanotexturing with a contact angle of 159 degrees and a roll-off angle of $<3^\circ$ (Fig. 11.5J and K). The butterfly (Fig. 11.5D and E) and dragonfly wings with microgrooves provide a rough surface with low adhesion characteristics and superhydrophobic characteristics which allow them to survive in constant rainfall environments [9].

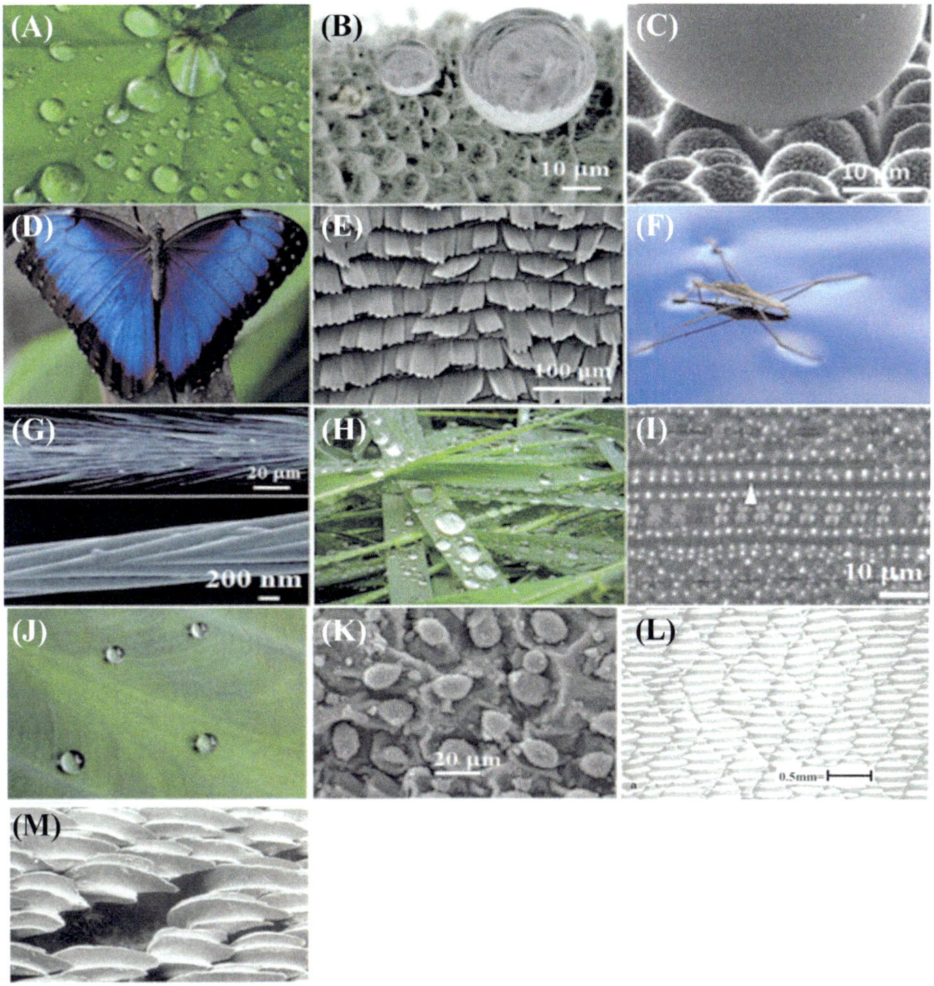

Figure 11.5 (A–C) Superhydrophobic lotus leaf surface, its nanoroughness imparting self-clean ability with scanning electron microscopy images; (D) butterflie's wings; (E) scanning electron microscopy image of the butterflie's wings with directional adhesion having radial outward direction; (F and G) water striders' legs over the water surface and scanning electron microscopy images of legs indicating the oriented micro setae covered with needle morphology and grooved nanostructures; (H and I) superhydrophobic rice leaf surface structure with water droplets on the surfaces, and scanning electron microscopy images of rice leaves with longitudinal grooves and transverse sinusoidal structure (J and K); taro leaf (*Colocasia*) and its scanning electron microscopy image showing nanogrooves; (L) silky shark skin; and (M) scales of great white shark. From D.W. Bechert, M. Bruse W. Hage, R. Meyer, Fluid mechanics of biological surfaces and their technological application, Naturwissenschaften 87(4) (2000), 157–171. https://doi.org/10.1007/s001140050696; S.S. Mohammed, Superhydrophobic self-cleaning surfaces in nature, Nanoarchitectonics 1 (1) (2000) 26–37. https://doi.org/10.37256/nat.112020121.26-37.

A water strider's legs (Fig. 11.5F and G) include grooved nanostructures and orientated microstate coated in needle-shaped formations. Water-repellent structure and microroughness are produced by these surface characteristics. The air is trapped in the surface nanogrooves of the legs, resulting in their superhydrophobicity (contact angle of 167 degrees). Water striders can survive on the water's surface because of this characteristic which allows the insect to walk on water. According to Bechert et al., the shark's skin is covered in tiny tooth-shaped dermal surfaces with longitudinal grooves and ribbed features. These grooves allow for superhydrophobicity by reducing the whirlpools that occur on the ultrasmooth shark skin. In this sense, a shark skin-like nanostructured surface made of silicone showed good fouling prevention against various microorganisms [23].

11.4 Designing of superhydrophobic surfaces

11.4.1 Surface chemical composition

In the 1990s, scientists engineered the first artificial superhydrophobic surface, drawing inspiration from the micro- and nanoscale roughness observed on lotus leaves. The fundamental principle of superhydrophobicity revolves around facilitating the effortless separation of droplets from solid surfaces with minimal external force. Hysteresis in the contact angle dictates how droplets respond to forces applied for removal, a critical factor in achieving easy detachment. Even a slight tilt in the surface can exert a force on the droplet due to its weight, influencing its equilibrium shape. The key to swift droplet removal lies in minimizing hysteresis and reducing the actual contact area between the droplet and the solid surface. Hence, the primary objective of design is to minimize both the wetted area and CAH. This goal, akin to natural phenomena, is achieved by promoting the drop to settle at the Cassie-Baxter condition. In this state, if the design is apt, the majority of the drop's base remains in contact with air rather than the solid surface, thus minimizing contact. The uniformity of air distribution also reduces hysteresis. Additionally, air, being highly hydrophobic in the Cassie-Baxter state, contributes to reducing the wetted area. Consequently, selecting an adequate roughness level to place the drop in the Cassie-Baxter condition satisfies the requirement for minimal wetted area and hysteresis.

Contrasted with the Wenzel state, the Cassie-Baxter state can exhibit thermodynamic stability, metastability, or instability. Stability occurs when the Gibbs energy of the Cassie-Baxter state is lower than that of the Wenzel state, often associated with a smaller contact angle. Instability arises when the Gibbs energy function has a saddle point, a characteristic influenced by roughness design. Convex roughness types favor the Cassie-Baxter state, while concave indentations do not. Metastability occurs when the drop's energy lies between stability and instability, typically with a sufficient energy gap between the Cassie-Baxter and Wenzel states [24].

To obtain the best possible superhydrophobic surfaces an optimum combination of surface chemistry and morphology is required. In terms of surface chemistry, the goal is to make the surface as hydrophobic as feasible. But, obtaining ideal roughness is more intricate. There are various studies available for roughness. According to several theoretical simulations, protrusions with rounded tops seem to be preferable over those with flat tops with sharp edges. This appears to be consistent with the patterns found in nature. Another concern is whether superhydrophobicity requires multiscale or fractal roughness. In which multiscale roughness is reported to be more useful, that is also seen throughout nature [25]. The multiscale roughness provides more durability as well which is the main concern while developing and using superhydrophobic surfaces.

Various chemical compositions have been developed by researchers for the fabrication of superhydrophobic nanopaints. Ko et al. have developed hexadecyltrimethoxysilane-modified SiO_2 particles which exhibit a contact angle of 152 degrees, and a nearly zero-degree roll-off angle [26]. Ruiz-Cabello et al. prepared nanopaints where roughness provided by ZnO nanoparticle and superhydrophobicity by fluorocarbon polymer which gave advancing contact angle (θ_a) 158 degrees, receding contact angle (θ_r) 154 degrees and 0.6 degrees roll-off angle [27]. Nine et al. have made superhydrophobic nanopaints by mixing reduced graphene oxide flakes, TiO_2 nanoparticles, and DE powder in polydimethylsiloxane polymer to attain a water contact angle (WCA) of 155 degrees [28]. Li et al. have fabricated superhydrophobic coating with the copper (Cu) stearate (CS) and PDMS mixture to achieve a contact angle of $>150^\circ$ and roll-off angle of $<10^\circ$ while in another study Suyambulingam et al. have developed CS and ZnO nanoparticles-based superhydrophobic coatings of same conatct angle (CA) [29,30]. A nanocomposite comprised of high molecular weight polydimethyl siloxane (PDMS) and SiO_2-doped ZnO giving to get CA of 163 degrees [31], whereas, PDMS with ZnO nanoparticles exhibit CA of 155 degrees, advancing angle of 160 degrees CAH of 5 degrees and SA of $<10^\circ$ [32]. Hwang et al. prepared superhydrophobic paint with perfluorooctyltriethoxysilane and TiO_2 to obtain a CA of 163 degrees and a roll angle of almost 0 degrees [33,34]. The superhydrophobic is also synthesized by the growth of SiO_2 nanoparticles on hydroxylated carbon nanotubes to achieve CA of 160 degrees and roll-off angle of 5 degrees [35]. Yi et al. have prepared supramolecular silicone coating by reacting Bis-aminopropyl-terminated poly(dimethylsiloxane) with dopamine hydrochloride and isophorone diisocyanate to obtain supramolecular polysiloxanes and further form its ligand with Zn ion which gave CA of 157 degrees and roll of angle of 8 degrees [36,37].

There is a very interesting study by Zhou et al. in which they modified palygorskite clay with amino silicone oil & aminopropyltriethoxysilane. They have incorporated this modified clay into various traditional paints and converted those hydrophilic paints into superhydrophobic nanopaints by achieving a CA of 154 degrees [38]. Wu et al. have synthesized epoxy emulsion comprised of perfluorooctyltriethoxysilane modified SiO_2 to get superhydrophobic paint with CA of 152 degrees and roll-off angle of 6° [39]. Whereas, in another study nano-SiO_2/epoxy resin solution is coated on the sandblasted panel, and further surface is modified with low surface energy polymer solution to get a CA of 167 degrees and roll-off angle of 7 degrees [40]. By mixing dual-sized titanium dioxide (TiO_2) nanoparticles to epoxy resin and then grafting 1H, 1H, 2H, and 2H-perfluorooctyltriethoxysilan, Zhu et al. created organic–inorganic superhydrophobic paint which exhibits CA $>150^\circ$ and SA$<10^\circ$ [41]. A superhydrophobic nanocoating is developed by dispersing magnetic nanoparticles in a PDMS matrix and achieving a CA of 153 degrees [42]. Xue et al. have prepared superhydrophobic nanopaints by adding octyltriethoxysilane-modified TiO_2 and $CoAl_2O_4$ powders into styrene–acrylic emulsion binder and getting CA of 160 degrees roll-off angle of $<3^\circ$ [43].

Zhang et al. have developed water-based superhydrophobic nanopaints by dispersing hydrophobic fumed in water/ethanol with the help of polyvinylpyrrolidone polymer which is water soluble polymer and has the ability to excellently stabilize many kinds of hydrophobic nanoparticles and achieve CA of 151 degrees and SA of $<10^\circ$ [44]. Selim et al. have used silica (SiO_2) carbide nanowires and PDMS to prepare superhydrophobic nanocomposite coating and obtained CA 153 degrees CA hysteresis 9° [45]. By a novel method, Kuzina et al. have prepared a superhydrophobic coating with CA of 170 degrees and SA of $<10^{\circ}$ over a paint surface by the using laser treatment of a paint surface, which is followed by the hydrophobic agent (methoxy-{3[(2,2,3,3,4,4,5,5,6,6,7,7,8,8,8-pentadecafluiorooctyl)-oxy]-propyl}silane being chemisorbed [46]. Il'darkhanova has developed a superhydrophobic coating with volume modification by using fluorosilane-treated carbon nanotubes and carbon nanofibers and incorporating them into epoxy silicone resin [47]. Methyltrimethoxysilane-functionalized SiO_2–titania core–shell nanoparticles are used in polyurethane and obtain a superhydrophobic coating with a CA of 145 degrees [48]. There are a few patents also in which they have fabricated superhydrophobic paint by using fluorinated SiO_2 nanoparticles in fluoropolymer [49], hydroxy-functional silicone-modified polyacrylate with organically modified SiO_2 nanoparticles in SiO_2 oil [50], diatomaceous SiO_2, polysiloxane oil, and perfluorinated organic layer [51]. The deposition of these superhydrophobic nanopaints on various substrates is carried out by several application techniques, such as spray coating, spin coating, dip coating, drop-pipetting, and brush [32,37,45,52].

11.4.2 Surface roughness

As per Cassie-Baxter model, nano and microroughness play a vital role in superhydrophobicity. The surface morphology of fabricated superhydrophobic surfaces is studied by scanning electron microscopy [5,20,34], Field emission scanning electron microscopy [32], transmission electron microscopy, high-resolution transmission electron microscopy [45,53] and atomic force microscope (AFM) [35,45]. These characterizations give visual qualitative results of surface roughness. Some of the results are shown in Figs. 11.6–11.8. Apart from this qualitative morphological evaluation, some scientists reported surface roughness quantitatively. Li et al. have obtained surface roughness by confocal laser scanning microscope and measured in μm for voids. They found that roughness decreases for fabric and paper while increases for glass, iron, and Cu surfaces after treatment with superhydrophobic composition [29]. Selim et al. measured root-mean-square roughness with the help of AFM. As the root-mean-square roughness increases the surface energy reduces, enabling to entrap air within surface interstices and reducing the surface convexity which collectively makes it superhydrophobic [45]. Nine et al. have studied surface roughness profile and reported roughness height, area roughness height, root-mean-square of profile roughness height, root-

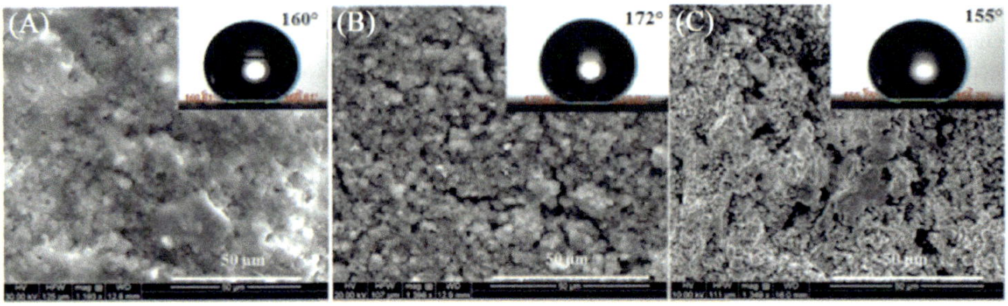

Figure 11.6 Scanning electron microscopy micrographs of (A) diatomaceous earth (DE)-based superhydrophobic coating; (B) DE/titanium dioxide (TiO_2)-based superhydrophobic coating; and (C) DE/TiO_2/reduced graphene oxide (rGO)-based superhydrophobic coating with their respecting static contact angle. From M.J. Nine, M.A. Cole, L. Johnson, D.N.H. Tran, D. Losic, Robust superhydrophobic graphene-based composite coatings with self-cleaning and corrosion barrier properties, ACS Applied Materials and Interfaces 7(51) (2015) 28482–28493. https://doi.org/10.1021/acsami.5b09611.

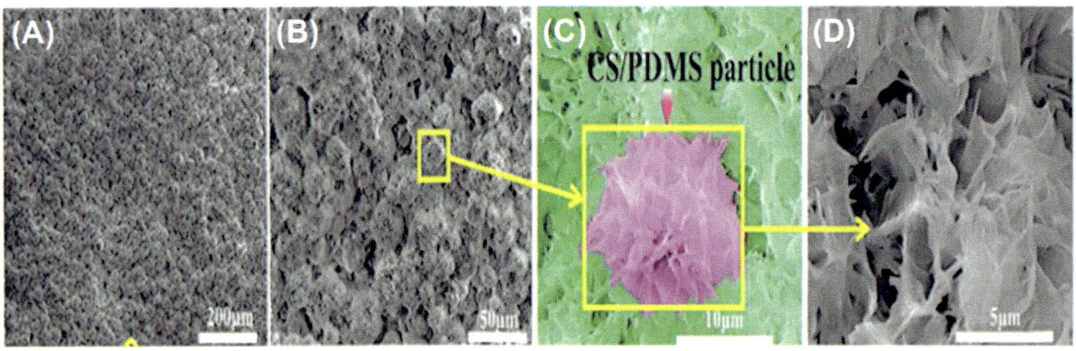

Figure 11.7 (A) Large area view of copper stearate (CS)/polydimethyl siloxane (PDMS) coating surface, (B, C and D) scanning electron microscopy image of CS/PDMS superhydrophobic coating. From L. Li, J. Zhu, Z. Zeng, A sky-blue superhydrophobic coating and applications, Progress in Organic Coatings 147 (2020) 105863. https://doi.org/10.1016/j.porgcoat.2020.105863.

mean-square of area roughness height, and mean spacing between peaks in μm or nm. By these values it can be concluded that a rough surface with large spacing may lead to a lower contact angle as it may follow the Marmur model and low roughness height is also not good enough to get superhydrophobic behavior. So, it can be said that only high roughness is not desirable, instead optimum roughness and spacing are required for superhydrophobicity [28].

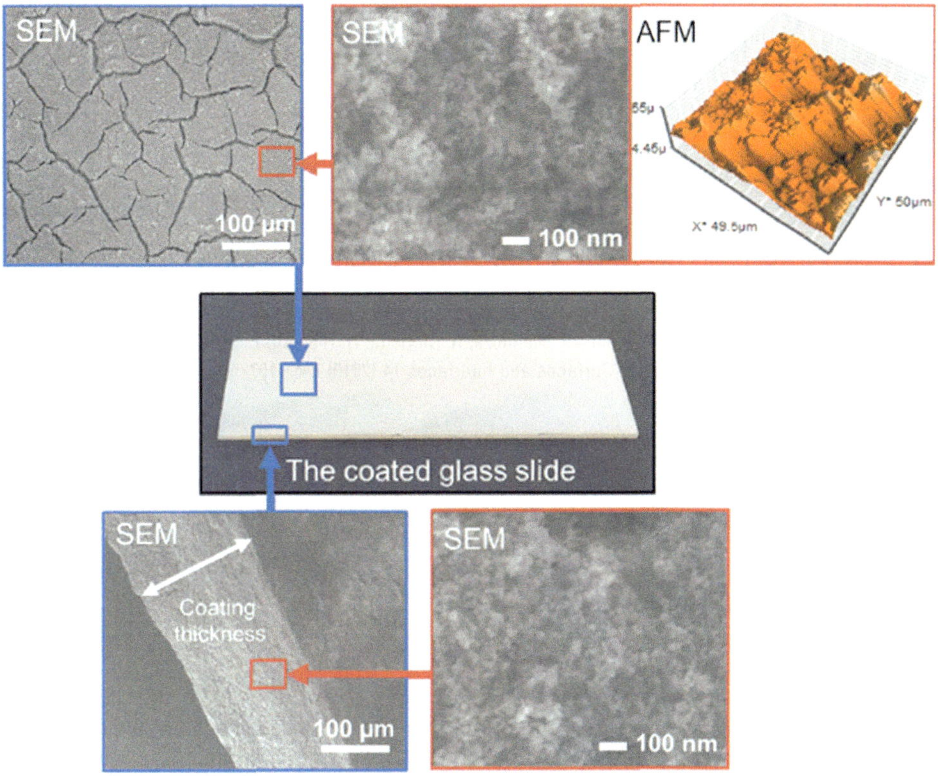

Figure 11.8 Scanning electron microscopy (SEM) and atomic force microscope (AFM) images of perfluorooctyltriethoxysilane/titanium dioxide nanoparticles-coated glass slide. From G.B. Hwang, A. Patir, K. Page, Y. Lu, E. Allan, I.P. Parkin, Buoyancy increase and drag-reduction through a simple superhydrophobic coating, Nanoscale 9(22) (2017) 7588–7594. https://doi.org/10.1039/c7nr00950j.

11.5 Properties and performance of superhydrophobic coatings

After the fabrication of a superhydrophobic surface by application of superhydrophobic nanopaints, the properties need to be checked to verify the superhydrophobic phenomenon of the coated substrate. In this regard, various parameters need to be tested. These tests provide qualitative as well as quantitative results.

11.5.1 Static and dynamic contact angles measurements

11.5.1.1 Measurement of static contact angle

Static contact angle is measured by using a contact angle meter on a flat surface to characterize the wettability of the

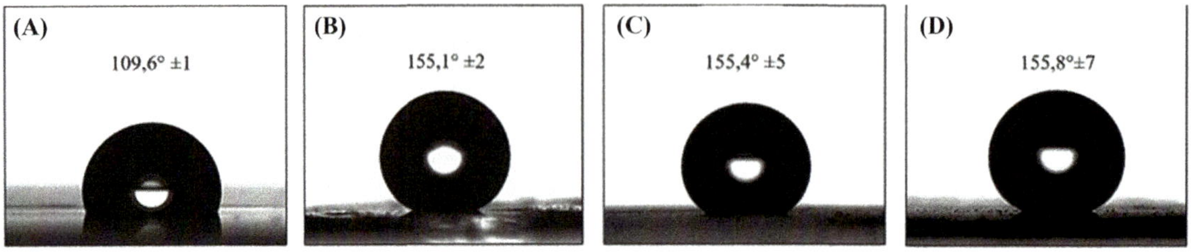

Figure 11.9 Static water contact angle images generated by CCD camera (CAM200—KSV Instruments, Helsinki, Finland) on (A) blank polydimethyl siloxane (PDMS) substrate; (B) hydrophilic ZnO nanoparticles sprayed on PDMS substrate; (C) hydrophobic ZnO nanoparticles sprayed on PDMS substrate; and (D) hydrophobic ZnO nanoparticles on PDMS substrate by dipping method. From Ā.I. Mendoza, R. Moriana, H. Hillborg, E. Strömberg, Super-hydrophobic zinc oxide/silicone rubber nanocomposite surfaces, Surfaces and Interfaces 14 (2019) 146–157. https://doi.org/10.1016/j.surfin.2018.12.008.

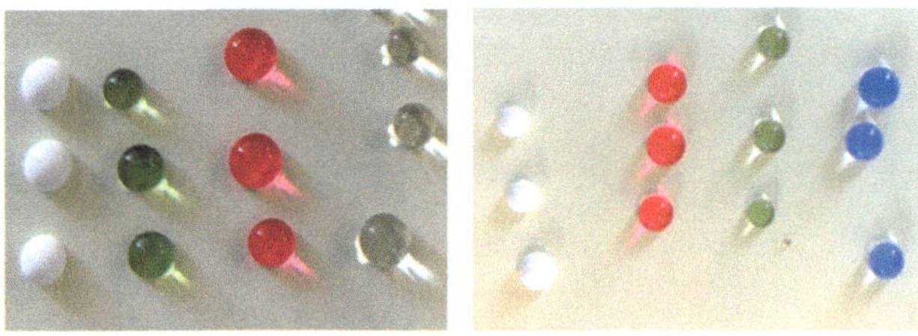

Figure 11.10 Visual appearance of superhydrophobic water droplets for static contact angle measurements of transparent pure water, blue dyes water, emulsion, methyl violet indicator with hydrochloric acid added green water, and phenolphthalein indicator with sodium hydroxide added in water on emery paper-roughened titanium dioxide/$CoAl_2O_4$/octyltriethoxysilane-loaded acrylic emulsion coating. From X. Xue, Z. Yang, Y. Li, P. Sun, Y. Feng, Z. He, T. Qu, J.G. Dai, T. Zhang, J. Qin, L. Xu, W. Zhang, Superhydrophobic self-cleaning solar reflective orange-gray paint coating, Solar Energy Materials and Solar Cells 174 (2018) 292–299. https://doi.org/10.1016/j.solmat.2017.09.014.

surface [26]. In this test a certain volume of deionized (DI) water droplet ranging from 3 to 10 μL is placed on different locations of the sample and average is reported. The static contact is used to record after 60 seconds of droplet stability [28]. In other experiments, scientists reported contact angles within the first 3 seconds of the deposition of droplets to avoid the drop stabilization effect as shown in Fig. 11.9A [32]. To be superhydrophobic a static contact angle of 150 degrees is required (Fig. 11.10).

11.5.1.2 Measurements of advancing and receding contact angle

The variable drop volume technique has been applied to determine the advancing (θ_a) and receding (θ_a) angles [42]. In this technique, generally, a small volume of the droplet was initially positioned on the surface, and then its volume increased and decreased at a certain rate. During this change in volume, the advancing and receding angles have been measured with the help of the camera. Mendoza et al. have carried out this testing as explained here. They placed a 3 μL volume of droplet initially on the surface. Then increasing the volume up to 25 μL using a stainless-steel needle centered in the drop which is also called "needle in drop" technique. Subsequently, the volume is continually raised by infusing water into the drop at a rate of 0.5 $\mu L\ s^{-1}$. Once the volume reaches 40 μL, the liquid is continuously withdrawn until 10 μL of liquid remains. The average $\theta_a°$ and $\theta_r°$ angles are calculated by recording images in the camera and using the Young–Laplace method for the data analysis and CAH is found to be around 7 degrees [32]. Similarly, Xue et al. have determined average advancing ($\theta_a°$) and receding ($\theta_r°$) contact angles by taking the initial volume of a drop of 7 μL and a rate of increasing/decreasing the droplet volume of 1 $\mu L\ s^{-1}$ and obtained CAH of $<5°$ [43]. Selim et al. determined CAH of 9 degrees for PDMS/SiC nanowired composite coatings by same method [45].

11.5.1.3 Measurements of roll-off angle

The roll-off (θ_R) angle or sliding angle (SA) of superhydrophobic surfaces is used to determine by using a tilted platform attached to the contact angle meter. The surface is superhydrophobic when it exhibits a roll-off angle of less than 10 degrees. The self-cleaning ability is linked to both the roll-off angle and CAH, which are complimentary characterization methods. Additionally, the roll-off angle gives a clear picture of the surface's real capacity for self-cleaning. A certain amount of water is placed on a horizontal surface, and the surface is gradually angled until the droplet begins to roll down. This process was used to determine the water droplet roll-off angle [46]. The droplet started to travel down the surface at an inclination angle that was measured using a goniometric scale or by recording images by the camera and measuring the angle with the help of imageJ software.

The Furmidge equation may be used to define the interaction strength between liquid and solid substrate in terms of the liquid droplet's θ_a and θ_r contact angles, which determines the roll-off angle [54].

$$m\,g\,\mathrm{Sin}\,\theta_R = \gamma\,w(\mathrm{Cos}\,\theta_r - \mathrm{Cos}\,\theta_a) \tag{11.6}$$

where g is the gravitational acceleration, m is the mass of the drop, w is the width of the contact circle of the liquid droplet, θ_R is the roll-off angle, and γ is the surface tension of the liquid drop. The capillary force is represented by the right side of this equation and gravity forces by the left side. The phrase included in the bracket refers to the strength of the interaction between a solid and water at the contact. A modest SA will arise from what seems to be a very small real water/solid contact area on the Cassie surface as well as a very low value for this term. Since the real contact area on the Wenzel surface is far bigger than the apparent size, the interface will experience a much stronger interaction. It is feasible for a surface with a very high contact angle to have a large SA at this regime [40].

Kuzina et al. placed 15 μL of the droplet and obtained less than 4.5 degrees roll-off angle for laser-treated fluoro polyurethane enamel-coated substrate [46]. Zhang et al. have reported a roll-off angle of less than 10 degrees for water-based paint prepared by SiO_2 nanoparticles and polyvinylpyrrolidone in the protected polymer [44]. Mendoza et al. deposited 10 μL water droplet but found a roll-off angle of 6 degrees for zinc oxide (ZnO)/silicone rubber nanocomposite surfaces [32]. Similarly, Xue et al. reported a roll-off angle of only 0.3 degrees for emery paper roughened coating surface by placing water droplets of 10 μL [43].

11.5.2 Dynamic droplet impact test/bouncing drop test

The purpose of the bouncing drop test is to qualitatively determination of the degree of water repellency. The foundation of these studies is the observation of a drop that, when thrown from a certain height, bounces on a repellent surface using a high-speed camera. The number of bounces is controlled by the dissipation brought on by the drop-surface tensile adhesion. Therefore the adhesion is lesser the more bounces there are [27,28,55].

Li et al. discussed the bouncing phenomenon of water droplets over uncoated and superhydrophobic paint-coated surfaces and named it the dynamic droplet impact test [29]. Whenever water droplets get in contact with the substrate adhesive forces are generated. The droplet is stretched when dropped from a certain distance and the adhesive forces increase with the increase in stretching length. When a water droplet is separated from the substrate the droplet restores its original deformation due to its surface tension. The water droplet starts shrinking and its shape deforms and water flows inside the droplet. When this internal flow and shrinkage reaches a balance, a short-term force balance appears but water still does not reach its original shape and this shrinkage phenomenon continues until the force value reaches zero. When this force becomes equal to or higher than the coated substrate, the droplet can easily roll off from an inclined superhydrophobic surface. On untreated surfaces when droplets dropped from a certain height it did not rebound off. The droplet stuck in various surfaces, deformed, and pulsed for a long time until kinetic energy dissipated by capillary force as shown in Fig. 11.11A [30]. In the case of the coated surface, droplets deformed, then flattened then the droplets' kinetic energy changed into vibrational energy, further retracting and rebounding off the treated surfaces as depicted in Fig. 11.11B. In this process, droplets remain intact and leave the coated surface completely without wetting. This rebounding behavior of droplets ensures the superhydrophobicity of the coated surface.

11.5.3 Self-cleaning test

Superhydrophocity can be easily assesed by self-cleaning efficiency [26]. The self-cleaning action of the prepared coatings is carried out and compared to a noncoated substrate. Superhydrophobic nanopaint dip-coated glass slides and uncoated glass slides are positioned at an inclination angle of less than 10 degrees and are compared by Nine et al. for self-cleaning behavior. Fine graphite powder as a debris is placed above all glass slides and then the surface is pipetted with water droplets. Water droplets on the coated surfaces effectively removed debris from their rolling and bouncing routes, demonstrating their acute self-cleaning function (Fig. 11.12). The uncoated control glass slide exhibited a build-up of graphite particles and no discernible removal of the particles when exposed to water droplets [28]. Similarly, CS/PDMS-coated various surfaces, such as iron, glass, fabric, paper, and Cu are tested for soil contamination [29] and emery paper roughened $TiO_2/CoAl_2O_4$/octyltriethoxysilane-loaded acrylic emulsion coating for fly ash contamination (Fig. 11.13) [43]. This self-cleaning behavior of superhydrophobic nanopaints also helps to remove marine foul as depicted by Selim et al. [42].

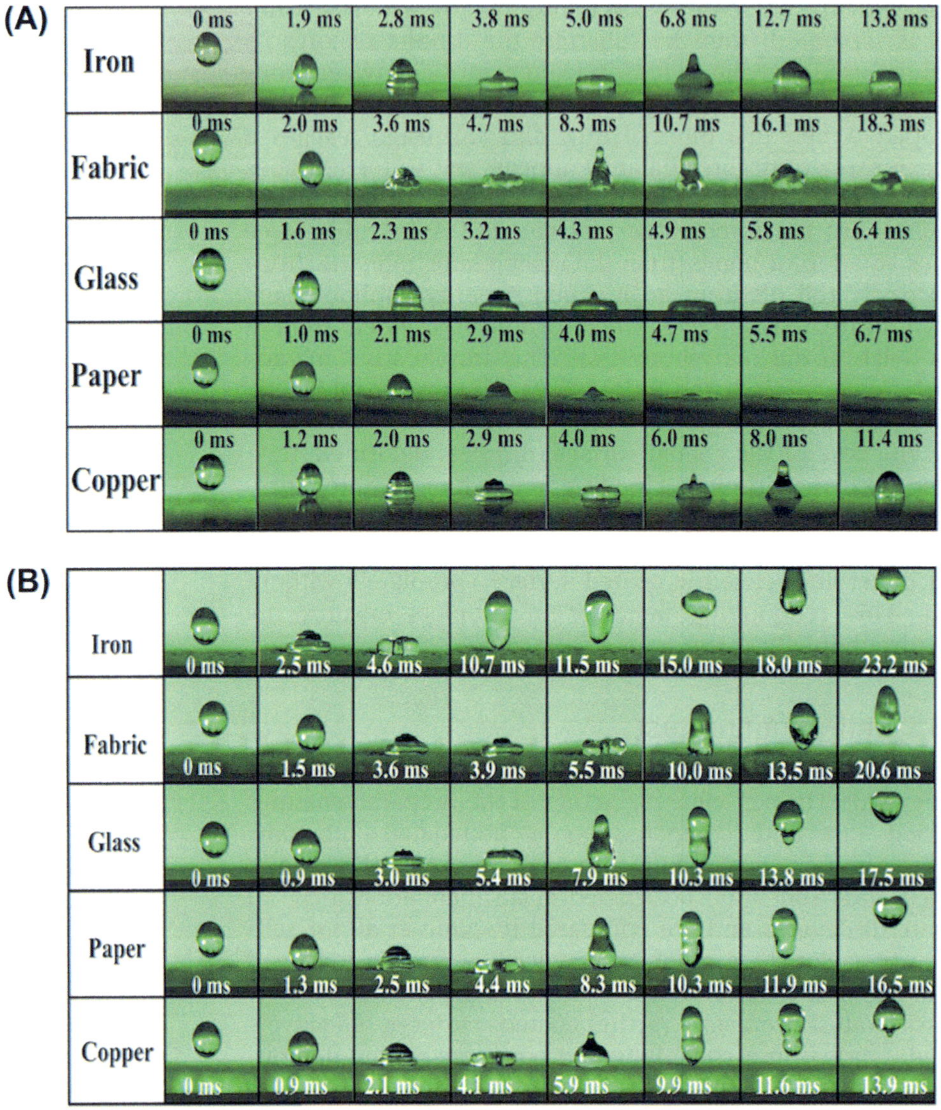

Figure 11.11 (A). Dropping of water droplet on the bare substrate, such as iron, fabric, glass, paper, and copper surfaces; and (B) water droplet bouncing phenomenon on the coated iron, fabric, glass, paper, and copper surfaces. Reproduced with permission from G.R.T. Suyambulingam, K. Jeyasubramanian, V.K. Mariappan, P. Veluswamy, H. Ikeda, K. Krishnamoorthy, Excellent floating and load bearing properties of superhydrophobic ZnO/copper stearate nanocoating, Chemical Engineering Journal 320 (2017) 468–477. https://doi.org/10.1016/j.cej.2017.03.052. Copyright 2020 Elsevier.

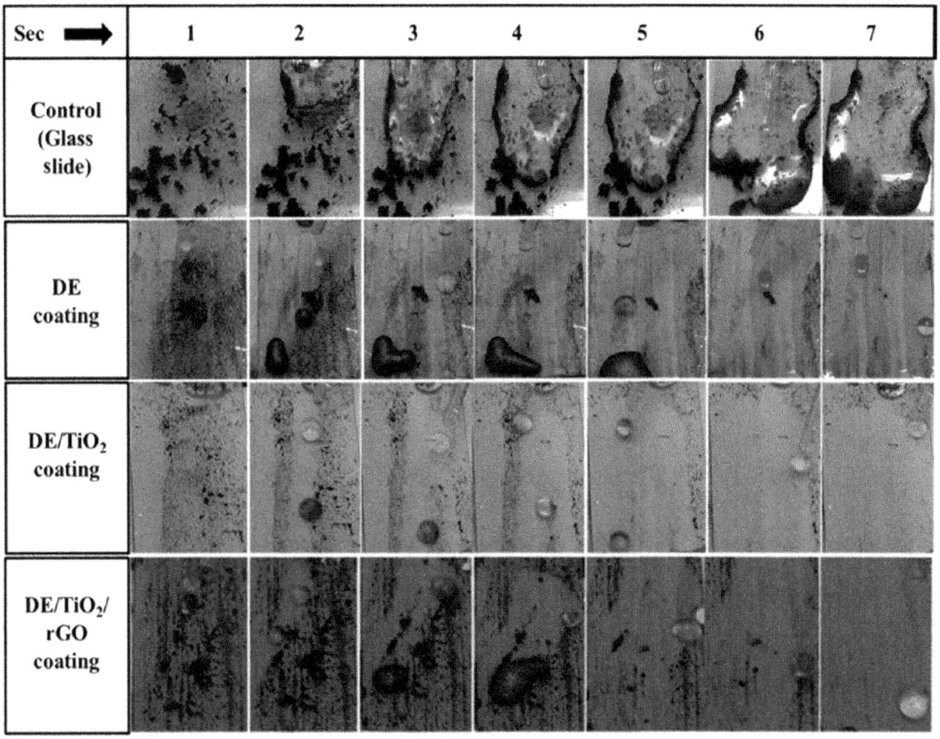

Figure 11.12 Depiction of the self-cleaning ability of diatomaceous earth (DE)-based superhydrophobic coating; DE/titanium dioxide (TiO_2)-based superhydrophobic coating; and DE/TiO_2/reduced graphene oxide(rGO)-based superhydrophobic coating contaminated with graphite powder. From M.J. Nine, M.A. Cole, L. Johnson, D.N.H. Tran, D. Losic, Robust superhydrophobic graphene-based composite coatings with self-cleaning and corrosion barrier properties, ACS Applied Materials and Interfaces 7(51) (2015) 28482–28493. https://doi.org/10.1021/acsami.5b09611.

11.5.4 Sailing test of boat

A test for superhydrophobicity that is qualitative is the boat sailing test. The superhydrophobic coating's ability to reduce surface energy is primarily responsible for the process underlying the decrease in drag. To ascertain the decrease in water and drag force caused by the superhydrophobic coating, Hwang et al. test-dives treated and untreated boats as shown in Fig. 11.14 [34]. By dipping the boat in the superhydrophobic nanopaints and allowing it to dry, they created a superhydrophobic covering. The boat, which weighed 14 grams and had a capacity of $8.5 \times 8.5 \times 5$ cm, was filled with 100 grams of glass beads and set in a 120-centimeter water tank. The boat was

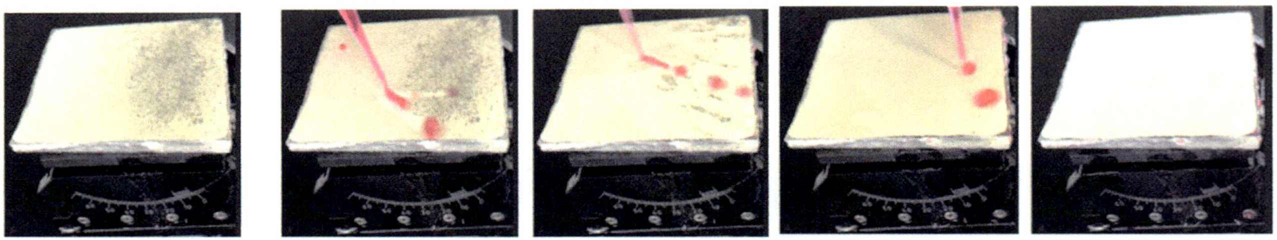

Figure 11.13 Testing of self-cleaning effect on emery paper roughened titanium dioxide/$CoAl_2O_4$/octyltriethoxysilane-loaded acrylic emulsion coating for fly ash contamination at 2 degrees tilted panel. From X. Xue, Z. Yang, Y. Li, P. Sun, Y. Feng, Z. He, T. Qu, J.G. Dai, T. Zhang, J. Qin, L. Xu, W. Zhang, Superhydrophobic self-cleaning solar reflective orange-gray paint coating, Solar Energy Materials and Solar Cells 174 (2018), 292–299. https://doi.org/10.1016/j.solmat.2017.09.014.

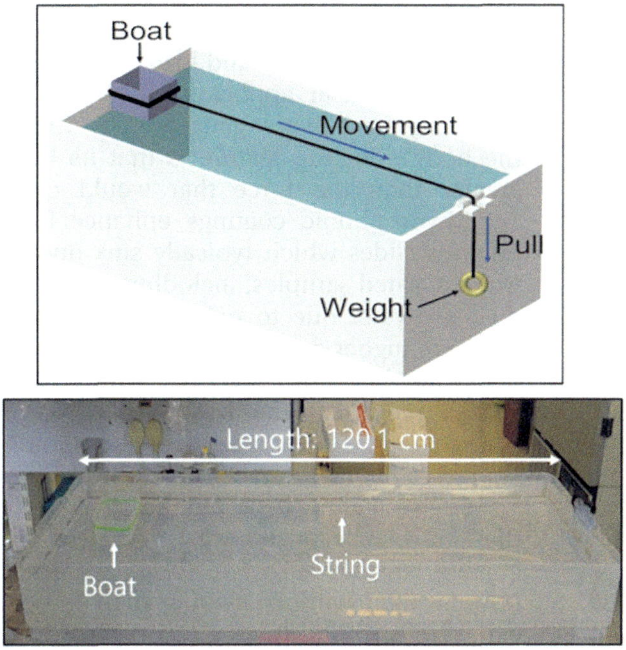

Figure 11.14 Sailing boat test to check reduction in drag forces. From G.B. Hwang, A. Patir, K. Page, Y. Lu, E. Allan, I.P. Parkin, Buoyancy increase and drag-reduction through a simple superhydrophobic coating, Nanoscale 9(22) (2017) 7588–7594. https://doi.org/10.1039/c7nr00950j.

partially buried in the water, with the remaining part still above the water's surface. A string was used to link the boat to the weights. The boat was able to go on the sea in a straight line because of the weight under gravity's pushing power. The boat with the superhydrophobic covering always had a greater sailing velocity than the boat without the coating, even when subjected to applied forces between 17.6 and 47 mN. The two boats' velocities were compared, and it was found that the superhydrophobic covering decreased water drag force by up to 12.7% at 47 mN. The water drag was significantly reduced by the microbubbles trapped at the solid/water contact. A superhydrophobic coating's ability to reduce surface energy also reduced the adhesion force of water molecules on the boat, resulting in less water drag for the sailing vessel. There are a number of another factor that also plays a role that may change increase the drag resistance, such as roughness morphology, viscosity of liquid, and menisci of microbubbles on the boundary of the water and the surface [56,57].

11.5.5 Buoyancy and loading capacity tests

The floating ability and load-bearing capacity are crucial factors for practical applications of superhydrophobic coatings, especially in aquatic gadgets. The fundamental idea behind the body's floating feature is that its buoyant force should be greater than the force that would cause it to drown [30]. Superhydrophobic coatings enhance buoyancy, demonstrated by glass slides which typically sink due to higher density than water. Coated samples, including edge, top, and whole surfaces, float on water due to reduced interaction with it. Despite the coating's higher density, air entrapped by the coating is not sufficient for buoyancy when only the bottom is coated. Instead, surface tension primarily supports the samples. Glass's higher surface energy attracts water but the coating lowers the surface energy, reducing interaction with water (Fig. 11.15).

Weight-loading tests on the treated glass samples revealed that the load capacity was influenced by the position of the coating on the substrate. Edge-coated samples supported a lesser weight without sinking than top-coated samples due to exposure to the untreated top surface while carrying the maximum bearable load. Samples coated on all sides had the highest load capacity, attributed to water surface tension and entrapped air bubbles at the treated surface [34].

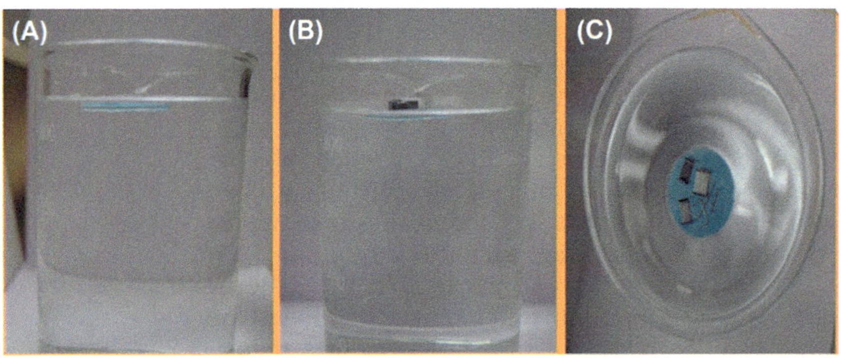

Figure 11.15 Load bearing capability of ZnO nanoparticle and copper stearate-loaded superhydrophobic nanopaint-coated substrate. http://doi.org/10.1016/j.cej.2017.03.067 G.R.T. Suyambulingam, K. Jeyasubramanian, V.K. Mariappan, P. Veluswamy, H. Ikeda, K. Krishnamoorthy, Excellent floating and load bearing properties of superhydrophobic ZnO/copper stearate nanocoating, Chemical Engineering Journal 320 (2017) 468–477. https://doi.org/10.1016/j.cej.2017.03.052. Copyright 2017 Elsevier.

11.5.6 Condensate transfer ability

Condensates, such as dew and frost, can pose significant challenges in both production and daily life if not promptly addressed. Altering surface wettability has emerged as a promising strategy to mitigate this issue. The coalescence-induced droplet jumping (CIDJ) effect presents a favorable mechanism for minimizing the accumulation of condensate.

Superhydrophobic surfaces have garnered significant attention for their ability to impede condensate nucleation, thereby slowing down condensate buildup. The CIDJ effect, observed on various superhydrophobic surfaces with diverse structures, operates on the principle where merging dewdrops transition into a state of reduced contact area, leading to diminished adhesion forces. When dewdrops coalesce, the kinetic energy of the resulting jumping droplet is determined by a coupling effect involving surface energy release, viscous flow-induced energy, and interfacial adhesion energy.

This CIDJ effect offers a practical means of maintaining condensate coverage at a minimal level over extended periods, thereby mitigating thermal resistance issues and related phenomena. Researchers, such as Wang et al., have depicted the CIDJ effect and its implications for condensate transfer (Fig. 11.16), particularly highlighting the efficacy of superhydrophobic surfaces. Utilizing superhydrophobic nanopaints can thus offer potential applications for antifrosting and antiicing effects [55].

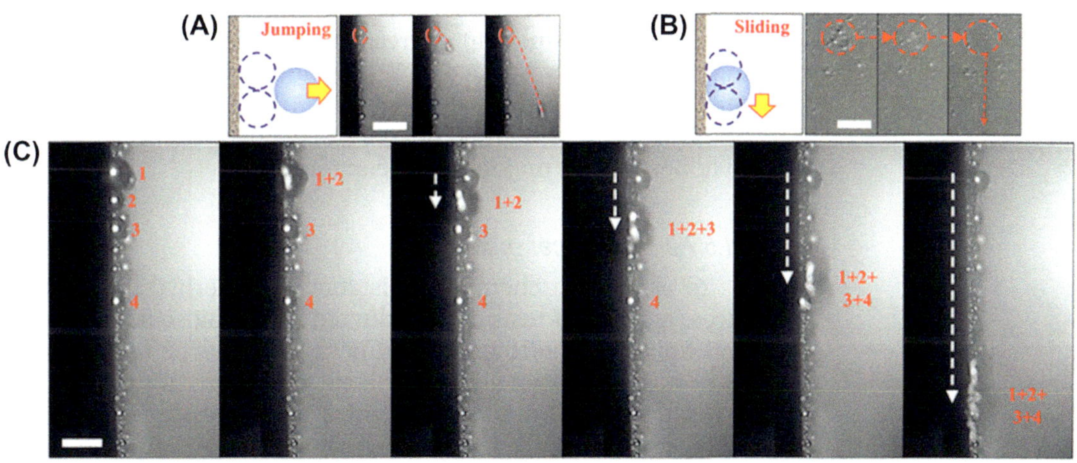

Figure 11.16 (A) Coalescence-induced droplet jumping; (B) sliding; (C) coalescence-induced condensate transfer ability of water droplet by superhydrophobic surface. From S. Wang, X. Zhao, Y. Teng, X. Chen, R. Ahuja, Large-scale fabrication of wettability-controllable coatings for optimizing condensate transfer ability, Langmuir 37(7) (2021) 2476–2484. https://doi.org/10.1021/acs.langmuir.0c03544.

11.5.7 Durability tests

Less durability and poor robustness are the two major barriers of superhydrophobic nanopaints to their widespread use. To check robustness and durability below test methods are applied on superhydrophobic nanopaints. It is expected that even after these harsh tests the surface should maintain its superhydrophobicity.

11.5.7.1 Water scouring test

According to Cui et al., a water scouring test may be used to assess the endurance of the generated superhydrophobic surface. In this test, coated samples are subjected to extremely severe water conditions designed to replicate a torpedo in the sea. A test for high-speed scouring is carried out in a water bath that has a 16-face rotor. After attaching the prepared sample to one face of the rotor, it revolved at a speed of $10\ \mathrm{m\ s^{-1}}$ (3700 rpm) in distilled water. Following the scouring process, the sample is dried at 100°C in an oven before the static contact angle is measured once more at room temperature. According to Cui et al., there is a noticeable 12-degree drop in the contact angle within the first 5 minutes of the test. It might have happened because the stream carried the loose particles off the surface, making the surface somewhat less abrasive [40]. The superhydrophobic property's strong durability against water scouring was then demonstrated by the contact angle, which only very little decreased as the scouring period grew and finally persisted at around 140 degrees. Nonetheless, when the scouring time rose, the surface's SA grew noticeably. Within five minutes, the SA rose to 20 degrees, and 4 hours later, the scouring SA grew to more than 90 degrees, allowing 10 μL water droplets to adhere to the majority of the surface.

11.5.7.2 Chemical resistance

To assess the chemical resistance of superhydrophobic surfaces, a series of tests are conducted involving soaking, heating, and boiling in both acidic and basic aqueous solutions as well as in water. Additionally, the solvent resistance of these surfaces is evaluated by immersing samples in various organic solvents for several days. The change in water contact angle observed during these tests is directly influenced by the solvency of the solvent for the coating material. Solvents with stronger solvating properties tend to cause greater swelling of the polymer and can potentially damage the very thin coating layer of nanoparticles. Similarly, samples sensitive to acidic solutions exhibit a rapid decrease in contact angles when soaked, with stronger acidic solutions leading to a more pronounced decrease. This sensitivity highlights the vulnerability of superhydrophobic surfaces to the acidity of the solution, underscoring the importance of chemical resistance testing in evaluating their performance and durability [25].

An acid barrier test is conducted using the color-changing method. To evaluate molecular permeability through the coating layer to the substrate qualitatively, a droplet of dilute mineral acid solution is applied to a variety of coated surfaces. Digital recording and visual examination of any color changes in the droplet—which signify the creation of reaction products—are used to assess the outcomes. For instance, the testing over a Cu substrate with 2 M hydrochloric acid, the formation of Cu chloride can lead to observable changes in the droplet's color which shows the reduction in water-repelling behavior. This method provides valuable insights into the effectiveness of the coating in preventing acid penetration and protecting the substrate from chemical damage [25].

An immersion test is also used to assess the stability of water-repellent coatings. It is determined by immersing the prepared samples in solutions of varying pH, including DI water. These pH values are selected to mimic diverse environmental conditions, such as washing liquor, tap water, and acid rain. Despite observing a slight reduction in contact angles after immersion, all samples are expected to retain their superhydrophobic properties. This evaluation method helps determine the coatings' resilience to different environmental factors and their ability to maintain water repellency under realistic conditions. Despite minor changes in contact angles, the coatings are anticipated to uphold their superhydrophobic nature, ensuring continued effectiveness in repelling water and protecting underlying surfaces [31].

11.5.7.3 Shear strength test

Cao et al. have tested the shear strength of superhydrophobic nanopaints to ensure their mechanical strength [38]. They first deposited the superhydrophobic nanopaint onto glass slides with different SiO_2 nanoparticle concentrations, that is, 0, 10, 20, 30, and 40 wt.%, Then pressed another coated glass slide to create a consistent and sturdy contact between them after curing. Shear strength tests are conducted on a universal testing machine at a specific rate following curing. As the roughness increases with the inclusion of nanoparticles the shear strength increases due to good anchoring and coordination bonds between the coating and substrates [37].

11.5.7.4 Tape peeling test or adhesion test

A peel-off test is used to evaluate the durability and strength of the superhydrophobic coating layers. This test is aimed at evaluating the mechanical resilience of the coatings [27]. A standard Elcometer 99 tape, known for its high adhesion to steel (642 N m^{-1}), is utilized to simulate severe peel-off conditions. During the test, a weight of 1 kg is applied to the tape, and then it was peeled off from the substrate as shown in Fig. 11.17A. Following each tape peel, the static contact angle and roll-off angle of the coating are measured. These measurements are crucial in determining the robustness of the coating layers under mechanical stress. By tracking changes in these angles after each tape peel, insights into the coating's ability to withstand peeling forces and maintain its superhydrophobic properties over time are gained [38]. Wu et al. have performed tape peeling tests and the superhydrophobic nanopaint is mechanically robust as shown in Fig. 11.18D1–D3 [39].

11.5.7.5 Abrasion test

To evaluate the superhydrophobic coating's mechanical stability, abrasion tests are carried out. The objective is to determine the coating's resistance to abrasion. The sample is positioned below a 1 kg weight on sandpaper during the test, as shown in Fig. 11.17B. After that, the sandpaper is dragged steadily in one direction. To assess the performance of the coating, measurements are made after a certain set of abrasion cycles. This involved weighing the composite covering and determining the water contact and slide angles. These measurements served as indicators of the coating's robustness and its ability to maintain superhydrophobic behavior under mechanical stress. By monitoring changes in these parameters

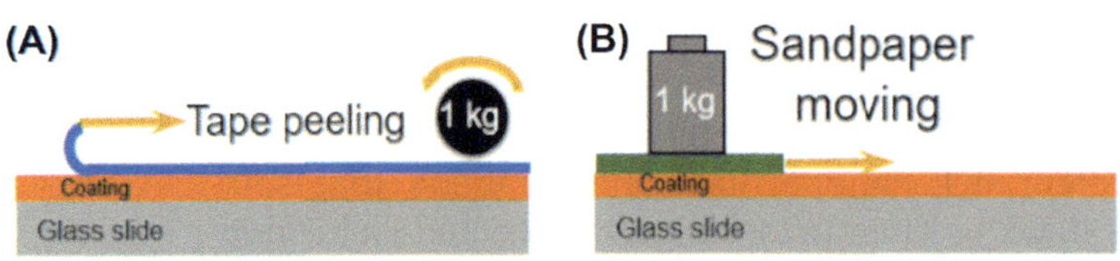

Figure 11.17 (A) Schematic presentation of tape peeling test; and (B) abrasion test using sandpaper. From C. Cao, B. Yi, J. Zhang, C. Hou, Z.Wang, G. Lu, X. Huang, X. Yao, Sprayable superhydrophobic coating with high processibility and rapid damage-healing nature, Chemical Engineering Journal 392 (2020). https://doi.org/10.1016/j.cej.2020.124834.

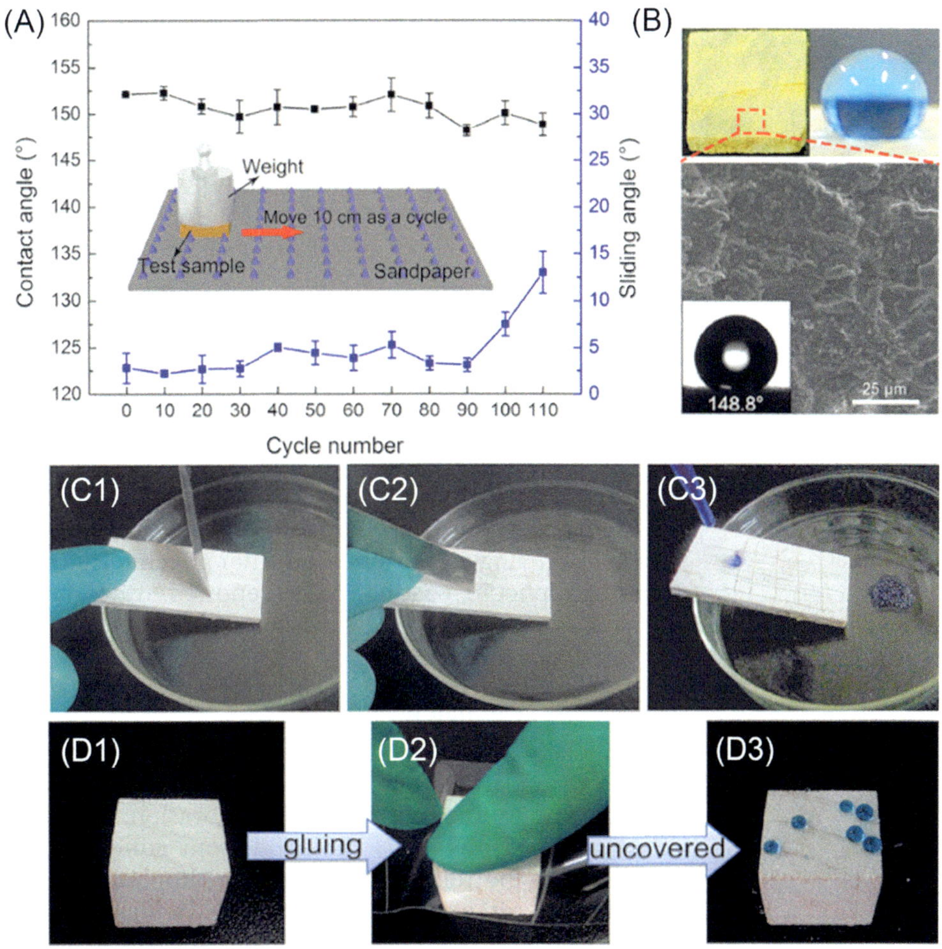

Figure 11.18 (A) Sandpaper abrasion test results after every 10 cycles; (B) water droplet image and its contact angle after 100 abrasion cycles; (C1–C3) knife scratching and roll-off results after scratching; (D1–D3) tape peeling test results on emulsion composed of SiO_2 nanoparticles and epoxy resin. From Y. Wu, S. Jia, S. Wang, Y. Qing, N. Yan, Q. Wang, T. Meng, A facile and novel emulsion for efficient and convenient fabrication of durable superhydrophobic materials, Chemical Engineering Journal 328 (2017) 186–196. https://doi.org/10.1016/j.cej.2017.07.023.

throughout the abrasion test, insights into the coating's durability and longevity are obtained [38]. Wu et al. have reported that SiO_2 or TiO_2 nanoparticles-induced epoxy paint is able to maintain its superhydrophobicity even after 100 abrasion cycles as shown in Fig. 11.18A and B [39].

The ASTM F735 standard was utilized in a different approach to test the surface structures' abrasion resistance. A calibrated 20 mm-high layer of sand is placed over a sample that was secured to the bottom of a vessel. A vibration shaker platform with a frequency and amplitude of 1050 min^{-1} and 3 mm, respectively, is used to move the vessel back and forth. The entire quantity of sand is propelled within the vessel by the forces of inertia, which also significantly abrasively acted on the surface of the sample that is secured to the bottom of the vessel [46].

11.5.7.6 Scratch resistance

To gauge the superhydrophobic surfaces' mechanical resilience, knife scratching is also done. The grid-like impression is produced using the knife on a superhydrophobic nanopaint-coated wood substrate. Drops of water could roll off the surface with ease and leave no trace, indicating that the superhydrophobicity remained as shown in Fig. 11.18C1–C3 [39]. Similarly, a cross-cut tester has been used by Nine et al. to make scratches and contact angles are checked after mechanical damage on superhydrophobic coating. They reported water repellence is maintained even after severe mechanical damage [28].

11.5.7.7 Accelerated weathering

Artificially fabricated superhydrophobic surfaces are susceptible to degradation when exposed to outdoor conditions, particularly due to rain and mechanical abrasion. Additionally, the effect of ultraviolet (UV) radiation from sunlight can further compromise the robustness of superhydrophobicity. According to certain research, organic compounds on the superhydrophobic surface with low SFE can be broken down by UV light, which decreases the material's efficiency. Notably, materials, such as TiO_2 and ZnO, originally hydrophobic, can undergo a transition to hydrophilicity when irradiated with UV light. However, these materials tend to revert to their initial hydrophobic state upon removal of UV exposure. Therefore ensuring the long-term performance of superhydrophobic nanopaints requires developing formulations that are regenerative and easily repairable. By creating superhydrophobic coatings that possess self-repairing capabilities, it becomes possible to mitigate the effects of UV degradation and extend the lifespan of these coatings in outdoor environments. Such advancements are essential for ensuring the sustained functionality and effectiveness of superhydrophobic surfaces in real-world applications.

In artificially accelerated weathering testing, the coated specimens' upper surfaces were exposed to either UV or sunlight, and water was sprayed using a predetermined spray cycle (18 minutes of water spraying and 102 minutes of dryness, for example). During the dry phase, the chamber temperature, black panel temperature, and relative humidity must all be kept within specified limits. The samples should be free of blistering, peeling, cracking, chalking, and color change/fading after the allotted time, which is typically 400 hours. They should also not have lost any contact angle or roll-off angle, nor changed surface morphology [36,43].

11.6 Conclusion

This chapter delves into a comprehensive examination of the fundamental parameters necessary for achieving superhydrophobic surfaces. It explores various aspects, such as wetting phenomena, contact angles, SFE, surface chemistry, and morphology, all of which play crucial roles in determining the hydrophobic properties of a surface. Additionally, the chapter discusses methods for determining these properties and evaluating the performance of superhydrophobic surfaces.

Researchers are actively engaged in addressing the limitations associated with superhydrophobic nanopaints, with the goal of moving closer to commercializing these innovative coatings. Of particular concern in the commercialization process is ensuring the durability and robustness of these coatings to meet real-world application demands. Overcoming these challenges is essential for the widespread adoption and successful integration of superhydrophobic nanopaints in various industries and applications.

Nanoparticles are identified as key players in advancing the commercialization of superhydrophobic coatings. These nanomaterials offer unique properties and functionalities that contribute to enhancing the performance and longevity of superhydrophobic surfaces. Their ability to modify surface characteristics and impart desired properties makes them indispensable in overcoming the hurdles associated with commercialization and ensuring the viability of superhydrophobic coatings in practical applications.

AI disclosure

During the preparation of this work the author(s) used Quilbot for rephrasing. After using this tool/service, the author (s) reviewed and edited the content as needed and take(s) full responsibility for the content of the publication.

References

[1] A. Kumar, D. Nanda, Methods and fabrication techniques of superhydrophobic surfaces, Superhydrophobic Polymer Coatings: Fundamentals, Design, Fabrication, and Applications, Elsevier, India, 2019, pp. 43–75. Available from: https://doi.org/10.1016/B978-0-12-816671-0.00004-7.

[2] L. Li, Q. Yang, Advanced Coating Materials, Wiley, 2018. Available from: https://doi.org/10.1002/9781119407652.

[3] L. Zhang, A.G. Zhou, B.R. Sun, K.S. Chen, H.Z. Yu, Functional and versatile superhydrophobic coatings via stoichiometric silanization, Nature Communications 12 (1) (2021). Available from: https://doi.org/10.1038/s41467-021-21219-y.

[4] E.R. Sadiku, O. Agboola, I.D. Ibrahim, P.A. Olumambi, B. Avabaram, Coating Materials Nanotechnology (2019) 175–233.

[5] T. Onda, S. Shibuichi, N. Satoh, K. Tsujii, Super-water-repellent fractal surfaces, Langmuir: the ACS Journal of Surfaces and Colloids 12 (9) (1996) 2125–2127. Available from: https://doi.org/10.1021/la950418o.

[6] S. Shibuichi, T. Onda, N. Satoh, K. Tsujii, Super water-repellent surfaces resulting from fractal structure, Journal of Physical Chemistry 100 (50) (1996) 19512–19517. Available from: https://doi.org/10.1021/jp9616728.

[7] P.C. Okafor, E.E. Ebenso, A.Y. El-Etre, M.A. Quraishi, Green approaches to corrosion mitigation, International Journal of Corrosion (2012).

[8] K.T. Rahman, Emerging Nano-Enable Materials in the Sports Industry, Materials Research Forum, LLC, 2023, pp. 75–100. Available from: https://doi.org/10.21741/9781644902295-4.

[9] M.S. Selim, S.A. El-Safty, M.A. Shenashen, S.A. Higazy, A. Elmarakbi, Progress in biomimetic leverages for marine antifouling using nanocomposite coatings, Journal of Materials Chemistry B 8 (17) (2020) 3701–3732. Available from: https://doi.org/10.1039/c9tb02119a.

[10] J.T. Simpson, S.R. Hunter, T. Aytug, Superhydrophobic materials and coatings: a review, Reports on Progress in Physics 78 (8) (2015). Available from: https://doi.org/10.1088/0034-4885/78/8/086501.

[11] W.L. Wisan, Stillman Drake, cause experiment and science: a Galilean dialogue incorporating a new english translation of Galileo's "bodies that stay atop water, or move in it, University of Chicago Press Journal 75 (1984) 614.

[12] J. Drelich, A. Marmur, Physics and applications of superhydrophobic and superhydrophilic surfaces and coatings, Surface Innovations 2 (4) (2014) 211–227. Available from: https://doi.org/10.1680/si.13.00017.

[13] A. Marmur, Wetting on hydrophobic rough surfaces: to be heterogeneous or not to be? Langmuir: the ACS Journal of Surfaces and Colloids 19 (20) (2003) 8343–8348. Available from: https://doi.org/10.1021/la0344682.

[14] J.H. Kim, P. Kavehpour, J.P. Rothstein, Dynamic contact angle measurements on superhydrophobic surfaces, Physics of Fluids 27 (3) (2015). Available from: https://doi.org/10.1063/1.4915112.

[15] M. Żenkiewicz, Methods for the calculation of surface free energy of solids, Journal of Achievements in Materials and Manufacturing Engineering 24 (2007) 137–145.

[16] C. Rulison, Measure Surface Energy: A Tutorial Designed to Provide Basic Understanding of the ConcEpt of Solid Surface Energy, Krus, USA, 1999, p. 40.

[17] F. Hejda, P. Solař, Surface free energy determination by contact angle measurements – a comparison of various approaches, WDS'10 Proceedings of Contributed Papers (2010) 25–30.

[18] G.H. Meier, Introduction to Surface Quantities, Cambridge University Press (CUP), 2014, pp. 40–72. Available from: https://doi.org/10.1017/cbo9781139047029.003.

[19] E.M. Nadgorny, D. Jaroslaw, Laser-based deposition technique: patterning nanoparticles into microstructures, Dekker Encyclopedia of Nanoscience and Nanotechnology (2004) 1565–1579. Available from: https://doi.org/10.1081/E-ENN.

[20] Y. Pan, S. Huang, F. Li, X. Zhao, W. Wang, Coexistence of superhydrophilicity and superoleophobicity: theory, experiments and applications in oil/water separation, Journal of Materials Chemistry A 6 (31) (2018) 15057–15063. Available from: https://doi.org/10.1039/c8ta04725a.

[21] J. Thus, How Material of Construction Relates to Drift Elimination, *Brentwood, California*, 2022.

[22] R.E. Johnson, R.H. Dettre, Contact Angle Hysteresis, American Chemical Society (ACS), 2009, pp. 112–135. Available from: https://doi.org/10.1021/ba-1964-0043.ch007.

[23] D.W. Bechert, M. Bruse, W. Hage, R. Meyer, Fluid mechanics of biological surfaces and their technological application, Die Naturwissenschaften 87 (4) (2000) 157–171. Available from: https://doi.org/10.1007/s001140050696.

[24] E. Bormashenko, V. Starov, Impact of surface forces on wetting of hierarchical surfaces and contact angle hysteresis, Colloid and Polymer Science 291 (2) (2013) 343–346. Available from: https://doi.org/10.1007/s00396-012-2785-9.

[25] Z. Cui, L. Yin, Q. Wang, J. Ding, Q. Chen, A facile dip-coating process for preparing highly durable superhydrophobic surface with multi-scale structures on paint films, Journal of Colloid and Interface Science 337 (2) (2009) 531–537. Available from: https://doi.org/10.1016/j.jcis.2009.05.061.

[26] K. Ko, D. Yoon, S.C. Yang, H.S. Lee, Brush-painted superhydrophobic silica coating layers for self-cleaning solar panels, Journal of Industrial and Engineering Chemistry 106 (2022) 460–468. Available from: https://doi.org/10.1016/j.jiec.2021.11.023.

[27] F.J.M. Ruiz-Cabello, S. Fusco, P. Ibáñez-Ibáñez, G. Guerrero-Vacas, M.A. Cabrerizo-Vílchez, M.A. Rodríguez-Valverde, Water-repellent galvanized steel surfaces obtained by sintering of zinc nanopowder, Langmuir: the ACS Journal of Surfaces and Colloids 39 (15) (2023) 5469–5476. Available from: https://doi.org/10.1021/acs.langmuir.3c00182.

[28] M.J. Nine, M.A. Cole, L. Johnson, D.N.H. Tran, D. Losic, Robust superhydrophobic graphene-based composite coatings with self-cleaning and corrosion barrier properties, ACS Applied Materials and Interfaces 7 (51) (2015) 28482–28493. Available from: https://doi.org/10.1021/acsami.5b09611.

[29] L. Li, J. Zhu, Z. Zeng, A sky-blue superhydrophobic coating and applications, Progress in Organic Coatings 147 (2020)105863. Available from: https://doi.org/10.1016/j.porgcoat.2020.105863.

[30] G.R.T. Suyambulingam, K. Jeyasubramanian, V.K. Mariappan, P. Veluswamy, H. Ikeda, K. Krishnamoorthy, Excellent floating and load bearing properties of superhydrophobic ZnO/copper stearate nanocoating, Chemical Engineering Journal 320 (2017) 468–477. Available from: https://doi.org/10.1016/j.cej.2017.03.052.

[31] M.S. Selim, M.A. Shenashen, A. Elmarakbi, N.A. Fatthallah, S.i Hasegawa, S. A. El- Safty, Synthesis of ultrahydrophobic and thermally stable inorganic–organic nanocomposites for self-cleaning foul release coatings, Chemical Engineering Journal 320 (2017) 653–666. Available from: https://doi.org/10.1016/j.cej.2017.03.067.

[32] A.I. Mendoza, R. Moriana, H. Hillborg, E. Strömberg, Super-hydrophobic zinc oxide/silicone rubber nanocomposite surfaces, Surfaces and Interfaces 14 (2019) 146–157. Available from: https://doi.org/10.1016/j.surfin.2018.12.008.
[33] A. Anus, V.C.T. Le, M. Sheraz, S. Kim, W.R. Lee, Visible light-activated photocatalytic activity of TiO_2–crystal violet based superhydrophobic paint against VOCs in indoor air, Environmental Science: Advances 2 (1) (2022) 69–77. Available from: https://doi.org/10.1039/d2va00120a.
[34] G.B. Hwang, A. Patir, K. Page, Y. Lu, E. Allan, I.P. Parkin, Buoyancy increase and drag-reduction through a simple superhydrophobic coating, Nanoscale 9 (22) (2017) 7588–7594. Available from: https://doi.org/10.1039/c7nr00950j.
[35] M.S. Selim, A. Elmarakbi, A.M. Azzam, M.A. Shenashen, A.M. EL-Saeed, S.A. El-Safty, Eco-friendly design of superhydrophobic nano-magnetite/silicone composites for marine foul-release paints, Progress in Organic Coatings 116 (2018) 21–34. Available from: https://doi.org/10.1016/j.porgcoat.2017.12.008.
[36] C. Cao, B. Yi, J. Zhang, et al., Sprayable superhydrophobic coating with high processibility and rapid damage-healing nature, Chemical Engineering Journal 392 (2020). Available from: https://doi.org/10.1016/j.cej.2020.124834.
[37] B. Yi, P. Liu, C. Hou, et al., Dual-cross-linked supramolecular polysiloxanes for mechanically tunable, damage-healable and oil-repellent polymeric coatings, ACS Applied Materials and Interfaces 11 (50) (2019) 47382–47389. Available from: https://doi.org/10.1021/acsami.9b17199.
[38] L. Zhou, S. Xu, G. Zhang, D. Cai, Z. Wu, A facile approach to fabricate self-cleaning paint, Applied Clay Science 132-133 (2016) 290–295. Available from: https://doi.org/10.1016/j.clay.2016.06.015.
[39] Y. Wu, S. Jia, S. Wang, et al., A facile and novel emulsion for efficient and convenient fabrication of durable superhydrophobic materials, Chemical Engineering Journal 328 (2017) 186–196. Available from: https://doi.org/10.1016/j.cej.2017.07.023.
[40] Z. Cui, Q. Wang, Y. Xiao, C. Su, Q. Chen, The stability of superhydrophobic surfaces tested by high speed current scouring, Applied Surface Science 254 (10) (2008) 2911–2916. Available from: https://doi.org/10.1016/j.apsusc.2007.09.062.
[41] H. Zhu, Y. Huang, S. Zhang, S. Jin, X. Lou, F. Xia, A universal, multifunctional, high-practicability superhydrophobic paint for waterproofing grass houses, NPG Asia Materials 13 (1) (2021). Available from: https://doi.org/10.1038/s41427-021-00315-x.
[42] M.S. Selim, H. Yang, F.Q. Wang, X. Li, Y. Huang, N.A. Fatthallah, Silicone/Ag@SiO_2 core-shell nanocomposite as a self-cleaning antifouling coating material, RSC Advances 8 (18) (2018) 9910–9921. Available from: https://doi.org/10.1039/c8ra00351c.
[43] X. Xue, Z. Yang, Y. Li, et al., Superhydrophobic self-cleaning solar reflective orange-gray paint coating, Solar Energy Materials and Solar Cells 174 (2018) 292–299. Available from: https://doi.org/10.1016/j.solmat.2017.09.014.
[44] K. Zhang, X. Yang, N. Zhu, Z.C. Wang, H. Yan, Environmentally benign paints for superhydrophobic coatings, Colloid and Polymer Science 295 (4) (2017) 709–714. Available from: https://doi.org/10.1007/s00396-017-4053-5.
[45] M.S. Selim, H. Yang, F.Q. Wang, et al., Superhydrophobic silicone/SiC nanowire composite as a fouling release coating material, Journal of Coatings Technology and Research 16 (4) (2019) 1165–1180. Available from: https://doi.org/10.1007/s11998-019-00192-8.
[46] E.A. Kuzina, K.A. Emelyanenko, A.G. Domantovskii, A.M. Emelyanenko, L.B. Boinovich, Preparation of stable superhydrophobic coatings on a paint surface with the use of laser treatment followed by hydrophobizer

deposition, Colloid Journal 84 (4) (2022) 445–455. Available from: https://doi.org/10.1134/S1061933X22040093.
[47] F.I. Il'darkhanova, G.A. Mironova, K.G. Bogoslovsky, V.V. Men'Shikov, E.D. Bykov, Development of paint coatings with superhydrophobic properties, Protection of Metals and Physical Chemistry of Surfaces 48 (7) (2012) 796–802. Available from: https://doi.org/10.1134/S2070205112070076.
[48] J. Verma, D. Kumar, B.S. Sikarwar, Fabrication of highly efficient nano core–shell structure for the development of super-hydrophobic polymeric coating on mild steel, Polymers and Polymer Composites 30 (2022). Available from: https://doi.org/10.1177/09673911221087835.
[49] W. Stephen, Z. Hong, S. Wang, Superhydrophobic Coatings, US 2014/0287243 A1, Sep 25 (2014).
[50] J.M. Sunder, S.A. Kapole, S. Deshpande, Super Hydrophobic Coatings, US20140208978A1, 31 Jul (2014).
[51] J.T. Simpson, S.R. Hunter, O. Ridge, et al., Method of making superhydrophobc/superoleophilc paints, epoxies, and composites, US 2014/0090578 A1, Apr. 3, (2014).
[52] D. Kumar, G. Babu, S. Krishnan, Study on mechanical & thermal properties of PCL blended graphene biocomposites, Polimeros 29 (2) (2019). Available from: https://doi.org/10.1590/0104-1428.05318.
[53] J. Zhang, J. Chen, A. Zhu, The novel hyperdispersant for CNT in waterborne paint and its metal corrosion protection behavior, Journal of Materials Science 58 (27) (2023) 10986–10999. Available from: https://doi.org/10.1007/s10853-023-08674-2.
[54] C.G.L. Fumridge, Studies at phase interfaces I. The sliding of liquid drops on solid surfaces and a theory for spray retention, Journal of Colloid Science 17 (1962) 309–324.
[55] S. Wang, X. Zhao, Y. Teng, X. Chen, R. Ahuja, Large-scale fabrication of wettability-controllable coatings for optimizing condensate transfer ability, Langmuir: the ACS Journal of Surfaces and Colloids 37 (7) (2021) 2476–2484. Available from: https://doi.org/10.1021/acs.langmuir.0c03544.
[56] M.S. Mohamad, C.M.M. Dover, K. Sefiane, Effect of surface treatment on drag coefficient of free-falling solid sphere in water, IOP Conference Series: Materials Science and Engineering 670(1) (2019). Available from: https://doi.org/10.1088/1757-899X/670/1/012067.
[57] A. Steinberger, C. Cottin-Bizonne, P. Kleimann, E. Charlaix, High friction on a bubble mattress, Nature Materials 6 (9) (2007) 665–668. Available from: https://doi.org/10.1038/nmat1962.

PART III

Applications and emerging trends

Nanopaint application in residential buildings. A study of the effect of cool paint on cooling energy demand

Noelia L. Alchapar and Julieta Balter
Institute of Environment, Habitat and Energy (INAHE) Scientific and Technological Centre Mendoza, National Council for Scientific and Technological Research (CONICET), Buenos Aires, Argentina

12.1 Introduction

Climate change is leading to the intensification of extreme weather events, such as heavy rainfall, floods, droughts, and heatwaves [1]. Heatwaves are often caused by large-scale high-pressure systems that provoke a temporary temperature anomaly over an entire region [2]. There is no unified definition of a heat wave in the international literature. However, studies agree that a heat wave is identified as a sustained period of excessively hot days during which the air temperature is above the average climatological mean [3].

Urban areas are more susceptible to climate vulnerability due to the overlap of heat waves with the urban heat island (UHI) effect, resulting in higher air temperatures in the cities compared to the surrounding rural areas [4,5]. This, in turn, leads to a permanent spatially concentrated temperature anomaly. Although heat waves increase air and surface temperature in both urban and rural areas [2]. The UHI exacerbates these effects within urban areas. This combination of anthropogenic and natural factors threatens the physical conditions and health of the urban dwellers [6].

A report by the World Bank of the United Nations (https://datos.bancomundial.org/indicator/SP.URB.TOTL.IN.ZS?end = 2022&start = 1960&view = chart) shows a growing worldwide migration toward urban centers [7]. Currently, 57% of the global population lives in urban regions, a figure expected to rise to 68% by

Nanopaints. DOI: https://doi.org/10.1016/B978-0-443-26512-9.00013-6

2050. This shift is a projected influx of 2.5 billion individuals from 2018 to 2050 [8]. In Latin America, the percentage is significantly higher, with 80% of the population living in urban centres. The adaptation to both heat stress and the mitigation challenges related to cooling is heavily influenced by future socioeconomic and climatic conditions [9].

In this context, the IPCC's 2022 [10] report on Climate Change, in Chapter 12, highlights the profound vulnerability of Latin America to climate change, a situation which is worsened by local factors such as inequality, poverty, and escalating deforestation. Argentina's climate, for instance, showed a marked trend with 2021 becoming the seventh hottest year since 1961. The mean temperature of the country was above the usual value for the climatological reference period. Moreover, four summer heat waves affected Argentina's central and southern regions, among them Mendoza, where this research is conducted. Further details can be found at the SMN (https://repositorio.smn.gob.ar/handle/20.500.12160/1707) Institutional Repository.

The urgent need to improve climate comfort has highlighted the need for cities to prepare for more severe climate conditions through climate adaptation processes [11,12]. City surfaces are predominantly made up of substances with high solar absorption and low solar reflectivity, which significantly intensify the development of the UHI effect [13]. The adoption of "cool" coatings with high solar reflectivity (also known as "albedo") and high thermal emissivity enables the management of solar radiation affecting on urban envelope surfaces [14,15]. This management consequently supports radiative cooling, leading to reduced surface temperatures and lowered heat input into buildings. Overall, this mitigates the urban heat island effect on an environmental scale. The international literature shows a great interest in the application of cool materials as a strategy to mitigate urban overheating [16]. However, the evaluation of the energy impact provoked by the application of cool coatings at building scale is less well documented [17], especially in Latin American cities where environmental policies are in their infancy and socio-economic vulnerability is high.

Research conducted by Alchapar et al. [18] showed that in the metropolitan area of Mendoza, Argentina, roofs stand for 35% of the total envelope of a dwelling. Moreover, due to their relative position in a dwelling, roofs are the element with the highest exposure to incident solar radiation, making them the main influential component in the energy balance of a building [19,20].

Studies in temperate cities in Argentina and Brazil have shown that the use of cool roofs is an effective strategy for reducing urban temperatures and mitigating some of the warming effects of climate change [21,22]. Their studies observed reductions in urban air temperature of up to 3.5°C in the city of Mendoza, Argentina (dry warm climate) and up to 4.5°C in the city of Campinas, Brazil (humid warm climate). For every 10% increase in the albedo of roofs and pavements [23], there is a 0.75°C decrease in outdoor air temperature and a 2.5% decrease in summer energy consumption [24].

The hypothesis of this research suggests that the application of a cool paint, with highly reflective nanocomposites, to a roof can reduce the amount of heat entering a building, thereby regulating indoor temperatures and leading to energy savings in summer cooling loads. Considering this, this chapter aims to evaluate the effect of using nanopaints with cool properties on roofs to assess their impact on a building's thermal performance and energy demand for cooling. The research is conducted in the Metropolitan Area of Mendoza, which is situated in the Cuyo Region of Argentina, on the South American dry diagonal. Regarding the climatic context, the area is characterized by a temperate continental climate and an abundant solar resource. Within the four most populated urban agglomerates in the country, Mendoza has the highest solar resource (25.4 MJ m^{-2} in December and 9.10 MJ m^{-2} in June). These conditions, coupled with the increase in UHI and heatwaves in the city, supply an opportunity to apply innovative paint technologies to investigate their impact throughout the summer season and during heatwaves.

To achieve this objective, this chapter is structured along two axes, the first one is linked to the analysis of the thermal performance of cool roof technology through experimental testing in a residential dwelling (Section 12.3.1), and the second axis, in which progress is made in the construction of a theoretical model (simulated) to predict energy demand for refrigeration under different scenarios that modify the radiative properties of the roof of the house under study.

12.2 Background and context

12.2.1 Building energy consumption

The building industry is a major energy consumer and one of the biggest emitters of black carbon at a global scale, responsible for roughly 32% of the global final energy use and over 30% of the emissions worldwide [25]. Due to population growth, fast urbanization, and augmented housing electricity access, energy demand in this sector is predicted to surge in the succeeding years [12,26]. Cool roofs decrease the radiative heat gain at the opaque envelope of the building, thereby reducing the heat flow into the conditioned space [27].

Santamouris [28], Biardeau et al. [29], and Bezerra et al. [30] analyzed global cooling energy requirements under different consumption scenarios. The results showed that 3.4 billion people in developing countries (Latin America and the Caribbean, Asia, and Africa) currently lack access to basic cooling (Sustainable Energy for All 2019). However, energy consumption scenarios in urban areas in developing countries are still largely unexplored. For different socio-economic development scenarios, Andrijevic et al. [31] projected an electricity shortage gap in 2050 in the order of 2 to 5 billion people, who will not be able to cope with the minimum energy requirement for cooling due to the energy crisis, low grid investments, and high costs [17].

Reducing emissions is imperative especially in energy-intensive sectors. The building sector has a substantial potential to deliver fast services and cost-effective greenhouse gases (GHG) reductions. The energy consumption in buildings is set up by the relationship between energy demand and the performance of the thermal systems. Energy demand varies according to the climatic conditions, the envelope characteristics and the operational and functioning conditions of the building. It is projected that by 2050, up to 61% of the total emissions could be reduced not only by retrofitting existing buildings, but also by designing new energy-efficient buildings, and effectively implementing renewable energy measures [10]. Among the passive solutions for building energy savings, the radiative properties of building envelope materials play a key role in reducing dependence on heating and cooling sources [32].

In this context, energy-efficient building material technologies used for heating, cooling, and building retrofitting have been increasingly in demand not only from the public sector, but also the construction industry and the homeowners. The application of double-glazed windows and doors, solar panels and insulating materials are being widely developed. However, paints and coatings incorporating reflective nanocomposites, i.e., cool materials, offer great opportunities for an easy installation and low costs. In many cases, the costs can be zero if the replacement of a conventional paint by one with cool characteristics is considered during maintenance periods. Cool coating technology has driven studies to improve the solar reflectance performance of pigments in a joint effort between industry and academia [33,34].

One of the main drivers for the expansion of cooling technology is the progressive increase in the cost of energy. Furthermore, a variety of incentives and policies have arisen within the construction industry to promote measures that enhance the energy efficiency of buildings and decrease carbon emissions in cities. For example, the Cool Roof Rating Council (CRRC) (https://coolroofs.org/) is a non-profit organization that devises precise and trustworthy techniques for evaluating and categorizing the radiative characteristics of roofing and exterior wall items. The CRRC's goal is to offer objective scientific information on cool surfaces for weighing crucial deliberations and making enlightened judgments regarding heat islands, scorching heat and energy consumption in infrastructure.

12.2.2 Optical properties and radiative cooling

As a general definition, cool paints are coatings distinguished by high solar reflectance (from now on SR) and high thermal emissivity (from now on TE) [35]. SR determines a surface material's ability to reflect incoming solar radiation, including the direct and diffused component. The TE of a cool coating refers to its capacity to re-radiate absorbed heat when compared to a black body operating at the same temperature [36]. Therefore, when a cool cladding is applied to a building envelope, it can reflect solar radiation and emit heat, which in turn lowers the surface temperature when exposed to sunlight, in consequence, less heat enters the building [37]. Both dimensions are typically assessed on a scale ranging from 0 to 1 or 0 to 100%.

High-performance cool paints show SRs exceeding 85% in contrast to standard black coatings like bituminous-based coatings, which reflect less than 10% of solar radiation. Moreover, cool paints are characterised by thermal emittance values ranging between 0.80 and 0.90. These properties are receptive to the surface's outer layer exposed to solar radiation, which explains why they are known as coatings designed for surface applications. Thus, they have a significant impact on the thermal balance of a horizontal or low-slope surface which is exposed to solar radiation, such as a roof or the pavement in a city.

The standard methods for evaluating the solar reflectance (SR) of a surface are

- ASTM E903-20 (2020): calculation from laboratory experimental data obtained using an integrating sphere spectrophotometer [38].
- ASTM C1549-02 (2017): a portable solar reflectometer is used on the in-situ target [39].

- ASTM E1918-06 (2015): used on large horizontal or low-slope surfaces using an albedometer in the field (two pyranometers). Applicable when the angle of the sun is less than 45 degrees regarding the normal of the observed surface [40].
- ASTM G173-03 (2008): provides the solar spectral irradiance values of direct normal and hemispherical reference with varying coated surface tilt angle, air mass, etc [41].

The standard methods to measure TE by portable emissometer are

- ASTM E1933-14 (2022): determines emissivity in a compensated manner, using thermocouples and infrared imaging radiometers [42].
- ASTM C1371-15 (2022): used for materials with significant thermal resistance and a surface temperature close to ambient [43].
- ASTM E408-13: determination of the total hemispheric emission, at 60° and 20° from the normal of a surface [44].

12.3 Assessing the impact of cool paint

This section will examine the effects of the application of cool paint technology on indoor air temperature reduction, as well as on indoor and outdoor surface temperature, and its potential impact on air conditioning energy demand in a house situated in the city of Godoy Cruz, Mendoza, Argentina (latitude 32°54'55.12" south, longitude 68°52'18.76" west and altitude of 750 meters). The chosen typology for the study is a local social housing unit with a covered area of 103 square meters and an exposed envelope area of 70.2 square meters, considering the surfaces of both the roof and the façade. The climate of Mendoza City is classified as BWh, which corresponds to a warm semi-arid climate according to the Köppen climatic classification [45]. This climate is characterised by a high annual heliophany with a maximum daily solar radiation of 1006 W m^{-2}, a wind speed at 10 m of 1.9 m s^{-1} with a predominant southeast direction, and low annual precipitation of 218 millimeters (mm). The average annual air temperature is 17.8°C, with average high temperatures of 31.5°C during the summer months. Mendoza's urban heat island, caused mainly by the morphological characteristics of the city, the dense forestation of the urban canyons and the materials used in the building envelope, has an average temperature of 6.5°C and can reach a maximum intensity of 10°C at night [46].

In terms of material characteristics, the roofs in Mendoza city are predominantly aluminised membranes (74%), while metal sheets and ceramic tiles are less commonly used [18]. The extensive prevalence of membranes in Mendoza supports the prospect of implementing high-performance reflective coatings. Particularly, the house under study presents low-cost and low energy-efficient materials with a thermal transmittance index (U-value) of 0.55 W m^{-2} K^{-1} in roofs, 2.08 W m^{-2} K^{-1} in external walls, and 5.82 W m^{-2} K^{-1} in windows. The roof structure makes up a 3 mm aluminised membrane, 10 mm concrete mortar, 50 mm expanded polystyrene, and 3/4" pine tongue and groove. In addition, the exterior walls consist of bricks (thickness: 280 mm) with a fine cement cladding, and the windows are single-glazed wooden frames. The principal front of the house is oriented to the west, while the secondary front is adjoined to the north-south perimeter walls with an eastern direction (Fig. 12.1).

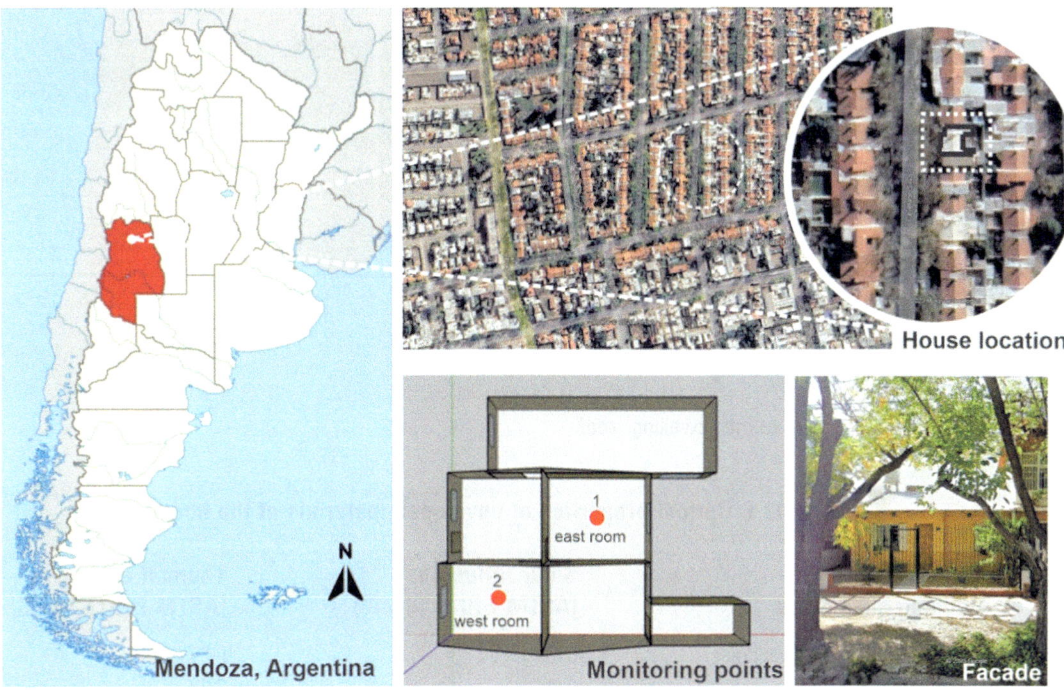

Figure 12.1 Location of the study house and measurement points. Map lines delineate study areas and do not necessarily depict accepted national boundaries.

12.3.1 Experimental analysis of the thermal performance of the house roof

12.3.1.1 Methodology and instrumental

The roof is the largest surface exposed to solar radiation in a dwelling, therefore evaluating its thermal performance under different material technologies can significantly improve the thermo-energetic performance of a dwelling. Therefore, the hygrothermal response of a detached house is analyzed under two roofing solutions: basic (aluminised membrane) and cool (nanopaint) (Fig. 12.2).

The following table describes the optical characteristics of the assessed roof and the external wall surfaces of the dwelling according to the reference standards (Table 12.1).

On-site audits were carried out for thirty consecutive days starting January 1, 2021 (Fig. 12.3). According to the functional layout of the dwelling, two rooms were selected for the study: one facing east and one facing west. The air temperature (Ta) and humidity sensors (Rh) were placed at an average height of

Figure 12.2 Applying cool paint to the dwelling roof.

Table 12.1 Optical properties of enveloped materials of the house.

Building component	Solar reflectance (ASTM E1918-06 [40])	Thermal emissivity (ASTM E1933-14 [43])
Base roof (aluminium membrane)	0.49	0.25
Cool roof (nano paint)	0.88	0.90
Exterior walls	0.55	0.80

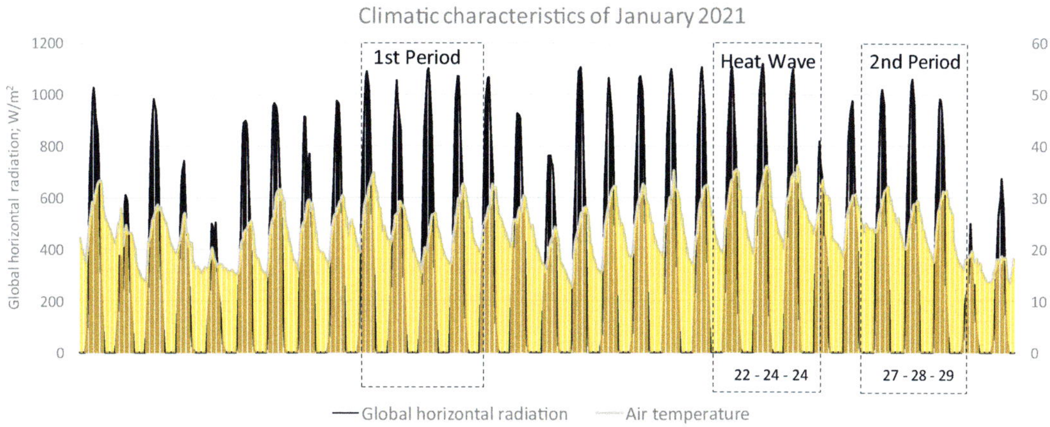

Figure 12.3 Global hourly radiation and air temperature during the month of January 2021.

2 meters inside the rooms and at a sufficient distance from the walls to avoid data distortion [47]. The outdoor air sensor (mobile weather station) was placed in the courtyard of the house. The surface temperature sensors were placed both on the outside of the roof (Tso) and on the inside roofing (Tsi). The instrumentation used at the monitoring sites is described in Table 12.2.

The global horizontal radiation (W m^{-2}) and air temperature (°C) curves monitored during January 2021 are plotted in Fig. 12.3 below. In this research, three periods of analysis are considered. The first and second periods correspond to days of standard climatic conditions for the summer season in the city of Mendoza (see Table 12.3). The third period corresponds to a series of consecutive excessively hot days in which maximum and minimum temperatures exceed the annual average for the city of Mendoza (18°C).

To assess the efficacy of the application of cool paint, we selected the following dates: 10th, 11th, and 13th January 2021 for the base roof and 27th, 28th, and 29th January 2021 for the cool roof due to their standard climatic features. Fig. 12.4 illustrates the distribution of climatic factors, including solar radiation (W m^{-2}), wind speed (ms^{-1}), air temperature (°C), and relative humidity (%) for both observation periods. The analysis shows that the four analysed climatic parameters exhibit a significant similarity between the periods under comparison, with an R^2 of 0.95 to 0.99. The high similarity presented by the selected days when comparing the two roof technologies strongly supports the analysis (Fig. 12.4).

For each climatic parameter, Table 12.3 outlines the corresponding calculated statistics. In the first monitoring period, with the base roof, the highest measured solar radiation

Table 12.2 Instrument description.

Parameter evaluated	Instrumental	
Outdoor environmental characteristics with a mobile weather station (temperature and relative humidity; wind direction and speed; direct and diffuse solar radiation and pressure).	ONSET Weather HOBO®, model H21 -001. The station has: HOBO S-THB-M002 61, HOBO S-WSA-M003, HOBO S-WDA-M003, HOBO S-LIB-M003 silicone pyranometer and HOBO S-BPA-CM10 barometric pressure sensor.	
Outside roof surface temperature (Tso)	Data logger type Onset HOBO® UX120 associated with type T thermocouples.	
Indoor air temperature (Ta) and relative humidity (Rh)	Data logger tipo Onset HOBO® H08-003-02	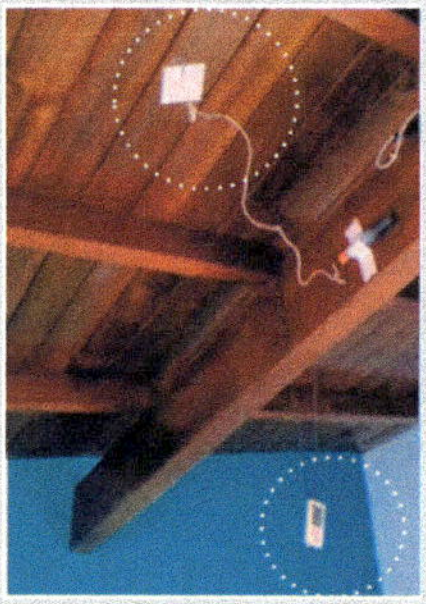
Indoor roof surface temperature (Tsi)	Data loggers type LASCAR® EL-USB-TC associated with type T thermocouples.	

was 1069 W m^{-2}, while during the second period, comparatively similar readings of 1047 W m^{-2} were recorded. Wind speed remained constant at an average of 1.7 ms^{-1} during both periods. Air temperature followed an average curve of 26°C over the two days of analysis, with a maximum variation of 2.5°C and 0.4°C during the lowest temperatures. Relative humidity displays the slightest similarity between the periods since the average humidity difference is 9%.

Table 12.3 Climate data monitored with base roof and cool roof.

	Base roof				Cool roof			
	Solar radiation; W m^{-2}	Outside air temperature; °C	Outside relative humidity; %	Wind speed; m/seg	Solar radiation; W m^{-2}	Outside air temperature; °C	Outside relative humidity; %	Wind speed; m/seg
Mean	342.5	25.7	48.0	0.7	327.4	25.8	39.2	0.6
Max	1069.4	34.8	69.7	1.7	1046.9	32.3	58.4	1.7
Min	0.6	17.0	32.1	0.0	0.6	17.4	29.1	0.0

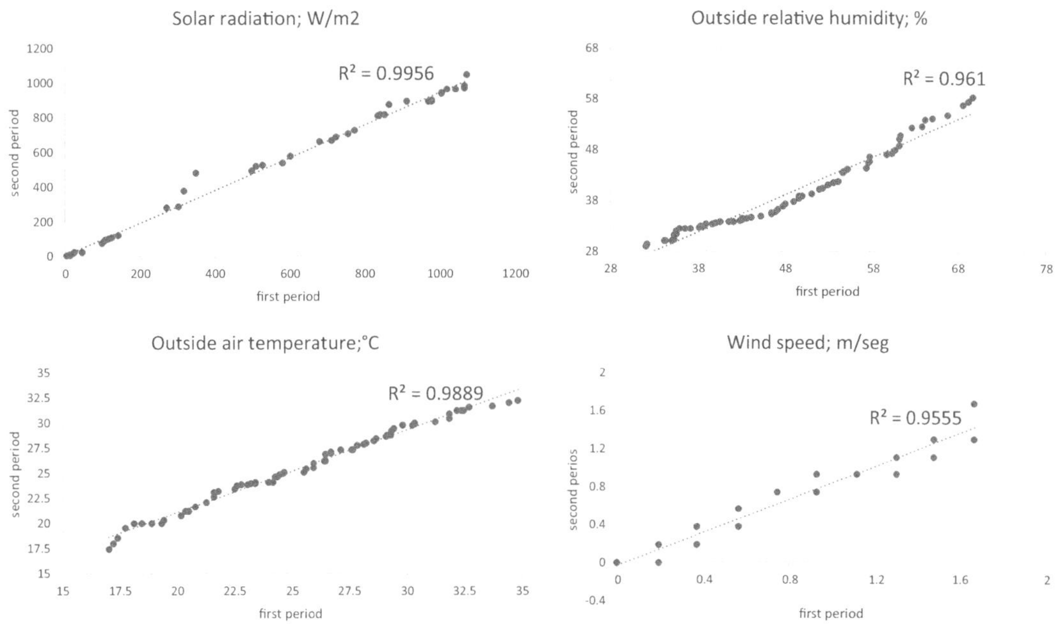

Figure 12.4 Scatter plot of the outdoor climatic conditions during the monitoring period of the base roof (first period) and the cool roof (second period).

12.3.1.2 Results

This section analyses the thermal responses when retrofitting the base roof with cool paint. Generally, the thermal performance of the dwelling with the base roof is considerably higher than that of the dwelling retrofitted with cool paint. Next, the daily thermal performances are individually analysed (day–night).

– *During maximum daytime temperatures,* the data demonstrates that the cool roof reduces external surface temperatures by up to 30°C for the west orientation, with a maximum base roof temperature of 73°C. In the east room, external surface temperatures decrease by 18°C. Likewise, interior roof surface temperatures also decrease by up to 5°C in the east room roofing. A decrease in indoor air temperature was seen, with a decline of 2.9°C in the east room and 1.4°C in the west room.

– *During the coldest night-time hours,* the data demonstrates that outdoor roof surface temperatures decrease by as much as 5°C, while indoor surface temperatures decrease by as much as 2.5°C. The reduction in indoor air temperature is more modest, with a decrease of 0.5°C (as presented in Table 12.4).

The study indicates that the most significant reduction in temperature (Tai, Tso and Tsi) occurs during periods of maximum temperature, particularly in the west room, with lower differences observed during minimum temperatures. The average temperature difference between outdoor surfaces reaches up to 8.7°C, meanwhile the indoor surface temperature variation is up to 3.6°C, with the maximum average indoor air temperature of 1.8°C for the east room (Fig. 12.5).

Further analysis of the behavior of the dwelling (Fig. 12.6) reveals the typical value range (between Q1 and Q3) for each roofing solution. The box plot shows that 50% of the air temperature occurrences are concentrated in a range between 31.3°C and

Table 12.4 Thermal behavior of the dwelling with base and cool roof and differences between the two solutions. Indoor air temperature (Tai.), outside (Tso.), and indoor (Tsi) surface temperature, and differences between average, maximum (max), and minimum (min) roof solutions.

Climate variables		Base roof			Cool roof			Differences		
		Mean	Max	Min	Mean	Max	Min	Mean	Max	Min
Point 1 East room	Indoor air temperature (Tai), °C	29.9	33	26.2	28.1	30.1	25.7	1.8	2.9	0.5
	Outside surface temperature (Tso), °C	34.1	57.5	17.5	25.4	39.3	12.6	8.7	18.2	5
	Indoor surface temperature (Tsi), °C	31.1	37	25	27.5	32	22.5	3.6	5	2.5
Point 2 Weast room	Indoor air temperature (Tai), °C	28.8	31.2	26.2	28.1	29.9	26	0.8	1.4	0.1
	Outside surface temperature (Tso), °C	31.9	73.5	15	26.4	42.8	14	5.5	30.7	1
	Indoor surface temperature (Tsi), °C	30	36.2	25.1	27.8	32	24	2.2	4.2	1.1

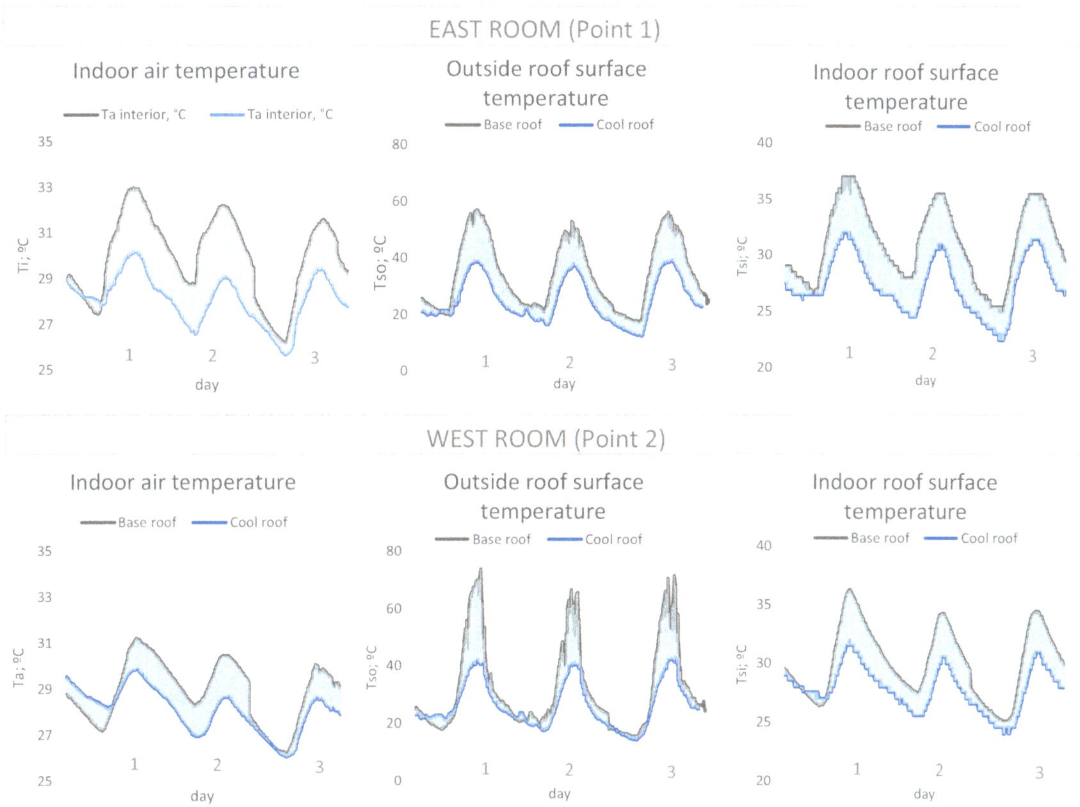

Figure 12.5 Comparison of the temperature distribution at the east room (point 1) and the west room (point 2) for the base and cool roof technological solutions.

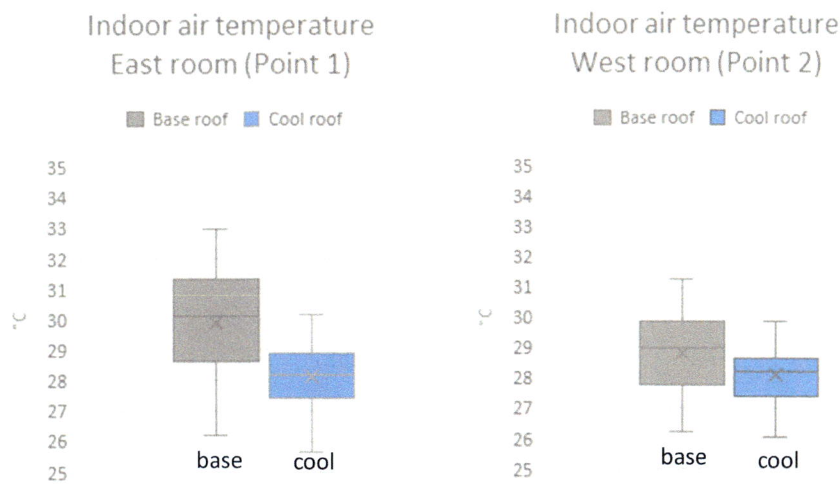

Figure 12.6 Range of indoor air temperature distribution within the monitored rooms.

28.6°C at the base roofing of the eastern room and between 28.9°C and 27.41°C in the cool roofing. Comparable values were seen in the western room, with a range of 29.9°C and 27.7°C at the base roofing, while in the cool roofing, a typical indoor air temperature range between 28.6°C and 27.4°C was seen. These values suggest that mechanical air conditioning equipment are necessary to reach comfortable temperatures within the house (25°C).

12.3.2 Theoretical analysis of the energy demand

12.3.2.1 Simulation model construction

In this section, we will analyze the effects of roof renovation on the auxiliary energy needed for indoor air conditioning (HVAC) during the summer. For this purpose, we developed a dynamic simulation model using the Energy Plus software, validating the thermal measurements and calculating the energy power needed (kW) for cooling the entire house with a thermostat set at 24°C. Energy demand was calculated for various solar reflectance values of different roof surfaces during a summer day with typical climatic conditions (mean temperature: 25.6°C) and during a more thermally challenging day, such as during a heatwave, where the air temperature is over 20°C during the event (mean temperature: 28.9°C). Fig. 12.7 displays a comparison of the temperature ranges during both periods.

For the energy analysis, eight scenarios were devised by determining four levels of solar reflectivity (SR = 0.3, 0.5, 0.7,

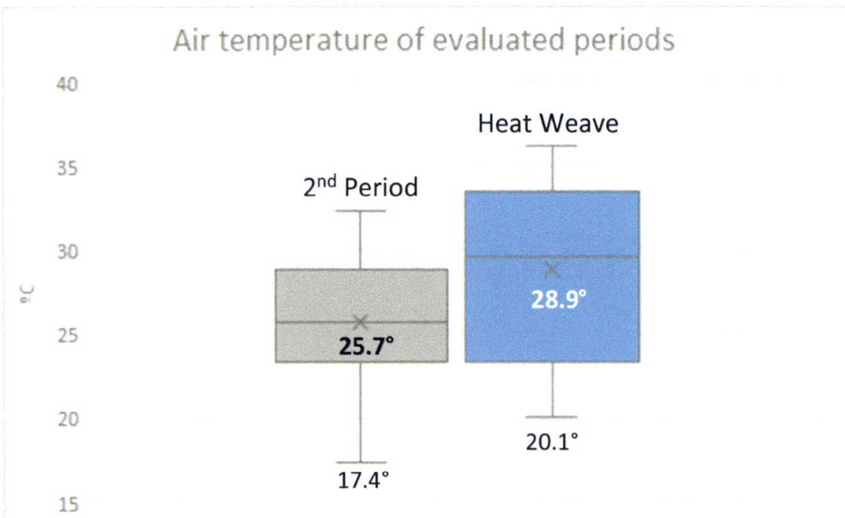

Figure 12.7 Range of outdoor air temperature distribution on a summer day and a heat wave day.

and 0.9) for the two climate periods under evaluation. These are shown in Table 12.5.

The measurements are used to validate the dynamic simulation model via Energy Plus software, version 23.1. This thermal and energy simulation software for buildings was developed by DOE (https://www.energy.gov/eere/buildings/articles/energy-plus), based on BLAST and DOE2 codes. It runs dynamically and enables the energy system performance to be interlinked with the significant material characteristics of the building envelope. The validation method for audit-diagnostic tools has been extensively studied as well as their adjustment through dynamic simulations. This method allows for the testing and evaluation of the state-of-the-art in the building under study, thereby generating several proposals for laboratory-designed improvements. A study of 158 articles scrutinizing different techniques for the evaluation of a buildings' thermal efficiency through energy simulations shows Energy Plus as the most widespread software (employed in 65% of the studies). In addition, 62% of those incorporate some kind of validation or calibration [48].

The geometric model of the house was defined using the Open Studio Plug-in 3.6.1 for Google Sketchup Pro 2021. Given the importance of selecting an adequate distribution of the thermal zones, the dwelling was divided into six thermal zones according to the following spaces: kitchen-dining room, bedroom 1, bedroom 2, bedroom 3, entrance hall, and service area (Fig. 12.8). A detailed description of the construction data assumed in the model is presented in Table 12.6. Regarding the windows, 3 mm single glazed windows (Solar Transmittance = 0.837) were considered.

Using the climatic variables measured during the period under study, an EPW climate file was created for model simulation. The necessary data includes global radiation on a horizontal surface, diffuse radiation on a horizontal surface, direct radiation normal to the beam, outdoor dry-bulb temperature, outdoor relative

Table 12.5 Scenarios determined according to four levels of RS.

Roof solar reflectance (SR)			
During summer period		**During heat wave**	
Case S1	SR = 0.3	Case H1	SR = 0.3
Case S2	SR = 0.5 (base)*	Case H2	SR = 0.5 (base)*
Case S3	SR = 0.7	Case H3	SR = 0.7
Case S4	SR = 0.9 (cool)*	Case H4	SR = 0.9 (cool)*

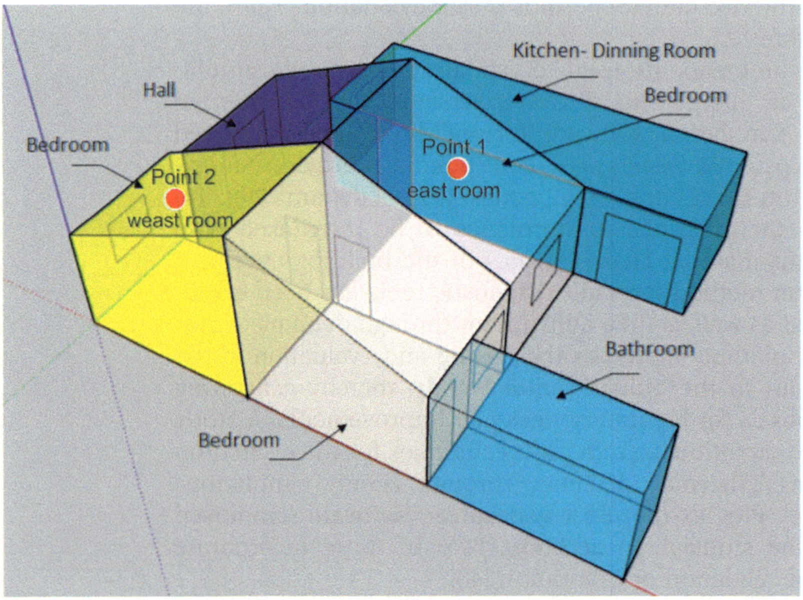

Figure 12.8 Thermal zones of the studied house.

Table 12.6 Detailed data construction of the house under study.

Layers	Roughness	Thickness [m]	Conductivity [W m^{-1}°C^{-1}]	Density [Kg m^{-3}]	Specific heat [J Kg^{-1}°C^{-1}]
WALLS					
Exterior plaster	Medium rough	0.025	0.81	1800	1000
Brick	Medium rough	0.2	0.91	2000	835
Interior plaster	Smooth	0.025	0.93	1800	1000
FLOOR					
Cement mortar	Rough	0.1	1.63	2400	800
Ceramic floor	Smooth	0.005	0.7	1800	920
ROOF					
Aluminium membrane	Medium smooth	0.003	0.17	200	1000
Lightweight mortar	Medium rough	0.01	0.12	500	1000
Expanded polystyrene (EPS)	Smooth	0.05	0.035	20	1300
¾" Wooden Planks	Smooth	0.018	0.17	1200	1500

humidity, air pressure, wind speed, and direction. Missing variables were gathered from climate databases(Repository of free climate data for building performance simulation (2019). https://climate.onebuilding.org/).

The calibration took place in room 1 (East), where no mechanical air conditioning was used during the measurement period. The Design Flow Rate Infiltration model was employed, as defined in the Energy Plus reference manual, to signify the unintended flow of air from the outdoor environment directly into a thermal zone. Infiltration is generally caused by the opening and closing of exterior doors, cracks around windows, and even in very small amounts through building elements. For the model to function effectively, schedules must be used for each zone under analysis. In the validated room, three air changes per hour throughout the entire day were decided.

The simulation was conducted from January 1 to 31. Figs. 12.9 and 12.10 illustrate the curves derived from the settings of the second measurement period (January 27–29). The model was calibrated using the optical and thermal specifications of the cool roof technology applied to the residence. The simulated models show a reasonable fit with differences of around 0.10°C for the indoor air temperature case (Fig. 12.9) and 0.65°C for the outdoor surface temperature case (Fig. 12.10).

12.3.2.2 Forecast of the energy demand

This section explores the outcomes of retrofitting a roof with cool paint during the summer and during a heat wave, with a focus on the external surface temperature and the cooling energy requirements regarding indoor air temperature. To assess the effectiveness

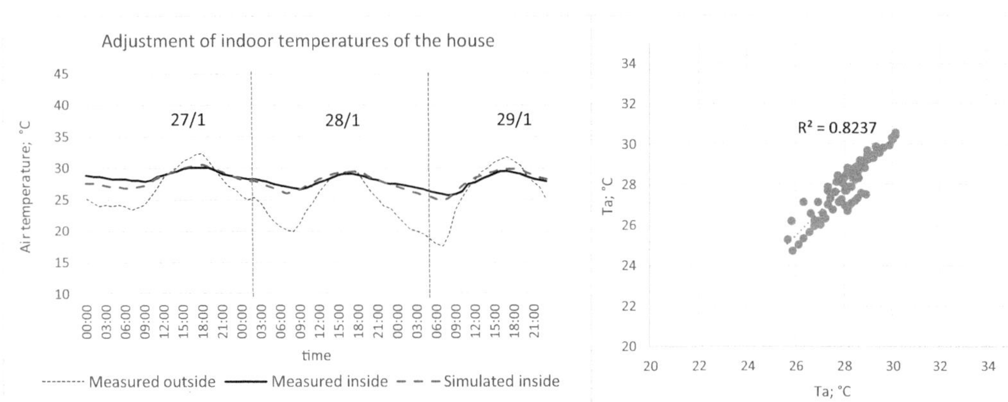

Figure 12.9 The relationship of measured and simulated interior air temperatures in room 1.

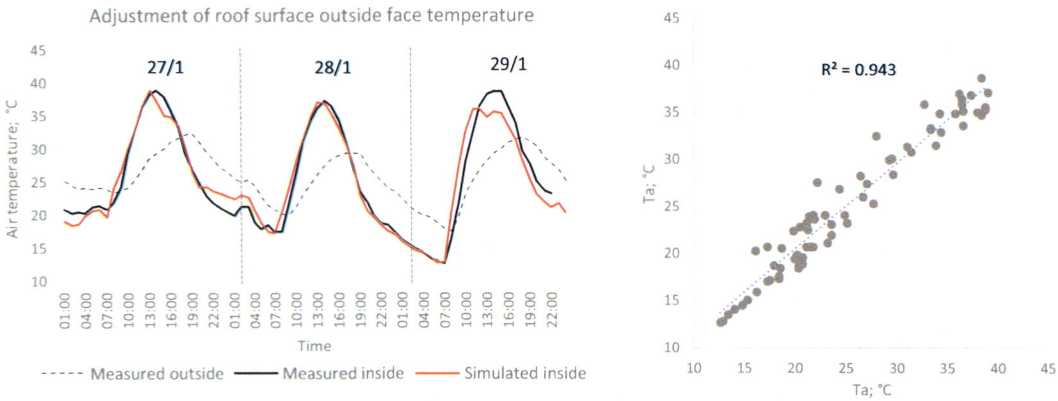

Figure 12.10 The relationship of measured and simulated roof surface temperatures in room 1.

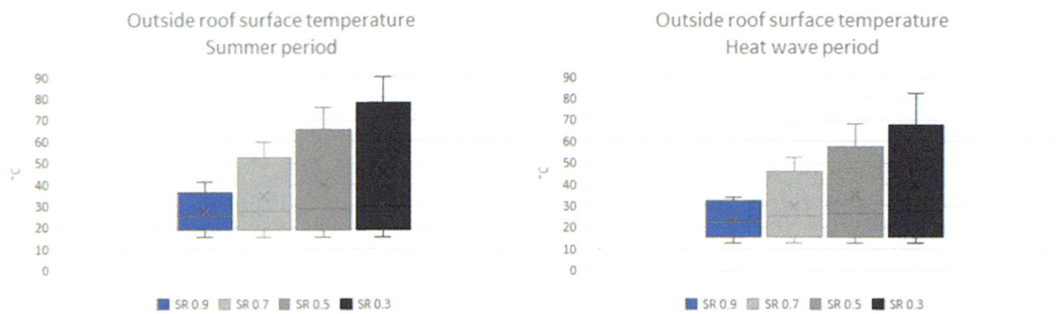

Figure 12.11 Outside roof surface temperatures.

of cool paint in reducing peak power demand and its impact on the energy load on the power grid, it is crucial to note that maximum surface temperatures serve as a suitable measure. Such demands are linked to unusual bursts of energy use within a brief period that can lead to power overload and, so, power outages. Average air temperatures are useful indicators when assessing the impact of technology on HVAC energy demand. Furthermore, both minimum and average temperatures significantly affect users' thermal comfort.

Upon the analysis of the thermal behavior of the roof surface for the cases proposed in Table 12.5, it is apparent that the most significant reductions in surface temperature occur at the maximum values, typically during the periods of higher solar radiation (Fig. 12.11). Notably, there is a recorded maximum decline of 48°C in roof surface temperature between case 1 (SR = 0.3) and case 4 (SR = 0.9), with an average difference exceeding 16°C in both periods. No differences have been found between the analysed period and the occurrence of heatwaves as solar radiation is the determining factor regarding surface thermal behavior and remains constant throughout both periods.

These results indicate the superior effectiveness of high SR and high ET paints in reducing surface temperatures compared to conventional paints. The use of a cool paint reduces the amount of heat transfer from the roof to the building, leading to a reduced energy demand for cooling during the summer. Nonetheless, cool roofs may lead to an increased necessity for heating energy during colder periods. In hot and warm climates, as the case of the city under research, cool roofs are often a more cost-effective and energy-efficient solution for reducing a building's energy requirements in the annual balance sheet as the heating penalty is lower than the cooling benefits [49].

To assess the effect of cool paint on energy savings for the interior thermal conditioning of the dwelling, a scatter diagram for the eight simulated cases is presented in Fig. 12.12. The graph reveals an inverse relationship between roof solar reflectance and power energy demand during both periods. The energy saving trend based on the application of cool roof technology is slightly greater during the regular summer period when compared to heat waves. Between the two extremes evaluated cases (SR = 0.3 and SR = 0.9), the daily energy demand decreases from 519.1 to 372.3 KW during the summer period, and from 910.9 to 721 KW during heatwave events. This implies that an increase of 1.5 in the SR level leads to a 10% reduction in energy demand during the standard period and a 6% reduction during unusual

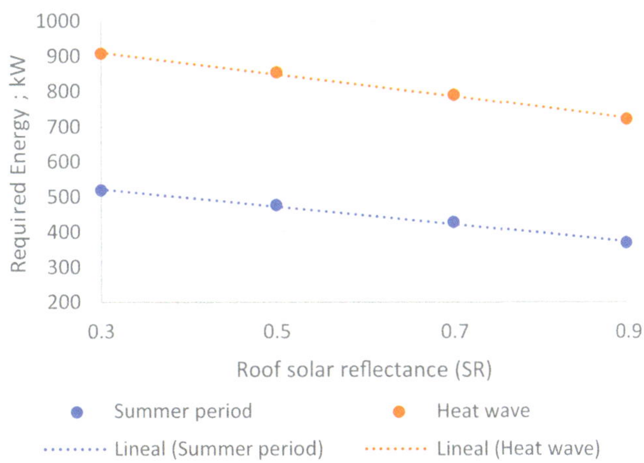

Figure 12.12 Power energy requirements for cooling on a summer day and on a heatwave day.

Table 12.7 Daily requirements (kW) for each of the scenarios.

SR scenario	Energy requirement, kW Summer day	Energy requirement, kW Heat wave day
0.3	519.1	910.9
0.5	474.4	855
0.7	426.4	793.2
0.9	372.3	721
Decreasing trend, %	28.28	20.28

circumstances. In a standard summer period, the energy demand can be reduced by 28.3%, and during a heat wave period, by 20.9% based on the analysis of SRs (Fig. 12.12 and Table 12.7).

Table 12.7 displays the daily power demands (in kW) for the different scenarios, each one altering the reflective features of the roof through nanopaint application.

12.4 Conclusions and discussion

According to several climate projections for the upcoming years, there is an expected rise in the frequency and severity of extreme weather events, like heat waves, in the coming decades [10,50]. In addition, low housing construction rates and an increasing urban population depict an urgent requirement to adapt/rehabilitate the existing housing stock. This study is framed in this context.

Experimental and theoretical tests, which use simulation models, have shown the effectiveness of cool paint technology in reducing roof surface temperatures, cooling energy demand and indoor air temperature in dwellings located in cities with a temperate climate. As a case study, the city of Mendoza was monitored during a standard summer weather period and a heat wave scenario.

The experimental test results showed that by substituting a conventional technology roofing with one having cool paint covering it, reductions in the roofing surface temperature of 18°C on the exterior side and 5°C on the interior side (ceiling) could be achieved. A decrease in indoor air temperature of 3°C was achieved, supplying significant benefits for cooling homes. These findings have a direct impact on the energy requirements for homes. Simulation modeling showed that a cool roof can reduce energy demand by over 20% during heat waves and 28% during the standard summer periods.

These findings are in line with a global survey conducted by Rawat et al. [19], which showed that cool roofs can reduce air temperature by 3°C throughout the day in a hot and arid region of central India. Furthermore, Macintyre [51] concluded that energy savings in cooling can be achieved, ranging from 20% to 70%, by increasing the relative humidity of a roof, depending on the perfect combination of surface relative humidity and roof thermal insulation.

The findings above show that these technologies are highly effective in reducing building energy demand for warm climate cities. However, to expand their benefits for enhancing the environment and the health of residents, an urban scale extension is necessary. To reach this objective, a collaborative effort between stakeholders from the public sector, responsible for the development and application of sustainability policies in urban areas, and representatives of the private sector, industry and commerce, responsible for product development, distribution and application, becomes paramount. In future studies, we plan to investigate the long-term energetic behavior of this technology when exposed to external weather conditions, including wear and tear. The impact projections of cool paint depend largely on the resilience of its optical properties (SR and ET). The scientific and academic community plays a crucial role at this point.

Research data

The data that support the findings will be available in "Repositorio Institucional CONICET Digital" at: http://hdl.handle.net/11336/256954.

Acknowledgements

This work was supported by the National Council for Scientific and Technological Research (PIP 2021–2023 GI 11220200100900CO and TE_2023_1_1012613 Projects).

References

[1] M.P. McCarthy, M.J. Best, R.A. Betts, Climate change in cities due to global warming and urban effects, Geophysical Research Letters 37 (9) (2010) 1–5.

[2] D. Li, E. Bou-Zeid, Synergistic interactions between urban heat islands and heat waves: The impact in cities is larger than the sum of its parts, Journal of Applied Meteorology and Climatology 52 (9) (2013) 2051–2064.

[3] D.A. Dominguez, M.M. Poggi, J.L. Stella, M. de los Milagros Skansi, Ola de calor y temperaturas extremas de enero de 2022 en Argentina [Internet], 7

(2022). Available from: https://repositorio.smn.gob.ar/bitstream/handle/20.500.12160/2426/0036ID2023.pdf?sequence = 1&isAllowed = y.
[4] T.R. Oke, The energetic basis of the urban heat island, Quarterly Journal of the Royal Meteorological Society 108 (455) (1982) 1–24.
[5] S. Grimmond, Cities and global environmental change. Urbanization and global environmental change: Local effects of urban warming, Geographical Journal 173 (1) (2007) 83–88.
[6] M. Hendel, K. Azos-Diaz, B. Tremeac, Behavioral adaptation to heat-related health risks in cities, Energy and Buildings [Internet] 152 (2017) 823–829. Available from: https://doi.org/10.1016/j.enbuild.2016.11.063.
[7] GlobalABC/IEA/UNEP. GlobalABC Regional Roadmap for Buildings and Construction in Latin America 2020–2050. 2020.
[8] K. Gi, F. Sano, A. Hayashi, T. Tomoda, K. Akimoto, A global analysis of residential heating and cooling service demand and cost-effective energy consumption under different climate change scenarios up to 2050, Mitigation and Adaptation Strategies for Global Change 23 (1) (2018) 51–79.
[9] United Nations Department of Economic and Social Affairs PD. The World's Cities in 2018 - The World's Cities in 2018—Data Booklet (ST/ESA/ SER.A/ 417). 2018.
[10] IPCC. Climate Change 2022_ Impacts, Adaptation and Vulnerability_Working Group II contribution to the Sixth Assessment Report of the Intergovernamental Panel on Climate Change. Working Group II contribution to the Sixth Assessment Report of the Intergovernamental Panel on Climate Change. 2022. 37 p.
[11] J. Yang, Z.-H. Wang, K.E. Kaloush, Environmental impacts of reflective materials: Is high albedo a 'silver bullet' for mitigating urban heat island? Renewable and Sustainable Energy Reviews [Internet] 47 (2015) 830–843. Available from: http://linkinghub.elsevier.com/retrieve/pii/S1364032115002452.
[12] K. Bamdad, Cool roofs: A climate change mitigation and adaptation strategy for residential buildings, Building and Environment [Internet] 236 (October 2022) (2023) 110271. Available from: https://doi.org/10.1016/j.buildenv.2023.110271.
[13] E. Morini, A.G. Touchaei, F. Rossi, F. Cotana, H. Akbari, Evaluation of albedo enhancement to mitigate impacts of urban heat island in Rome (Italy) using WRF meteorological model, Urban Climate [Internet] 24 (August 2017) (2018) 551–566. Available from: https://doi.org/10.1016/j.uclim.2017.08.001.
[14] M. Santamouris, A. Synnefa, T. Karlessi, Using advanced cool materials in the urban built environment to mitigate heat islands and improve thermal comfort conditions, Solar Energy [Internet] 85 (12) (2011) 3085–3102. Available from: https://linkinghub.elsevier.com/retrieve/pii/S0038092X10004020.
[15] Z. Jandaghian, H. Akbari, The effect of increasing surface albedo on urban climate and air quality: A detailed study for Sacramento, Houston, and Chicago, Climate 6 (2) (2018).
[16] H. Akbari, D. Kolokotsa, Three decades of urban heat islands and mitigation technologies research, Energy and Buildings [Internet] 133 (2016) 834–842. Available from: https://doi.org/10.1016/j.enbuild.2016.09.067.
[17] A. Mastrucci, E. Byers, S. Pachauri, N. Rao, B. van Ruijven, Cooling access and energy requirements for adaptation to heat stress in megacities, Mitigation and Adaptation Strategies for Global Change [Internet] 27 (8) (2022) 1–16. Available from: https://doi.org/10.1007/s11027-022-10032-7.
[18] N. Alchapar, E. Correa, M. Colli, Cool materials in buildings. Roofs as a measure for urban energy rehabilitation, in: V. Costanzo, G. Evola, L. Marletta (Eds.), Urban Heat Stress and Mitigation Solutions An Engineering

Perspective [Internet], 1st Editio, Routledge Taylor & Francis Group, London, Reino Unido, 2021, p. 20. Available from: https://www.taylorfrancis.com/chapters/edit/10.1201/9781003045922-12-15/cool-materials-buildings-roofs-measure-urban-energy-rehabilitation-noelia-liliana-alchapar-maría-florencia-colli-erica-norma-correa?context = ubx&refId = 65b9ca8d-acda-4f84–9821-4bc3b4b0a.

[19] M. Rawat, R.N. Singh, A study on the comparative review of cool roof thermal performance in various regions, Energy [Internet] 3 (3) (2022) 327–347. Available from: https://doi.org/10.1016/j.enbenv.2021.03.001.

[20] C.H. Ben, A. Bouchair, Experimental Studies of a Passive Cooling Roof in Hot Arid Areas, The Open Energy & Fuels Journal 1 (1) (2008) 1–6.

[21] C.C. Pezzuto, N.L. Alchapar, E.N. Correa, Urban cooling technologies potential in high and low buildings densities, Solar Energy Advances [Internet] 2 (September 2021) (2022) 100022. Available from: https://linkinghub.elsevier.com/retrieve/pii/S2667113122000109.

[22] N.L. Alchapar, C.C. Pezzuto, E.N. Correa, Parámetros urbanos morfo-materiales y su correlación con las temperaturas de aire en verano, Ambiente Construído [Internet 18 (4) (2018) 199–213 [cited 2020 Mar 25]. Available from: http://www.scielo.br/scielo.php?script = sci_arttext&pid = S1678–86212018000400199&lng = es&tlng = es.

[23] N.L. Alchapar, E.N. Correa, The use of reflective materials as a strategy for urban cooling in an arid "OASIS" city, Sustainable Cities and Society [Internet] 27 (2016) 1–14 Nov 1 [cited 2020 Mar 25]. Available from: https://linkinghub.elsevier.com/retrieve/pii/S2210670716302256.

[24] N.L. Alchapar, C.C. Pezzuto, E.N. Correa, L. Chebel Labaki, The impact of different cooling strategies on urban air temperatures: the cases of Campinas, Brazil and Mendoza, Argentina, Theoretical and Applied Climatology [Internet] 130 (1–2) (2017) 35–50. [cited 2020 Mar 25]. Available from: http://link.springer.com/10.1007/s00704-016-1851-5.

[25] P. Eickemeier, S. Schlömer, E. Farahani, S. Kadner, S. Brunner, I. Baum, et al., IPCC, 2012: Summary for Policymakers, in: C.B. Field, V. Barros, T.F. Stocker, Q. Dahe, D.J. Dokken, K.L. Ebi, et al. (Eds.), Planning for Climate Change [Internet], Routledge, 2018, pp. 111–128. Available from: https://www.taylorfrancis.com/books/9781351201100/chapters/10.4324/9781351201117-15.

[26] F. Tariku, S. Hagos, Performance of green roof installed on highly insulated roof deck and the plants' effect: An experimental study, Building and Environment [Internet] 221 (Complete) (2022) 109337. Available from: http://resolver.scholarsportal.info/resolve/03601323/v221icomplete/nfp_pogriotpeaes.

[27] Akbari H., Konopacki S.J. The impact of reflectivity and emissivity of roofs on building cooling and heating energy use. In: Thermal Performance of the Exterior Envelopes of Whole Buildings [Internet]. Building Technologies Department, 1998. p. 29–39. http://www.ornl.gov/sci/buildings/2012/1998 B7 papers/003.pdf.

[28] M. Santamouris, Cooling the buildings – past, present and future, Energy and Buildings [Internet] 128 (2016) 617–638. Available from: https://doi.org/10.1016/j.enbuild.2016.07.034.

[29] L.T. Biardeau, L.W. Davis, P. Gertler, C. Wolfram, Heat exposure and global air conditioning, Nature Sustainability [Internet] 3 (1) (2019) 25–28. Available from: https://doi.org/10.1038/s41893-019-0441-9.

[30] P. Bezerra, F. da Silva, T. Cruz, M. Mistry, E. Vasquez-Arroyo, L. Magalar, et al., Impacts of a warmer world on space cooling demand in Brazilian

households, Energy and Buildings [Internet] 234 (2021) 110696. Available from: https://doi.org/10.1016/j.enbuild.2020.110696.
[31] M. Andrijevic, E. Byers, A. Mastrucci, J. Smits, S. Fuss, Future cooling gap in shared socioeconomic pathways, Environmental Research Letters [Internet] 16 (9) (2021) 094053. Available from: https://iopscience.iop.org/article/10.1088/1748–9326/ac2195.
[32] H. Akbari, S. Konopacki, M. Pomerantz, Cooling energy savings potential of reflective roofs for residential and commercial buildings in the United States, Energy [Internet] 24 (1999) 391–407.
[33] R. Levinson, H. Akbari, S. Konopacki, S. Bretz, Inclusion of cool roofs in nonresidential Title 24 prescriptive requirements. 2010.
[34] R. Levinson, H. Akbari, J.C. Reilly, Cooler tile-roofed buildings with near-infrared-reflective non-white coatings, Building and Environment 42 (7) (2007) 2591–2605.
[35] H. Taha, D. Sailor, H. Akbari, High-Albedo Materials for Reducing Building Cooling Energy Use. California Institute for Energy Efficiency (CIEE). 1992.
[36] A.R. Gentle, J.L.C. Aguilar, G.B. Smith, Optimized cool roofs: Integrating albedo and thermal emittance with R-value, Solar Energy Materials and Solar Cells [Internet] 95 (12) (2011) 3207–3215. Available from: https://doi.org/10.1016/j.solmat.2011.07.018.
[37] A.L. Pisello, State of the art on the development of cool coatings for buildings and cities, Solar Energy 144 (2017) 660–680.
[38] ASTM E903-20: Standard Test Method for Solar Absorptance, Reflectance, and Transmittance of Materials Using Integrating Spheres. 2020.
[39] ASTM C1549-02: Standard Test Method for Determination of Solar Reflectance Near Ambient Temperature Using a Portable Solar Reflectometer. 2017.
[40] ASTM E1918-06: Standard Test Method for Measuring Solar Reflectance of Horizontal and Low-Sloped Surfaces in the Field. 2015.
[41] ASTM G173-03: Standard Tables for Reference Solar Spectral Irradiances: Direct Normal and Hemispherical On 37° Tilted Surface. 2008.
[42] ASTM E1933-14: Standard Practice for Measuring and Compensating for Emissivity Using Infrared Imaging Radiometers. 2022.
[43] ASTM C1371-15: Standard Test Method for Determination of Emittance of Materials Near Room Temperature Using Portable Emissometers. 2022.
[44] ASTM E408-13: Standard Test Methods for Total Normal Emittance of Surfaces Using Inspection-Meter Techniques. 2019.
[45] M. Kottek, J. Grieser, C. Beck, B. Rudolf, F. Rubel, J. Grieser, et al., World map of the Köppen-Geiger climate classification updated, Meteorologische Zeitschrift 15 (3) (2006) 259–263.
[46] E.N. Correa, Isla de calor urbana. El caso del Área Metropolitana de Mendoza, Universidad Nacional de Salta, 2006.
[47] T.R. Oke, Initial guidance to obtain representative meteorological observations at urban sites, World Meteorological Organization [Internet] (81)(2004) 51. Available from: http://www.geog.ubc.ca/~toke/IOM-81-UrbanMetObs.pdf.
[48] V.F. Mendes, A.S. Cruz, A.P. Gomes, J.C. Mendes, A systematic review of methods for evaluating the thermal performance of buildings through energy simulations, Renewable and Sustainable Energy Reviews [Internet] 189 (PA) (2024) 113875. Available from: https://doi.org/10.1016/j.rser.2023.113875.
[49] M. Hosseini, H. Akbari, Heating energy penalties of cool roofs: the effect of snow accumulation on roofs, Advances in Building Energy Research [Internet] 8 (1) (2014) 1–13. Available from: http://www.tandfonline.com/doi/abs/10.1080/17512549.2014.890541.

[50] IPCC. Foreword Technical and Preface. Climate Change and Land: an IPCC special report on climate change, desertification, land degradation, sustainable land management, food security, and greenhouse gas fluxes in terrestrial ecosystems. 2019, 35–74 p.
[51] H.L.L. Macintyre, C. Heaviside, Potential benefits of cool roofs in reducing heat-related mortality during heatwaves in a European city, Environment International [Internet] 127 (May) (2019) 430–441. Available from: https://doi.org/10.1016/j.envint.2019.02.065 [cited 2019 Jul 18].

13

Nanopaints in food packaging applications

Md. Sohel Rana[1], Most. Afroza Khatun[1], K.M. Anis-Ul-Haque[2] and Md. Wasikur Rahman[1,3,4]

[1]*Department of Chemical Engineering, Jashore University of Science and Technology, Jashore, Bangladesh* [2]*Department of Chemistry, Jashore University of Science and Technology, Jashore, Bangladesh* [3]*Department of Physics and Astronomy, University of Texas Rio Grande Valley, Edinburg, TX, United States* [4]*Department of Mechanical Engineering, University of Texas Rio Grande Valley, Edinburg, TX, United States*

13.1 Nanopaints

Nanopaints, often termed nano-enabled coatings, represent an innovative application of nanotechnology in surface engineering. These coatings are created by dispersing nanoparticles, which are minuscule particles with dimensions ranging from 1 to 100 nanometers, into a liquid medium such as a polymer matrix or solvent. The addition of nanoparticles to these coatings imparts unique and superior properties compared to conventional paints, making them suitable for a wide range of applications across various industries.

The nanoparticles used in nanopaints can be categorized based on their material composition. Metallic nanoparticles such as silver (Ag), gold (Au), and titanium dioxide (TiO_2) are popular for their antimicrobial, UV-resistant, and photocatalytic properties. Clay nanoparticles, like montmorillonite, are widely used for their exceptional barrier properties, effectively reducing the permeability of gases and moisture in packaging applications. Polymeric nanoparticles, which are often biodegradable, are designed for enhanced flexibility and environmentally friendly applications.

The incorporation of nanoparticles into coatings enables significant functional improvements. Nanopaints exhibit enhanced barrier performance by creating denser and less porous films, which are particularly advantageous for applications in packaging where protection against environmental factors is critical. For instance, in food packaging, barrier nanopaints prevent the

Nanopaints. DOI: https://doi.org/10.1016/B978-0-443-26512-9.00014-8

ingress of oxygen and moisture, preserving food quality and extending shelf life. Similarly, in industrial applications, these coatings protect machinery and equipment from corrosion and environmental degradation.

Another key advantage of nanopaints lies in their antimicrobial activity. Coatings infused with nanoparticles like silver or zinc oxide (ZnO) exhibit broad-spectrum antimicrobial properties, which are particularly beneficial in healthcare and food industries. These coatings inhibit bacterial and fungal growth on surfaces, reducing the risk of contamination and enhancing product safety. Furthermore, the nanoscale dispersion ensures that these properties are uniformly distributed across the coated surface, maximizing effectiveness.

Nanopaints are also highly valued for their surface functionalization capabilities. By tailoring the properties of the nanoparticles, these coatings can be engineered to exhibit specific functionalities such as self-cleaning, anti-fogging, and scratch resistance. For example, self-cleaning nanopaints, often incorporating hydrophobic or photocatalytic nanoparticles, reduce the need for frequent maintenance and cleaning, making them ideal for architectural and automotive applications.

In addition to their functional benefits, nanopaints offer aesthetic enhancements. The nanoscale particles interact with light in unique ways, creating effects such as vibrant colors, iridescence, and color-shifting, which are highly desirable in decorative and cosmetic applications. These coatings also improve durability by enhancing mechanical properties like hardness, scratch resistance, and adhesion to surfaces.

Overall, nanopaints represent a groundbreaking advancement in coating technologies, merging nanoscale innovations with practical applications. Their ability to provide superior barrier properties, antimicrobial activity, surface functionality, and aesthetic appeal positions them as an essential tool for industries ranging from food packaging and healthcare to electronics, construction, and automotive sectors. However, the development and widespread adoption of nanopaints require addressing challenges such as cost, regulatory compliance, and environmental impact, paving the way for future innovations in this transformative field.

13.2 Types of nanopaints

Nanopaints are revolutionizing the packaging industry by enhancing the functionality, safety, and durability of packaging materials. Depending on their specific characteristics and applications, nanopaints can be categorized into several types, each addressing distinct challenges in packaging. Following is a detailed discussion of the main types of nanopaints used in this domain.

13.2.1 Barrier nanopaints

Barrier nanopaints are specifically designed to enhance the barrier properties of packaging materials. These coatings mitigate the transmission of gases (such as oxygen and carbon dioxide) and moisture, which are critical for maintaining product freshness, particularly in food and beverage packaging.

- **Composition:** Nanoparticles such as clay nanocomposites and metal oxides (e.g., silicon oxide [SiO] and aluminum oxide [Al_2O_3]) are commonly used in barrier nanopaints. These materials form a dense, impermeable layer when integrated into coatings, which significantly reduces the permeability of the packaging surface.
- **Applications:** Barrier nanopaints are used in food packaging to preserve flavor, prevent spoilage, and extend shelf life. In addition, they are applied in pharmaceutical packaging to protect sensitive drugs from moisture and air exposure, ensuring their efficacy.

13.2.2 Antimicrobial nanopaints

Antimicrobial nanopaints are gaining increasing prominence due to the growing need to ensure the safety and hygiene of packaged goods. These coatings prevent the growth and spread of harmful microorganisms such as bacteria and fungi on packaging surfaces.

- **Composition:** Nanoparticles with antimicrobial properties, such as silver nanoparticles (AgNPs) and zinc oxide (ZnO) nanoparticles, are incorporated into these paints. AgNPs, in particular, are highly effective due to their ability to disrupt microbial cell membranes and inhibit reproduction. ZnO nanoparticles are valued for their broad-spectrum antimicrobial activity and low toxicity.
- **Applications:** These coatings are extensively used in packaging for perishable food items, where microbial contamination can lead to spoilage. By preventing the growth of pathogens, antimicrobial nanopaints help in maintaining product freshness and safety, especially for dairy products, meat, and ready-to-eat meals.

13.2.3 UV-resistant nanopaints

UV-resistant nanopaints protect packaged goods from the harmful effects of ultraviolet (UV) radiation, which can degrade packaging materials and negatively impact the quality of the product.

- **Composition:** Titanium dioxide (TiO_2) and ZnO nanoparticles are the primary components of UV-resistant nanopaints. These nanoparticles absorb and scatter UV radiation, shielding the contents of the package from UV-induced damage.
- **Applications:** These coatings are particularly important for food, beverage, and pharmaceutical packaging, where exposure to UV light can lead to the deterioration of color, flavor, and nutritional value. For example, UV-resistant nanopaints prevent color fading in beverages and maintain the nutritional integrity of UV-sensitive vitamins in packaged foods.

13.2.4 Functional nanopaints

Functional nanopaints offer specialized properties such as self-cleaning, anti-fogging, and scratch resistance, which enhance the performance and usability of packaging materials.

- **Composition:** These nanopaints often incorporate metal oxides (e.g., TiO_2, ZnO) or other nanomaterials that are engineered to exhibit smart properties. The hydrophobic or hydrophilic nature of these coatings enables functionalities such as self-cleaning or anti-fogging.
- **Properties and applications:**
  - **Self-cleaning:** Coatings with photocatalytic properties (e.g., TiO_2) break down organic contaminants on the surface under UV light, keeping the package clean.
  - **Anti-fogging:** These coatings prevent condensation of water droplets, maintaining visibility and aesthetic appeal. They are often used in transparent packaging for food products like fresh produce and baked goods.
  - **Scratch resistance:** Nanopaints with enhanced hardness prevent scratches, preserving the visual and structural integrity of the packaging, especially in luxury and high-value goods.

13.3 Properties of nanopaints

Nanopaints are celebrated for their unique and enhanced properties, which make them invaluable in diverse applications, particularly in the packaging industry. These properties arise due to the integration of nanoparticles that offer improved functionality, durability, and aesthetic appeal compare

13.3.1 Improved mechanical properties

Nanopaints significantly enhance the mechanical performance of coatings, ensuring that packaging materials can withstand the rigors of transportation, handling, and storage.

- **Hardness and scratch resistance:**
  The incorporation of nanoparticles such as AgNPs in a reinforced polymeric matrix greatly improves the hardness and scratch resistance of coatings. This ensures that packaging materials maintain their integrity and appearance despite mechanical stresses.
- **Tensile strength and durability:**
  Nanopaints offer superior tensile strength and resistance to mechanical damage, making them ideal for packaging applications where durability is crucial. This is particularly important during shipping, where packages are exposed to physical impacts and pressure.
- **Thermal and glass transition improvements:**
  Research has shown that when nanoparticles like AgNPs are mixed with polymeric matrices, properties such as thermal stability, hardness, and glass transition temperature (Tg) are enhanced [1]. These improvements ensure the structural integrity of packaging under various temperature conditions.

13.3.2 Heat-tolerant properties

Nanopaints are designed to improve the thermal stability of packaging materials, making them resilient to extreme temperatures.

- **High and low-temperature resistance:**
  Packaging with nanopaint coatings can withstand significant temperature fluctuations, making it suitable for products stored in extreme environments. This ensures that the texture, flavor, and overall quality of the packaged product remain intact.
- **Protection against sunlight:**
  Exposure to sunlight can degrade the quality of packaging and its contents. Nanopaints with heat-resistant properties protect products from UV-induced temperature changes, ensuring the preservation of sensitive goods, such as food and beverages.

13.3.3 Antimicrobial properties

The antimicrobial properties of nanopaints are among their most sought-after features, especially for food and medical packaging.

- **Silver and gold nanoparticles (AgNPs and AuNPs):**
  Silver is considered one of the noblest metals, known for its broad-spectrum antimicrobial activity. Silver nanoparticles disrupt the cell membranes of microorganisms, effectively inhibiting their growth and reproduction [2]. Gold nanoparticles also exhibit similar properties and are widely used in biomedical applications.
- **Applications in food packaging:**
  AgNPs were among the first additives used in food packaging due to their ability to combat foodborne diseases. When combined with polymer matrices, these nanoparticles enhance the antibacterial, antioxidant, and barrier properties of the packaging material [3].
- **Titanium dioxide nanoparticles:**
  The addition of titanium dioxide nanoparticles (TiO_2NPs) to polymer matrices improves microbial resistance, ensuring food safety during storage and transportation [4].

13.3.4 Barrier properties

Nanopaints enhance the barrier capabilities of packaging materials, protecting products from environmental factors that can compromise quality.

- **Gas and moisture resistance:**
  The dense and compact coating created by nanoparticles reduces the transmission of gases (e.g., oxygen and carbon dioxide) and moisture. This is vital in preventing spoilage, maintaining flavor, and extending the shelf life of perishable goods.
- **Environmental protection:**
  By preventing the exchange of odors, colors, and other environmental variables, nanopaints ensure that products remain unaltered. This is particularly important for food packaging, as inappropriate interactions with external factors can degrade product quality [5].

13.3.5 Aesthetic and sensory enhancement

Nanopaints contribute significantly to the visual and sensory appeal of packaging, improving customer satisfaction and brand perception.

- **Visual effects:**
  Nanopaints can create stunning visual effects such as iridescence, color-shifting, and glossy finishes. These features make packaging more attractive and noticeable on store shelves.

- **Textural improvements:**
  Nanopaints can provide unique textures to packaging materials, enhancing their tactile appeal. These sensory enhancements can influence consumer perceptions, making the product feel premium and sophisticated.

13.3.6 Smart packaging

Nanopaints are integral to the development of smart packaging technologies, which enhance the functionality and interactivity of packaging materials.

- **Intelligent packaging:**
  Intelligent packaging (IP) systems use sensors and indicators integrated with nanopaints to monitor the condition of packaged products. Examples include:
  - **Time-temperature indicators (TTIs):** These detect and display temperature variations, ensuring that products remain within safe storage conditions.
  - **Gas indicators:** These monitor the levels of gases such as oxygen and carbon dioxide inside the packaging, indicating spoilage or leakage.
- **Active packaging:**
  Active packaging (AP) incorporates nanoparticles to actively interact with the packaged product or its environment. For example, nanopaints with antimicrobial properties can directly inhibit microbial growth, ensuring product safety.
- **Advanced technologies:**
  Nanopaints facilitate cutting-edge innovations such as biological sensors, humidity sensors, barcoding techniques, and radio-frequency identification (RFID) systems. These technologies enhance traceability, quality control, and consumer engagement, making packaging smarter and more effective.

13.4 Production process of nanopaints

The production of nanopaints involves a series of carefully engineered steps, starting from the synthesis of nanoparticles to their dispersion and the final formulation of the paint. Each stage is critical to achieving the desired properties and ensuring the optimal performance of the final product. Following is a detailed exploration of each step.

13.4.1 Synthesis of nanoparticles

The first step in nanopaint production is the synthesis of nanoparticles. This process ensures that nanoparticles possess the specific size, shape, and properties required for their intended application. Various synthesis methods are employed, depending on the type of nanoparticles needed:

1. **Chemical vapor deposition (CVD):**
   - In this method, gaseous reactants are used to form a solid material on a substrate.
   - **Process:** Gases are introduced into a reaction chamber containing a substrate. Under specific temperature and pressure conditions, a chemical reaction occurs, depositing a thin film of nanoparticles on the substrate.
   - **Applications:** CVD is commonly used to synthesize metallic and ceramic nanoparticles with high purity and controlled particle size.
2. **Sol-gel process:**
   - This technique involves converting a solution (sol) into a gel-like phase, which is then used to produce nanoparticles, particularly metal oxides.
   - **Process:** Metal alkoxides or chlorides are hydrolyzed and condensed to form a gel. Upon drying and calcination, the gel transitions into nanoparticles.
   - **Applications:** The sol-gel method is widely used for producing silica, titanium dioxide, and other metal oxide nanoparticles.
3. **Hydrothermal synthesis:**
   - This method uses high-temperature and high-pressure conditions to grow nanoparticles from precursors dissolved in water.
   - **Process:** Precursors are mixed in an autoclave and subjected to elevated temperatures and pressures, resulting in the crystallization of nanoparticles.
   - **Applications:** Hydrothermal synthesis is ideal for producing nanoparticles with uniform size and shape, such as ZnO and TiO_2.
4. **Mechanical milling:**
   - This technique involves grinding bulk materials into nanoparticles using a high-power ball mill.
   - **Process:** Materials are placed in a milling chamber containing grinding media (e.g., balls). The chamber is rotated, and the mechanical force breaks the material into nanoparticles.
   - **Applications:** Mechanical milling is often used for producing metal and ceramic nanoparticles.

13.4.2 Dispersion of nanoparticles

Once synthesized, nanoparticles must be evenly dispersed within the paint matrix. This step is essential to ensure that the nanoparticles impart their unique properties uniformly throughout the coating. Several methods are employed to achieve effective dispersion.

1. **Solvent dispersion:**
   - In this method, nanoparticles are dispersed in a liquid solvent compatible with the paint matrix.
   - **Process:** The solvent aids in breaking up nanoparticle agglomerates, allowing them to spread evenly throughout the mixture.
2. **Ultrasonication:**
   - Ultrasonic waves are used to break up nanoparticle aggregates and disperse them homogeneously.
   - **Process:** A probe or bath ultrasonicator generates high-frequency sound waves, creating cavitation (microscopic bubbles) in the liquid. The bubbles collapse, generating localized high energy that disperses the nanoparticles.
   - **Applications:** Ultrasonication is particularly effective for dispersing carbon nanotubes, graphene, and metal oxide nanoparticles.
3. **Mechanical stirring (high-shear mixing):**
   - This method uses mechanical force to achieve uniform dispersion.
   - **Process:** Nanoparticles are mixed with the paint matrix using high-speed blades or impellers, breaking up agglomerates and distributing the particles evenly.
   - **Applications:** Mechanical stirring is often combined with other methods, such as ultrasonication, for optimal results.

13.4.3 Formulation of nanopaints

The final step in the production process is the formulation of nanopaints. This involves blending nanoparticles with binders, solvents, and additives to create a coating with the desired properties and performance characteristics.

1. **Binders:**
   - Binders are the primary component of the paint matrix, providing adhesion to the substrate and mechanical strength to the coating.
   - **Examples:** Epoxy resins, polyurethane, and acrylic binders are commonly used.
   - **Role:** The choice of binder significantly affects the durability, flexibility, and mechanical properties of the nanopaint.

2. **Solvents:**
   - Solvents help in dispersing nanoparticles and ensuring smooth application of the paint.
   - **Examples:** Organic solvents (e.g., ethanol, acetone) or water-based solvents are used, depending on the paint type.
   - **Role:** Solvents evaporate during the drying process, leaving behind a uniform coating of nanoparticles.
3. **Additives:**
   - Additives are incorporated to enhance specific properties of the nanopaint, such as UV resistance, antimicrobial activity, and color stability.
   - **Examples:**
     - **UV resistance:** TiO_2 and ZnO nanoparticles protect against ultraviolet radiation.
     - **Antimicrobial activity:** AgNPs and ZnO nanoparticles inhibit microbial growth.
     - **Color and gloss:** Pigments and surfactants improve the aesthetic appeal and surface finish.
   - **Role:** Additives allow customization of nanopaints for specific applications, such as food packaging, medical devices, or automotive coatings.

13.5 Application strategy

The application of nanopaints involves various techniques, each suited to specific types of packaging materials, surfaces, and desired finishes. Choosing the appropriate application method is critical for achieving uniform coatings, optimal performance, and aesthetic appeal. Following is a detailed description of the primary application strategies for nanopaints:

1. **Spraying**

   Spraying is one of the most widely used techniques for applying nanopaints, particularly in large-scale applications.

- **Process:**
  - A spray gun or automated spray system is used to atomize the nanopaint into fine droplets, which are then directed onto the target surface.
  - The process can be conducted using compressed air, airless systems, or electrostatic spray systems.
- **Advantages:**
  - **Uniform coverage:** Spraying ensures an even distribution of nanopaint across large surfaces.

    - **Scalability:** It is ideal for industrial-scale applications, such as coating large packaging materials or machinery.
    - **Flexibility:** Can be used on flat, curved, or irregular surfaces.
- **Applications:**
    - Spraying is commonly used for food packaging materials, automotive parts, and consumer goods packaging, where consistent coating thickness and high-quality finishes are required.
    - For instance, nanopaints with antimicrobial properties can be sprayed onto food containers to enhance shelf life and reduce contamination risks.
- **Limitations:**
    - Overspray can lead to material wastage, especially when precise control is lacking.
    - Requires proper ventilation and safety precautions due to potential exposure to aerosolized nanoparticles.

2. Dip coating

Dip coating is a straightforward and effective method for achieving uniform coatings on small to medium-sized objects.

- **Process:**
    - The object to be coated is submerged in a bath of nanopaint, then withdrawn at a controlled speed to allow excess liquid to drip off.
    - The coated object is then dried or cured to set the nanopaint layer.
- **Advantages:**
    - **Uniform coating:** Ensures a consistent thickness over the entire surface, including hard-to-reach areas.
    - **Simple and cost-effective:** Requires minimal equipment and is easy to implement for small-scale applications.
    - **Versatility:** Suitable for a variety of substrates, including glass, plastics, and metals.
- **Applications:**
    - Dip coating is frequently used in food and beverage packaging, such as coating bottles, cans, or small containers with nanopaints to improve barrier properties or UV resistance.
    - In electronics, it is used to coat components with nanopaints for enhanced durability and thermal stability.
- **Limitations:**
    - Not ideal for large-scale production due to its slower processing speed.
    - Excess nanopaint must be managed to minimize waste.

3. Roll coating

Roll coating is a highly efficient method for applying nanopaints to large, flat surfaces or continuous rolls of packaging materials.

- **Process:**
  - In this technique, nanopaint is transferred to the substrate using a rotating roller.
  - The process can involve one or multiple rollers to control the thickness and uniformity of the coating.
- **Advantages:**
  - **High speed and efficiency:** Ideal for large-scale industrial applications, such as coating flexible packaging films or sheets.
  - **Precision:** Provides excellent control over the thickness and quality of the coating.
  - **Automation-friendly:** Easily integrated into automated production lines.
- **Applications:**
  - Roll coating is commonly used in food packaging to create oxygen and moisture barriers.
  - It is also utilized in flexible packaging for snacks, pharmaceuticals, and personal care products, where lightweight yet durable coatings are essential.
- **Limitations:**
  - Limited to flat or slightly curved surfaces.
  - Requires precise calibration of rollers and nanopaint viscosity for optimal results.

4. **Electrostatic coating**

Electrostatic coating is a sophisticated technique used to apply nanopaints to complex shapes and surfaces.

- **Process:**
  - Nanopaint is atomized and charged with an electric current, while the target object is grounded.
  - The electrostatic attraction causes the paint particles to adhere uniformly to the surface.
  - The coated object is then cured to solidify the nanopaint layer.
- **Advantages:**
  - **Uniform coverage:** Ideal for complex geometries, including corners, edges, and intricate designs.
  - **Efficiency:** Minimizes material waste due to precise application and reduced overspray.
  - **Improved adhesion:** The electrostatic charge enhances the bond between the nanopaint and the substrate.
- **Applications:**
  - Widely used in the automotive and electronics industries for creating durable, high-quality finishes on components.
  - In packaging, electrostatic coating is used for specialty applications, such as coating metal cans or intricate packaging designs.

- **Limitations:**
  - Requires specialized equipment and expertise.
  - Conductive substrates are preferred, limiting its use with certain materials unless pre-treated.

Factors influencing the choice of application method

The selection of an application strategy depends on several factors, including:

1. **Substrate material:**
   - Different methods are suited to specific substrates, such as plastics, metals, glass, or paper.
   - For example, roll coating works well for flat films, while electrostatic coating is better for metal cans.
2. **Desired coating properties:**
   - The method chosen should achieve the desired thickness, uniformity, and performance characteristics.
3. **Production scale:**
   - Large-scale industrial applications often use spraying or roll coating, while small-scale or specialized applications may rely on dip coating or electrostatic techniques.
4. **Cost and equipment availability:**
   - Methods like spraying and roll coating are more cost-effective for large volumes, while electrostatic coating may require significant initial investment.
5. **Safety and environmental considerations**
   - Proper handling of nanopaints and minimizing material waste are essential, especially in techniques like spraying.

13.6 Application of nanopaints in packaging

Nanopaints have revolutionized the packaging industry by enhancing the functional, aesthetic, and protective properties of packaging materials. The integration of nanoparticles into paints provides unique features, such as antimicrobial properties, UV resistance, barrier capabilities, and mechanical strength. These characteristics make nanopaints an indispensable tool in various packaging sectors, including food, pharmaceuticals, cosmetics, and electronics. Following is a detailed exploration of these applications.

13.6.1 Food packaging

Food packaging is one of the most significant applications of nanopaints due to its direct impact on food safety, quality, and shelf life. Nanopaint-based coatings ensure the protection of food products against environmental factors, microbial contamination, and spoilage.

13.6.1.1 Seafood packaging

- **Challenges in seafood packaging:**
  Seafood such as fish, shrimp, crab, and lobster is highly perishable due to its high moisture content and susceptibility to bacterial growth.
- **Nanopaint solutions:**
  - Barrier nanopaints reduce the transmission of oxygen and moisture, which helps in maintaining freshness.
  - Antimicrobial nanopaints containing AgNPs or ZnO nanoparticles inhibit the growth of bacteria and fungi, extending the shelf life of seafood.
  - Intelligent packaging can be integrated with colorimetric sensors to detect spoilage through pH changes or the release of volatile compounds.

13.6.1.2 Meat packaging

- **Intelligent packaging for quality monitoring:**

Meat products are prone to degradation, releasing volatile compounds that indicate spoilage. Nanopaint-based intelligent packaging with pH-sensitive dyes can monitor changes in meat quality.

- For example, a packaging material with a dye that changes color as the meat deteriorates provides a visual indication of its freshness.
- This technology can also be applied to fish products, providing similar benefits.

13.6.1.3 Dairy product packaging

- **Nanopaints for dairy stability:**
  Dairy products like milk, cheese, and yogurt are sensitive to microbial contamination and environmental factors.
  - Nanopaints with antimicrobial properties help in preventing the growth of spoilage-causing bacteria such as *Lactobacillus* and *Escherichia coli*.
  - UV-resistant nanopaints protect dairy products from light-induced degradation, maintaining their quality and nutritional value.

13.6.1.4 Perishable food packaging

- **Fruits and vegetables:**
  Perishable food items like fruits and vegetables are highly sensitive to environmental factors such as oxygen and moisture.
  - Nanopaints enhance the barrier properties of packaging materials, preventing moisture loss and gas exchange.

- Example: A colorimetric film embedded with bromophenol blue was used to monitor the freshness of guava fruit. The sensor indicated spoilage by changing color, aiding in the timely consumption or disposal of the fruit.
- Such technologies reduce food waste and enhance consumer satisfaction.

13.6.2 Pharmaceutical packaging

Nanopaints are increasingly used in pharmaceutical packaging to ensure the safety, stability, and efficacy of drugs.

- **Barrier properties:**
  - Nanopaints create an effective barrier against moisture, oxygen, and light, which are known to degrade the potency of pharmaceutical products.
  - This is particularly important for drugs with a high sensitivity to environmental factors, such as antibiotics, vaccines, and vitamins.
- **Antimicrobial protection:**
  - Antimicrobial nanopaints, incorporating nanoparticles like silver or titanium dioxide, prevent microbial contamination of drug packaging.
  - This ensures product safety and reduces the risk of infections caused by contaminated medications.
- **Innovative applications:**
  - Smart pharmaceutical packaging uses nanopaints to incorporate features like tamper evidence, real-time monitoring of storage conditions, and authentication technologies to combat counterfeiting.

13.6.3 Cosmetic packaging

Nanopaints are widely used in the cosmetic industry to enhance the appeal, durability, and functionality of packaging materials.

13.6.3.1 Aesthetic enhancements

- Nanopaints provide high gloss, vibrant colors, and unique textures, making the packaging visually appealing to consumers.
- Color-shifting nanopaints and iridescent coatings add a luxurious feel to cosmetic products, increasing their marketability.

13.6.3.2 *Barrier properties*

- UV-resistant nanopaints protect cosmetic formulations from photodegradation caused by sunlight exposure, preserving their color and efficacy.
- They also prevent oxidation and contamination, ensuring the longevity of cosmetic products.

13.6.3.3 *Durability*

- Functional nanopaints provide scratch resistance, anti-fogging properties, and enhanced mechanical strength.
- These features maintain the pristine appearance of packaging materials even during transportation and handling.

13.6.4 Electronics packaging

Nanopaints play a crucial role in the packaging of electronic devices, offering protection, functionality, and aesthetic enhancements.

13.6.4.1 *Anti-static coating*

- Nanopaints with anti-static properties prevent the buildup of static electricity, which can damage sensitive electronic components.
- This is particularly important for packaging materials used in semiconductors, microchips, and circuit boards.

13.6.4.2 *Conductive coating*

- Conductive nanopaints containing materials like graphene or carbon nanotubes facilitate electromagnetic interference shielding.
- These coatings are essential for maintaining the performance and reliability of electronic devices in environments with high electromagnetic activity.

13.6.4.3 *Mechanical protection*

- Nanopaints enhance the scratch and impact resistance of electronic packaging materials, ensuring the safety of delicate components during shipping and handling.
- They also provide corrosion resistance, protecting electronic devices from environmental factors such as humidity and salt spray.

13.7 Advantages of nanopaints in packaging

Nanopaints offer significant advantages in packaging applications, enhancing protection and extending product shelf life. With their enhanced barrier properties, nanopaints reduce gas and moisture permeability, keeping packaged goods fresher for longer. Their antimicrobial properties inhibit bacterial and fungal growth on surfaces, reducing contamination risks for perishable items like food and pharmaceuticals. In addition, nanopaints with UV-blocking features protect light-sensitive products from degradation and color fading, especially in cosmetics and certain foods. They also provide advanced functionalities, such as scratch resistance, anti-fogging, and self-cleaning, adding durability and convenience to packaging. Eco-friendly nanopaint formulations support sustainability by reducing harmful chemical use and facilitating recyclable packaging options, meeting both regulatory standards and consumer demand for greener products.

13.7.1 Enhanced barrier properties

Nanopaints improve packaging barrier properties by embedding nanoparticles that act as barriers to gas and moisture. Traditional packaging materials, such as plastics, often permit some permeability to gases like oxygen and carbon dioxide, which can accelerate product spoilage, especially for foods and beverages. By applying nanopaints, packaging materials gain a dense, nanoparticle-rich layer that fills the microscopic gaps that might allow oxygen or moisture to penetrate. This creates a packaging solution that drastically reduces gas permeability and moisture transmission, extending product freshness and shelf life.

- **Applications in food packaging**: With better barrier properties, food packaging benefits from extended freshness and reduced spoilage. For instance, oxygen-sensitive products like chips, dried fruits, or pre-cooked meals stay fresher for longer. Nanopaints can help maintain the product's texture, flavor, and nutritional value, thereby meeting consumer demand for longer-lasting foods.
- **Impact on pharmaceuticals**: Pharmaceuticals are often sensitive to moisture, which can degrade their effectiveness. Nanopaint coatings provide an ideal barrier to moisture, ensuring the potency and safety of medications, particularly for those that require consistent humidity levels.

- **Preservation in electronics packaging**: Many electronics need protection from moisture and oxygen to avoid corrosion. Nanopaints with enhanced barrier properties help shield sensitive components, thereby improving the lifespan and reliability of electronic devices.

13.7.2 Antimicrobial activity

Nanopaints formulated with antimicrobial nanoparticles, such as silver or copper, inhibit the growth of bacteria, fungi, and other harmful microorganisms on packaging surfaces. This feature is especially valuable in food packaging, where contamination can lead to spoilage, illness, and significant economic losses. Nanopaints provide a proactive defense against microbes by creating a surface that is inhospitable to bacterial and fungal growth.

- **Food safety**: Incorporating antimicrobial nanopaints into packaging for perishable goods like dairy, meats, and fresh produce prevents bacterial growth, reducing the risk of foodborne illness and spoilage. This is crucial for extending product shelf life and preserving food quality.
- **Applications in pharmaceuticals**: For packaging pharmaceuticals, especially sterile products, antimicrobial nanopaints ensure added safety by maintaining sterility and preventing microbial contamination, which is critical in healthcare settings.
- **Sustainability benefits**: The antimicrobial properties can potentially reduce the need for preservatives in food products, allowing for cleaner labels with fewer additives. This aligns with consumer preferences for more natural products and supports sustainable packaging solutions.

13.7.3 UV Protection

Ultraviolet (UV) radiation can degrade various packaged products, including foods, medicines, and cosmetics. Nanopaints with UV-blocking nanoparticles absorb or reflect UV rays, shielding the contents from harmful light exposure. This protective layer prevents degradation, discoloration, and loss of effectiveness, which is vital for maintaining product integrity over extended periods.

- **Benefits for light-sensitive foods**: Many foods, such as oils, dairy products, and beverages, are sensitive to UV light, which can trigger oxidation, spoilage, and nutrient loss. By incorporating UV-blocking nanopaints, these foods maintain their quality and flavor for longer.

- **Pharmaceutical preservation**: Some medicines are light-sensitive, and exposure to UV light can reduce their potency. Nanopaints with UV protection offer a solution to preserve the efficacy of drugs by minimizing their degradation over time.
- **Cosmetics and personal care**: Cosmetics, especially those with natural ingredients, are prone to color fading and ingredient breakdown when exposed to UV light. UV-blocking nanopaints help maintain the appearance, texture, and effectiveness of these products, enhancing their appeal to consumers.

13.7.4 Functionality

Nanopaints can be tailored to provide specific functional properties that traditional packaging cannot offer. These functionalities include self-cleaning, anti-fogging, scratch resistance, and even conductive properties, opening new possibilities in "smart" and multifunctional packaging solutions.

- **Self-cleaning**: Using nanoparticles that create a hydrophobic (water-repelling) surface, nanopaints make it difficult for dirt and contaminants to adhere to the packaging. This is particularly useful for packaging that needs to maintain a clean appearance or withstand exposure to various environmental conditions.
- **Anti-fogging**: Nanopaints with anti-fogging properties prevent the accumulation of condensation on packaging surfaces. This feature is especially valuable in refrigerated foods, where fogging can obscure product visibility and reduce shelf appeal.
- **Scratch resistance**: Nanopaints can enhance the durability of packaging by making it more resistant to scratches and abrasions, which helps maintain product aesthetics and reduces damage in transit.
- **Smart packaging**: Some nanopaints are conductive and can be integrated with sensors or data tags, enabling functionalities like temperature monitoring, freshness indicators, or interactive features. This paves the way for smart packaging, which can alert consumers about product conditions or even interact with mobile devices for added information.
- **Applications in electronics**: Conductive nanopaints are useful in electronic packaging, allowing packages to perform functions such as signal shielding, static dissipation, or even providing on-package displays for branding or product information.

13.7.5 Environmentally friendly

Nanopaints can be formulated to avoid the use of harmful chemicals, heavy metals, or other pollutants, making them suitable for sustainable packaging initiatives. Environmentally friendly nanopaints align with the growing emphasis on eco-conscious products and packaging, especially in markets where regulatory pressures and consumer expectations for green packaging are high.

- **Reduced environmental impact**: Nanopaints designed without volatile organic compounds and other harmful chemicals minimize environmental impact and health risks. This makes them an appealing option for industries aiming to reduce emissions and improve sustainability.
- **Compostable and biodegradable options**: Some nanopaints are compatible with compostable or biodegradable packaging materials, ensuring the entire package can decompose without releasing harmful substances, addressing the problem of long-term plastic waste.
- **Support for circular economy**: Nanopaints that are environmentally friendly contribute to a circular economy by enhancing the recyclability of packaging materials. Their durability can prolong product life, reducing the frequency of packaging replacement and waste generation.
- **Consumer appeal**: Eco-friendly nanopaints meet consumer demand for sustainable packaging solutions, improving brand reputation and marketability, especially in sectors like food, beverages, and personal care, where consumers are increasingly conscious of environmental impact.

These advantages illustrate why nanopaints have gained attention across diverse packaging sectors. By combining enhanced protective features, antimicrobial and UV shielding, added functionalities, and environmentally friendly formulations, nanopaints help meet modern packaging demands for safety, sustainability, and smart functionality.

13.8 Challenges and limitations

Despite their many advantages, nanopaints in packaging come with several challenges and limitations. The high cost of production, due to expensive nanoparticles and specialized manufacturing processes, makes them less accessible for many industries. Regulatory hurdles add another layer of complexity, as nanopaints must pass extensive health, safety, and environmental assessments before they can be widely used. There are also potential health and environmental concerns around nanoparticles, as their small size and reactivity raise questions about their impact on human health and ecosystems. Ensuring uniform dispersion of nanoparticles within the paint matrix is challenging, as nanoparticles tend to aggregate, which can compromise the coating's effectiveness. In addition, the durability of nanopaints must be thoroughly evaluated to withstand diverse environmental and mechanical stresses, and compatibility with various packaging materials requires careful testing. Overcoming these challenges is essential to unlocking the full potential of nanopaints in packaging applications.

13.8.1 Cost

Nanopaints are generally more expensive than traditional coatings, largely because of the cost of nanoparticles and the specialized processes required for their production. Producing nanoparticles often involves advanced equipment, sophisticated synthesis methods, and precision control, all of which add to the material and production expenses. Furthermore, ensuring a high level of purity and quality consistency in nanopaints involves additional quality control measures and testing, contributing to overall costs.

- **Manufacturing costs**: The production of nanoparticles involves complex methods such as chemical vapor deposition, laser ablation, or sol-gel techniques. These methods require high precision and significant investment in machinery and skilled labor, raising the cost of nanopaint production.
- **Research and development**: Nanopaints are often the result of extensive research and development to achieve the desired properties, such as antimicrobial resistance or UV protection. The R&D phase incurs significant expenses, which further increase the final cost.
- **Limited scalability**: While nanopaints offer unique benefits, scaling up their production from lab to industrial scale can be challenging. This difficulty in scalability impacts the cost-effectiveness of nanopaints, making them less accessible for widespread use in packaging, especially in cost-sensitive markets like food and consumer goods.
- **Potential solutions**: To make nanopaints more affordable, ongoing research aims to develop simpler and more efficient synthesis methods for nanoparticles. In addition, scaling up production facilities and improving yield efficiency may reduce costs, as could the development of cheaper alternative nanoparticles.

13.8.2 Regulatory approvals

Nanopaints must undergo rigorous regulatory assessments due to concerns about their potential health and environmental impacts. The novelty and size of nanoparticles introduce challenges in assessing safety, making it necessary to perform extensive testing to meet regulatory standards for packaging materials. Regulatory requirements can vary widely between countries, adding complexity for companies operating on a global scale.

- **Health and safety concerns**: Regulatory agencies such as the FDA and the European Food Safety Authority mandate that nanopaints be tested for any potential health effects, especially in food contact materials. The unique properties of nanoparticles, such as their increased surface area and reactivity, mean they could behave differently than bulk materials, raising concerns about toxicity.
- **Environmental impact**: Agencies require that companies demonstrate that their nanopaint formulations do not negatively impact the environment during production, use, or disposal. This includes evaluating the potential for nanoparticles to leach into soil and water, as well as the effect of disposal or recycling.
- **Time and cost of compliance**: The process of obtaining regulatory approval is both time-consuming and expensive, adding to the overall cost and slowing down the commercialization of nanopaint technologies.
- **Strategies for addressing regulatory challenges**: Companies and researchers are focusing on developing biodegradable and non-toxic nanoparticles that meet regulatory standards more easily. Furthermore, establishing standardized testing protocols for nanopaints may streamline approval processes.

13.8.3 Health and environmental concerns

The potential health and environmental risks associated with nanoparticles are a significant limitation. Nanoparticles can be small enough to penetrate biological membranes, raising concerns about their impact if inhaled, ingested, or absorbed through the skin. Similarly, their behavior in the environment remains poorly understood, with concerns that they might accumulate in ecosystems or interfere with natural processes.

- **Health risks**: Research indicates that certain nanoparticles, if inhaled or ingested, could pose respiratory or gastrointestinal risks. For workers involved in the manufacturing process, occupational exposure to nanoparticles is a concern that necessitates proper protective measures.

- **Environmental concerns**: Nanoparticles, once released into the environment, may persist and accumulate in soil, water, or living organisms. Their small size and chemical reactivity raise concerns about potential ecotoxicity, such as the disruption of aquatic ecosystems or harmful effects on soil organisms.
- **Developing safer alternatives**: Scientists are exploring biodegradable nanoparticles that break down more easily in the environment. Non-toxic materials, such as naturally derived or bio-based nanoparticles, are being developed as safer alternatives that minimize health and environmental risks.
- **Risk management and safety protocols**: Companies are implementing safety protocols to minimize exposure to nanoparticles during production and use. These include the use of personal protective equipment and adherence to strict handling, storage, and disposal procedures.

13.8.4 Uniform dispersion

Achieving uniform dispersion of nanoparticles within the paint matrix is essential for consistent performance. However, nanoparticles have a strong tendency to aggregate, clumping together due to van der Waals forces or electrostatic interactions. This aggregation can compromise the effectiveness of the nanopaint, leading to uneven coatings and reduced protective properties.

- **Impact of aggregation on performance**: When nanoparticles are not evenly distributed, the nanopaint's barrier, antimicrobial, or UV-blocking properties can be compromised. An uneven distribution may also result in weaker mechanical properties, such as reduced scratch resistance or durability.
- **Challenges in dispersion techniques**: Dispersing nanoparticles uniformly requires advanced techniques, such as high-shear mixing, ultrasonic dispersion, or the use of dispersants. These techniques add complexity and cost to the production process.
- **Solutions to improve dispersion**: Researchers are exploring methods to improve dispersion, such as functionalizing the surface of nanoparticles to make them more compatible with the paint matrix. Another approach is to develop advanced mixing techniques that can prevent aggregation and ensure a more uniform nanopaint application.

13.8.5 Durability

The durability of nanopaints, particularly in diverse environmental conditions, is another challenge. Packaging materials are exposed to various stresses, including temperature fluctuations, humidity, and mechanical handling. Ensuring that nanopaints retain their protective and functional properties under these conditions is crucial for their long-term reliability.

- **Environmental stresses**: Temperature changes, humidity, and UV exposure can degrade nanopaints over time. For example, excessive heat may cause nanopaints to lose their adhesive strength, while high humidity may affect their barrier properties.
- **Mechanical stresses**: Packaging is often subjected to bending, friction, and other mechanical stresses. Nanopaints must be resilient enough to withstand these stresses without cracking, peeling, or losing their effectiveness.
- **Evaluating durability**: Comprehensive testing of nanopaints under simulated environmental and mechanical conditions is essential to assess their durability. However, such testing can be time-intensive and expensive.
- **Research on enhanced durability**: Scientists are working to develop formulations that improve nanopaint resilience, such as reinforcing the paint matrix with tougher materials or developing nanoparticles that are inherently resistant to environmental degradation.

13.8.6 Compatibility

Nanopaints may interact differently with various packaging materials, leading to compatibility issues. Packaging materials like plastics, metals, glass, and paper each have unique properties, and nanopaints must be tailored to ensure they adhere properly and maintain functionality on these different surfaces.

- **Material-specific challenges**: For example, nanopaints designed for plastics may not bond as effectively to metals or glass. Incompatible nanopaints may experience poor adhesion, reduced durability, or compromised barrier properties.
- **Testing and customization**: Compatibility testing is essential to ensure that nanopaints perform as expected on the target packaging material. This may involve adjusting the nanopaint formulation to optimize adhesion, flexibility, or other properties specific to the substrate.
- **Complexity in multi-material packaging**: Some packaging solutions use multiple materials (e.g., plastic and metal in the same package), making it challenging to find a nanopaint that is compatible with each component. Customized formulations may be required for these applications, adding to production complexity and cost.

- **Ongoing innovations**: Researchers are working to develop more versatile nanopaints that can be used on a range of substrates, as well as surface treatments that enhance nanopaint adhesion and compatibility across materials.

These challenges highlight the need for ongoing research and development to overcome the limitations of nanopaints in packaging. Addressing these issues will be key to maximizing the benefits of nanopaints while minimizing potential risks and making them more commercially viable for widespread adoption in the packaging industry.

13.9 Future vision

The future of nanopaints in packaging is poised for groundbreaking advancements as researchers explore new compositions and functionalities that enhance the safety, sustainability, and versatility of packaging materials. Future developments in nanopaints for packaging focus on improving barrier properties, antibacterial capabilities, UV resistance, and sustainability. By integrating nanoparticles with polymer matrices, researchers are creating multifunctional nanocomposites that provide a broader range of protective features for packaged goods while supporting environmental goals.

13.10 Multifunctional nanopaints with enhanced barrier, antibacterial, and UV-resistant properties

In recent years, researchers have made significant strides in developing nanopaints that offer barrier, antibacterial, and UV-resistant properties in a single coating. The integration of these multiple functionalities provides a protective solution that can extend product shelf life, reduce contamination risks, and preserve product quality. For instance, nanopaints incorporating barrier-enhancing nanoparticles can restrict the movement of oxygen and moisture, two elements that contribute to food spoilage. By keeping out gases and moisture, nanopaints help maintain the texture, flavor, and freshness of food products, reducing waste and increasing consumer satisfaction.

Antibacterial properties in nanopaints are another promising area of development. By incorporating antibacterial nanoparticles, such as silver or zinc oxide, nanopaints prevent the growth of harmful bacteria and fungi on the surface of packaging, reducing contamination risks. These coatings are especially useful for food and pharmaceutical packaging, where maintaining sterility and freshness is crucial. Furthermore, adding UV-resistant properties to nanopaints offers additional protection for products that are sensitive to light. UV exposure can lead to chemical degradation, discoloration, and loss of effectiveness in products such as medicines, cosmetics, and certain foods. Nanopaints that can block UV radiation help extend the longevity of these products by shielding them from harmful light exposure.

13.11 Advances in polymer matrices and nanocomposite packaging materials

The combination of nanoparticles and polymer matrices is revolutionizing the performance of packaging materials, creating nanocomposites that are stronger, more durable, and more versatile. Nanocomposite materials can exhibit superior mechanical, thermal, and barrier properties compared to traditional packaging materials. For instance, polymers reinforced with nanoparticles like ZnO, TiO_2, or montmorillonite provide excellent strength, elasticity, and heat resistance. This allows for thinner, lighter packaging materials that still offer the necessary protection and durability, potentially lowering material costs and reducing waste.

In addition, polymer matrices can enhance the dispersion of nanoparticles within the paint, ensuring consistent properties across the coating. Improved nanoparticle dispersion reduces the risk of aggregation, a common challenge that can impair coating quality and functionality. This uniform distribution ensures that the barrier, antibacterial, and UV-resistant features of the nanopaints remain effective across the entire packaging surface, maximizing protection and performance.

13.12 Exploring bionanocomposites for innovative packaging solutions

Bionanocomposites, which combine biodegradable polymers with nanoparticles, represent a new frontier in sustainable packaging. Materials such as alginate-nanocomposites are being studied for their potential to create eco-friendly packaging solutions that meet the demands of both performance and environmental responsibility. Alginate, derived from seaweed, is biodegradable and naturally sourced, making it an attractive choice for packaging materials. By reinforcing alginate with nanoparticles, researchers aim to improve its strength, barrier properties, and thermal stability, enabling it to compete with traditional plastics in packaging applications.

The development of bionanocomposites aligns with the growing trend of reducing plastic waste and supporting sustainable practices. As consumers and regulatory bodies increasingly prioritize eco-friendly packaging, bionanocomposites offer an innovative way to meet these demands. This approach not only reduces the environmental impact of packaging waste but also supports a circular economy by enabling materials to break down more naturally after use.

13.13 Incorporating zinc oxide, titanium dioxide, montmorillonite, and silver nanoparticles

The use of nanoparticles like ZnO, TiO_2, montmorillonite, and silver in packaging films is promising for creating materials with outstanding mechanical, barrier, and thermal properties. ZnO nanoparticles provide strong antibacterial effects, which can be critical in packaging environments where hygiene is essential. TiO_2 offers both antibacterial and UV-blocking properties, making it an excellent choice for protecting products from both microbial contamination and light-induced degradation. Montmorillonite, a type of clay nanoparticle, significantly enhances barrier properties by reducing gas permeability, helping to keep out moisture and oxygen.

These nanoparticles also improve the thermal stability of packaging materials, lowering the risk of fire and increasing durability under high temperatures. Enhanced thermal properties allow for safer and more versatile packaging solutions, which is especially valuable in applications where products might be exposed to extreme conditions during storage or transport. In addition, nanomaterials such as AgNPs offer antimicrobial properties, further improving the hygienic qualities of packaging and extending the shelf life of perishable goods.

13.14 Smart packaging with nanomaterials to inform consumers about product freshness

Another exciting development in nanopaints is the potential for smart packaging, which can communicate information about product freshness to consumers. Nanomaterials can be engineered to interact with environmental factors such as temperature or gas levels within the packaging. These materials can be incorporated into labels or coatings that change color based on specific conditions, alerting consumers when a product is nearing spoilage. For instance, packaging with gas-sensitive nanoparticles could detect the buildup of gases associated with food spoilage and signal this change by altering the color of the packaging.

Smart packaging has significant implications for reducing food waste, as it allows consumers to make informed decisions about the freshness of their products. In addition, it provides manufacturers with valuable data on product shelf life and storage conditions, helping them optimize packaging and logistics for perishable goods.

13.15 Developing sustainable and eco-friendly nanopaints

As the demand for sustainable packaging grows, researchers are focusing on creating nanopaints that incorporate eco-friendly nanoparticles, biodegradable polymers, and environmentally conscious manufacturing methods. Eco-friendly nanoparticles, such as naturally derived or bio-based materials, provide a safer alternative to traditional nanoparticles that may have uncertain environmental impacts. In addition, biodegradable polymers can replace conventional plastics in nanopaint formulations, allowing the entire packaging to break down more easily after disposal.

To further reduce the environmental impact, sustainable nanopaint production methods are being explored. These methods aim to minimize the use of energy and hazardous chemicals in the manufacturing process, making nanopaint production cleaner and more efficient. Such developments align with global sustainability goals and respond to consumer and regulatory pressures for packaging solutions that are not only effective but also environmentally responsible.

13.16 Conclusion

Exciting advancements are underway in nanopaint formulations, with a focus on enhancing antimicrobial, barrier, and resilience properties for packaging applications. Alginate-based nanocomposites offer transformative potential in material development, paving the way for innovative, eco-friendly packaging solutions. The integration of nanoparticles like silver,

TiO_2, ZnO, and montmorillonite enhances packaging films, giving them superior mechanical strength, barrier protection, and thermal stability. In addition, sustainability efforts are driving the creation of more environmentally friendly nanopaints, incorporating eco-friendly nanoparticles, biodegradable polymers, and greener production methods. These developments position nanopaints as a vital component in the future of sustainable and high-performance packaging.

Acknowledgment

The authors acknowledge the support of Laboratory of Materials, Energy and Nanotechnology (MENTECH Labs).

References

[1] Z.H. Mbhele, M.G. Salemane, C.G.C.E. Van Sittert, J.M. Nedeljković, V. Djoković, A.S. Luyt, Fabrication and characterization of silver-polyvinyl alcohol nanocomposites, Chemistry of Materials 15 (26) (2003) 5019–5024.

[2] C.V. Garcia, G.H. Shin, J.T. Kim, Metal oxide-based nanocomposites in food packaging: applications, migration, and regulations, Trends in Food Science and Technology (2018). Available from: https://doi.org/10.1016/j.tifs.2018.09.021.

[3] H. Althues, J. Henle, S. Kaskel, Functional inorganic nanofillers for transparent polymers, Chemical Society Reviews 36 (9) (2007) 1454–1465.

[4] Y. Xing, W. Li, Q. Wang, X. Li, Q. Xu, X. Guo, H. Yang, Antimicrobial nanoparticles incorporated in edible coatings and films for the preservation of fruits and vegetables, Molecules (Basel, Switzerland) 24 (9) (2019) 1695.

[5] G.I. Olivas, G.V. Barbosa-Cánovas, Edible coatings for fresh-cut fruits, Critical reviews in food science and nutrition 45 (7–8) (2005) 657–670. Available from: https://doi.org/10.1080/10408690490911837.

14

Nanopaints applications in energy sector

Hani Nasser Abdelhamid
Department of Chemistry, College of Science, Imam Mohammad Ibn Saud Islamic University (IMSIU), Riyadh, Saudi Arabia

14.1 Introduction

Surface modification with chemical materials, such as paints, has become essential for many applications [1,2]. The word "paint" refers to the liquid form of the substance. However, once it is spread over a surface, it is called a "coating." Paint is a suspension of tiny insoluble solid particles called "pigments" in an organic medium called "resin." The light absorption and reflection of paint can be controlled using pigments. Paints may be applied to a surface using many processes, including coating via roll (i.e., coil coating), dip, spin, spray, and brush painting.

Nanopaints are paints containing nanoparticles (NPs) (particle size of 1−100 nm) in the composition of traditional paints. They may be customized for particular purposes by selecting appropriate NPs and formulations. The NPs confer distinct characteristics to the paint. They are usually added to improve conductivity, offer self-cleaning capabilities, and increase light absorption. Nanopaints are widely used for applications such as antibacterial agents [3], protection [4], and antifouling paint for marine [5]. Nanopaints containing metallic and carbon nanomaterials offer conductive coating [6]. Silver nanoparticles (AgNPs) and copper oxide (CuO) NP-based nanopaints can be used as effective antimicrobial paints [7]. Adding TiO_2 NPs to polymer paint enhances the photoactivity, but it can also lead to polymer breakdown and the emission of hazardous volatile organic chemicals [8]. Several NPs were reported for nanopaints including superparamagnetic NPs [9], carbon nanotubes (CNTs), for example, multiwall carbon nanotubes (MWCNTs) [10], titanium oxide (TiO_2) [8], zinc oxide (ZnO), and iron oxide (Fe_2O_3) nanopaints [5]. NPs, such as CuO enabled selective absorption with low infrared reflectance in thermal emittance

Nanopaints. DOI: https://doi.org/10.1016/B978-0-443-26512-9.00015-X

[11]. Nanopaints offer multifunctional properties, such as anti-corrosive, self-cleaning, and antibacterial agents [12].

This chapter offers an overview of nanopaints in the energy sector. It covers an introduction to the topic, synthesis, and preparation procedure for nanopaints. Nanopaint applications for energy management, storage, and production are covered.

14.2 Nanopaints materials and preparation

Paint consists of solid particles known as "pigments" dispersed in an organic material called "resin." Pigments are generally in-soluble particles used to paint primarily to provide color by the absorption and reflection of light. Resin is a substance that forms a film. The bulk of resins consists of organic silicones, which exhibit prominent absorption bands in the mid-infrared range. Commonly used resins include polyethylene, polyester, and propylene (Table 14.1). The nanopaint consists mainly of (1) NPs as a pigment, (2) alkyd resin as a binder, (3) thickener, (4) wetting agent, (5) inner coat drier, (6) outer coat drier, (7) thinner, and (8) stabilizer (Table 14.1).

Carbon nanomaterials are widely used for nanopaints. They are present in different allotropes with two-dimensional morphologies, such as graphene oxide (GO) [[4]], and one-dimensional, such as carbon nanotubes (CNTs) [13]. Carbon nanomaterials, such as CNTs showed high thermal conductivity reaching 3000–3500 $W\,mK^{-1}$ [13]. The electrophoretic deposition method was reported for three different types of CNT (e.g., T-CNT, N-CNT, and P-CNT) onto aluminum substrates [14]. The authors investigated the effect of several parameters, including the thickness of the CNT coating, the morphology, and the reflectance of the CNT absorber in the spectral range. The materials exhibited solar absorbance (α) values of 0.79, 0.90, and 0.90, respectively with thermal emittance (ε) values 0.14, 0.14, and 0.13, respectively [14].

There are several methods for nanopaints preparation. The suitable synthesis procedures depend on several parameters, including the nature and phase of pigment NPs used for nanopaint preparation. Graphene oxide (GO)-based nanopaint was synthesized by integrating GO sheets into alkyd resin via a ball-milling method [4]. TiO_2 NPs were also prepared using the same resin and method of preparation, that is, ball-milling [3]. Simple stirring can be also used for nanopaints preparation. NPs were combined with the other components, for example, a dispersant, wetting agent, and water in a tank [15]. The components are

Table 14.1 Composition of nanopaints and their purposes.

Additive	Purpose	Weight percentage (%)	Examples
Resin or binder	Film-forming	64.0	Poly(acrylamide), polyacrylic acid, methyl methacrylate (acryloid), alkyd, cellulose acetate butyrate, cellulose acetate, methylcellulose, nitrocellulose, epoxy, resin, polyester polymers, polyethylene glycol, polyvinyl ethers, silicone, dimethylpolysiloxane, sodium methyl siliconate, polysiloxane, and polyvinyl alcohol
Thinner or solvent	Dispersing agent	17.0	Benzyl alcohol, butanol, ethanol, isobutanol, methanol, ethyl acetate, cyclohexane, *N*-pentane, benzene, carbon tetrachloride, chloroform, turpentine, tripolene, acetone, and pentoxone
Pigment	Providing the color of the paint	16.0	Fe_2O_3, zinc chromate, asbestos, Cu, Cr (black), Nichem, TiO_2, silica (SiO_2), and nickel oxide (NiO)
Stabilizer	Reducing the color-fading effect	0.6	ZnO
Antisettling agent	Preventing pigment and binder settling	0.5	ZnO nanoparticles
Defoamer or antifoaming	Reduces the formation of foam	1–2	Insoluble oils, polydimethylsiloxanes, certain alcohols, stearates, and glycols
Wetting agent	Improve dispersion uniformity	0.6	Aluminum stearate
Inner coat drier	Chemical cross-linking agent of unsaturated fatty acids	0.6	Soya lecithin
Outer coat drier	Active catalyst for the lipid autoxidation process	0.6	Zironium dioxide (ZrO_2)
Thickener	Tune the viscosity	0.1	Cobalt napthenate

Fe_2O_3, Iron oxide; *TiO_2*, titanium oxide; *ZnO*, zinc oxide.

added successively using agitation at high speed for a certain time [15].

Nanopaint using silver NPs as a plasmonic metasurface was reported [16]. It offered shallow nanoapertures that can control the color, saturation, and brightness [16]. The principle of the fabrication is based on adjusting the size and orientation of each nanoaperture independently. Initially, a silicon wafer underwent a series of ultrasonic cleaning steps utilizing organic solvents (e.g., acetone and chloroform), and water. Then, the wafer underwent a boiling process in a solution consisting of H_2SO_4:H_2O_2 (in a ratio of 3:1) for one hour. It was then sequentially cleaned using organic solvents and water. Next, the AR-3170 positive photoresist (PR) was applied onto the silicon substrate using spin coating at a rate of 4000 r.p.m. It was then subjected to a baking process at a temperature of 100°C for a duration of 10 minutes. The thickness of the PR was about 120 nm. The PR was fabricated using electron-beam lithography. Following exposure, the PR undergoes periodic patterning upon development in an AR 300–35 solution. Next, the grating was transferred to the silicon substrate, which had a thickness of 40 nm, using reactive-ion etching. Subsequently, the PR on the silicon with patterns was removed using acetone and subjected to an O_2 plasma treatment. The silicon surface was covered with a self-assembled layer of 1H,1H,2H,2H-perfluorooctyl-triethoxy silane molecules using a physical vapor deposition process. An electron-beam evaporation process was used to deposit a 1 μm thick coating of silver onto the silicon wafers. Subsequently, the uncovered side of the silver film was affixed to a glass slide using an epoxy resin adhesive (NOA63) with ultraviolet-curable properties and then subjected to 15 minutes of ultraviolet radiation to solidify. Ultimately, the silver film with a pattern, along with the glass slide, was manually removed from the silicon mold because of the limited bonding strength between the silver and silicon. Consequently, a polished silver layer adorned with a nanoaperture array (thickness of 40 nm) was achieved. These procedures offered miniature artworks by using a reusable template-stripping technology that produces very accurate and three-dimensional (3D) imprints. Furthermore, by carefully organizing nanoapertures with varying orientations, it is possible to decipher kaleidoscopic information states using certain combinations of incoming and reflected polarized light [16].

Several tips were reported to improve the nanopaint's preparation. To improve the quality of the substrate, MWCNTs can be included in the paint during the preparation of the nanopaint using ultrasonication and magnetic stirrer techniques [10]. Advanced technologies could help to improve the nanopaint's preparation. For example, they can be applied for organic photovoltaic devices using aqueous solar NP (ASNP) inks or solar nanopaints. These methods solve the difficulty due to the conflicting demands of preserving emulsion stability and promoting surface wettability [17].

Surfactants are usually used to improve the production of high-quality films. However, the high surfactant content hindered the material's performance. Advanced methods, such as surface ozone treatments may help to avoid surfactant usage.

14.3 Nanopaints for enhancing energy efficiency

Nanopaints advanced thermal management using various NPs (Fig. 14.1). Buildings may benefit from the use of nanopaints that include reflective NPs that can reduce heat absorption and decrease the need for air conditioning. On the other hand, in cold areas, the use of nanopaints with insulating qualities may reduce the amount of heat that is lost. Nanopaints can be used to control thermal radiation emissions [18].

Nanopaints have been applied for solar air heaters and cooling systems. Nanopaints with superhydrophobic and photocatalytic capabilities may be used to coat solar panels and wind turbines, resulting in self-cleaning surfaces. Before paint application, a coating of brilliant nickel-based NPs was electrochemically deposited onto copper substrates [19]. The infrared reflectance of the paint was measured to be about 0.4 in the absence of a brilliant nickel coating. However, when a nickel layer with a thickness of 750 nm was added, the reflectance rose to 0.6 [19].

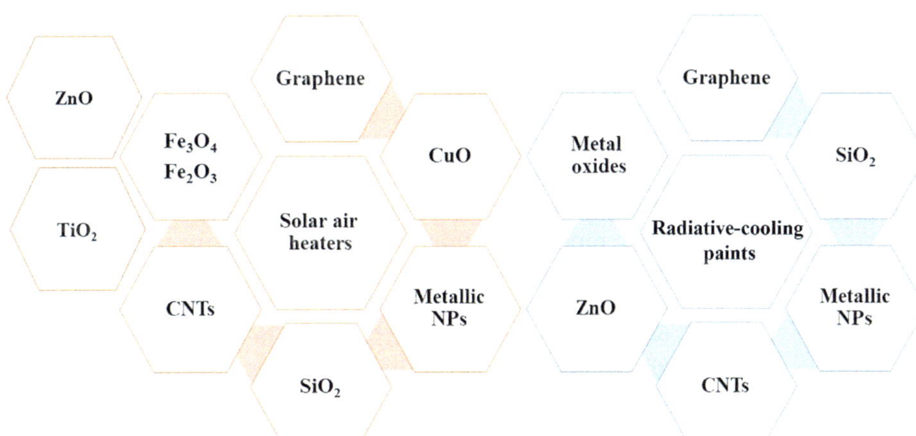

Figure 14.1 Common nanoparticles used for solar air heaters and radiative-cooling paints.

14.3.1 Nanopiants for solar distillers

Solar distillers are simple desalination devices that use sun energy to generate freshwater from saltwater water. It is a promising device for obtaining pure drinking water; however, its main downside is its relatively low output. Several NPs were reported including TiO_2 NPs [20] and carbon fiber-CNT epoxy [21]. NPs can be added directly to the black paint and applied to the absorber plate of a pyramid solar still.

The thermal performance of commercial thermal reflection paints was evaluated by comparing them with paints containing TiO_2 and nano-ilmenite [15]. The color was assessed under ISO 1164–4. The measurement of reflectance in the ultraviolet (UV)–visible (vis)–near-infrared (NIR) range was conducted under the ASTM E 903 standard. The thermal performance was assessed by a simulation test utilizing an infrared light, as well as a field test. The findings indicated that the materials were transformed into a nanofiber arrangement, with a higher concentration of titanium dioxide (62.1%), and exhibited the existence of hematite, metal oxides, and ceramics. The thermal reflection paint containing nano-ilmenite exhibited the greatest NIR and total solar reflectance reflectances, measuring 56.8% and 59.8%, respectively. Thermal performance attributes were assessed using an IR camera for both samples, that is, TiO_2 and nano-ilmenite. The results revealed that both substances had low surface temperatures, suggesting their ability to obstruct IR radiation. Thus, the use of nano-ilmenite in thermal reflection paints not only serves as a barrier but also effectively lowers the surface temperature. Enhancing the thermal efficiency of a structure by minimizing heat transfer and limiting the entry of sunlight may effectively decrease the amount of heating required [15].

Nanopaints improve the low limit of water output for conventional solar distillers. The internal heat transfer coefficients (HTCs) and the amount of distilled water produced are low in the solar desalination system (Fig. 14.2). A solar distiller design was reported using lanthanum cobalt oxide ($LaCoO_3$) NPs doped in black paint [22]. The system was called a double slope wick-type solar distiller (DSWSD). The $LaCoO_3$ NPs/black paint was applied to the surface of the jute wick to enhance the absorption of solar radiation, hence improving the efficiency of desalination. The effect of water flow rates was investigated using different rate values. The daily productivity values with and without $LaCoO_3$ (20 wt.%)/black paint were 5.40 and 3.85 kg m^{-2} day^{-1}, respectively, when the saline water flow rate

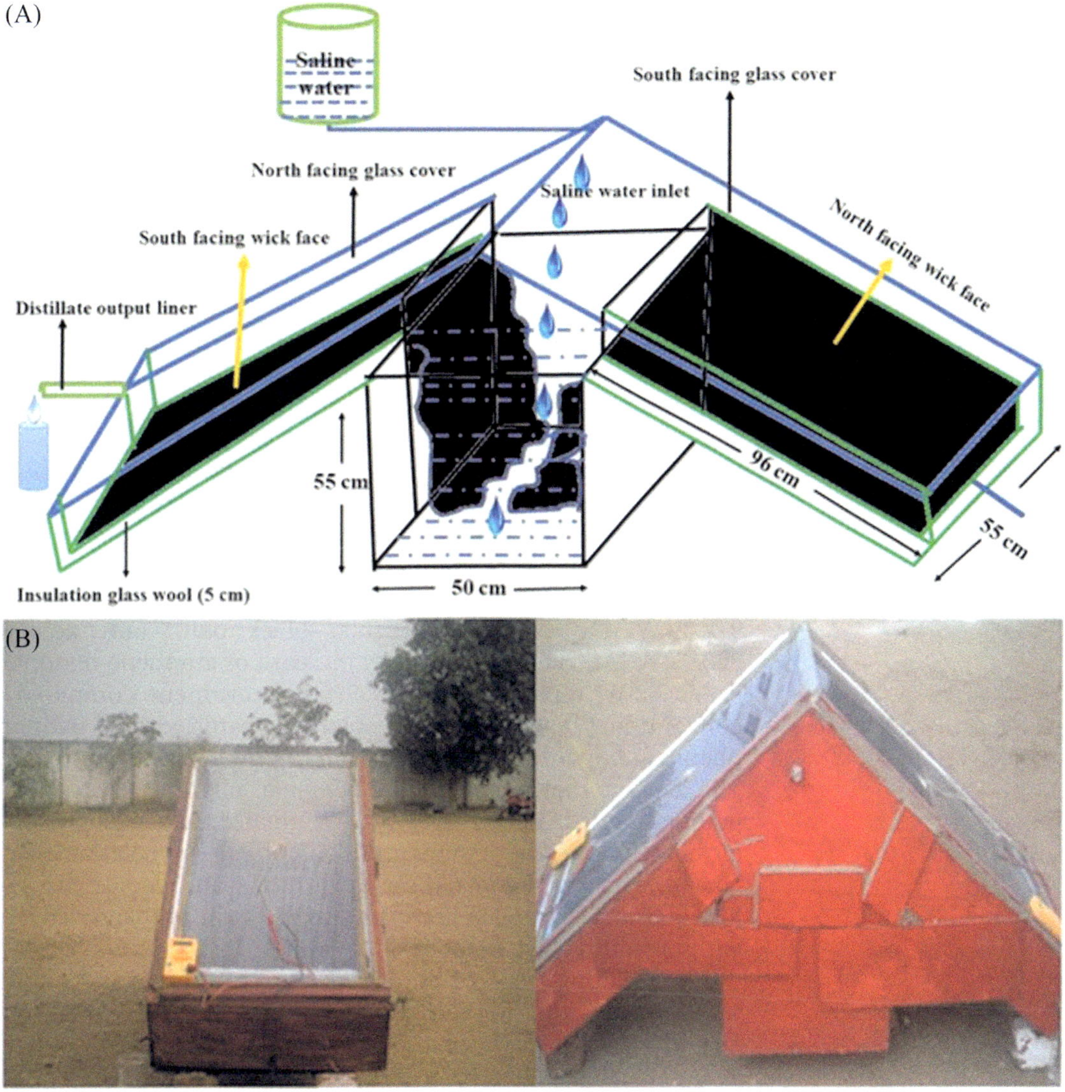

Figure 14.2 (A) The schematized layout of the double slope wick-type solar distiller and (B) photographic view of double slope wick-type solar distiller. Reproduced with permission from the open access article of E.I. Ghandourah, A. Sangeetha, S. Shanmugan, M.E. Zayed, E.B. Moustafa, A. Tounsi et al., Performance assessment of a novel solar distiller with a double slope basin covered by coated wick with lanthanum cobalt oxide nanoparticles, Case Studies in Thermal Engineering 32 (2022) 101859. http://doi.org/10.1016/j.csite.2022.101859.

was 0.05 kg min^{-1}. In addition, the average values of evaporative, convective HTCs, and energy efficiency were higher by 11.20%, 17.54%, and 24.86%, respectively, than for the solar distiller without $LaCoO_3$/black paint [22].

14.3.2 Nanopiants for solar air heaters

Solar air heaters (SAHs) are a renewable energy heating device that attempts to provide warm and fresh air by converting solar energy into a usable form. They capture sunlight via an absorption surface, which converts it into thermal energy and transfers it to the heating fluid. They offered several advantages, such as affordability, adaptability, simple design, and ease of maintenance. They are used in diverse industries for several functions, including preheating, provision of fresh and hot air, dehydration of fruits and vegetables, and energy retention. They exhibit low value of the thermodynamic characteristics of air in SAHs. There are several geometries for SAHs, including vertical solar air heaters without and with baffles and vertical solar air heaters with perforated baffles (Fig. 14.3).

Graphene-based nanopaint improved the thermal efficiency of solar air heaters. The absorber plate of a solar air heater was covered with conventional black paint and graphene-based nanopaint using a standard coating process [24]. The graphene-coated solar air heater achieves a maximum energy gain of 972.84 W day^{-1}, whereas regular black paint only achieves 808.02 W [24]. The mean energy increase of graphene nanopaint is 655.85 W, representing a 12.9% improvement compared to conventional black paint. The solar air heater covered with graphene nanopaint has a maximum thermal efficiency of 81%. The average thermal efficiency of a solar air heater coated with graphene is 72.5%, which is 13.24% greater than that of a solar air heater coated with standard black paint. The solar air heater coated with graphene nanopaint exhibits an average top heat loss that is 8.48% lower compared to the solar air heater covered with ordinary black paint [24].

Two heating systems equipped with perforated baffles heaters were reported for comparison [23]. One of the systems was coated with a conventional matte black paint, while the second solar heater was coated with black nanopaint containing CuO NPs. The comparison was performed at three distinct flow rates. The thermal efficiency values for the heaters, both with and without NPs, were determined to be in the range of 58.10%–76.22% and 54.96%–72.05%, respectively [23]. The application of CuO-embedded coating resulted in a significant improvement in exergy efficiency, ranging from 9.25% to 10.58% [23]. CuO NPs were blended with CNTs [25]. The findings indicated that the solar selective coating with 4% CNTs/CuO-black paint had the maximum solar absorbance (i.e., 0.964) and thermal emittance (0.124). The energy and exergy performances of the

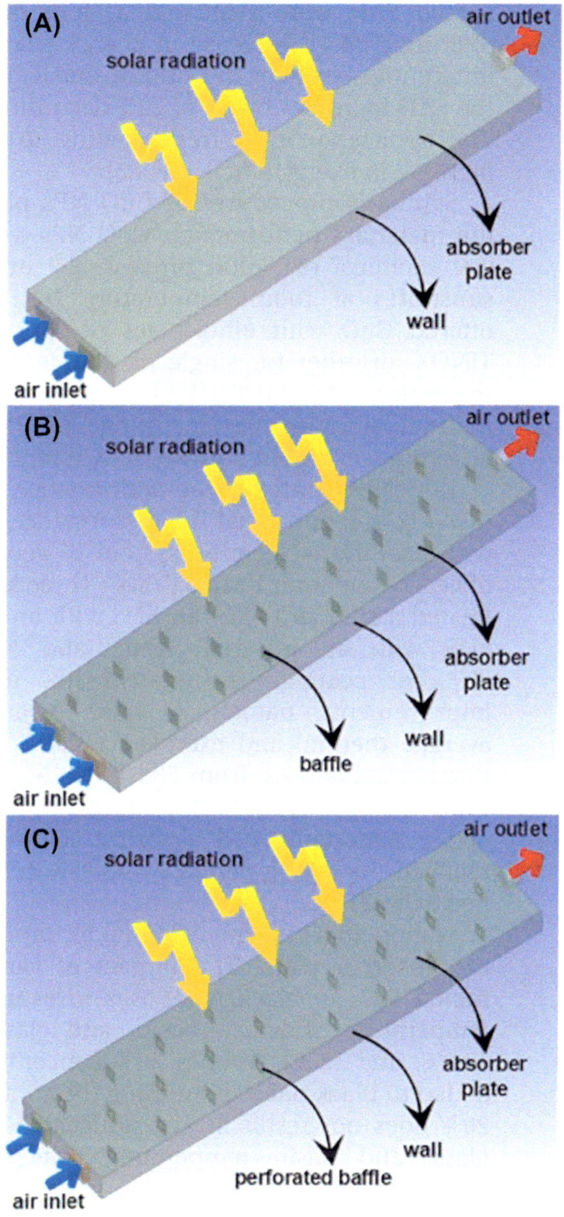

Figure 14.3 The geometries of vertical solar air heaters. (A) Vertical solar air heater without baffles; (B) vertical solar air heater with baffles; and (C) vertical solar air heater with perforated baffle [23]. Reproduced with permission from A. Khanlari, A.D. Tuncer, A. Sözen, İ. Aytaç, E. Çiftçi, H.İ. Variyenli, Energy and exergy analysis of a vertical solar air heater with nano-enhanced absorber coating and perforated baffles, Renewable Energy 187 (2022) 586–602. http://doi.org/10.1016/j.renene.2022.01.074. Copyright belongs to Elsevier 2022.

coated SAH were evaluated at four different airflow speeds, offering a 24.4% increase in energy efficiency. The temperature differential between the output and entrance of the air across the SAH increases by up to 22% according to the average values. The application of a fresh coating to the SAH results in an increase in exergy efficiency [25].

Synthesis procedures of CuO NPs play a significant role in the material's performance. CuO NPs were synthesized using a wet-chemical oxidation process [26] and the oxidation of Cu substrates at room temperature [11]. The former methods offered CuO with efficiencies of 95% and 97% of the used TiNOX absorber for single or double glazing, respectively. On the other side, the CuO films of the second method offered thermal emittance in the range of 6%–7%, while the solar absorptances varied between 84% and 90% [11].

ZnO NPs improved the performance of solar drying systems (SDS) [27]. It enhanced the performance of the absorber coating with an infrared heating system of geometrical configurations of vertical solar air heater (VSH). Three kinds of SDS were investigated a standard SDS, an SDS with an infrared heater, and an SDS with an infrared heater and a ZnO nano-enhanced absorber coating. The system has undergone two further improvements, namely, of nano-enhanced black paint. The average thermal and exergetic efficiencies of the VSH for the three systems were from 53.54% to 65.12% and from 9.94% to 14.32%, respectively. Furthermore, using both an infrared heater and a nano-enhanced absorber coating material in the VSH reduced the drying period by 43.75% compared to the unmodified SDS [27].

Nanopaints become essential to improve the performance of the absorber plate. The impact of varying concentrations of fumed SiO_2 in black paint was investigated [28]. The average temperature of water, basin, and glass increased by 10.4%, 12.2%, and 12.3% when a 20% concentration of SiO_2 NPs was added to black paint. Increasing the concentration of NPs above 20% does not result in any noticeable improvement in water, glass, and basin temperature. The solar still's output is enhanced by 27.2% and 34.3% when employing 10% and 20% NPs in black paint, respectively [28].

14.3.3 Nanopaints for cooling systems

Nanopaints have been widely used as radiative cooling paint (RCP) materials. An RCP system was reported by incorporating polymethylpentene into acrylic resin blended with SiO_2

microparticles [29]. At the system's optimal condition, the materials offered an emissivity of 0.91 in the atmospheric window and a reflectivity of 92% in the solar spectrum when backed with aluminum foil. The temperature decreases by an average of 1.5°C (during the evening) with a cooling power of 53.3 W m^{-2} and 5.2°C (during the daytime) with a cooling power of 46.8 W m^{-2} [29].

The cooling properties of barium sulfate ($BaSO_4$) NP films and paints over the whole daylight period [30]. The $BaSO_4$ NP coating achieves an exceptionally high solar reflectance and sky window emissivitiy of 97.6% and 0.96%, respectively due to its suitable particle size and wide particle size distribution (Fig. 14.4). The $BaSO_4$ film demonstrates a temperature reduction of more than 4.5°C compared to the surrounding environment or reaches an average cooling power of 117 W m^{-2}. The $BaSO_4$-acrylic paint is formulated with a 60% volume concentration to improve its durability when used outdoors. This formulation results in solar reflectance, sky window emissivity, and a figure of merit of 98.1%, 0.95%, and 0.77%, respectively. The field experiments demonstrate comparable cooling efficiency to the $BaSO_4$ films. $BaSO_4$-acrylic paint offers excellent dependability, a practical paint form, convenience of use, and compatibility with the commercial paint fabrication process (Fig. 14.4) [30]. Polyvinylpyrrolidone-modified $BaSO_4$ NPs (20 nm), can be easily and consistently distributed in ethanol for a period exceeding 6 months [31]. These NPs were then used as fundamental components to fabricate spherical $BaSO_4$ clusters (0.9 μm), using a scalable spray-drying process. The $BaSO_4$ NPs/clusters, with a mass ratio 1:1, were used to fabricate an epoxy resin film for radiative cooling. The film exhibited a

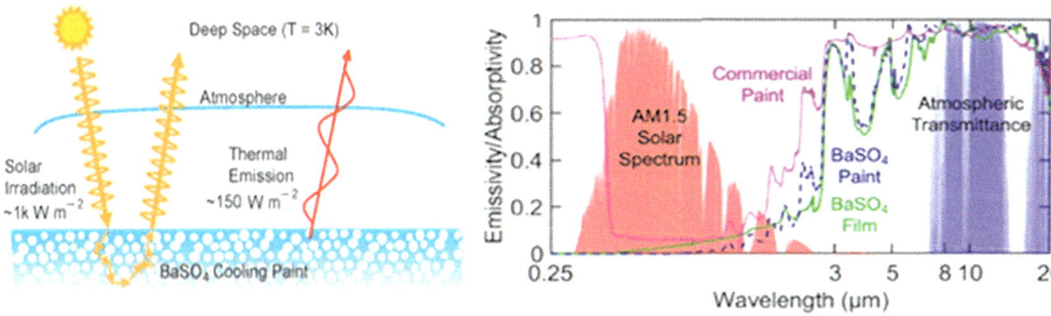

Figure 14.4 Schematic representation of barium sulfate cooling paints and their emission/absorption performance compared to commercial paint. Reproduced with permission from X. Li, J. Peoples, P. Yao, X. Ruan, Ultrawhite $BaSO_4$ paints and films for remarkable daytime subambient radiative cooling, ACS Applied Materials & Interfaces 13 (2021) 21733–21739. http://doi.org/10.1021/acsami.1c02368.

solar reflectance and sky window emissivity of 71% and 0.94%, respectively. It showed exceptional radiative cooling capabilities, resulting in a reduction in the surrounding temperature by 13.5°C and 7°C during indoor testing and outdoor testing, respectively [31]. $BaSO_4$-epoxy resin composite film outperforms the commercial $BaSO_4$-filled film in both radiative cooling and mechanical properties (Fig. 14.4).

A water-based porous composite made of calcium silicate ($CaSiO_3$) and SiO_2 aerogel (SiO_2–CS-coating) was reported for the cooling system [32]. It showed excellent passive radiative cooling and high thermal insulation capabilities. It has a remarkable solar reflectance of 94%, a high sky window emissivity of 96%, and a thermal conductivity of 0.0854 W m · K. A method is suggested to improve the emissivity of the atmospheric window on the SiO_2/radiative cooling coating by using lattice resonance. This resonance is caused by the eight-membered ring structure of porous $CaSiO_3$, increasing the emissivity of the atmospheric window. During the daylight test, with a solar irradiation of 900 W m^{-2}, an ambient temperature of 43°C, a wind speed of 0.53 m s^{-1}, and a humidity of 25%, the temperature within the box may be lowered by 13°C compared to the environment, which was at 30°C. The theoretical cooling power is calculated to be 96 W m^{-2}. In comparison to the commercial white coating, the SiO_2–CS coating is capable of conserving 70 kW h^{-1} of electric energy during a span of 1 month, resulting in a reduction of energy usage by 36%. It offered a versatile and scalable radiative-cooling coating that may effectively lower interior temperatures for building surfaces [32].

The selection of NP materials for cooling is a state of art. Materials, such as $BaSO_4$ have a significant difference in energy levels between their valence and conduction bands, resulting in a large electron band gap. This property makes it very resistant to absorbing solar radiation. Additionally, $BaSO_4$ shows strong phonon resonance at a wavelength of 9 μm, which contributes to its ability to emit thermal radiation effectively via the sky window. These properties ensure promising potential for applications in building air conditioning.

14.4 Nanopaints for renewable energy technologies

Nanopaints have the potential to be used in the production of antireflective coatings for solar panels, which would enhance their ability to absorb light and improve their efficiency in converting sunlight into energy. In addition, luminous nanopaints may be used in solar concentrators to augment light absorption and increase energy production.

Nanopaints with hydrophobic qualities may be used on wind turbine blades to decrease ice buildup and enhance energy generation in winter.

14.4.1 Solar energy

The process of creating organic photovoltaic devices using aqueous solar nanoparticle (ASNP) inks or solar nanopaints is difficult due to the conflicting demands of preserving emulsion stability and promoting surface wettability [17]. This challenge can be solved using the implementation of roll-to-roll (R2R) printing of ASNP inks. The method can enhance the film's quality by adjusting the concentration of free SDS in the ASNP inks. The study examined the impact of surface ozone treatment on several properties of poly(3,4-ethylenedioxythiophene) polystyrene sulfonate (PEDOT:PSS) hole-transport layer (HTL) electrodes on small-scale devices, including the work function, contact angle, surface composition, and surface morphology. The use of optimized surface treatments using ozone resulted in NP organic photovoltaic (NP-OPV) device efficiencies of 1.5%. Based on the surface characterization, the ozone treatment of the HTL films primarily led to an increase in PSS surface concentration. Applying surface ozone treatments to medium SDS content ASNP films can effectively coat them and produce high-quality large-area devices with an efficiency of 0.45%. This emphasizes the importance of controlling surface energy to determine the performance of R2R OPV devices made from solar paints [17].

The thickness of nanopaints on the subtract affects the performance of the solar absorbers [14]. The effect of CNTs thickness was investigated [14]. By increasing the thickness of CNT coating, the point at which the CNT absorbers transition from low to high reflectance moves towards longer wavelengths resulting in a significant increase in solar absorptance and a decrease in thermal emittance [14].

14.4.2 Nanopaints for energy storage

Nanopaints have been applied for energy storage systems, including batteries and supercapacitors. Nanopaints with enhanced conductivity may be used as a coating on battery electrodes, enabling more rapid charging and discharging processes. In addition, nanopaints can prolong the lifetime of batteries by inhibiting corrosion.

Supercapacitors are created using nanopaints that include conductive NPs with a large surface area [33,34]. These supercapacitors can store and release energy quickly, making them ideal for applications such as electric cars. NPs are usually prepared as a slurry or paint. They are usually mixed with polymer, such as polyvinylidene difluoride and carbon blacks [35–40].

14.5 Challenges and Considerations

Although nanopaints show great potential, several obstacles must be overcome. These challenges involved scalability and the environmental impact of NPs. It is essential to devise economical techniques to mass manufacture nanopaints to facilitate their broad use.

The assessment and management of the possible environmental consequences of NPs used in paints throughout their entire life cycle and disposal need meticulous examination and acceptable methods. Ensuring the long-term efficacy and longevity of nanopaint characteristics under real-world situations is essential for its durability and long-term performance. A study for TiO_2 added into paints was reported to investigate the material transformation and release [41]. TiO_2 NPs added to paints may undergo physicochemical changes but no observation of release indicates that Ti after incineration becomes highly immobilized inside a glass matrix [41]. The authors observed no emission of TiO_2 NPs into the environment [41]. The presence of photocatalytic NPs, such as TiO_2 may offer self-cleaning properties; however, it can also lead to polymer breakdown and the emission of hazardous volatile organic chemicals [8].

Theoretical and experimental investigations should be included. In a study of CuO-based nanopaint, the highest discrepancy between the calculated values and experimentally obtained outlet temperature values was determined to be 4.74% [23]. The numerical calculation based on Mie theory and Monte Carlo simulation were reported for RCP using SiO_2-based nanopaint [29]. Data analysis showed that the optimal performance can be obtained using a volume fraction of 7% of SiO_2 microparticles. The optimal particle size and thickness of SiO_2 were 8 and 200 μm, respectively. Furthermore, the RCP showed an emissivity and reflectivity of 0.91% and 92% (backed with aluminum foil), respectively [29].

Nanopaints are an innovative technology that can greatly boost energy efficiency, improve renewable energy production, and optimize energy storage systems. Advancing research and development will enable us to overcome the existing obstacles .

14.6 Conclusions

Nanopaints offer significant advantages in the energy sector. NPs enhance energy-based technologies, e.g., temperature regulation and energy storage devices. They can enhance the efficiency of energy production in solar cells. Nanopaints may be applied on the plate absorber of solar heaters to augment the evaporation and heat transfer rate. NPs can amplify the light absorption in solar panels, reinforce the durability and strength of wind turbine blades, and improve the catalytic efficiency in fuel cells. Nanopaints in solar stills provide a financially efficient and environmentally beneficial substitute for conventional desalination methods. They improve the durability of vital elements in power plants, pipelines, and transmission lines. They may be designed with inherent self-cleaning properties, allowing them to resist the accumulation of dust and debris effectively. Nanopaints can enhance sustainability and cost-efficiency in the energy sector by tackling efficiency and maintenance issues.

References

[1] F. Fanicchia, S.N. Karlsdottir, Research and development on coatings and paints for geothermal environments: a review, Advanced Materials Technologies 8 (2023). Available from: https://doi.org/10.1002/admt.202202031.

[2] Z. Ranjbar, D. Pourhadadi, S. Montazeri, M. Roshanzamir Modaberi, Lead compounds in paint and coatings: a review of regulations and latest updates, Progress in Organic Coatings 174 (2023) 107247. Available from: https://doi.org/10.1016/j.porgcoat.2022.107247.

[3] G. Subbiah, M. Premanathan, S.J. Kim, K. Krishnamoorthy, K. Jeyasubramanian, Preparation of TiO_2 nanopaint using ball milling process and investigation on its antibacterial properties, Materials Express 4 (2014) 393–399. Available from: https://doi.org/10.1166/mex.2014.1185.

[4] K. Krishnamoorthy, K. Jeyasubramanian, M. Premanathan, G. Subbiah, H.S. Shin, S.J. Kim, Graphene oxide nanopaint, Carbon 72 (2014) 328–337. Available from: https://doi.org/10.1016/j.carbon.2014.02.013.

[5] M.V.K. Mohan, T.V.K.B. Prakash, A. Mukherjee, Antifouling paint formulations with ZnO and Fe_2O_3 nano-paints for marine applications, Sustainable Chemistry and Pharmacy 30 (2022) 100858. Available from: https://doi.org/10.1016/j.scp.2022.100858.

[6] M.N. de Queiroz, A.M. de Oliveira Lima, M.E.G. Winkler, V.H. Fragal, A.F. Rubira, T. Sequinel, et al., Conductive nanopaints: a remarkable coating, Nanotechnology in the Automotive Industry (2022) 429–449. Available from: https://doi.org/10.1016/B978-0-323-90524-4.00020-7.

[7] A. Alali, A. Hosseini-Abari, A. Bahrami, M. Yazdan Mehr, Biosynthesis of copper oxide and silver nanoparticles by *Bacillus* spores and evaluation of the feasibility of their use in antimicrobial paints, Materials 16 (2023) 4670. Available from: https://doi.org/10.3390/ma16134670.

[8] Q. Maqbool, O. Favoni, T. Wicht, N. Lasemi, S. Sabbatini, M. Stöger-Pollach, et al., Highly stable self-cleaning paints based on waste-valorized PNC-doped TiO_2 nanoparticles, ACS Catalysis 14 (2024) 4820–4834. Available from: https://doi.org/10.1021/acscatal.3c06203.

[9] H. Zhang, H. Li, W. Li, J. Wang, W. Wang, Y. Zhang, et al., Application of superparamagnetic nanoparticle (SPM-NP) heating in wax removal during crude oil pipeline pigging, Energies 15 (2022) 6464. Available from: https://doi.org/10.3390/en15176464.

[10] R.V.S.K. Jakkula, P. Sethuramalingam, Analysis of coatings based on carbon-based nanomaterials for paint industries – a review, Australian Journal of Mechanical Engineering 21 (2023) 1008–1036. Available from: https://doi.org/10.1080/14484846.2021.1938953.

[11] S. Karthick Kumar, S. Murugesan, S. Suresh, Preparation and characterization of CuO nanostructures on copper substrate as selective solar absorbers, Materials Chemistry and Physics 143 (2014) 1209–1214. Available from: https://doi.org/10.1016/j.matchemphys.2013.11.023.

[12] R. Solano, D. Patiño-Ruiz, A. Herrera, Preparation of modified paints with nano-structured additives and its potential applications, Nanomaterials and Nanotechnology 10 (2020). Available from: https://doi.org/10.1177/1847980420909188.

[13] A.A. Balandin, Thermal properties of graphene and nanostructured carbon materials, Nature Materials 10 (2011) 569–581. Available from: https://doi.org/10.1038/nmat3064.

[14] Z. Chen, T. Boström, Electrophoretically deposited carbon nanotube spectrally selective solar absorbers, Solar Energy Materials and Solar Cells 144 (2016) 678–683. Available from: https://doi.org/10.1016/j.solmat.2015.10.016.

[15] P. Chaisaenrith, N. Srisawat, C. Phuangsuwan, V. Chobpattana, N. O-charoen, S. Pavasupree, Development and characterization of a reflection paint with nano-ilmenite material for energy-saving in buildings, Case Studies in Construction Materials 20 (2024) e02894. Available from: https://doi.org/10.1016/j.cscm.2024.e02894.

[16] M. Song, L. Feng, P. Huo, M. Liu, C. Huang, F. Yan, et al., Versatile full-colour nanopainting enabled by a pixelated plasmonic metasurface, Nature Nanotechnology 18 (2023) 71–78. Available from: https://doi.org/10.1038/s41565-022-01256-4.

[17] F. Almyahi, T.R. Andersen, A. Fahy, M. Dickinson, K. Feron, W.J. Belcher, et al., The role of surface energy control in organic photovoltaics based on solar paints, Journal of Materials Chemistry A 7 (2019) 9202–9214. Available from: https://doi.org/10.1039/C8TA09521C.

[18] S. Wijewardane, D.Y. Goswami, A review on surface control of thermal radiation by paints and coatings for new energy applications, Renewable and Sustainable Energy Reviews 16 (2012) 1863–1873. Available from: https://doi.org/10.1016/j.rser.2012.01.046.

[19] J. El Nady, A.B. Kashyout, S. Ebrahim, M.B. Soliman, Nanoparticles Ni electroplating and black paint for solar collector applications, Alexandria Engineering Journal 55 (2016) 723–729. Available from: https://doi.org/10.1016/j.aej.2015.12.029.

[20] A.E. Kabeel, R. Sathyamurthy, S.W. Sharshir, A. Muthumanokar, H. Panchal, N. Prakash, et al., Effect of water depth on a novel absorber plate of pyramid solar still coated with TiO_2 nano black paint, Journal of Cleaner Production 213 (2019) 185–191. Available from: https://doi.org/10.1016/j.jclepro.2018.12.185.

[21] N. Abdelal, Y. Taamneh, Enhancement of pyramid solar still productivity using absorber plates made of carbon fiber/CNT-modified epoxy composites, Desalination 419 (2017) 117–124. Available from: https://doi.org/10.1016/j.desal.2017.06.012.

[22] E.I. Ghandourah, A. Sangeetha, S. Shanmugan, M.E. Zayed, E.B. Moustafa, A. Tounsi, et al., Performance assessment of a novel solar distiller with a double slope basin covered by coated wick with lanthanum cobalt oxide nanoparticles, Case Studies in Thermal Engineering 32 (2022) 101859. Available from: https://doi.org/10.1016/j.csite.2022.101859.

[23] A. Khanlari, A.D. Tuncer, A. Sözen, İ. Aytaç, E. Çiftçi, H.İ. Variyenli, Energy and exergy analysis of a vertical solar air heater with nano-enhanced absorber coating and perforated baffles, Renewable Energy. 187 (2022) 586–602. Available from: https://doi.org/10.1016/j.renene.2022.01.074.

[24] S.K. Dharmaraj, H. Ramasubbu, V. Rajendran, P. Ravichandran, Effect of graphene nanopaint on performance of solar air heater attached with inclined and winglet baffles, Environmental Science and Pollution Research 30 (2023) 87330–87342. Available from: https://doi.org/10.1007/s11356-023-28646-y.

[25] T.K. Abdelkader, Y. Zhang, E.S. Gaballah, S. Wang, Q. Wan, Q. Fan, Energy and exergy analysis of a flat-plate solar air heater coated with carbon nanotubes and cupric oxide nanoparticles embedded in black paint, Journal of Cleaner Production 250 (2020) 119501. Available from: https://doi.org/10.1016/j.jclepro.2019.119501.

[26] D. Jeong, J. Lee, H. Hong, D. Choi, J.-W. Cho, S.-K. Kim, et al., Absorption mechanism and performance characterization of CuO nanostructured absorbers, Solar Energy Materials & Solar Cells 169 (2017) 270–279. Available from: https://doi.org/10.1016/j.solmat.2017.05.029.

[27] A.D. Tuncer, A. Amini, A. Khanlari, Developing an infrared-assisted solar drying system using a vertical solar air heater with perforated baffles and nano-enhanced black paint, Solar Energy. 263 (2023) 111958. Available from: https://doi.org/10.1016/j.solener.2023.111958.

[28] R. Sathyamurthy, A.E. Kabeel, M. Balasubramanian, M. Devarajan, S.W. Sharshir, A.M. Manokar, Experimental study on enhancing the yield from stepped solar still coated using fumed silica nanoparticle in black paint, Materials Letters 272 (2020) 127873. Available from: https://doi.org/10.1016/j.matlet.2020.127873.

[29] K. Jiang, K. Zhang, Z. Shi, H. Li, B. Wu, O. Mahian, et al., Experimental and numerical study on the potential of a new radiative cooling paint boosted by SiO_2 microparticles for energy saving, Energy 283 (2023) 128473. Available from: https://doi.org/10.1016/j.energy.2023.128473.

[30] X. Li, J. Peoples, P. Yao, X. Ruan, Ultrawhite $BaSO_4$ paints and films for remarkable daytime subambient radiative cooling, ACS Applied Materials & Interfaces 13 (2021) 21733–21739. Available from: https://doi.org/10.1021/acsami.1c02368.

[31] T. Wu, Q. Zou, Z. Li, B. Chen, W. Gao, Q. Sun, et al., $BaSO_4$-epoxy resin composite film for efficient daytime radiative cooling, Langmuir 40 (2024) 638–646. Available from: https://doi.org/10.1021/acs.langmuir.3c02827.

[32] D. Han, C. Wang, C.B. Han, Y. Cui, W.R. Ren, W.K. Zhao, et al., Highly optically selective and thermally insulating porous calcium silicate composite SiO_2 aerogel coating for daytime radiative cooling, ACS Applied Materials & Interfaces 16 (2024) 9303–9312. Available from: https://doi.org/10.1021/acsami.3c18101.

[33] S. Onajah, R. Sarkar, M.S. Islam, M. Lalley, K. Khan, M. Demir, et al., Silica-derived nanostructured electrode materials for ORR, OER, HER, CO_2RR electrocatalysis, and energy storage applications: a review, Chemical Record 24 (2024). Available from: https://doi.org/10.1002/tcr.202300234.

[34] M. Ibrahim, Z. Wen, X. Sun, H.N. Abdelhamid, In situ polymerization of a melamine-based microsphere into 3D nickel foam for supercapacitors, RSC Advances 14 (2024) 5566–5576. Available from: https://doi.org/10.1039/D3RA08489B.

[35] H.N. Abdelhamid, Three-dimensional (3D) printed supercapacitor, Handbook Energy Materials, Springer Nature, Singapore, 2023, pp. 1–21. Available from: https://doi.org/10.1007/978-981-16-4480-1_77-1.

[36] S.A. Al Kiey, H.N. Abdelhamid, Metal-organic frameworks (MOFs)-derived Co_3O_4@N-doped carbon as an electrode materials for supercapacitor, Journal of Energy Storage 55 (2022) 105449. Available from: https://doi.org/10.1016/j.est.2022.105449.

[37] M. Ibrahim, H.N. Abdelhamid, A.M. Abuelftooh, S.G. Mohamed, Z. Wen, X. Sun, Covalent organic frameworks (COFs)-derived nitrogen-doped carbon/reduced graphene oxide nanocomposite as electrodes materials for supercapacitors, Journal of Energy Storage 55 (2022) 105375. Available from: https://doi.org/10.1016/j.est.2022.105375.

[38] M. Ibrahim, M.G. Fayed, S.G. Mohamed, Z. Wen, X. Sun, H.N. Abdelhamid, High-performance lithium-ion battery and supercapacitors using covalent organic frameworks (COFs)/graphitic carbon nitride (g-C_3N_4)-derived hierarchical N-doped carbon, ACS Applied Energy Materials 5 (2022) 12828–12836. Available from: https://doi.org/10.1021/acsaem.2c02415.

[39] H.N. Abdelhamid (Ed.), Covalent organic frameworks-based nanomaterials as electrode materials for supercapacitors, Covalent organic frameworks: Chemistry, properties, and energy applications for a sustainable future, CRC Press, 2022.

[40] H.N. Abdelhamid, S.A. Al Kiey, W. Sharmoukh, A high-performance hybrid supercapacitor electrode based on ZnO/nitrogen-doped carbon nanohybrid, Applied Organometallic Chemistry (2021). Available from: https://doi.org/10.1002/aoc.6486.

[41] A. Massari, M. Beggio, S. Hreglich, R. Marin, S. Zuin, Behavior of TiO_2 nanoparticles during incineration of solid paint waste: a lab-scale test, Waste Management 34 (2014) 1897–1907. Available from: https://doi.org/10.1016/j.wasman.2014.05.015.

15

Nanopaints: revolutionizing catalysis for enhanced efficiency and sustainability

Rishi Raj, Nancy Sharma, Rohit Verma, Tejendra K. Gupta and Sheenam Thatai

Department of Physics, Amity Institute of Applied Sciences, Amity University Uttar Pradesh, Noida, Uttar Pradesh, India

15.1 Introduction to nanopaints in catalysis

Catalysis plays a pivotal role in enhancing chemical reactions, and the integration of nanopaints further expands the horizons of catalytic processes. The evolution of nanomaterials has led to the development of nanopaints tailored for catalytic applications, opening new avenues in the field.

15.1.1 Evolution and significance

The evolution of nanopaints in catalysis is accompanied by the progress of nanotechnology and material science. Nanopaints that are composed of nanomaterials show unmatched characteristics which can affect the catalytic performance. The addition of nanoparticles, for instance metal and metal oxide nanomaterials, has given positive outcomes in terms of catalytic activity and selectivity [1].

The value lies in the fact that nanopaints can function as catalytic coatings, providing better surface interactions and stability. These coatings not only inhibit fouling but also offer a controlled environment for catalytic reactions, leading to improved functionality [1,2].

15.1.2 Scope and objectives of the chapter

This chapter encompasses the study of nanopaints role in catalysis. Its focus is to explore all types of nanomaterials employed in nanopaint formulations, their catalytic actions,

Nanopaints. DOI: https://doi.org/10.1016/B978-0-443-26512-9.00016-1

and their applications. The chapter aims at giving a brief history as well as the importance of nanopaints, and discusses the current state of research and future prospects.

Objectives:

- Classify the types of nanomaterials used in nanopaint formulations for catalysis.
- Study the catalytic mechanisms mediated by nanopaints.
- Gain insights into nanopaints in catalysis.
- Examine the challenges and probable solutions for nanopaints in catalytic processes [1,2].

This chapter acts as a useful reference for researchers, scientists, and practitioners interested in the interface of nanotechnology and catalysis as the chapter delves into the evolution, importance, and future trends of nanopaints in catalysis.

Catalysis constitutes a cornerstone of the field of chemistry that operates as a driving force, guiding the paths of chemical conversions through time. The development of the material—highly entangled with the scientific advances—testifies to its paramount significance in multiple industrial applications.

15.2 Catalysis fundamentals

Catalysis stands as a fundamental pillar in the realm of chemistry, representing a dynamic force that has shaped (as in Fig. 15.1) the course of chemical transformations throughout history. Its evolution, deeply entwined with scientific progress, underscores its paramount importance across various industrial applications.

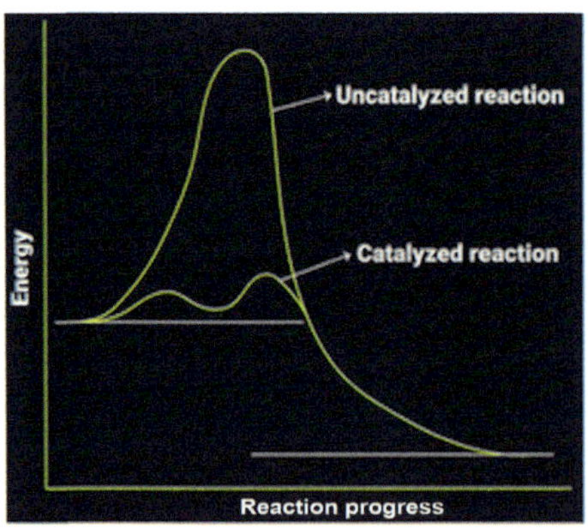

Figure 15.1 Depicted difference between catalyzed and noncatalyzed reaction pathway.

15.2.1 Historical overview

The historical narrative of catalysis spans epochs, with its origins rooted in ancient practices, notably observed in the fermentation processes employed for food and beverage production. However, it was in the 18th century that catalysis emerged as a distinct field of study. Pioneering chemists, including Jöns Jacob Berzelius and Marcellin Berthelot, laid the groundwork for a systematic exploration of catalytic phenomena.

The 20th century marked a watershed moment for catalysis, witnessing transformative breakthroughs. Among these milestones, the elucidation of enzyme catalysis provided profound insights into the intricate workings of biological catalysts. Simultaneously, the development of heterogeneous catalysis, exemplified by the ground-breaking work of Haber and Bosch in ammonia synthesis, revolutionized industrial processes. This era witnessed the convergence of theoretical understanding and practical applications, laying the foundation for modern catalytic science.

As we delve into the annals of catalysis, this historical overview serves as a portal to comprehend the diverse facets that have shaped its trajectory, making it an indispensable field that continues to redefine the frontiers of chemistry.

15.2.2 Role of catalysts in chemical transformations

Catalysts in chemical changes exhibit a complex dance of molecular interactions which play a determining role in reaction rates and success. Catalysts, acting as molecular catalysts, exquisitely regulate the trajectory of chemical reactions, thus implicating them in a multitude of industrial and biological systems. Catalysts, strictly speaking, are those substances which, while influencing chemical reactions, do not undergo any permanent changes themselves.

This catalytic action is shown through an offering other reaction pathway with lower activation energies. Therefore catalysts speed up the conversion of reactants into products, leading to faster and energy-saving reactions. In catalyzed reactions there is a dynamic interaction between a catalyst and reactant to generate an intermediate complex. This molecular dance plays a role in alteration of the reaction pathway thus enhancing the process efficiency. Importantly, what characterizes catalysts is their capacity to be regenerated, returning to their original state at the end of the reaction.

Such distinctive properties guarantee the same catalyst is available for the next series of reactions thus constituting their sustainability and economic livelihood. Yet, this catalytic power finds usefulness in numerous chemical processes from the intricate refining of oil to the accurate manufacturing of medicinal substances. The capacity of catalysts to accelerate rates of reactions is a major pillar in achieving industrial process optimization and demonstrates their critical position in modern chemical expeditions [3,4].

15.2.3 Types of catalysts (homogeneous and heterogeneous)

Studying the taxonomy of catalysts exposes a complex reality, being classified into homogeneous and heterogeneous.

Homogeneous catalysts: reaction in the same phase as reactants, homogeneous catalysts are versatile across chemical reactions as depicted in Fig. 15.2. Many examples can be cited, such as acids, bases, and particular transition metal complexes. The characteristic to adhere firmly with the reaction medium makes them an essential part of any synthetic or catalytic process.

Heterogeneous catalysts: on the other hand, heterogeneous catalysts move across different phases compared to the reactants and oftentimes become solid materials, like metals or metal oxides. The unique phase segregation allows simple separation of catalyst from the reaction mixture especially in heterogeneous catalysis enabling reuse and making it dominant in

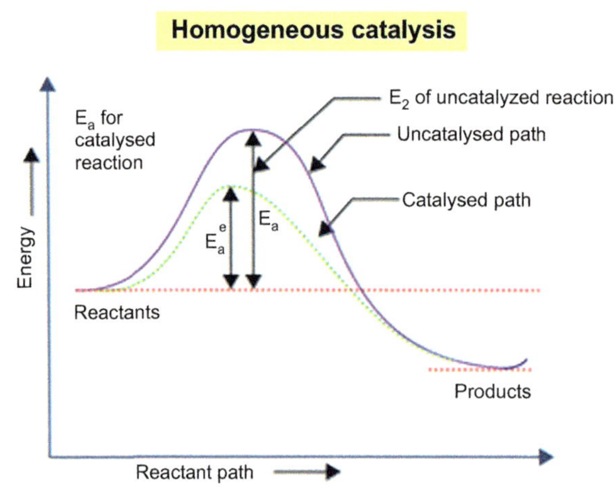

Figure 15.2 Plot demonstrating homogeneous catalysis.

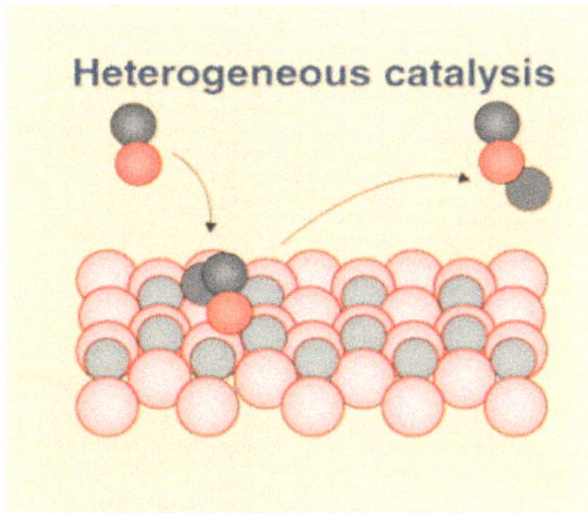

Figure 15.3 Molecular depiction of heterogeneous catalysis.

many industries as seen in Fig. 15.3. This also includes, but is not limited to, the production of chemicals, environmental cleanup, and catalytic converters in automotive systems [5].

Understanding the subtleties of the function and processes of catalysts, not only enriches our view of catalysis but also lays the foundation for unveiling novel and advanced catalysis applications and trends.

15.3 Key concepts in catalytic processes: unveiling the intricacies

Catalysis, a multifaceted discipline at the heart of chemical transformations, unveils a rich tapestry of key concepts shaping its applications in natural and industrial realms. Delving into these fundamental principles is paramount for deciphering the complexity inherent in catalytic processes.

15.3.1 Enzyme catalysis in biological systems: the maestros of biochemical symphony

In the fabric of catalysis, the catalysis of enzymes in biological systems as in Fig. 15.4 comes out as a symphony master-minded by the molecular masters. Enzymes, the master biological catalysts, take the spotlight in speeding up biochemical reactions crucial for life. The extraordinary selectivity in substrate recognition is what results in an unprecedented catalytic activity. The enzymatic flexibility stems from their capacity to construct amiable surroundings, stabilize transition states, and precisely orient the reactants. At a mechanistic level, enzymes participate in an intricate dance of conformational changes

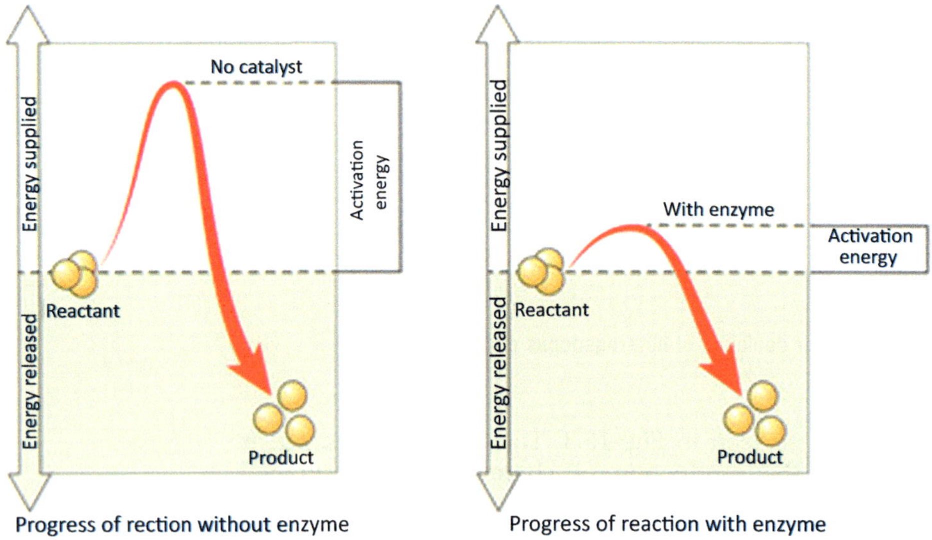

Figure 15.4 Pictorial depiction of enzyme catalysis.

during catalysis, an exquisitely designed choreography that leads to the coordinated movements of molecules. These examples, including deoxyribonucleic acid (DNA) polymerase, the guardian of genetic replication and amylase, the architect of starch hydrolysis, describe the diverse roles that enzymes play in biochemical symphony [6,7].

In essence, understanding these key concepts opens a gateway to the profound world of catalysis, where molecules dance to the orchestrated tunes of enzymes, shaping the very fabric of life and industry.

15.3.2 Heterogeneous catalysis: unveiling the role of metal surfaces and nanoparticles

Heterogeneous catalysis, the key element in the catalysis theory, consists of catalysts which are in distinct phases away from reactants usually involving metal surfaces or nanoparticles. The functionalities of metal catalysts are extremely dependent on the size and the structure, which leads to more functionalities. Nanoparticles—having amplified surface area—have added novelty to reactivities within catalytic processes. Metal surfaces predominantly act as adsorption sites, triggering subsequent reaction steps which ultimately determine the reaction selectivity and yield.

Examples of this are the use of platinum in catalytic converters, a technology employed in reducing automotive emissions. Moreover, metal nanoparticles are used in a number of green chemistry programmes thus demonstrating their multipurpose nature and greener profile [8–10].

Understanding these principles in catalytic reactions goes beyond theological aspect; they form the basis of outstanding discoveries. The spread of impact is interdisciplinary, with medical technology advancements and environmental innovations. Marrying theory with practice makes the path for innovation although it broadly shapes the movement through different fields.

15.3.3 Unraveling the complexities: mechanisms, kinetics, and thermodynamics of catalysis

Catalytic events do not stop to outside physical features, rather it goes to the depth of the complicated realms of mechanism, kinetics, and thermodynamics. Like molecular detectives, mechanistic studies show stepwise sequence of reactions. Such a trip exposes intermediate species and transition states, revealing the molecular richness that drives the catalytic process. This subtle perception provides scientists with a means to study catalysis at the molecular level, thus being a guide for molecular engineering. Parallel to this, the study of kinetics becomes crucial, considering the rate of reactions. Information gained from kinetics is indispensable for the process optimization, attaining the reliable results in various scales. Such diligent observance of reaction rates makes scientists manage to improve the catalytic processes by performance. In tandem, the study of kinetics acquires paramount importance, examining the speed at which reactions occur. Information derived from kinetics is invaluable for condition optimization, guaranteeing the same efficiency across different applications. Such a careful inspection of reaction rates enables scientists to optimize the catalytic processes for better performance.

On the side, thermodynamics comes into prominence, looking into the energetic changes linked to catalysis. This discipline of research offers crucial understanding of the spontaneity and equilibrium of reactions which reveal the energetic terrain that catalysis negotiates. These three pillars—mechanistic understanding, kinetic exploration, and thermodynamic scrutiny—thus form the skeleton of designing highly efficient catalytic systems. Such systems, equipped with a deep knowledge of mechanisms, kinetics, and thermodynamics, find applications across a broad spectrum of disciplines. Catalysis plays a key role from complex industrial processes to the urgent needs of environmental remediation [9,11].

Through the exploration of these basic issues researchers create a pathway for ground-breaking changes in catalysis, smoothly connecting the areas of academia and practice. The gained knowledge from these studies not only goes beyond just theoretical insight but also fuels breakthroughs that lead to sustainable and practical solutions in various research and industrial fields.

15.4 Applications of catalysis: a multifaceted landscape

Catalysis, a versatile and indispensable tool, unfolds its multifaceted applications across diverse domains, playing a pivotal role in driving essential processes. Let's delve into the three major facets of catalysis applications.

15.4.1 Industrial applications

In the intricate landscape of industrial settings, catalysis stands as a linchpin, contributing to the manufacturing of over 80% of products. Catalysts, acting as facilitators, orchestrate a myriad of chemical transformations, optimizing reaction conditions, and elevating overall efficiency. Industries strategically employ catalysts in pivotal processes, exemplified by catalytic cracking in petroleum refining, a fundamental step in the generation of valuable fuels. The synthesis of ammonia, crucial in fertilizer production, heavily relies on heterogeneous catalysis [12].

Moreover, catalytic converters in automobiles serve as exemplars of catalysis mitigating environmental impact. In this context, catalysts actively convert harmful emissions into less detrimental substances, showcasing their profound role in environmental sustainability. These applications underscore the indispensability of catalysis in industrial landscapes not only optimizing chemical processes but also contributing significantly to environmental stewardship and sustainable practices [13].

The broad impact of catalysis in industrial applications resonates not only in the efficiency of manufacturing processes but also in the conscientious efforts towards a greener and more sustainable industrial ecosystem.

15.4.2 Environmental catalysis: a catalyst for sustainable solutions

Catalysis takes center stage in addressing urgent environmental concerns, playing a pivotal role in crafting sustainable solutions as in Fig. 15.5. Through the utilization of advanced

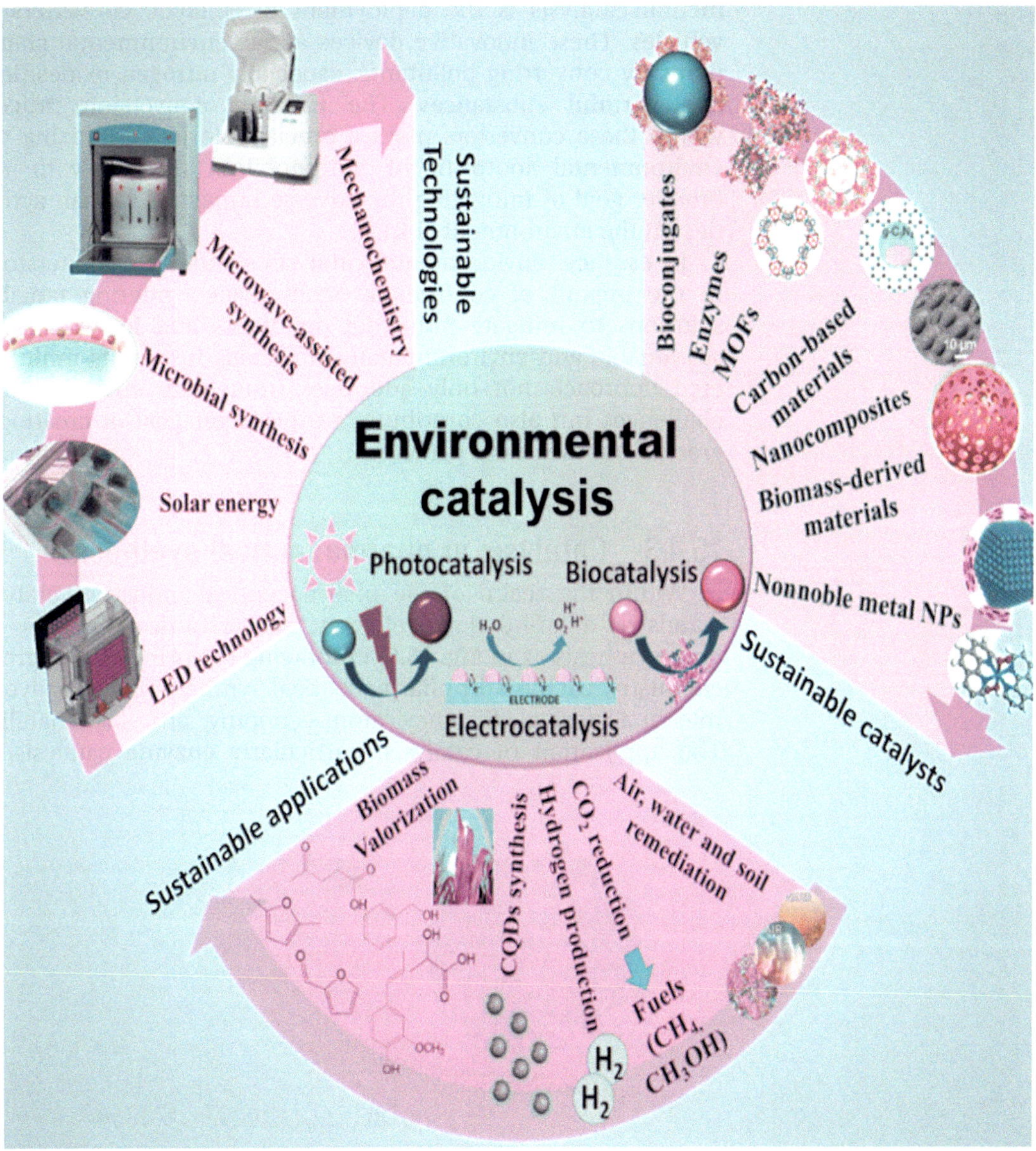

Figure 15.5 Applications of environmental catalysis.

catalyst technologies, environmental catalysis emerges as a potent tool in combating pollutants stemming from industrial processes. The impact of environmental catalysis extends across various dimensions, contributing to the reduction of harmful emissions, efficient waste treatment, and the promotion of sustainable energy production.

A notable example illustrating the effectiveness of environmental catalysis is the deployment of catalytic converters in vehicles. These innovative devices act as environmental champions by converting pollutants, especially nitrogen oxides, into less harmful substances. The catalytic conversion process within these converters plays a crucial role in minimizing the environmental footprint of automobiles, aligning with the broader goal of mitigating the adverse impact of human activities on the environment [14,15].

In essence, environmental catalysis stands as a cornerstone in the pursuit of eco-friendly technologies, offering tangible solutions to mitigate industrial pollutants and foster a more sustainable and environmentally conscious future. Its multifaceted approach not only addresses immediate environmental challenges but also contributes to the larger goal of creating a greener and healthier planet.

15.4.3 Catalysis in pharmaceutical synthesis

Within the realm of the pharmaceutical industry, catalysis stands as a foundational element in the intricate process of drug synthesis as in Fig. 15.6. Embracing the principles of green chemistry, catalysis in pharmaceutical synthesis plays a pivotal role in fostering efficiency, atom-economy, and sustainability. The application of catalysis, particularly enzyme catalysis, is

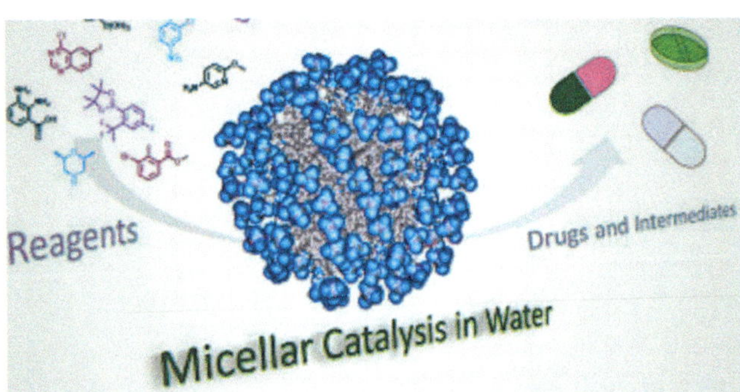

Figure 15.6 Application of micellar catalysis in drug synthesis.

widespread, ensuring the maintenance of chirality in a diverse array of substrates. This not only enhances the precision of drug synthesis but also aligns with the industry's pursuit of environmentally conscious practices.

Enzyme catalysis takes center stage in pharmaceutical synthesis, contributing to the synthesis of complex molecules with specific stereochemistry. This application of catalysis ensures the creation of pharmaceuticals with the desired properties, essential for their efficacy and safety. Moreover, both chemical synthesis and bioprocessing, processes inherently dependent on catalysis, are integral components of the manufacturing pipeline for pharmaceuticals [16].

Recognizing the broad spectrum of applications in pharmaceutical synthesis highlights catalysis as an indispensable factor in shaping industrial processes. By marrying efficiency with sustainability, catalysis not only advances pharmaceutical synthesis but also underscores its significant contribution to addressing environmental challenges. In essence, catalysis in pharmaceutical synthesis represents a harmonious integration of scientific innovation, industrial necessity, and eco-friendly practices [17].

15.5 Mitigating fouling with nanopaints in catalysis: revolutionizing efficiency and reliability

The emergence of nanopaints stands as a revolutionary innovation in catalysis, specifically designed to combat the pervasive challenge of fouling within catalytic systems. Nanopaints introduce a novel paradigm not only addressing fouling issues but also ushering in a multitude of benefits that reshape the operational dynamics and longevity of catalytic systems.

15.5.1 Addressing fouling challenges in catalytic systems

The deposition of undesirable materials onto the surface of the catalyst leads to fouling as one of the most dangerous threats to catalytic performance in the long-term. The novel proposal offered by nanopaints is comprised of the formation of protective coating on catalytic surfaces which decreases the impact of fouling. Such coatings thus function as tough barriers resisting fouling of all kinds from softening of the organic matter to hardening of minerals.

The use of nanopaints in fighting fouling problems is ultra important in industrial world. In the industrial processes where efficiency dominates, fouling may lead to a drop in the overall performance which results in decreased efficiency and higher maintenance costs. The protective layers induced by nanopaints act as barriers to the destructive effect of fouling, hence the long life and sustained efficacy of catalytic systems [18,19].

This novel method, however, additionally brings out the multifunctional character of nanopaints and, consequently, represents a landmark point in the way of improving the trustworthiness and productivity of catalytic processes—above all in those industries that still suffer the incidences of fouling.

The incorporation of nanopaints into the catalytic systems implies an important progress towards fettered fouling and the enhancement of the performance of these relevant industrial processes. Nanopaints acting as a shield make the field of catalysis eyewitness the rebirth, greatly tackling the problem of fouling.

15.5.2 Nanopaints: a catalyst for long-term catalytic excellence

Incorporation of nanopaints into catalytic systems makes a revolutionary stride in the history of long-term catalytic performance.

The deliberate use of nanopaints goes beyond just preventing fouling, significantly influencing the long-term performance and durability of catalytic processes as seen in Fig. 15.7.

15.5.2.1 Lessening fouling for maintained catalytic efficiency

Nanopaints serve a vital function in maintaining catalytic systems operation through enhancement of fouling. Nanopaints make a significant contribution as thickets due to the formation of super covering on surfaces of catalysts, preventing the accumulation of undesired substances. The proactive fouling prevention approach attains continued high catalytic efficiency during extended operations, preventing the inevitable performance deterioration triggered by fouling.

15.5.2.2 Durability and maintenance

The inherent protection of nanopaints safeguards catalytic activity as well as enhances the overall lifespan of the catalytic surfaces. The nanopaint coatings, therefore act as protective shields, decreasing catalytic materials' wear and tear. This decrease in surface degradation leads to a reduced frequency in maintenance and

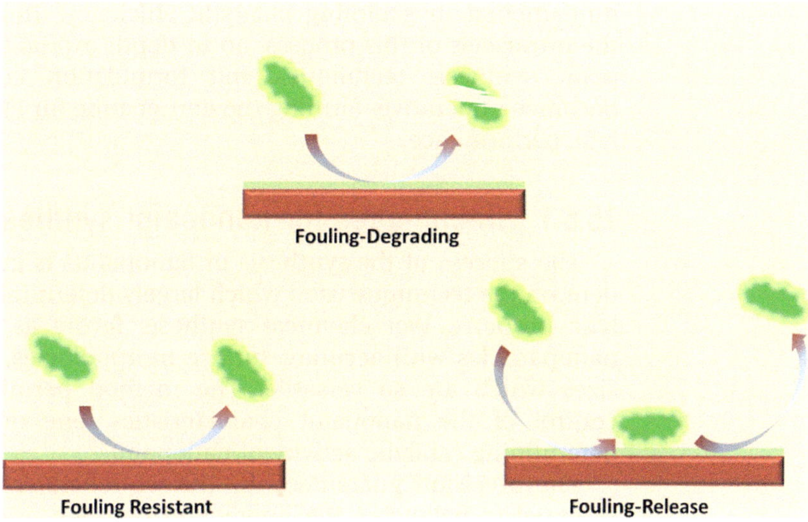

Figure 15.7 Mechanism of antifouling nanopaints.

replacement which is a cost-effective solution for industries which depend on catalytic processes.

15.5.2.3 Wide compatibility with different catalytic systems

Nanopaints differ from each other by their own composition that facilitates compatibility with different catalytic materials. Such versatility, therefore makes nanopaints compatible with virtually all the catalytic systems making them an attractive choice as a universal solution to the long-term catalytic performance. Gradually, the importance of nanopaints as a fouling remover stands out, and their tremendous effect on the durability of catalytic processes becomes explicit. Nanopaints go beyond immediate fouling reduction to become key players in the maintained efficiency, durability, and economic life of catalytic systems. This dynamic terrain regards nanopaints as essential actors in the endeavor of maximizing catalytic endurance [20].

15.6 Synthesis and formulation strategies: pioneering catalytic efficiency with nanopaints

The incorporation of nanopaints into catalysis requires a strategic orchestration of synthesis and formulation strategies, fundamental in sculpting catalytic efficiency. To comprehend the intricacies of this process, an in-depth exploration of nanopaint synthesis techniques and formulation considerations becomes imperative, forming the cornerstone for elevating catalytic performance.

15.6.1 Techniques for nanopaint synthesis

The success of the synthesis of nanopaints is greatly dependent on the technique used which largely determines their catalytic property. Wet chemical synthesis favors as it fabricates nanoparticles with accurate surface morphologies, shapes, and sizes which are so versatile. The method permits a precise control of the nanopaint characteristics, one of the factors determining catalytic activity of nanopaints.

The precision guaranteed by the wet chemical synthesis is essential in balancing the interaction between nanopaints and catalytic surfaces. Broaching unconventional paths, the input of biological substances gives an ecologically sound feature to the nanopaints. Employing biological bodies during synthesis is a sustainable method. Research studies the application of biological species to define the architectonic features of nanoparticles [21,22].

This bioinspired synthesis also improves the features of nanopaints while in line with the trend of eco-friendly and green catalysis. With the development of nanopaint generation, the gathering of wet chemistry methods and biomimetic processes shows potential for breaking novel avenues of catalytic performance. The potential for tuning nanopaints with precision and in a sustainable fashion paves the way for their powerful place in catalysis—defining how the future of catalysis will be characterized by effective synthesis schemes aimed at improved catalytic performance [23,24].

15.6.2 Factors influencing formulation for catalytic efficiency

In the intricate world of nanopaint formulation for catalytic efficiency, multiple factors come into play as seen in Fig. 15.8, shaping the composition and characteristics of these minute particles to ensure optimal performance. At the nanoscale, particles exhibit extraordinary properties, including heightened catalytic, magnetic, and chemical attributes. The challenge lies in navigating these properties effectively to create nanopaints that act as catalytic powerhouses.

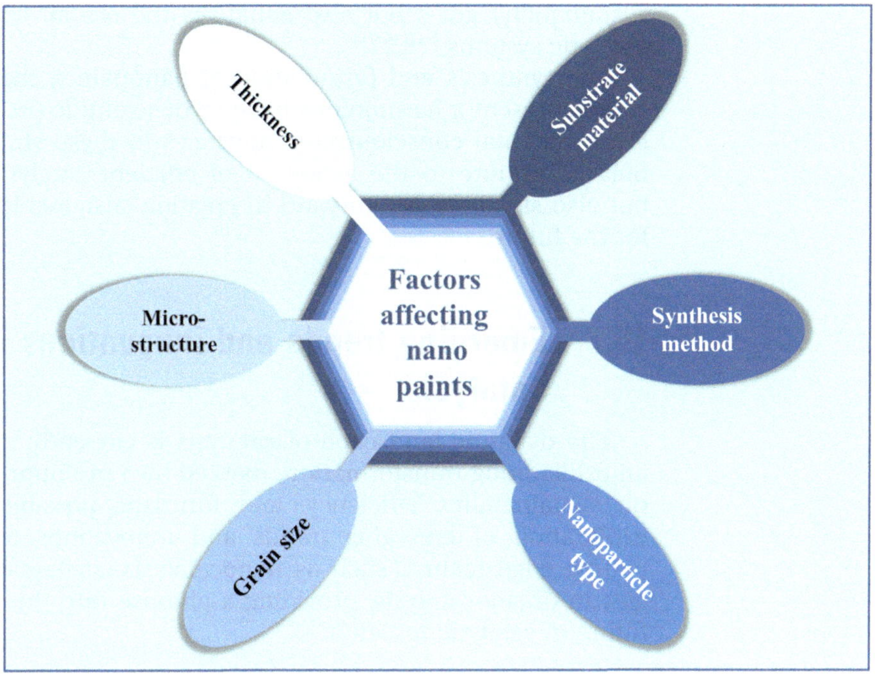

Figure 15.8 Factors affecting nanopaints.

The first crucial consideration is the high surface-to-volume ratio inherent at the nanoscale. This ratio accentuates the impact of material choice and their respective ratios in nanopaint formulation. The delicate balance between different materials becomes paramount, influencing the catalytic behavior of the nanopaints. Achieving the right mix is akin to orchestrating a symphony of elements, where each component plays a unique role in defining the catalytic prowess [25].

Moreover, the compatibility of nanopaints with specific catalytic surfaces adds another layer of complexity. Tailoring nanopaints to seamlessly integrate with diverse catalytic systems ensures their effectiveness in a variety of applications. This compatibility factor not only influences the overall performance but also expands the versatility of nanopaints across different catalytic scenarios.

In the quest for sustainability, green synthesis methods have emerged as a driving force. The emphasis on eco-friendly practices in nanopaint formulation addresses both environmental concerns and catalytic performance. As the field progresses, the integration of green synthesis not only aligns with global sustainability goals but also enhances the overall efficiency of catalytic systems [26,27].

The synthesis and formulation of nanopaints, though intricate, represent a harmonious balance of scientific precision and environmental consciousness. Advances in these strategies not only contribute to the evolution of efficient catalytic systems but also signify a step forward in creating sustainable solutions for the future.

15.7 Emerging trends and innovations in catalysis

The dynamic landscape of catalysis is currently undergoing an exhilarating transformation, marked by a profound emphasis on sustainability, efficiency, and futuristic possibilities. This exploration of emerging trends and innovations delves into cutting-edge realms, such as nanocatalysis, sustainable strategies in organocatalysis, providing a glimpse into the promising future of catalytic research.

15.7.1 Nanocatalysis: revolutionizing catalytic evolution

Recent advancements in nanocatalysis are at the forefront of catalytic evolution, bringing about a revolution in green organic synthetic methodologies. This progressive approach is pivotal in navigating the complexities of synthesizing diverse heterocyclic scaffolds. Nanocatalysts are emerging as trailblazers, playing a central role in enhancing the efficiency and selectivity of reactions, marking a paradigm shift in catalytic research.

Within the realm of nanocatalysis, innovation is taking center stage, extending beyond catalytic processes to the very core of catalysts. Nanocatalysts, characterized by their minute size, wield unprecedented control over reactions, providing enhanced efficiency and selectivity. Strategies employed in nanocatalysis involve meticulous tailoring of surface morphologies, shapes, and sizes. This precision engineering empowers scientists with unparalleled control, ushering in a new era of mastery in synthetic processes.

The synthesis of complex organic compounds is evolving into an intersection of art and science, where nanocatalysis emerges as the brushstroke painting the future of catalytic advancements. This transformative journey in catalytic research holds the promise of reshaping industries and paving the way for sustainable, efficient, and innovative chemical processes [28,29].

15.7.2 Organocatalysis: pioneering sustainable approaches

The extensive world of chemical transformations experiences a revolutionary impetus from organocatalysis, an approach that fields small organic molecules as catalysts for sophisticated reactions. This blazing field places a definite focus on redox-neutral catalytic processes, providing new ways to handle intricacies of synthetic problems. Fundamentally, sustainable organocatalysis is more than just scientific investigation; it is about responsible chemistry.

This methodology involves development of techniques that not only progresses catalytic efficiency but also be the green developments and economically viable. It turns into the material incarnation of the green chemistry principles in which the production of knowledge and the quest for environmental sustainability attain a perfectly balanced coexistence. Organocatalysis being treated by researchers, they cling upon a fine balance between innovative intellectuality and environmental accountability. Under the given economic paradigm, the methodologies that are developed are a notable shift in the previously employed traditional technologies, thus heralding a new age where the sustainability of chemical processes is at the forefront [30,31].

This undertaking is not only a scientific endeavor; it's an intentional shift toward a greener future in the field of catalytic research. With the field developing, sustainable organocatalysis appears to not only handle the synthetic challenges but also bring chemistry closer to green and cost-effective processes.

15.7.3 Future prospects in catalytic research

As the landscape of catalytic research undergoes dynamic transformations, the future unfolds with a keen focus on accelerating innovations, particularly in CH activation/functionalization. This underscores the imperative to develop catalytic materials that can orchestrate regioselective and enantioselective synthesis, unlocking new frontiers in chemical transformations.

The trajectory of future research in catalysis is distinctly marked by a commitment to the principles of green chemistry. This commitment manifests in the relentless exploration of novel catalytic processes that align with sustainability goals.

Researchers are not merely pursuing scientific breakthroughs; they are actively shaping a future where catalysis is synonymous with environmental responsibility.

Key research directions on the horizon include the harnessing of selective catalytic methods, delving into uncharted territories of catalytic materials, and confronting the challenges associated with sustainable catalysis. These forward-looking endeavors aim not only to expand the scientific understanding of catalysis but also to contribute practical solutions to real-world challenges [32].

The synergy between burgeoning fields like nanocatalysis, the sustainable approaches of organocatalysis, and the relentless exploration of new catalytic frontiers paints a promising picture for the future of catalytic research. It is a future where innovation, sustainability, and the pursuit of greener alternatives converge to redefine the boundaries of what catalysis can achieve.

15.8 Challenges and solutions in catalysis

Catalysis, an important element of many industrial fields, faces a variety of issues of different levels of complexity which requires careful analysis and remedy. From these challenges, catalyst deactivation, environmental problems, and economic issues are the most quoted. In the following part, we go into the different aspects of these challenges and offer a glimpse of modern approaches trained mostly on green catalysis.

15.8.1 Catalyst deactivation and regeneration

Catalyst deactivation, a permanent form of catalyst activity loss where the activity slowly diminishes over time, is a primary problem in industrial applications. The driving force due to fouling, sintering, or poisoning, requires strategic approaches to ensure sustainability. Catalyst regeneration becomes a critical point which allows breathing new life into deactivated catalysts, restoring their activity and selectivity.

While in several cases, regeneration appears to be rather simplistic and requires minimal input, its intricacies can add additional problems that surpass economic feasibility of the entire catalytic process. Researchers have started working in search of novel regeneration methods proactively. Their explorations range from materials to catalysts with higher stability. These efforts seek to disentangle the intricacies of catalyst deactivation, giving rise to solutions not only for restoring catalytic activity but also prolonging catalytic processes, and consequently, their economic feasibility [33,34]. The dynamic scenery of catalyst revitalization is captivating, presenting an avenue for the creation of regenerative and efficient catalytic systems.

15.8.2 Navigating environmental and economic challenges in catalytic processes

Catalytic processes though promising benefits in terms of industrial operations, bring a tricky balance between advantages and pressing ecological and economic issues. While the catalysis becomes a beast of burden which drives industrial processes, these concerns about catalyst synthesis arise due to such undesirable side effects as greenhouse gas emissions and resource-intensive nature of process. The green chemistry principles will serve as a beacon while facing the challenges directly. An environmental considerate approach to catalysis is the route to address urging environmental concerns associated with production processes. This whole new way of thinking includes a systemic approach, from the switch to renewables to the careful design of ensembles of highly efficient catalysts that minimize the waste generated.

The interplay between environmental impact and economic viability becomes a pivotal consideration. Striking a harmonious balance between these two aspects is crucial to ensuring the sustainability of catalytic processes. Embracing green chemistry principles propels catalysis into a domain where environmental conservation and economic viability seamlessly converge. This evolution aligns with the global momentum towards the adoption of eco-friendly industrial practices, ushering in a transformative phase in catalytic processes that signifies a departure from conventional practices [35,36].

15.8.3 Pioneering green catalysis for a sustainable future

In the dynamic world of catalysis, recent advances in green catalysis represent an epoch-making development, placing sustainability and environmental awareness at the cradle. This development is tantamount to a paradigm shift introducing chemical processes with pronounced ecological concerns as the primary attention. For the green catalysis, the holistic approach is used which unites the key elements of industry to transform it.

The use of renewable feedstocks results in utilizing renewable resources which reduces dependence on finite resources. Environmentally benign solvents favorable for their low ecological footprint replace conventional counterparts, thus increasing the green performance of catalytic processes. As well, the advent of catalytic methodologies which reduce waste fits squarely within the bigger picture of environmental protection. Beyond obeying environmental laws, green catalysis improves process efficiency, and sustainability greatly. The instrumentalization of such green principles reveals the green research community's effort of engineering eco-friendly and economically viable technologies. The dedication to the relentless search of innovative approaches to problems, such as catalyst deactivation as well as wider environmental concerns expresses a certain unmistakable commitment to a better sustainable and efficient future for catalytic processes.

This commitment is reflected in the current research projects that are placing green catalysis as a marching force toward a more green- and economical-based catalytic environment [37].

15.9 Case studies and examples

15.9.1 Unveiling the impact of notable catalytic reactions

A detailed analysis of the catalytic reactions of great moments illuminates the wide utilization and deep impartation of catalysis by numerous industries. An illustrative case is the catalytic hydrogenation of octanal, a reaction that reveals the complex processes of catalyst deactivation and regeneration associated with hydrogenation processes. This case study emphasizes the complex issues that arise to sustain a hydrogenation efficiency, central to a vital function from food production to pharmaceuticals.

Beyond that example, catalytic converters in automobiles present one further convincing instance of catalysis at work. These converters are of paramount significance in decreasing the pollution from the internal combustion engines. Nitrogen oxides are among harmful pollutes that catalytic conversion turns into less toxic products; catalysis is thus an unavoidable instrument that helps car industry to relieve its environmental burden.

These illustrations cumulatively highlight the robustness of catalytic reactions in bringing about different outcomes, thereby, solving a wide range of diverse application problems in different sectors of the society. The endeavor in these industrially significant catalytic reactions besides expanding our knowledge of catalysis, also reveals its tremendous impact on various industries. Thus with these vigorous reactions, catalysis stands out to be a pivot forming a conscientious and eco-friendly present for many sectors.

15.9.2 Illuminating success stories in catalytic processes

The ambiance of catalytic processes is illuminated by remarkable success stories which depict accomplishments realized through refinement and imagination. These victories are usually borne from the sensitive balancing and perfecting of reaction conditions, the tactical design of catalysts, and the wise application of resources. Case studies in catalysis are signifying illuminating narratives that capture the role played by a profound understanding of reaction kinetics, mechanisms, and pathways in getting optimum catalytic performance.

The essence of these success stories is rooted in the implementation of innovative techniques, as exemplified by discussions in catalysis case studies. These discussions intricately delve into atomic to microscale characterizations of catalytic materials and in situ studies, offering invaluable insights into the complexities of catalytic processes. By embracing these cutting-edge approaches, both researchers and industries propel catalysis into new realms of enhanced efficiency and performance [38,39].

These success stories are not merely isolated triumphs; rather, they constitute integral contributions to the collective knowledge base. Serving as benchmarks, they play a crucial role in refining catalytic processes, fostering a continuous cycle of improvement and innovation. With each unfolding success story, another layer is added to our understanding of catalysis, contributing significantly to the evolution of this dynamic field. These narratives collectively shape the landscape of catalytic processes, driving progress and innovation in the pursuit of more sustainable, efficient, and impactful solutions.

15.9.3 Navigating the landscape of catalyst development: lessons learned

Embarking on the intricate journey of catalyst development is akin to navigating uncharted waters—a narrative teeming with challenges, revelations, and invaluable lessons that illuminate the path forward. These acquired insights stand as beacons, guiding future endeavors toward the creation of more effective and resilient catalysts.

One pivotal aspect in the realm of catalyst development is the dance with catalyst deactivation mechanisms. The meticulous study of catalyst regeneration, an exploration into reviving deactivated catalysts not only addresses immediate challenges but also furnishes profound insights for crafting catalysts with enduring efficacy. These lessons, stemming from a nuanced understanding of regeneration dynamics, contribute significantly to the knowledge base, allowing the formulation of robust and durable catalysts.

Environmental and economic challenges, acting as catalysts in their own right, propel the evolution of green catalysis. Lessons learned in this domain orbit around the delicate equilibrium required to design processes that are not only efficient but also eco-friendly. The adoption of sustainable practices becomes a pivotal lesson, underscoring the importance of aligning industrial efficiency with environmental responsibility.

The on-going exploration of emerging trends and innovations in catalysis adds layers to the repository of lessons learned. This continuous quest for knowledge actively shapes the future landscape of catalyst development, ensuring that every challenge surmounted becomes a stepping stone for subsequent advancements.

In a broader context, delving into case studies and examples within catalysis provides a comprehensive education. This exploration transcends understanding the intricacies of individual reactions; it constitutes a collective effort to foster continuous improvement, contribute to the refinement of catalytic processes, and pave the way for a future where sustainability and efficiency seamlessly coexist. The lessons learned in this dynamic field not only enhance our understanding but serve as catalysts for transformative progress.

15.10 Conclusion: unveiling insights from nanopaints in catalysis

15.10.1 Synthesizing key insights

A synopsis of critical insights comes to maturity as the final pages of the book chapter that deal with nanopaints in catalysis are reached.

The domain of nanopaints and their catalytic applications have not only been illuminating but also extensive. The chapter captures salient points that traverse different aspects. It commences with a deep dive into the many methods applied in nanopaint preparation, peeling back the curtains on the intricate procedures that create them. The detailed grasping of the involvement of nanopaints in dealing with fouling as far as catalytic systems are concerned makes the central theme.

In the exegesis of this issue, the exploration is anchored on the study of how these nanopaints become the innovations counteracting and preventing biofouling, thus enriching the catalytic efficiency. Moreover, the evaluation spans over the long-term catalytic performance influenced by nanopaints [40,41].

This stage entails an incisive scrutiny of the influence imparted by these nanomaterials, which results in improved catalytic stability and efficacy along the extended periods. The obtained findings not only deepen the perception of nanopaintings but also guide their smart employment in the catalytic processes. At the end of the chapter, a lake of knowledge is left that reverberates with the possibility for world changing inventions in the field of catalysis.

15.10.2 Implications for future research and applications

With a look to the future, the consequences from gathered insights are weighty. The novelty in nano baits for catalysis created opens for studies to carry out research processes. Novel synthesis techniques, influencing agents on formulation and application of nanopaints in catalysis are avenues that need to be explored. The use of nanopaints in catalysis is likely to have well beyond the lab scope. The practical implications for the industries from automotive to environmental catalysis are huge. Leveraging the catalytic power of nanopaints also solves existing barriers and gives way to new horizons of green and effective processes.

15.10.2.1 An echo from the challenges

After all, we had to face several bumps in the road. Catalyst deactivation, the ecological consequences of catalytic processes, and the economic aspect of the matter manifest as the dark side of the landscape. Yet, in facing these issues frontally, the human spirit became evident as novel regeneration technologies, green catalysis rules, and sustainable practices emerged.

Our response to the challenges we face has turned out to be the crucible for catalytic innovation.

15.10.2.2 In the tapestry of case studies

The case studies and examples interwoven in our story provided us with compasses, exposing the practical applications of catalysts. Starting from the remarkable catalytic hydrogenation of octanal to the catalytic converters in cars, each case study bears witness to the broad range of applications of catalysis. These stories do not merely refer to the chronographs of scientific inquiry; they have live illustrations of the practices that apply in the catalytic arena.

Ultimately, we have come to the end of the narration revealing nuances of nanopaints in catalysis and the story is full of knowledge and potentials. This chapter builds foundation for upcoming research efforts, calling for more investigations of nanopaints as potential catalytic transformative agents.

References

[1] U. Chadha, et al., Complex nanomaterials in catalysis for chemically significant applications: from synthesis and hydrocarbon processing to renewable energy applications, Advances in Materials Science and Engineering 2022 (2022). Available from: https://doi.org/10.1155/2022/1552334.

[2] K. Krishnamoorthy, K. Jeyasubramanian, M. Premanathan, G. Subbiah, H.S. Shin, S.J. Kim, Graphene oxide nanopaint, Carbon 72 (2014) 328–337. Available from: https://doi.org/10.1016/J.CARBON.2014.02.013.

[3] M. Garcia-Diaz, K. Bebenek, Multiple functions of DNA polymerases, Critical Reviews in Plant Sciences 26 (2) (2007) 105–122. Available from: https://doi.org/10.1080/07352680701252817.

[4] Amylase – PubMed. https://pubmed.ncbi.nlm.nih.gov/32491670/ (accessed 11.02.24).

[5] H.S. Taylor, Catalysis | Chemistry, Classification, & Chemical Reactions | Britannica, 2023. https://www.britannica.com/science/catalysis (accessed 10.02.24).

[6] V.L. Arcus, M.W. van der Kamp, C.R. Pudney, A.J. Mulholland, Enzyme evolution and the temperature dependence of enzyme catalysis, Current Opinion in Structural Biology 65 (2020) 96–101. Available from: https://doi.org/10.1016/J.SBI.2020.06.001.

[7] G.M. Cooper, The Central Role of Enzymes as Biological Catalysts, 2000. Available from: https://www.ncbi.nlm.nih.gov/books/NBK9921/ (accessed 10.02.24) [Online].

[8] 7.1: Catalytic Converters – Chemistry LibreTexts. https://chem.libretexts.org/Bookshelves/Physical_and_Theoretical_Chemistry_Textbook_Maps/Supplemental_Modules_(Physical_and_Theoretical_Chemistry)/Kinetics/07%3A_Case_Studies-_Kinetics/7.01%3A_Catalytic_Converters (accessed 11.02.24).

[9] A. Cao, R. Lu, G. Veser, Stabilizing metal nanoparticles for heterogeneous catalysis, Physical Chemistry Chemical Physics: PCCP 12 (41) (2010) 13499–13510. Available from: https://doi.org/10.1039/C0CP00729C.
[10] A. Alshammari, V.N. Kalevaru, A. Martin, A. Alshammari, V.N. Kalevaru, A. Martin, Metal nanoparticles as emerging green catalysts, Green Nanotechnology – Overview and Further Prospects, 2016. Available from: https://doi.org/10.5772/63314.
[11] L. Liu, A. Corma, Metal catalysts for heterogeneous catalysis: from single atoms to nanoclusters and nanoparticles, Chemical Reviews 118 (10) (2018) 4981–5079. Available from: https://doi.org/10.1021/ACS.CHEMREV.7B00776.
[12] Industrial Catalysts – Catalyst Analysis – Advancing Materials. https://www.thermofisher.com/blog/materials/characterizing-the-effectiveness-of-industrial-catalysts/ (accessed 11.02.24).
[13] Environmental impact – Energy Education. https://energyeducation.ca/encyclopedia/Environmental_impact (accessed 11.02.24).
[14] Catalysis Applications | PIDC. https://www.pidc.com/applications/catalysis (accessed 11.02.24).
[15] T. Islam, J. Ryan, Mitigation rules and regulations, Hazard Mitigation in Emergency Management (2016) 37–68. Available from: https://doi.org/10.1016/B978-0-12-420134-7.00002-3.
[16] S. Kar, H. Sanderson, K. Roy, E. Benfenati, J. Leszczynski, Green chemistry in the synthesis of pharmaceuticals, Chemical Reviews 122 (3) (2022) 3637–3710. Available from: https://doi.org/10.1021/ACS.CHEMREV.1C00631/ASSET/IMAGES/MEDIUM/CR1C00631_0213.GIF.
[17] D. Sharma, C.M. Hussain, Smart nanomaterials in pharmaceutical analysis, Arabian Journal of Chemistry 13 (1) (2020) 3319–3343. Available from: https://doi.org/10.1016/J.ARABJC.2018.11.007.
[18] S. Kumar, F. Ye, S. Dobretsov, J. Dutta, Nanocoating is a new way for biofouling prevention, Frontiers in Nanotechnology 3 (2021) 771098. Available from: https://doi.org/10.3389/FNANO.2021.771098/BIBTEX.
[19] H.E. Yong, K. Krishnamoorthy, K.T. Hyun, S.J. Kim, Preparation of ZnO nanopaint for marine antifouling applications, Journal of Industrial and Engineering Chemistry 29 (2015) 39–42. Available from: https://doi.org/10.1016/J.JIEC.2015.04.020.
[20] C.I. Idumah, C.M. Obele, E.O. Emmanuel, A. Hassan, Recently emerging nanotechnological advancements in polymer nanocomposite coatings for anti-corrosion, anti-fouling and self-healing, Surfaces and Interfaces 21 (2020) 100734. Available from: https://doi.org/10.1016/J.SURFIN.2020.100734.
[21] N. Abid, et al., Synthesis of nanomaterials using various top-down and bottom-up approaches, influencing factors, advantages, and disadvantages: a review, Advances in Colloid and Interface Science 300 (2022). Available from: https://doi.org/10.1016/j.cis.2021.102597.
[22] V. Harish, et al., Nanoparticle and nanostructure synthesis and controlled growth methods, Nanomaterials 12 (18) (2022). Available from: https://doi.org/10.3390/NANO12183226.
[23] Y. Pan, W.J. Paschoalino, A. Szuchmacher Blum, J. Mauzeroll, Recent advances in bio-templated metallic nanomaterial synthesis and electrocatalytic applications, ChemSusChem 14 (3) (2021) 758–791. Available from: https://doi.org/10.1002/cssc.202002532.
[24] C. Pandit, et al., Biological agents for synthesis of nanoparticles and their applications, Journal of King Saud University – Science 34 (3) (2022) 101869. Available from: https://doi.org/10.1016/J.JKSUS.2022.101869.

[25] R. Chaudhary, K. Nawaz, A.K. Khan, C. Hano, B.H. Abbasi, S. Anjum, An overview of the algae-mediated biosynthesis of nanoparticles and their biomedical applications, Biomolecules 10 (11) (2020) 1–35. Available from: https://doi.org/10.3390/BIOM10111498.
[26] P.K. Dikshit, et al., Green synthesis of metallic nanoparticles: applications and limitations, Catalysts 11 (8) (2021) 902. Available from: https://doi.org/10.3390/CATAL11080902.
[27] L. Xu, Y.Y. Wang, J. Huang, C.Y. Chen, Z.X. Wang, H. Xie, Silver nanoparticles: synthesis, medical applications and biosafety, Theranostics 10 (20) (2020) 8996. Available from: https://doi.org/10.7150/THNO.45413.
[28] K. Singh, S. Mehra, A. Kumar, Recent advances in catalytic conversion of lignin to value-added chemicals using ionic liquids and deep eutectic solvents: a critical review, Green Chemistry: an International Journal and Green Chemistry Resource: GC 26 (3) (2024) 1062–1091. Available from: https://doi.org/10.1039/D3GC03439A.
[29] S.B. Somwanshi, S.B. Somvanshi, P.B. Kharat, Nanocatalyst: a brief review on synthesis to applications, Journal of Physics: Conference Series 1644 (1) (2020) 012046. Available from: https://doi.org/10.1088/1742-6596/1644/1/012046.
[30] L.C. Branco, et al., Recent advances in sustainable organocatalysis, Recent Advances in Organocatalysis (2016). Available from: https://doi.org/10.5772/64279.
[31] A. Antenucci, S. Dughera, P. Renzi, Green chemistry meets asymmetric organocatalysis: a critical overview on catalysts synthesis, ChemSusChem 14 (14) (2021) 2785–2853. Available from: https://doi.org/10.1002/CSSC.202100573.
[32] S. Kumar, et al., Recent developments in nanocatalyzed green synthetic protocols of biologically potent diverse O-heterocycles—a review, Catalysts 12 (6) (2022) 657. Available from: https://doi.org/10.3390/CATAL12060657.
[33] C.H. Bartholomew, M.D. Argyle, Advances in catalyst deactivation and regeneration, Catalysts 5 (2) (2015) 949–954. Available from: https://doi.org/10.3390/CATAL5020949.
[34] C.V. Satyanarayana, D. Srikant, H.R. Gurav, Catalyst deactivation and regeneration, Industrial Catalytic Processes for Fine and Specialty Chemicals (2016) 187–219. Available from: https://doi.org/10.1016/B978-0-12-801457-8.00005-7.
[35] I.V. Yentekakis, F. Dong, Grand challenges for catalytic remediation in environmental and energy applications toward a cleaner and sustainable future, Frontiers in Environmental Chemistry 1 (2020) 554309. Available from: https://doi.org/10.3389/FENVC.2020.00005.
[36] D. Weber, et al., Recent advances in the mitigation of the catalyst deactivation of CO_2 hydrogenation to light olefins, Catalysts 11 (12) (2021) 1447. Available from: https://doi.org/10.3390/CATAL11121447.
[37] I.T. Horváth, P.T. Anastas, Innovations and green chemistry, Chemical Reviews 107 (6) (2007) 2169–2173. Available from: https://doi.org/10.1021/CR078380V.
[38] Catalysis Case Studies | Diamond Light Source. https://www.diamond.ac.uk/industry/Case-Studies/Catalysis-Case-Studies.html (accessed 11.0224).
[39] A. Govender, A.S. Mahomed, H.B. Friedrich, Water: friend or foe in catalytic hydrogenation? A case study using copper catalysts, Catalysts 8 (10) (2018). Available from: https://doi.org/10.3390/CATAL8100474.

[40] Nanotechnology in High Performance Paint Coatings. https://www.researchgate.net/publication/266269380_Nanotechnology_in_High_Performance_Paint_Coatings (accessed 1102.24).
[41] M.N. de Queiroz, et al., Conductive nanopaints: a remarkable coating, Nanotechnology in the Automotive Industry (2022) 429–449. Available from: https://doi.org/10.1016/B978-0-323-90524-4.00020-7.

16

Nanopaints in medical applications

Abhinay Thakur[1] and Ashish Kumar[2]

[1]Department of Research Impact and Outcome, Division of Research and Development Cell, Lovely Professional University, Phagwara, Punjab, India

[2]Department of Science, Technology and Technical Education, Nalanda College of Engineering, Bihar Engineering University Government of Bihar, Patna, Bihar, India

16.1 Introduction

Nanotechnology, defined as the manipulation of matter at the nanoscale, has emerged as a powerful tool with the potential to revolutionize multiple industries. In the biomedical sector, the unique properties exhibited by materials at the nanoscale have spurred the development of innovative solutions to longstanding challenges [1–4]. Historically, the concept of nanotechnology can be traced back to visionary physicist Richard Feynman, who, in his famous 1959 lecture, "There's Plenty of Room at the Bottom," laid the conceptual groundwork for manipulating individual atoms and molecules. The subsequent decades witnessed the gradual convergence of physics, chemistry, and biology, leading to the birth of nanotechnology as a distinct scientific discipline. The background section highlights key milestones, including the advent of scanning tunneling microscopy in the 1980s, which enabled scientists to visualize and manipulate individual atoms. The convergence of nanotechnology with biology opened up new avenues for exploration, giving rise to the field of nanobiotechnology [5,6]. Researchers recognized the potential for developing nanoscale materials tailored for biomedical applications, setting the stage for the emergence of nanopaints. These materials, characterized by their unique properties at the nanoscale, have since become a focal point of research, promising transformative advancements in medical diagnostics, imaging, and targeted drug delivery. The significance of nanotechnology in biomedical applications lies in its capacity to bridge the gap between the macroscopic world of medical treatments and the intricate,

Nanopaints. DOI: https://doi.org/10.1016/B978-0-443-26512-9.00017-3

subcellular landscapes of biological entities. At the nanoscale, materials exhibit novel physical, chemical, and biological behaviors that diverge from their bulk counterparts. One of the defining features is the high surface area-to-volume ratio of nanoparticles. This characteristic enhances their reactivity and enables efficient interactions with biological entities, making them ideal candidates for targeted applications. The introduction explores how nanoparticles can be designed to mimic biological structures or carry therapeutic payloads, paving the way for innovative diagnostic and therapeutic approaches. Moreover, the unique optical, magnetic, and electronic properties of nanomaterials empower them with functionalities that transcend traditional biomedical tools. Nanoparticles can serve as contrast agents in medical imaging, allowing for enhanced visualization of tissues and organs. Additionally, their ability to cross biological barriers offers a gateway for precisely delivering therapeutic agents to specific cellular targets [7,8].

The evolution of nanopaints is a testament to the dynamic and iterative nature of scientific discovery. This section chronicles the developmental journey of nanopaints, elucidating the key stages that have shaped their current state and potential future trajectories. The inception of nanopaints can be traced to the intersection of nanotechnology with the field of materials science. Early experimentation with nanomaterials, such as quantum dots and gold nanoparticles, paved the way for understanding their unique properties and potential applications. Researchers recognized the need for materials that could be manipulated not only for their physical attributes but also for their interactions with biological systems. The evolution of nanopaints reflects a deliberate shift from passive nanomaterials to actively engineered agents with tailored functionalities. Initially, the focus of nanopaint development centered around imaging applications. The introduction provides insights into the pivotal studies that laid the groundwork for using nanoparticles as contrast agents in various imaging modalities. For example, the exploration of magnetic nanoparticles for enhancing magnetic resonance imaging (MRI) contrast marked a significant breakthrough, offering a glimpse into the potential of nanopaints to revolutionize diagnostics [9–11]. The iterative process of refinement and innovation led to the diversification of nanopaint functionalities. This section explores how researchers began to engineer nanopaints not only for imaging but also for targeted drug delivery. The ability to encapsulate therapeutic payloads within nanoparticles opened up new avenues for precision medicine, where drugs could be delivered directly to diseased cells while sparing healthy tissues. The evolution of nanopaints is also intertwined with advancements in nanomaterial synthesis and fabrication techniques. These advancements not only enhanced the performance of nanopaints but also facilitated scalability, making them more feasible for practical applications in a clinical setting. The integration of nanotechnology with other emerging fields, such as artificial intelligence and machine learning, further propelled the evolution of nanopaints. Computational approaches enabled the rational design of nanoparticles with desired properties, accelerating the development process and expanding the scope of potential applications. Throughout the evolutionary timeline, the section emphasizes the collaborative nature of nanopaint research. Multidisciplinary teams of scientists, engineers, and medical professionals collaborated to overcome challenges and explore the full spectrum of possibilities offered by nanopaints. This collaborative spirit underscores the importance of a holistic approach in advancing the field and maximizing the societal impact of nanopaint technologies [12–14]. This evolutionary perspective sets the stage for the subsequent sections, where the current applications and future potentials of nanopaints in medical science will be systematically reviewed.

16.2 Nanopaint formulation principles

16.2.1 Nanoparticle engineering

Nanoparticle engineering forms the cornerstone of nanopaint formulation, wherein scientists manipulate the size, shape, composition, and surface characteristics of nanoparticles to achieve desired functionalities. The introduction of nanoscale materials into paints brings about a convergence of materials science, chemistry, and nanotechnology. Nanoparticle engineering allows for a high degree of customization, enabling researchers to design nanoparticles with specific interactions and behaviors crucial for biomedical applications. One key aspect of nanoparticle engineering is the choice of materials. Different materials exhibit distinct properties at the nanoscale, influencing how nanoparticles interact with biological systems. Metals, such as gold or iron oxide, and organic materials, like lipids or polymers, are common choices. The selection is often dictated by the intended application; for instance, magnetic nanoparticles might be chosen for imaging purposes due to their response to magnetic fields [15–17].

The size of nanoparticles is another critical parameter that researchers carefully engineer. The quantum effects that dominate at the nanoscale can significantly alter the properties of materials. Fine-tuning the size allows for control over optical, magnetic, and electronic characteristics, making nanoparticles versatile tools in the design of nanopaints. Surface modification is an essential component of nanoparticle engineering. The surface of nanoparticles plays a crucial role in their interactions with biological entities. Functionalizing the surface with specific molecules, such as antibodies or peptides, enhances the ability of nanoparticles to target specific cells or tissues. This tailored approach enables the development of nanopaints with a high degree of specificity, a crucial aspect for applications like targeted drug delivery.

16.2.2 Tailoring properties at the nanoscale

Tailoring properties at the nanoscale involves a meticulous adjustment of various parameters to achieve the desired characteristics in nanopaints. The unique behaviors of materials at this scale result from quantum effects, surface interactions, and the inherent nature of nanomaterials. This section explores how researchers tailor these properties to make nanopaints well-suited for diverse biomedical purposes. One significant property that researchers aim to control is the stability of nanopaints. Nanoparticles tend to agglomerate due to van der Waals forces and other interactions, which can compromise their performance [18,19]. Surface modifications and the choice of stabilizing agents are critical in preventing agglomeration, ensuring the stability and efficacy of nanopaint formulations. Optical properties, crucial for imaging applications, are often tailored through the manipulation of nanoparticle size and composition. Quantum dots, for instance, exhibit size-dependent fluorescence, allowing for the creation of vibrant and tunable labels for imaging. This property has profound implications for improving the contrast and sensitivity of imaging techniques such as fluorescence microscopy.

Magnetic properties are harnessed in nanopaints designed for MRI. The tailoring of magnetic nanoparticles enables enhanced contrast in imaging, providing detailed information about anatomical structures and pathological conditions. The ability to fine-tune these properties contributes to the versatility of nanopaints in different imaging modalities. Surface charge is another property that researchers manipulate for specific applications. The surface charge of nanoparticles influences their interaction with biological membranes and cells. Positively charged nanoparticles may have enhanced cellular uptake, while negatively charged ones might show improved stability in physiological environments. Tailoring the surface charge allows for optimizing the behavior of nanopaints based on their intended use.

16.2.3 Unique characteristics of nanopaints

Nanopaints, as a distinctive class of materials, possess unique characteristics that set them apart from conventional paints and have profound implications for their applications in the biomedical field. This section explores these distinctive features, shedding light on why nanopaints are particularly well-suited for diverse biomedical purposes. One of the defining characteristics is the ability of nanopaints to interact selectively with biological entities. This specificity is achieved through the careful engineering of nanoparticles, allowing for surface modifications that enable targeted interactions [20–22]. This selectivity is a crucial aspect in applications like medical imaging, where nanopaints can be designed to bind to specific cellular markers, providing enhanced contrast and diagnostic accuracy. The small size of nanoparticles is a feature that imparts unique advantages to nanopaints. Their nanoscale dimensions facilitate penetration through biological barriers, allowing for targeted delivery of therapeutic agents to specific sites within the body. This capability is pivotal in the development of nanopaints for targeted drug delivery, where precision and minimized side effects are paramount.

Furthermore, the high surface area-to-volume ratio of nanoparticles contributes to their reactivity and ability to carry a substantial payload. This feature is exploited in drug delivery applications, where nanoparticles can encapsulate and deliver therapeutic agents, ensuring controlled release and improved efficacy. The unique physicochemical properties at the nanoscale enable nanopaints to act as carriers for a diverse range of therapeutic payloads, from small molecules to nucleic acids. The adaptability of nanopaints is a notable characteristic that allows for their integration into multifunctional platforms. Nanopaints can be designed not only for imaging but also for therapeutic interventions. The concept of theranostics, combining diagnostics and therapeutics, is made feasible through the unique characteristics of nanopaints [23,24]. This multifunctionality opens new avenues for personalized medicine, where nanopaints can be tailored to individual patient profiles.

16.3 Nanopaints in medical imaging

16.3.1 Contrast enhancement in imaging

The significant contribution of nanopaints to medical imaging lies in their exceptional ability to enhance contrast, addressing a longstanding challenge faced by traditional imaging modalities. Conventional imaging techniques often encounter limitations in accurately distinguishing between different tissues or pinpointing specific pathological conditions due to inherent similarities in contrast among various structures [25,26]. Nanopaints, with their meticulously tailored properties, offer a solution to this challenge by introducing specialized contrast agents that greatly amplify the discernibility of anatomical structures and cellular components within medical images. The core mechanism behind the contrast enhancement achieved by nanopaints is rooted in the unique properties of the nanoparticles they encapsulate. These nanoparticles, meticulously engineered for specific biomedical applications, bring a level of precision and functionality not achievable with conventional contrast agents. When introduced into the imaging process, these nanoparticles interact selectively with the target tissues or structures of interest. This selective interaction is a result of surface modifications and functionalizations that enable the nanopaints to bind specifically to certain biomarkers or cell types. As the nanopaint-laden contrast agents selectively bind to their intended targets, they introduce a nuanced alteration in the signal intensity during the imaging process. This interaction is finely tuned to enhance or modify the contrast between different regions within the imaged area. Consequently, the resulting images exhibit a clearer differentiation between structures that may have been challenging to distinguish with traditional imaging alone. In an experiment, Wade et al. [27] introduced a novel approach to multiplexed super-resolution microscopy, aiming to efficiently detect hundreds of molecular species simultaneously. In this study, they designed the blinking behavior of targets with engineered binding frequency and duration in DNA-PAINT. Herein, the kinetic barcoding approach was assessed both in silico and in vitro using DNA origami structures, demonstrating its applicability for multiplexed RNA and protein detection in cells. The authors successfully achieved 124-plex super-resolution imaging within minutes using DNA-PAINT. The implementation of this approach was demonstrated in vitro and in situ, showcasing its potential for single-cell systems biology studies with super-resolution capabilities. The proposed method, termed "localizomics," envisions addressing systems biology questions by simultaneously imaging hundreds of DNA, RNA, and protein targets, resembling transcriptomics- and proteomics-style experiments through a straightforward localization microscopy approach.

Similarly, Gowtham et al. [28] revealed the transformative role of medical imaging in the diagnosis and treatment of various diseases, facilitated by advancements in science and engineering. The emergence of modern chemistries has further enhanced imaging tools, leading to the development of contrast agents crucial for localizing diseased tissues. This review focused on nanovectors as essential components for multimodal imaging due to their significant physicochemical properties. Initially, the review outlines the fundamental principles, key parameters, and limitations of prominent imaging modalities such as MRI, computed tomography (CT), and fluorescence imaging. It highlights the drawbacks of traditional contrast agents, emphasizing the need for novel agents. The review elaborated on the importance of multimodal imaging and advocated for a unified contrast agent for various imaging applications. Lastly, it delves into the advantages, limitations, and design considerations of nanovectors based on magnetic and metallic nanoparticles, emphasizing surface modifications to mitigate toxicity and enhance efficacy. Movahedi et al. [29] introduced an innovative approach to radiation protection by designing a lead-free paint based on nanopowder. The conventional method of intraoperative radiation protection involves lead-based shields, which, while effective, pose well-documented human lead toxicity risks, lack durability, are heavy, are difficult to launder, and are environmentally hazardous to dispose of. In response to these drawbacks, the study aimed to develop a lead-free alternative. Various types of nano metal powder were evaluated for their absorption properties, with nano tungsten trioxide and nano tin dioxide identified as the most suitable candidates for paint in the diagnostic photon energy range. A paint formulation combining 15% nano tungsten trioxide and 85% nano tin dioxide demonstrated optimal protection in both simulation and experiments. Subsequent efforts focused on producing this nanopowder-based paint, resulting in a material with a density of 4.2 g cm^{-3}. Monte Carlo simulations and experimental measurements confirmed the efficacy of the nano tungsten trioxide-nano tin dioxide paint, with HVL (half-value layer) values at 100 kVp measured at 0.25 and 0.23 mm, respectively. The findings suggest that this cost-effective, lead-free paint has the potential to effectively absorb

X-rays and gamma rays, providing a viable alternative to lead-based shields in radiation protection applications.

This heightened contrast is particularly advantageous in the realm of medical diagnostics, where early disease detection and precise localization of abnormalities are paramount. The ability of nanopaints to provide a more detailed and distinct representation of anatomical features enables healthcare professionals to identify subtle changes in tissues or cellular structures indicative of pathological conditions. The improved clarity in imaging facilitates a more accurate diagnosis, allowing for timely intervention and personalized treatment strategies. In the context of early disease detection, nanopaints play a crucial role in highlighting subtle anomalies that might go unnoticed with conventional imaging methods. By enhancing the contrast between normal and abnormal tissues, nanopaints contribute to the identification of irregularities at their incipient stages [30–32]. This has profound implications for improving diagnostic accuracy and patient outcomes, as early detection often correlates with more effective treatment options and improved prognoses. Moreover, nanopaints excel in enabling precise localization of abnormalities within the body. Whether it be identifying specific areas of inflammation, detecting tumors, or pinpointing the onset of certain diseases, the enhanced contrast provided by nanopaints aids healthcare professionals in precisely mapping the location and extent of abnormalities. This precision is invaluable in guiding subsequent medical interventions, ranging from targeted therapies to surgical procedures.

16.3.2 Role of nanopaints in magnetic resonance imaging

Nanopaints play a pivotal and transformative role in advancing MRI, a cornerstone of noninvasive medical imaging. The essence of this contribution lies in how nanopaints enhance the contrast in MRI images, leading to improvements in both anatomical and functional imaging. In the realm of MRI, nanopaints contribute significantly to contrast enhancement by manipulating the relaxation times of protons in surrounding tissues. The incorporation of paramagnetic or superparamagnetic nanoparticles into nanopaint formulations introduces agents that influence the local magnetic field. As a result, there is a modification in the relaxation times of protons, which enhances the signal intensity and overall contrast in the resulting MRI images [33–35]. This heightened contrast enables healthcare professionals to discern anatomical structures with greater clarity, providing a more detailed and accurate representation of tissues. The nuanced engineering of nanopaints tailored for MRI applications brings forth a range of benefits that contribute to their effectiveness in clinical settings. One notable advantage is the prolonged circulation time of nanopaints in the bloodstream. This characteristic is crucial for optimizing the imaging window, allowing for an extended period during which the contrast agents remain active. The prolonged circulation time enhances the opportunity for imaging procedures, ensuring that healthcare professionals have an adequate timeframe to capture detailed and comprehensive images. Reduced toxicity is another key benefit associated with the use of nanopaints in MRI. The carefully selected materials and surface modifications in nanopaint formulations contribute to minimizing potential adverse effects on the body. This is particularly important in the context of repeated imaging procedures or in patients with specific health considerations, where minimizing toxicity enhances the safety profile of MRI examinations [36,37].

The specificity of nanopaints for particular tissues further amplifies their impact on MRI applications. By engineering nanopaints with surface modifications that enable selective binding to specific biomarkers or tissues of interest, researchers and clinicians can achieve a higher degree of precision in imaging. This specificity is particularly valuable in neurological or oncological imaging, where distinguishing between normal and pathological tissues is critical for accurate diagnosis and treatment planning. Controlled release mechanisms embedded within nanopaint formulations contribute to sustained contrast enhancement. This aspect ensures that the contrast agents are released in a controlled manner over time, extending the period during which enhanced imaging is possible. This controlled release is strategically designed to align with the temporal requirements of MRI procedures, providing a more consistent and prolonged enhancement throughout the imaging process. The implications of these advancements in MRI technology, facilitated by nanopaints, are far-reaching, impacting a diverse array of medical disciplines. In neurology, nanopaint-enhanced MRI enables detailed imaging of the brain's intricate structures, aiding in the diagnosis and monitoring of neurological conditions. In oncology, nanopaints contribute to precise imaging of tumors and surrounding tissues, informing treatment strategies and assessing therapeutic responses. In an experiment, Zhen et al. [38] focused on the development of manganese-based nanoparticles as contrast probes for MRI. MRI is a crucial imaging tool in clinical settings, providing detailed delineation with high spatial and temporal resolutions by interrogating nuclei of atoms in living subjects. To enhance the inherent low sensitivity of MRI, contrast probes are utilized, typically composed of paramagnetic or superparamagnetic materials that reduce the relaxation times of nearby protons. Traditionally, gadolinium-based T1 contrast probes were commonly employed, but recent associations with nephrogenic system fibrosis (NSF) have raised concerns, necessitating the exploration of alternatives. Manganese-containing nanoparticles have emerged as promising candidates, offering comparable r1 relaxivities through careful engineering. Additionally, these nanoparticles can be designed to carry various functionalities, such as targeting motifs, therapeutic agents, or second imaging components, resulting in versatile and multifunctional nanoparticle platforms. This research contributes to addressing the challenges associated with conventional contrast probes in MRI, offering potential alternatives with enhanced safety and functionality. Similarly, Ashokan et al. [39] focused on the development of a biomineral nano-theranostic agent for the augmented radiofrequency (RF) ablation of liver tumors, guided by MRI. The study introduced a single-phase calcium phosphate biomineral nanoparticle (nCP) designed to exhibit dual-mode magnetic resonance (MR) contrast (T1–T2) and an RF-mediated thermal response for effective cancer therapy. The nanoparticles were engineered to achieve dual MR contrast by optimizing the doping concentration (4.1 at.%) of the paramagnetic ion, Fe^{3+}, which imparts a lossy dielectric character to nCP, resulting in a thermal response under RF exposure. To assess the in vivo dual-mode T1–T2 contrast properties of nCP:Fe, a dose of $10\,mg\,kg^{-1}$ was intravenously administered to healthy rats. Thirty minutes postinjection, an enhancement in both T1 and T2 contrast was observed, particularly in the liver and heart regions, as depicted in the white dotted box in Fig. 16.1(A and B). T2 mapping analysis indicated a reduction in T2 contrast in the liver region, corresponding to a decrease in T2 relaxation time from 40 ± 1 mS to 25 ± 1 mS. Axial T2-weighted images of liver sections before (Fig. 16.1C) and after (Fig. 16.1D) nCP:Fe injection demonstrated an increase in T2 contrast intensity. T2 mapping data obtained from the selected region of interest (white circles in Fig. 16.1C and D) reflected the heightened relaxivity changes in the liver post nCP:Fe injection. Biodistribution and clearance time of nCP:Fe were also evaluated after intravenous injection. Whole-body coronal (Fig. 16.1E) and axial (Fig. 16.1F) T2-weighted MRI exhibited an initial increase in

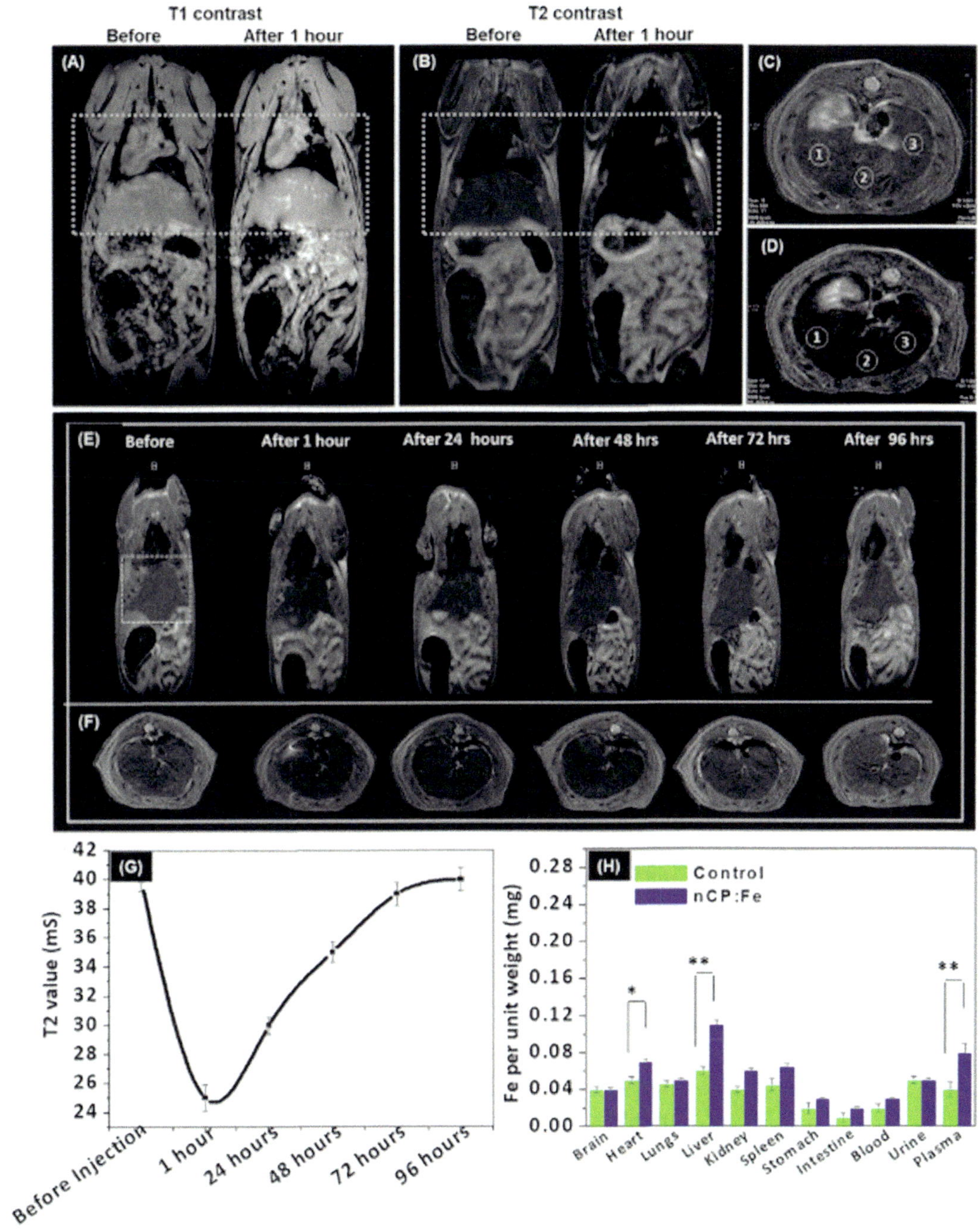

Figure 16.1 Illustration of the in vivo dual T1–T2 contrast and biodistribution of nCP:Fe. Coronal T1-weighted (A) and T2-weighted (B) MRI of a Wistar rat before and after nCP:Fe injection display enhanced contrast in the liver and heart region (white dotted box). Axial liver sections are shown before (C) and 30 minutes after (D) sample injection. T2-weighted MRI (E) and corresponding axial liver sections (F) depict changes over 96 hours post intravenous nCP:Fe injection. Variation in liver T2 time over this period is demonstrated (G), along with Fe content in different organs one hour after injection (H), compared to PBS control, as determined by ICP analysis. Adapted from A. Ashokan, V.H. Somasundaram, G.S. Gowd, et al., Biomineral nano-theranostic agent for magnetic resonance image guided, augmented radiofrequency ablation of liver tumor, Scientific Reports 7 (1) (2017) https://doi.org/10.1038/s41598-017-14976-8. under CCBY 4.0.

T2 contrast in the liver (highlighted in the white dotted box) within the first hour, gradually returning to the initial contrast level by 96 hours. This pattern was mirrored in T2 relaxation values, which significantly decreased in the first hour after injection (from 40 ± 1 mS to 25 ± 1 mS) and then gradually increased to the initial value of 40 mS by ~96 hours (Fig. 16.1G). This suggests the initial uptake of nanoparticles by the mononuclear phagocyte system in the liver, followed by a gradual clearance over a 96-hour period. Tissue-ICP analysis after 1 hour of injection confirmed an increase in Fe^{3+} content in all organs and plasma, with significantly higher concentrations in the heart, plasma, and liver (Fig. 16.1H). Importantly, our in vitro studies on nCP:Fe-treated RAW 264.7 macrophages demonstrated minimal degradation even after 12 hours of intracellular localization, supporting the conclusion that the nanoparticles circulated in the body, accumulated in the liver within 1 hour, and were slowly cleared by 96 hours.

Maeng et al. [40] focused on the development of a novel drug delivery system, YCC-DOX, comprising poly(ethylene oxide)-trimellitic anhydride chloride-folate (PEO-TMA-FA), doxorubicin (DOX), and superparamagnetic iron oxide (Fe_3O_4) with folate targeting. In rat and rabbit models of liver cancer, YCC-DOX demonstrated enhanced anticancer efficacy and specific targeting of folate receptor-expressing tumors, leading to increased bioavailability and DOX efficacy. Compared to free DOX (FD) and the commercial liposome drug DOXILÒ, the YCC-DOX group exhibited a two- to four-fold reduction in relative tumor volume. Additionally, YCC-DOX demonstrated MRI sensitivity comparable to ResovistÒ, a conventional MRI contrast agent, even with lower iron content. Immunohistochemical analysis revealed lower expression of angiogenesis and cell proliferation markers (CD34 and Ki-67) in the YCC-DOX group, while the TUNEL assay indicated a higher presence of apoptotic cells. These findings suggest that YCC-DOX holds promise for liver cancer treatment and MRI monitoring of cancer progression. Similarly, Kono et al. [41] devised liposomes integrating temperature-triggered drug release and MRI capabilities. These multifunctional liposomes were crafted by incorporating thermosensitive poly(2-ethoxy(ethoxyethyl)vinyl ether) chains (with a lower critical solution temperature around 40°C) and polyamidoamine G3 dendron-based lipids featuring Gd^{3+} chelate residues into pegylated liposomes. Below physiological temperature, these stable doxorubicin (DOX)-loaded liposomes retained DOX internally but promptly released it at temperatures surpassing 40°C. They demonstrated superior efficacy in shortening the longitudinal proton relaxation time for enhanced MRI sensitivity. In colon 26 tumor-bearing mice, intravenously administered liposomes exhibited increased accumulation in tumors over time, leveling off 8 hours postadministration as indicated by T1-weighted MRI signal intensity in tumors. The liposome size and tumor dimensions influenced accumulation efficiency, with approximately 100 nm diameter liposomes showing more efficient accumulation, particularly in larger tumors. Intravenous administration of DOX-loaded liposomes followed by mild tumor heating at 44°C for 10 minutes at 8 hours postadministration significantly suppressed tumor growth. These temperature-triggered drug release and MRI-functionalized liposomes present a potential avenue for personalized chemotherapy, offering optimized and efficient treatment strategies for individual patients.

16.3.3 Advancements in computed tomography with nanopaints

CT imaging, a staple in medical diagnostics, has also undergone significant improvements with the integration of nanopaints. In CT applications, nanopaints play a pivotal role in overcoming limitations associated with conventional imaging techniques, especially in the visualization of soft tissues and subtle anatomical features. The key mechanism driving this enhancement lies in the incorporation of high-density materials within nanopaint formulations [42–46]. Materials such as gold or iodine are often chosen for their ability to strongly absorb X-rays, a fundamental principle in CT imaging. The introduction of nanopaints with high-density materials significantly alters the interaction of X-rays with the imaged tissues. Traditional CT images may face challenges in distinguishing between tissues of varying densities, leading to limited contrast in certain regions. Nanopaints, equipped with high-density materials, enhance the contrast by absorbing X-rays more effectively [47,48]. This heightened contrast facilitates a clearer and more detailed differentiation between various tissues and structures within the imaged area. One of the primary contributions of nanopaints to CT imaging is their capacity to improve visualization in areas where conventional techniques may fall short. Soft tissues, blood vessels, and intricate anatomical structures, which might be challenging to discern with standard CT approaches, become more distinctly visible with the aid of nanopaint-enhanced contrast. This is particularly valuable in clinical scenarios where precise imaging is crucial for accurate diagnosis and treatment planning.

The choice of high-density materials, such as gold or iodine, in nanopaint formulations is strategic. These materials possess a pronounced ability to absorb X-rays, leading to a stark contrast in the resulting CT images. The effective absorption of X-rays by nanopaints enhances the clarity of the imaged structures, aiding radiologists in making more accurate and informed diagnoses. This heightened contrast allows for improved delineation of boundaries between different tissues, providing a comprehensive view of the anatomical landscape. The advancements in CT imaging brought about by nanopaints find particular relevance in cardiovascular imaging. In this context, the precise visualization of blood vessels and soft tissues is paramount for diagnosing and monitoring conditions such as arterial stenosis, aneurysms, or vascular malformations. The enhanced contrast provided by nanopaints allows for a more detailed and accurate representation of the vascular system, facilitating early detection and intervention in cardiovascular pathologies. Moreover, the benefits extend beyond cardiovascular imaging to various other clinical scenarios where improved visualization is critical. Nanopaint-enhanced CT imaging has implications in oncology, where the enhanced contrast aids in the detection and characterization of tumors. It also proves valuable in imaging studies involving the musculoskeletal system, where distinguishing between different soft tissues and structures is essential for identifying injuries or abnormalities.

For instance, in an experiment, Bonnin et al. [49] demonstrated that the impurity precipitation in atomized particles profoundly impacts the performance and physical properties of advanced materials, often influencing their initial microstructure and overall behavior. Accurate characterization of these impurities and their interactions with the surrounding matrix is crucial but typically involves destructive methods, posing a risk of misinterpretation. This study addressed these challenges by leveraging advancements in high-resolution X-ray diffraction computed tomography, coupled with an X-ray nanoprobe, enabling non-destructive crystallographic analysis of materials exhibiting microscopic heterogeneous microstructures (with a grain size ranging from 10 nm to 10 μm). The presented investigation focuses on the grain localization within a 2D slice of a 20 μm solidified atomized cU-Mo particle, revealing the presence of a minority U(C,O) phase (1 wt.%) characterized by submicrometer-sized grains. The findings provide compelling evidence indicating a precipitation mechanism during the onset of U(C,O) grain crystallization, where a single U-Mo grain exhibits a direct orientation relationship with multiple surrounding

U(C,O) grains. This innovative approach offers a non-destructive means to unravel the intricate details of impurity dynamics in atomized particles, contributing to a deeper understanding of material microstructures.

Chatzipetros et al. [50] assessed the capability of cone beam computed tomography (CBCT) for imaging critical-size defects (CSD) in rat calvaria filled with 75/25 w/w nano-hydroxyapatite/chitosan (nHAp/CS) scaffolds, with subsequent histological verification. Using 30 adult Sprague Dawley rats, bilateral CSDs (5 mm in diameter) were created in the parietal bone. The right CSD was filled with nHAp/CS scaffold, while the left CSD remained empty as the control. After euthanizing rats at 2, 4, and 8 weeks postsurgery, 28 specimens containing both CSDs were scanned using a NewTom VGi CBCT imaging unit. Grayscale values (in VGiHU) and the region of interest (ROI, in mm^2) were automatically calculated using the manufacturer's software. Histological examination was performed on all specimens. Results indicated increased VGiHU in the experimental group compared to the control ($P=.000$). The ROI of CSD (in mm^2) significantly decreased ($P=.001$) from the fourth to the eighth week in both groups. No significant difference between male and female rats ($P=.188$) was observed for VGiHU. The study demonstrates the effective visualization of nHAp/CS scaffolds using high-resolution CBCT, suggesting their contribution to new bone formation in rat calvarial CSD. The combined CBCT measurements and histological findings highlight the clinical relevance of nHAp/CS scaffold presence in promoting bone regeneration.

16.3.4 Fluorescence imaging applications

Nanopaints have found a niche in fluorescence imaging, a technique that utilizes the emission of light from fluorophores to visualize biological structures or processes. The unique optical properties of nanoparticles incorporated into nanopaints underpin their effectiveness in fluorescence imaging applications. Quantum dots, a noteworthy example, exhibit size-dependent fluorescence, allowing for tunable emission wavelengths. This tunability enables researchers and clinicians to tailor the emission characteristics of nanopaints according to specific imaging needs. Such precision in emission wavelengths contributes to the versatility of nanopaints in fluorescence imaging, allowing for targeted investigations at the cellular and molecular levels. The surface modifications on nanopaints play a pivotal role in enhancing the specificity of fluorescence imaging [51–53]. These modifications enable nanopaints to interact selectively with specific cells or tissues of interest. By incorporating ligands, antibodies, or peptides on the nanopaint surface, researchers can design formulations that bind specifically to biomarkers or receptors associated with certain pathological conditions. This targeted approach enhances the accuracy and precision of fluorescence imaging, enabling the visualization of specific cellular events and molecular processes [54–57].

The versatility of nanopaints in fluorescence imaging extends beyond basic research applications to intraoperative imaging. Real-time visualization of tissues or structures of interest during surgical procedures is paramount for surgeons to make informed decisions. Nanopaints, by providing this real-time imaging capability, contribute to improved precision during surgeries. The potential for fluorescence-guided surgery, facilitated by nanopaints, represents a significant advancement in the field. In intraoperative settings, nanopaints act as beacons, allowing surgeons to visualize specific tissues or structures that might be challenging to discern under conventional lighting. This is particularly crucial in procedures involving complex anatomical structures or those where precise localization is essential, such as tumor resections. By leveraging nanopaint-enhanced fluorescence, surgeons can navigate with enhanced precision, ensuring thorough and accurate interventions. The concept of fluorescence-guided surgery, propelled by nanopaints, addresses a longstanding challenge in surgical practice—the need for improved visualization and differentiation between healthy and diseased tissues. By highlighting specific structures, tumors, or abnormal cells with fluorescence, surgeons can minimize damage to healthy tissues, reduce the risk of leaving residual disease, and enhance overall surgical outcomes. This approach represents a significant stride toward personalized and minimally invasive surgical interventions. In an experiment, Park et al. [58] introduced a novel category of eco-friendly fluorescent CD paints (C-paints) generated from polyethylene glycol (PEG) through a straightforward ultrasound irradiation process at room temperatures. Unlike previously derived green CDs sourced from potentially harmful natural substances, these C-paints achieve quantum yields of approximately 14%. Notably, C-paints demonstrate a potent photoluminescent (PL) property responsive to both UV and visible light. Furthermore, when incorporated into a photocatalytic system, C-paints play an additional role in emitting fluorescent light to activate TiO_2 nanoparticles (NPs), facilitating the detoxification of various organic dyes. This multifaceted functionality positions C-paints as promising contributors to advanced water purification systems. Similarly, Mishra et al. [59] did the controlled carbonization of the P123 block copolymer in ethanol, which resulted in the formation of micelle-protected carbon nanodots (CNDs) at room temperature. Subsequently, a purified and concentrated CND solution, boasting a quantum yield of 10%, was integrated into the organically-modified SiO_2 (ORMOSIL) sol for the development of transparent fluorescent coatings with a thickness of approximately 3 mm on glass substrates. Notably, these films exhibit scratch resistance and possess a surface hardness value of 8H. Furthermore, the sol exhibits utility as a fluorescent ink/paint. Importantly, the films retain the inherent photophysical properties of CNDs, emphasizing their potential in various applications.

Taylor et al. [60] addressed the escalating challenge of bacterial antibiotic resistance, which is imperative, particularly in the context of medical devices inserted into the body, where the robust adherence of antibiotic-resistant biofilm infections poses a significant threat. Concerns about the proliferation of bacterial genetic tolerance, exemplified by multidrug-resistant strains like Staphylococcus aureus (MRSA), have intensified. In the realm of biomaterial design, the emergence of nanomaterials, defined by at least one dimension less than 100 nm, offers a promising avenue to mitigate bacterial functions leading to infections. A crucial aspect involves exploring various nanoparticles to enhance bacteria and biofilm penetration, induce the generation of reactive oxygen species, and eradicate bacteria. This approach presented a nondrug-related strategy for combatting infections. This review delves into the detailed mechanisms and applications of such nanoparticles, examining recent material designs employing nanotechnology to counteract medical device infections.

16.4 Targeted drug delivery with nanopaints

16.4.1 Precision in drug delivery

Targeted drug delivery with nanopaints represents a revolutionary approach that addresses one of the primary challenges faced by conventional drug delivery systems—achieving precise localization of therapeutic agents to specific cells or tissues while minimizing off-target effects. The hallmark of this innovation lies in the unparalleled precision achieved through careful engineering and surface modifications of nanopaints [61]. Conventional drug delivery often encounters limitations in ensuring that therapeutic compounds reach their intended destinations without affecting healthy tissues, leading to undesirable side effects. Nanopaints, however, navigate and overcome these challenges by offering a targeted approach, revolutionizing the landscape of drug delivery. The surface of nanopaints serves as a customizable canvas that can be strategically functionalized with ligands, antibodies, or peptides. These functional groups are meticulously chosen to recognize and bind to specific receptors on the surface of target cells. This selective interaction is akin to a molecular recognition system, ensuring that the nanopaints precisely hone in on the desired location [62–65]. The specificity achieved through these surface modifications enhances the precision of drug delivery, allowing the therapeutic payload to be delivered with unprecedented accuracy. For instance, in an experiment, Merisko-Liversidge et al. [66] employed wet media milling for the prominent particle size reduction approach successfully employed to formulate such problematic compounds. This water-based media milling process involves shear-fracturing micron-sized drug particles into nanometer-sized particles, resulting in stable nanoparticle dispersions with a mean diameter below 200 nm and 90% of particles below 400 nm. The formulation comprises water, drug, and generally recognized as safe (GRAS) excipients. Achieving drug concentrations of 300–400 mg g^{-1} with minimal stabilizer is feasible, with a typical drug-to-stabilizer ratio ranging from 2:1 to 20:1. These liquid nanodispersions demonstrate acceptable shelf life and can be postprocessed into various solid dosage forms. Nanoparticulate-based drug products have demonstrated enhanced bioavailability and increased drug exposure for both oral and parenteral formulations. This nanosuspension approach for formulating poorly water-soluble compounds has evolved from concept to realization in recent years, showcasing the platform's versatility and applicability in drug delivery. Similarly, Solano et al. [67] synthesized titanium dioxide (TiO_2) and zinc oxide (ZnO) nanoparticles via a green chemistry approach and thoroughly characterized them using various techniques. Commercial enamel-type paint was then modified with different concentrations (2, 3.5, and 5 w/v%) of these nanoparticles. The nanofilled paints were applied to carbon steel sheets, wood sheets, and aluminum disks, and their anticorrosive, self-cleaning, and antibacterial properties were assessed. Characterization results indicated that the synthesized TiO_2 and ZnO nanoparticles exhibited physicochemical properties comparable to those obtained through traditional methods.

Anticorrosion tests revealed that nanofilled paints formed an effective barrier even at low nanoparticle concentrations, minimizing surface agglomerates and reducing porosity. The proposed degradation mechanism demonstrated substantial methylene blue degradation in the presence of both nanoparticles, attributed to the high surface area provided by the nanoparticles. Under UV light, antibacterial activity was observed exclusively for ZnO nanoparticles, suggesting their diffusion into bacterial cell membranes and interference with normal cellular functions. These findings present a promising outlook for the modification of paints using TiO_2 and ZnO nanoparticles, showcasing potential applications in construction and even in textile fields.

This heightened level of precision has transformative implications for minimizing off-target effects. By guiding the nanopaints to the intended target cells, healthy tissues are spared from unnecessary exposure to the therapeutic agents. This not only reduces the risk of side effects but also improves the overall therapeutic index—the ratio of the desired therapeutic effect to the potential adverse effects. The ability to minimize off-target effects enhances the safety and tolerability of therapeutic interventions, a critical factor in the development of effective and patient-friendly treatments. Furthermore, the small size of nanoparticles is a pivotal factor in their success as carriers for targeted drug delivery. Their diminutive size allows them to navigate the intricate and dynamic biological environment with remarkable ease. This property is in stark contrast to larger drug carriers, which may encounter obstacles in reaching specific cellular or tissue targets. Nanopaints, with their optimized size, can traverse biological barriers more effectively, increasing the likelihood of reaching target sites that might be challenging for larger carriers. The combination of surface modifications and size optimization establishes nanopaints as versatile and efficient carriers for therapeutic agents. The tailored surface functionalization ensures a targeted approach, while the small size facilitates the efficient navigation through biological barriers [55,68,69]. This synergy allows nanopaints to deliver therapeutic payloads with unprecedented precision, opening new avenues for therapeutic strategies that were previously challenging or unattainable.

16.4.2 Navigating physiological barriers

Navigating physiological barriers presents a formidable obstacle in the realm of drug delivery, and overcoming these barriers is essential for the success of therapeutic interventions. The human body employs intricate defense mechanisms, such as the blood–brain barrier and various cellular barriers, which can impede the efficient delivery of therapeutic agents to specific sites. In this context, nanopaints emerge as innovative carriers equipped with unique properties to address these challenges [70–72]. The small size of nanoparticles is a key attribute that confers a distinct advantage in traversing biological barriers that would otherwise restrict larger drug carriers. Unlike conventional drug delivery systems, which may struggle to penetrate certain anatomical regions, nanopaints can navigate through these barriers with remarkable efficiency. This capability is particularly relevant in scenarios where the targeted delivery of therapeutic agents is crucial for achieving therapeutic efficacy. Surface modifications on nanopaints play a pivotal role in enhancing their ability to navigate physiological barriers. These modifications can be tailored to interact with specific transport mechanisms present at the barrier interfaces. For instance, nanopaints can be engineered to recognize and engage with receptor-mediated transport systems, facilitating their efficient crossing of barriers that would otherwise hinder drug delivery. This precision in surface engineering ensures that nanopaints interact selectively with the cellular and molecular components of physiological barriers, optimizing their passage to target sites.

The challenge of delivering therapeutic agents to the brain is emblematic of the difficulties posed by physiological barriers. The blood–brain barrier, a protective shield that regulates the passage of substances from the bloodstream to the brain, poses a significant hurdle in neurology. Nanopaints, with their small size and tailored surface modifications, offer a breakthrough in overcoming this challenge. By efficiently traversing the blood–brain barrier, nanopaints enable the delivery of therapeutic agents to the brain, opening new avenues for the treatment of neurological disorders. The ability of nanopaints to overcome physiological barriers has profound implications for enhancing the bioavailability of therapeutic agents at target sites [73–75]. This breakthrough not only improves the effectiveness of drug treatments but also broadens the spectrum of conditions that can be addressed through targeted drug delivery. Diseases that were historically challenging to treat due to barrier-related limitations, such as neurological disorders, can now be targeted more precisely, offering renewed hope for improved therapeutic outcomes.

16.4.3 Therapeutic payloads and precision medicine

Targeted drug delivery with nanopaints is intricately entwined with the transformative paradigm of precision medicine, a dynamic approach that tailors medical interventions to the individual characteristics of patients. The seamless alignment between nanopaints and precision medicine principles lies in their ability to deliver therapeutic payloads with unparalleled precision, ushering in a new era of personalized treatment strategies [76,77]. Nanopaints serve as versatile carriers capable of encapsulating a diverse range of therapeutic payloads, ranging from small molecules to nucleic acids. This versatility enables tailored interventions based on the specific characteristics of individual patients and the nuances of their diseases. In the context of cancer treatment, nanopaints exemplify the potential for precision medicine applications. These carriers can be meticulously engineered to target tumor-specific markers, facilitating the direct delivery of chemotherapeutic agents to cancer cells while sparing adjacent healthy tissues. The targeted approach enabled by nanopaints in cancer treatment holds significant promise for minimizing side effects associated with traditional chemotherapy. By precisely guiding therapeutic agents to malignant cells, nanopaints reduce the exposure of healthy tissues to cytotoxic drugs, mitigating adverse effects and improving the overall tolerability of the treatment. This targeted delivery not only enhances the safety profile of anticancer drugs but also amplifies their therapeutic efficacy.

Moreover, the adaptability of nanopaints offers a unique advantage in the integration of multiple therapeutic agents [78,79]. This capability aligns with the evolving understanding of disease complexity, especially in conditions where a single-agent approach may fall short. In diseases characterized by diverse molecular pathways or resistant subpopulations, nanopaints can be engineered to carry a combination of therapeutic payloads. This opens up avenues for innovative combination therapies that address the intricacies of diseases at multiple levels, reflecting the complexity inherent in conditions like cancer. The era of precision medicine emphasizes the individualized nature of healthcare, recognizing that patients exhibit unique responses to treatments based on their genetic makeup, physiological factors, and the specific characteristics of their diseases. In this context, nanopaints emerge as key contributors to the paradigm shift by offering a platform for tailoring drug delivery to the specific needs of each patient. The ability to encapsulate a diverse range of therapeutic agents and precisely deliver them to target sites aligns seamlessly with the core tenets of precision medicine. This personalized approach not only improves treatment outcomes but also fosters innovation in disease management. Nanopaints, as carriers of therapeutic payloads, empower healthcare professionals to navigate the intricacies of individual patient responses. By facilitating tailored interventions, nanopaints contribute to optimizing treatment strategies, enhancing therapeutic outcomes, and addressing diseases characterized by high variability among patients.

16.5 Current challenges and future directions

16.5.1 Limitations of nanopaint technology

While nanopaints present innovative solutions in various biomedical applications, there exist challenges that necessitate careful consideration. One primary limitation is the potential toxicity of certain nanoparticles. Ensuring the biocompatibility of nanopaints is critical, as unintended interactions with biological systems could lead to adverse effects. Moreover, the long-term fate and environmental impact of nanoparticles must be thoroughly investigated to assess their sustainability [80–82]. Another challenge lies in achieving consistent and reproducible nanopaint formulations. Variability in synthesis processes and the influence of external factors may result in batch-to-batch differences, hindering standardization. Overcoming these challenges requires stringent quality control measures and a deeper understanding of the factors influencing nanopaint properties.

Additionally, the complexity of biological environments poses hurdles to the precise targeting of nanopaints. Biological variability, dynamic physiological conditions, and the presence of interfering substances can affect the efficacy of targeted drug delivery and imaging applications. Enhancing the selectivity and stability of nanopaint interactions with biological entities is pivotal for overcoming these challenges.

16.5.2 Addressing safety and regulatory concerns

The integration of nanopaints into medical applications raises safety and regulatory concerns that warrant comprehensive scrutiny. Regulatory bodies must establish clear guidelines for the development, testing, and approval of nanopaint-based products [83–85]. Addressing safety concerns involves evaluating potential side effects, long-term toxicity, and the impact on different patient populations. Furthermore, the ethical considerations surrounding nanopaint technology should be carefully navigated. Transparent communication about potential risks, benefits, and uncertainties is essential, fostering informed decision-making by patients and healthcare professionals. Striking a balance between innovation and safety is crucial in establishing a robust regulatory framework that promotes the responsible development and deployment of nanopaint technologies.

16.5.3 Future prospects and emerging trends

Despite current challenges, the future of nanopaint technology holds promising prospects and is marked by emerging trends that could reshape biomedical applications. Continued research into novel nanoparticle formulations, such as multifunctional nanoparticles capable of simultaneous imaging and drug delivery, is anticipated [86–88]. These advanced formulations aim to enhance the versatility and efficacy of nanopaints in clinical settings. The integration of artificial intelligence (AI) and machine learning (ML) in nanopaint-based applications is an emerging trend with transformative potential. AI can analyze vast datasets generated by nanopaint-enhanced imaging techniques, aiding in pattern recognition, diagnostics, and treatment planning. This synergy between nanotechnology and AI holds the promise of more precise and data-driven medical interventions. Furthermore, advancements in nanotechnology research are expected to yield nanoparticles with enhanced functionalities, improved biocompatibility, and increased targeting specificity. Tailoring nanopaints for individualized medicine, where therapies are customized based on a patient's unique characteristics, represents a future direction that aligns with the principles of precision medicine.

16.6 Conclusion

The exploration of nanopaint technology in this chapter has unveiled its transformative potential in biomedical applications. Engineered with precision at the nanoscale, nanopaints exhibit unique attributes suitable for diverse purposes, spanning from medical imaging to targeted drug delivery. Key findings underscore their significance in enhancing contrast for medical imaging modalities, overcoming physiological barriers for precise drug delivery, and contributing to the evolving landscape of precision medicine. The ability to tailor nanopaint formulations, exemplified in applications like cancer treatment, showcases the versatility and promise of this technology. The implications for biomedical research and practice are profound, with nanopaints poised to revolutionize diagnostic capabilities, offer more effective and personalized therapeutic interventions, and usher in an era of patient-centric healthcare. However, as nanopaint technology progresses, addressing safety concerns and establishing guidelines for ethical integration into clinical practice are crucial.

References

[1] T.C. Dzogbewu, W.B. du Preez, Additive manufacturing of titanium-based implants with metal-based antimicrobial agents, Metals 11 (3) (2021) 1–12. Available from: https://doi.org/10.3390/met11030453.

[2] S.C. Liao, C.T. Chang, C.Y. Chen, C.H. Lee, W.L. Lin, Functionalization of pure titanium MAO coatings by surface modifications for biomedical applications, Surface and Coatings Technology 394 (2020). Available from: https://doi.org/10.1016/j.surfcoat.2020.125812.

[3] T. Hanawa, Titanium-tissue interface reaction and its control with surface treatment, Frontiers in Bioengineering and Biotechnology 7 (2019). Available from: https://doi.org/10.3389/fbioe.2019.00170.

[4] X. Fan, L. Yahia, E. Sacher, Antimicrobial properties of the Ag, Cu nanoparticle system, Biology 10 (2) (2021) 1–37. Available from: https://doi.org/10.3390/biology10020137.

[5] A. Kurup, P. Dhatrak, N. Khasnis, Surface modification techniques of titanium and titanium alloys for biomedical dental applications: a review, Materials Today: Proceedings 39 (2020) 84–90. Available from: https://doi.org/10.1016/j.matpr.2020.06.163.

[6] Q. Wang, P. Zhou, S. Liu, et al., Multi-scale surface treatments of titanium implants for rapid osseointegration: a review, Nanomaterials 10 (6) (2020) 1–27. Available from: https://doi.org/10.3390/nano10061244.

[7] M. Chen, H. Li, X. Wang, G. Qin, E. Zhang, Improvement in antibacterial properties and cytocompatibility of titanium by fluorine and oxygen dual plasma-based surface modification, Applied Surface Science 463 (2019) 261–274. Available from: https://doi.org/10.1016/j.apsusc.2018.08.194.

[8] S. Tortorella, V.V. Buratti, M. Maturi, L. Sambri, M.C. Franchini, E. Locatelli, Surface-modified nanocellulose for application in biomedical engineering and nanomedicine: a review, International Journal of Nanomedicine 15 (2020) 9909–9937. Available from: https://doi.org/10.2147/IJN.S266103.

[9] M.-K. Lee, H. Lee, H.-E. Kim, E.-J. Lee, T.-S. Jang, H.-D. Jung, Nano-topographical control of Ti-Nb-Zr alloy surfaces for enhanced osteoblastic response, Nanomaterials 11 (6) (2021). Available from: https://doi.org/10.3390/nano11061507.

[10] M. John, P.R. Kalvala, M. Misra, P.L. Menezes, Peening techniques for surface modification: processes, properties and applications, Materials 14 (14) (2021). Available from: https://doi.org/10.3390/ma14143841.

[11] C. Ungureanu, C. Dumitriu, S. Popescu, et al., Enhancing antimicrobial activity of TiO_2/Ti by torularhodin bioinspired surface modification, Bioelectrochemistry (Amsterdam, Netherlands) 107 (2016) 14–24. Available from: https://doi.org/10.1016/j.bioelechem.2015.09.001.

[12] A. Mahapatro, Bio-functional nano-coatings on metallic biomaterials, Materials Science and Engineering C 55 (2015) 227–251. Available from: https://doi.org/10.1016/j.msec.2015.05.018.

[13] P. Šugár, B. Ludrovcová, M.H. Kalbáčová, J. Šugárová, M. Sahul, J. Kováčik, Laser surface modification of powder metallurgy-processed Ti-graphite composite which can enhance cells' osteo-differentiation, Materials 14 (20) (2021). Available from: https://doi.org/10.3390/ma14206067.
[14] A. Woźniak, M. Adamiak, G. Chladek, M. Bonek, W. Walke, O. Bialas, The influence of hybrid surface modification on the selected properties of cp titanium grade ii manufactured by selective laser melting, Materials 13 (12) (2020) 1–22. Available from: https://doi.org/10.3390/ma13122829.
[15] A.A. John, S.K. Jaganathan, E. Supriyanto, A. Manikandan, Surface modification of titanium and its alloys for the enhancement of osseointegration in orthopaedics, Current Science 111 (6) (2016) 1003–1015. Available from: https://doi.org/10.18520/cs/v111/i6/1003-1015.
[16] Y. Zhao, K.W.K. Yeung, P.K. Chu, Functionalization of biomedical materials using plasma and related technologies, Applied Surface Science 310 (2014) 11–18. Available from: https://doi.org/10.1016/j.apsusc.2014.02.168.
[17] S. Kargozar, F. Kermani, S.M. Beidokhti, et al., Functionalization and surface modifications of bioactive glasses (BGs): tailoring of the biological response working on the outermost surface layer, Materials 12 (22) (2019). Available from: https://doi.org/10.3390/ma12223696.
[18] C. Celesti, T. Gervasi, N. Cicero, et al., Titanium surface modification for implantable medical devices with anti-bacterial adhesion properties, Materials 15 (9) (2022). Available from: https://doi.org/10.3390/ma15093283.
[19] X. Yan, W. Cao, H. Li, Biomedical alloys and physical surface modifications: a mini-review, Materials 15 (1) (2022). Available from: https://doi.org/10.3390/ma15010066.
[20] M.K. Sung, J. Rho, I.S. Choi, P.B. Messersmith, H. Lee, Norepinephrine: material-independent, multifunctional surface modification reagent, Journal of the American Chemical Society 131 (37) (2009) 13224–13225. Available from: https://doi.org/10.1021/ja905183k.
[21] Y. Kirmanidou, M. Sidira, M.E. Drosou, et al., New Ti-alloys and surface modifications to improve the mechanical properties and the biological response to orthopedic and dental implants: a review, BioMed Research International 2016 (2016). Available from: https://doi.org/10.1155/2016/2908570.
[22] J. Xu, J. Zhang, Y. Shi, et al., Surface modification of biomedical Ti and Ti alloys: a review on current advances, Materials 15 (5) (2022). Available from: https://doi.org/10.3390/ma15051749.
[23] H.S. Jung, T. Lee, I.K. Kwon, H.S. Kim, S.K. Hahn, C.S. Lee, Surface modification of multipass caliber-rolled ti alloy with dexamethasone-loaded graphene for dental applications, ACS Applied Materials and Interfaces 7 (18) (2015) 9598–9607. Available from: https://doi.org/10.1021/acsami.5b03431.
[24] Y. Okazaki, S.-i Katsuda, Biological safety evaluation and surface modification of biocompatible Ti–15Zr–4Nb alloy, Materials 14 (4) (2021). Available from: https://doi.org/10.3390/ma14040731.
[25] A. Fattahi, P. Koohsari, M.S. Lakmehsari, K. Ghandi, The impact of the surface modification on tin-doped indium oxide nanocomposite properties, Nanomaterials 12 (1) (2022). Available from: https://doi.org/10.3390/nano12010155.
[26] B. Chico, B.T. Pérez-maceda, S.S. José, M.L. Escudero, M.C. García-alonso, R.M. Lozano, Corrosion behaviour and j774a.1 macrophage response to hyaluronic acid functionalization of electrochemically reduced graphene oxide on biomedical grade cocr, Metals 11 (7) (2021). Available from: https://doi.org/10.3390/met11071078.

[27] O.K. Wade, J.B. Woehrstein, P.C. Nickels, et al., 124-Color Super-resolution imaging by engineering DNA-PAINT blinking kinetics, Nano Letters 19 (4) (2019) 2641–2646. Available from: https://doi.org/10.1021/acs.nanolett.9b00508.
[28] P. Gowtham, V. Haribabu, A.D. Prabhu, P. Pallavi, K. Girigoswami, A. Girigoswami, Impact of nanovectors in multimodal medical imaging, Nanomedicine Journal 9 (2) (2022) 107–130. Available from: https://doi.org/10.22038/NMJ.2022.63459.1664.
[29] M. Movahedi, A. Abdi, A. Mehdizadeh, et al., Novel paint design based on nanopowder to protection against X and gamma rays, Indian Journal of Nuclear Medicine 29 (1) (2014) 18–21. Available from: https://doi.org/10.4103/0972-3919.125763.
[30] I. Demetrescu, C. Dumitriu, G. Totea, C.I. Nica, A. Dinischiotu, D. Ionita, Zwitterionic cysteine drug coating influence in functionalization of implantable Ti50Zr alloy for antibacterial, biocompatibility and stability properties, Pharmaceutics 10 (4) (2018). Available from: https://doi.org/10.3390/pharmaceutics10040220.
[31] M. Delfi, M. Ghomi, A. Zarrabi, et al., Functionalization of polymers and nanomaterials for biomedical applications: antimicrobial platforms and drug carriers, Prosthesis 2 (2) (2020) 117–139. Available from: https://doi.org/10.3390/prosthesis2020012.
[32] M.B. Berger, P. Slosar, Z. Schwartz, et al., A review of biomimetic topographies and their role in promoting bone formation and osseointegration: implications for clinical use, Biomimetics 7 (2) (2022). Available from: https://doi.org/10.3390/biomimetics7020046.
[33] P. Biehl, M. von der Lühe, S. Dutz, F.H. Schacher, Synthesis, characterization, and applications of magnetic nanoparticles featuring polyzwitterionic coatings, Polymers 10 (1) (2018). Available from: https://doi.org/10.3390/polym10010091.
[34] O.A. Noqta, A.A. Aziz, I.A. Usman, M. Bououdina, Recent Advances in iron oxide nanoparticles (IONPs): synthesis and surface modification for biomedical applications, Journal of Superconductivity and Novel Magnetism 32 (4) (2019) 779–795. Available from: https://doi.org/10.1007/s10948-018-4939-6.
[35] L. Guo, R. Smeets, L. Kluwe, et al., Cytocompatibility of titanium, zirconia and modified PEEK after surface treatment using UV light or non-thermal plasma, International Journal of Molecular Sciences 20 (22) (2019). Available from: https://doi.org/10.3390/ijms20225596.
[36] R. Quiñones, D. Shoup, G. Behnke, et al., Study of perfluorophosphonic acid surface modifications on zinc oxide nanoparticles, Materials 10 (12) (2017). Available from: https://doi.org/10.3390/ma10121363.
[37] Z. Ma, Z. Mao, C. Gao, Surface modification and property analysis of biomedical polymers used for tissue engineering, Colloids and Surfaces B: Biointerfaces 60 (2) (2007) 137–157. Available from: https://doi.org/10.1016/j.colsurfb.2007.06.019.
[38] Z. Zhen, J. Xie, Development of manganese-based nanoparticles as contrast probes for magnetic resonance imaging, Theranostics 2 (1) (2012) 45–54. Available from: https://doi.org/10.7150/thno.3448.
[39] A. Ashokan, V.H. Somasundaram, G.S. Gowd, et al., Biomineral nano-theranostic agent for magnetic resonance image guided, augmented radiofrequency ablation of liver tumor, Scientific Reports 7 (1) (2017). Available from: https://doi.org/10.1038/s41598-017-14976-8.

[40] J.H. Maeng, D.H. Lee, K.H. Jung, et al., Multifunctional doxorubicin loaded superparamagnetic iron oxide nanoparticles for chemotherapy and magnetic resonance imaging in liver cancer, Biomaterials 31 (18) (2010) 4995–5006. Available from: https://doi.org/10.1016/j.biomaterials.2010.02.068.
[41] K. Kono, S. Nakashima, D. Kokuryo, et al., Multi-functional liposomes having temperature-triggered release and magnetic resonance imaging for tumor-specific chemotherapy, Biomaterials 32 (5) (2011) 1387–1395. Available from: https://doi.org/10.1016/j.biomaterials.2010.10.050.
[42] A. Thakur, S. Kaya, A. Kumar, Recent trends in the characterization and application progress of nano-modified coatings in corrosion mitigation of metals and alloys, Applied Sciences (Switzerland) 13 (2) (2023). Available from: https://doi.org/10.3390/app13020730.
[43] S. Kaya, H. Lgaz, A. Thakkur, et al., Molecular insights into the corrosion inhibition mechanism of omeprazole and tinidazole: a theoretical investigation, Molecular Simulation 49 (17) (2023) 1632–1646. Available from: https://doi.org/10.1080/08927022.2023.2256888.
[44] D. Sharma, A. Thakur, M.K. Sharma, et al., Effective corrosion inhibition of mild steel using novel 1,3,4-oxadiazole-pyridine hybrids: synthesis, electrochemical, morphological, and computational insights, Environmental Research 234 (2023). Available from: https://doi.org/10.1016/j.envres.2023.116555.
[45] A. Thakur, A. Kumar, Unraveling the multifaceted mechanisms and untapped potential of activated carbon in remediation of emerging pollutants: a comprehensive review and critical appraisal of advanced techniques, Chemosphere 346 (2024). Available from: https://doi.org/10.1016/j.chemosphere.2023.140608.
[46] A. Thakur, A. Kumar, A. Singh, Adsorptive removal of heavy metals, dyes, and pharmaceuticals: carbon-based nanomaterials in focus, Carbon 217 (2024). Available from: https://doi.org/10.1016/j.carbon.2023.118621.
[47] S. Demirci, R. Dalmış, T. Dikici, M.M. Tünçay, N. Kaya, A.N. Güllüoğlu, Effect of surface modifications of additively manufactured Ti-6Al-4V alloys on apatite formation ability for biomedical applications, Journal of Alloys and Compounds 887 (2021). Available from: https://doi.org/10.1016/j.jallcom.2021.161445.
[48] D.B. Panemangalore, R. Shabadi, M. Gupta, Corrosion behavior, microstructure and mechanical properties of novel mg-zn-ca-er alloy for bio-medical applications, Metals 11 (3) (2021) 1–16. Available from: https://doi.org/10.3390/met11030519.
[49] A. Bonnin, J.P. Wright, R. Tucoulou, H. Palancher, Impurity precipitation in atomized particles evidenced by nano x-ray diffraction computed tomography, Applied Physics Letters 105 (8) (2014). Available from: https://doi.org/10.1063/1.4894009.
[50] E. Chatzipetros, Z. Yfanti, P. Christopoulos, et al., Imaging of nano-hydroxyapatite/chitosan scaffolds using a cone beam computed tomography device on rat calvarial defects with histological verification, Clinical Oral Investigations 24 (1) (2020) 437–446. Available from: https://doi.org/10.1007/s00784-019-02939-4.
[51] Y. Park, J. Jung, M. Chang, Research progress on conducting polymer-based biomedical applications, Applied Sciences (Switzerland) 9 (6) (2019). Available from: https://doi.org/10.3390/app9061070.
[52] P. Szymczyk-Ziółkowska, V. Hoppe, M. Rusińska, et al., The impact of ebm-manufactured ti6al4v eli alloy surface modifications on cytotoxicity toward

eukaryotic cells and microbial biofilm formation, Materials 13 (12) (2020) 1–21. Available from: https://doi.org/10.3390/ma13122822.
[53] R. Ravichandran, S. Sundarrajan, J.R. Venugopal, S. Mukherjee, S. Ramakrishna, Applications of conducting polymers and their issues in biomedical engineering, Journal of the Royal Society Interface 7 (5) (2010) S559–S579. Available from: https://doi.org/10.1098/rsif.2010.0120.focus.
[54] A. Kumar, A. Thakur, Encapsulated nanoparticles in organic polymers for corrosion inhibition, Corrosion Protection at the Nanoscale, Elsevier, India, 2020, pp. 345–362. Available from: https://doi.org/10.1016/B978-0-12-819359-4.00018-0.
[55] A. Thakur, S. Sharma, R. Ganjoo, H. Assad, A. Kumar, Anti-corrosive potential of the sustainable corrosion inhibitors based on biomass waste: a review on preceding and perspective research, Journal of Physics: Conference Series 2267 (1) (2022). Available from: https://doi.org/10.1088/1742-6596/2267/1/012079.
[56] C. Verma, A. Thakur, R. Ganjoo, et al., Coordination bonding and corrosion inhibition potential of nitrogen-rich heterocycles: azoles and triazines as specific examples, Coordination Chemistry Reviews 488 (2023). Available from: https://doi.org/10.1016/j.ccr.2023.215177.
[57] A. Thakur, A. Kumar, Recent trends in nanostructured carbon-based electrochemical sensors for the detection and remediation of persistent toxic substances in real-time analysis, Materials Research Express 10 (3) (2023). Available from: https://doi.org/10.1088/2053-1591/acbd1a.
[58] S.Yo Park, H.Uk Lee, Y.C. Lee, et al., Eco-friendly carbon-nanodot-based fluorescent paints for advanced photocatalytic systems, Scientific Reports 5 (2015). Available from: https://doi.org/10.1038/srep12420.
[59] M.K. Mishra, A. Chakravarty, K. Bhowmik, G. De, Carbon nanodot-ORMOSIL fluorescent paint and films, Journal of Materials Chemistry C 3 (4) (2015) 714–719. Available from: https://doi.org/10.1039/c4tc02140a.
[60] E. Taylor, T.J. Webster, Reducing infections through nanotechnology and nanoparticles, International Journal of Nanomedicine 6 (2011) 1463–1473. Available from: https://doi.org/10.2147/ijn.s22021.
[61] S. Ferraris, S. Spriano, G. Pan, et al., Surface modification of Ti–6Al–4V alloy for biomineralization and specific biological response: part I, inorganic modification, Journal of Materials Science: Materials in Medicine 22 (3) (2011) 533–545. Available from: https://doi.org/10.1007/s10856-011-4246-2.
[62] C. Dhonchak, N. Agnihotri, Computational insights in the spectrophotometrically 4H-chromen-4-one complex using DFT method, Biointerface Research in Applied Chemistry 13 (2023).
[63] S. Kaya, A. Thakur, A. Kumar, The role of in Silico/DFT investigations in analyzing dye molecules for enhanced solar cell efficiency and reduced toxicity, Journal of Molecular Graphics and Modelling 124 (2023). Available from: https://doi.org/10.1016/j.jmgm.2023.108536.
[64] A. Thakur, A. Kumar, Ecotoxicity analysis and risk assessment of nanomaterials for the environmental remediation, Macromolecular Symposia 410 (1) (2023). Available from: https://doi.org/10.1002/masy.202100438.
[65] D. Sharma, A. Thakur, M.K. Sharma, et al., Synthesis, electrochemical, morphological, computational and corrosion inhibition studies of 3-(5-naphthalen-2-yl-[1,3,4]oxadiazol-2-yl)-pyridine against mild steel in 1 M HCl, Asian Journal of Chemistry 35 (5) (2023) 1079–1088. Available from: https://doi.org/10.14233/ajchem.2023.27711.

[66] E. Merisko-Liversidge, G.G. Liversidge, Nanosizing for oral and parenteral drug delivery: a perspective on formulating poorly-water soluble compounds using wet media milling technology, Advanced Drug Delivery Reviews 63 (6) (2011) 427–440. Available from: https://doi.org/10.1016/j.addr.2010.12.007.

[67] R. Solano, D. Patiño-Ruiz, A. Herrera, Preparation of modified paints with nano-structured additives and its potential applications, Nanomaterials and Nanotechnology 10 (2020). Available from: https://doi.org/10.1177/1847980420909188.

[68] A. Thakur, A. Kumar, S. Kaya, et al., Electrochemical and computational investigations of the *Thysanolaena latifolia* leaves extract: an eco-benign solution for the corrosion mitigation of mild steel, Results in Chemistry 6 (2023). Available from: https://doi.org/10.1016/j.rechem.2023.101147.

[69] A. Thakur, A. Kumar, S. Sharma, R. Ganjoo, H. Assad, Computational and experimental studies on the efficiency of *Sonchus arvensis* as green corrosion inhibitor for mild steel in 0.5 M HCl solution, Materials Today: Proceedings 66 (2022) 609–621. Available from: https://doi.org/10.1016/j.matpr.2022.06.479.

[70] P.C. Chiang, C.W. Chen, F.T. Tsai, C.K. Lin, C.C. Chen, Hard anodization film on carbon steel surface by thermal spray and anodization methods, Materials 14 (13) (2021). Available from: https://doi.org/10.3390/ma14133580.

[71] J. Quinn, R. McFadden, C.-W. Chan, L. Carson, Titanium for orthopedic applications: an overview of surface modification to improve biocompatibility and prevent bacterial biofilm formation, iScience 23 (11) (2020). Available from: https://doi.org/10.1016/j.isci.2020.101745.

[72] H. Chouirfa, H. Bouloussa, V. Migonney, C. Falentin-Daudré, Review of titanium surface modification techniques and coatings for antibacterial applications, Acta Biomaterialia 83 (2019) 37–54. Available from: https://doi.org/10.1016/j.actbio.2018.10.036.

[73] B. Arifa Farzana, N. Mujafarkani, A. Thakur, A. Kumar, A. Mushira Banu, M. Shifana, Evaluating (p-semidine-guanidine-formaldehyde) terpolymer resin efficiency as anti-corrosive agent for mild steel in 1 M H_2SO_4: an experimental and computational approach, Inorganic Chemistry Communications 158 (2023). Available from: https://doi.org/10.1016/j.inoche.2023.111572.

[74] A. Thakur, A. Kumar, Exploring the potential of ionic liquid-based electrochemical biosensors for real-time biomolecule monitoring in pharmaceutical applications: from lab to life, Results in Engineering 20 (2023) 101533. Available from: https://doi.org/10.1016/j.rineng.2023.101533.

[75] A. Thakur, S. Kaya, A. Kumar, Recent innovations in nano container-based self-healing coatings in the construction industry, Current Nanoscience 18 (2) (2022) 203–216. Available from: https://doi.org/10.2174/1573413717666210216120741.

[76] S. Chauhan, L.S.B. Upadhyay, Biosynthesis of iron oxide nanoparticles using plant derivatives of *Lawsonia inermis* (Henna) and its surface modification for biomedical application, Nanotechnology for Environmental Engineering 4 (1) (2019). Available from: https://doi.org/10.1007/s41204-019-0055-5.

[77] L.T. De Jonge, S.C.G. Leeuwenburgh, J.G.C. Wolke, J.A. Jansen, Organic-inorganic surface modifications for titanium implant surfaces, Pharmaceutical Research 25 (10) (2008) 2357–2369. Available from: https://doi.org/10.1007/s11095-008-9617-0.

[78] J. Kim, H. Lee, T.S. Jang, et al., Characterization of titanium surface modification strategies for osseointegration enhancement, Metals 11 (4) (2021) 618. Available from: https://doi.org/10.3390/met11040618.

[79] I. Miranda, A. Souza, P. Sousa, et al., Properties and applications of PDMS for biomedical engineering: a review, Journal of Functional Biomaterials 13 (1) (2022) 2. Available from: https://doi.org/10.3390/jfb13010002.

[80] S. Mohammadzadeh Asl, M. Ganjali, M. Karimi, Surface modification of 316L stainless steel by laser-treated HA-PLA nanocomposite films toward enhanced biocompatibility and corrosion-resistance in vitro, Surface and Coatings Technology 363 (2019) 236–243. Available from: https://doi.org/10.1016/j.surfcoat.2019.02.052.

[81] E. Saborowski, A. Dittes, T. Lindner, T. Lampke, Nickel-aluminum thermal spray coatings as adhesion promoter and susceptor for inductively joined polymer-metal hybrids, Polymers 13 (8) (2021) 1320. Available from: https://doi.org/10.3390/polym13081320.

[82] A.R. Mahdavian, M.A.S. Mirrahimi, Efficient separation of heavy metal cations by anchoring polyacrylic acid on superparamagnetic magnetite nanoparticles through surface modification, Chemical Engineering Journal 159 (1-3) (2010) 264–271. Available from: https://doi.org/10.1016/j.cej.2010.02.041.

[83] A. Saberi, H.R. Bakhsheshi-Rad, S. Abazari, et al., A comprehensive review on surface modifications of biodegradable magnesium-based implant alloy: polymer coatings opportunities and challenges, Coatings 11 (7) (2021). Available from: https://doi.org/10.3390/coatings11070747.

[84] C. Popescu, S. Alain, M. Courant, A. Vardelle, A. Denoirjean, M. Cavarroc, Thermal spray copper-based coatings against contamination of thermoplastic surfaces: a systematic review, Engineering Science and Technology, An International Journal 35 (2022). Available from: https://doi.org/10.1016/j.jestch.2022.101194.

[85] S. Abazari, A. Shamsipur, H.R. Bakhsheshi-Rad, S. Ramakrishna, F. Berto, Graphene family nanomaterial reinforced magnesium-based matrix composites for biomedical application: a comprehensive review, Metals 10 (8) (2020) 1–40. Available from: https://doi.org/10.3390/met10081002.

[86] K. Zhang, T. Liu, J.A. Li, J.Y. Chen, J. Wang, N. Huang, Surface modification of implanted cardiovascular metal stents: from antithrombosis and antirestenosis to endothelialization, Journal of Biomedical Materials Research - Part A 102 (2) (2014) 588–609. Available from: https://doi.org/10.1002/jbm.a.34714.

[87] S. Cai, C. Wu, W. Yang, W. Liang, H. Yu, L. Liu, Recent advance in surface modification for regulating cell adhesion and behaviors, Nanotechnology Reviews 9 (1) (2020) 971–989. Available from: https://doi.org/10.1515/ntrev-2020-0076.

[88] S. Yang, O.W. Dillon, D.A. Puleo, I.S. Jawahir, Enhancement of wear resistance for improved functional performance of Co-Cr-Mo hip implants through cryogenic surface treatment: a case study, Machining Science and Technology 25 (3) (2021) 455–476. Available from: https://doi.org/10.1080/10910344.2021.1903924.

17

Nanomaterials and nanotechnology based coatings for stealth applications

R.V. Lakshmi, Parthasarathi Bera and Harish C. Barshilia
Surface Engineering Division, CSIR-National Aerospace Laboratories, Bengaluru, Karnataka, India

17.1 Introduction

Stealth or low-observable technology is a military concept which reduces the visibility of aircrafts and makes them difficult to be detected by enemy sensors. This is often misunderstood in a way that aircraft incorporating this technology become invisible to the enemy. In fact, stealth only reduces aircraft visibility by making their appearance or signature small. The technology majorly caters to the military projects.

Stealth technology started with fighter jets. Later, this technology has been used in warships, submarines, and even military helicopters. The search for this technology began almost eight decades ago during World War II. Hitler's Germany built an experimental fighter plane made of transparent fiber to attack without being noticed by the enemy. Because of low weight and easy observability in the daytime, it was not implemented. Incidentally, it was during World War II that radio detection and ranging (RADAR) was first used to identify enemy aircraft. The development of radar, which can detect distant objects like ships and airplanes, drastically transformed military surveillance and defense. But as soon as radar was put into use, studies into ways to block or avoid its detection started. Nazi Germany was also a pioneer in this regard. Aeronautical engineers in Germany tested and found that the radio waves emitted by the radar bounced off the main body of the aircraft. After this, attempts were made to shorten the body of the aircraft and lengthen the wings. In the last phase of World War II, they built a fighter plane named 'Horten 229' to evade radar surveillance. But before it could be used in war, Hitler was defeated.

Nanopaints. DOI: https://doi.org/10.1016/B978-0-443-26512-9.00018-5

The rush to develop and combat radar technology has escalated after the war as improved knowledge and its tactical importance grew over the years. Researchers have been exploring the subtleties of "radar echoes" in several laboratories across the globe. However, a significant turning point in stealth technology occurred in the 1980s. After decades of study and testing, results were ultimately obtained, particularly in the USA during the Cold War with the Soviet Union. In this context, the USA unveiled the prototype airplanes that were nearly imperceptible to radar due to unique designs, cutting-edge materials, and sophisticated design procedures. Redefining the art of aerial warfare through a unique blend of science, engineering, and strategy, the prototype stealth bomber served as a symbol of this success. As a result, several modern stealth aircraft, warships, and helicopters have been developed in recent decades. Some examples are the combat-ready stealth aircraft (Northrop Grumman B-2 Spirit, the Lockheed Martin F-22 Raptor, the Lockheed Martin F-35 Lightning II, Lockheed F-117 Nighthawk, B-2 Spirit, Chengdu J-20, Sukhoi Su-57), stealth ships (American Sea Shadow (IX-529), Swedish Visby class corvette, Indonesian Klewang-class, Independence-class littoral combat ship of US Navy, French multirole ship design Gowind-class design, and Chinese Type 055 destroyer), helicopters (Hughes 500P, RAH-66 Comanche, and MH-60 Black Hawk stealth helicopter), and submarines (German Type 212A).

17.2 Stealth technologies

Stealth technology includes various methods to make objects less visible (detectable) from radar and other sensors. This technology, if on a high level, provides a significant advantage over an opponent by enabling the user to carry out a striking military attack with increased survivability. Stealth is not a single technology: there are various hidden options for making aircraft or other objects low-observable. Stealth is exploited to camouflage objects, absorb or reflect some of the incoming radar waves, reduce the IR (inherent) signature, or to avoid LIDAR (Light Detection and Ranging) [1,2]. The objects are classified as "Target" and "Clutter." Targets are the objects of interest for detection, like aircraft, missiles, and satellites, while clutters are the objects, like buildings, that are often confused as targets.

Stealth technologies have been under continuous development to impart maximum possible invisibility for a stealthy aircraft. They can be broadly categorized into five types, namely,

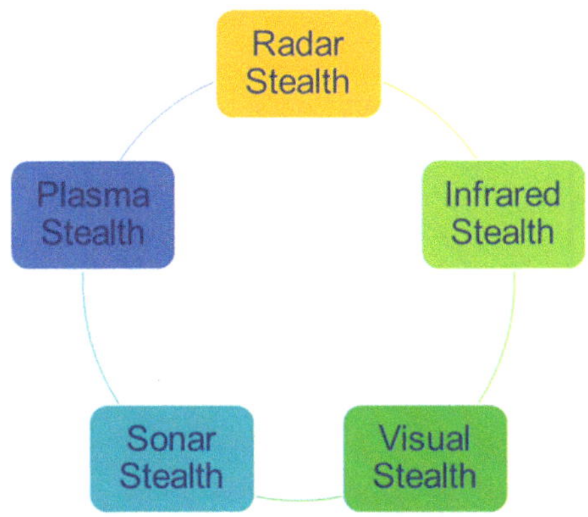

Figure 17.1 Classification of stealth technologies.

radar stealth, infrared stealth, visual stealth, sonar stealth, and plasma stealth, which are shown in Fig. 17.1. This section focuses on the basic types of stealth and the prime methods of making them low-observable.

17.2.1 Radar stealth

RADAR is an acronym for "radio detection and ranging." Microwave (MW) which is a form of electromagnetic (EM) wave is used in radar system. The main working principle of radar is that it sends electromagnetic pulses in the direction of the target at the speed of light. Upon encountering an object, which could be a target, they are reflected back. Radar computes the time taken by the pulse of energy to hit the target and return back to the radar. With this, the distance between the target and radar is estimated. To counter this technique, radar stealth is employed. The main mission of this technology is to reduce radar cross-section (RCS). Two techniques are primarily used to minimize RCS. One is based on vehicle shaping, restructuring the frame, and changing geometric design to achieve the desired RCS level. Subsequently, effective structures, known as "radar absorbent structures" (RAS), are created [3]. The other refers to radar-absorbing materials (RAM) and is linked with designing materials capable of reducing the reflectivity of the airframe. RAMs are generally coated on the external or internal reflective metal surfaces of military vehicles, aircraft, and ships, which reduces the RCS of these targets, thus making them difficult to detect by enemy radars. The typical RCS values of certain objects or targets, such as aircraft or ships, are indicated in

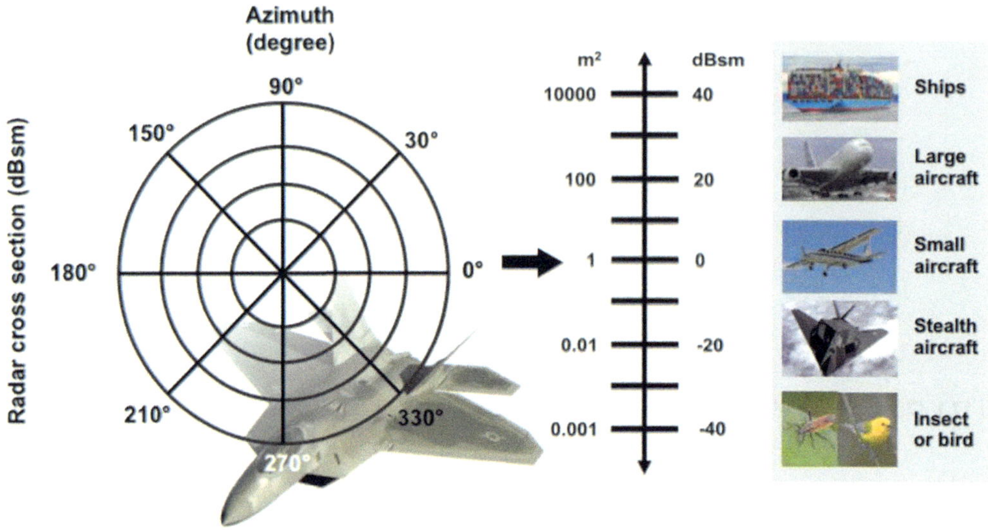

Figure 17.2 The radar cross-section values of various objects or targets in square meter and decibel square meter. Reproduced with permission from S.-H. Kim, S.-Y. Lee, Y. Zhang, S.-J. Park, J. Gu, Carbon-based radar absorbing materials toward stealth technologies, Advanced Sciences 10 (32) (2023) 2303104. https://doi.org/10.1002/advs.202303104, http://onlinelibrary.wiley.com/journal/10.1002/(ISSN)2198-3844.

Fig. 17.2. When the RCS is low, there is a lower chance of being detected by the enemy during a combat mission and a greater chance of infiltrating enemy territory.

Vehicle shape is the basic aspect of every construction. It turns out that it also has a huge impact on low-observable technology. Since the invention of radars, they have had a huge impact on the extensiveness of war. For example, in the Battle of Britain, engineers recognized the need to reduce the RCS of aircraft. Since the 1960s, it has been known that vehicle shape changes detectability. The British Avro Vulcan bomber had an incredibly small appearance on radar, despite its large size. Besides eliminating propellers in a stealth airplane, the design of the internal construction is vital.

The idea of re-entrant triangles under the skin of aircraft was first used in the F-117, as shown in Fig. 17.3(A). Radar waves penetrating the "skin" of an airplane get trapped in these structures and bounce inside, thus losing energy. Tilting the tail surface was also reported to reduce corner reflections that occur between them. Going further, the best effect is achieved when the tail is eliminated completely, as in the B-2 Spirit aircraft presented in Fig. 17.3(B). Furthermore, stealth-designed aircraft must bury elements like the engine and fuselage, and external

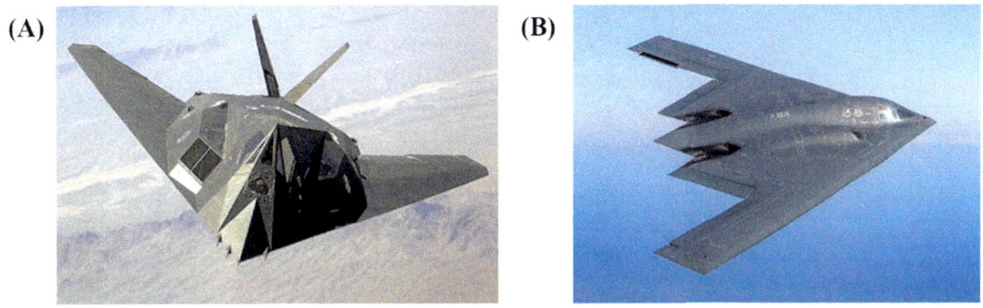

Figure 17.3 Aircraft with stealth achieved by shaping (A) F-117 and (B) B-2 Spirit. From Wikipedia.

weapons or fuel tanks cannot be attached to the undercarriage because such elements reflect radar waves back very well.

Use of radar-absorbing materials and radar-absorbing structures is the second possible solution to achieve radar stealth. When electromagnetic radiation impacts an object, it can be transmitted, absorbed, or reflected. The geometric shape and material characteristics determine the different magnitudes of these quantities. RAMs are a special class of materials, which are chemically designed to absorb incident radio frequency signals. The MW radiation energy is converted to heat, which gets dissipated. This attenuates the reflected radar wave and alters the wavelength. Fillers can be classified as dielectric and magnetic RAMs. The higher the effectiveness of RAM, the lower the resulting reflected EM radiation level will be. RAM is also applied in the form of a paint termed radar-absorbing paint (RAP), comprising an insulating binder (generally a polymeric matrix) and fillers that are electrically conductive and magnetic in nature (Fig. 17.4). The polymers used in these materials are generally EM transparent. They are used to diffuse the energy of the radar wave, preventing the signal from returning to the antenna. More information on RAP will be provided in the later section. It is to be noted that the global radar-absorbing materials market was $470 million in 2021, which is expected to grow to $630 million by 2030 with a compound annual growth rate (CAGR) of 3.8%. This growth is attributed to rising national security demands in terms of advanced aerial combat capabilities and the making of cutting-edge electronic warfare equipment. This growth is further going to see an increase due to recent rising demands for stealth unmanned aerial vehicles (UAVs) in the military market.

Another interesting technique used in radar stealth technology is the use of a water curtain. This is only for warships.

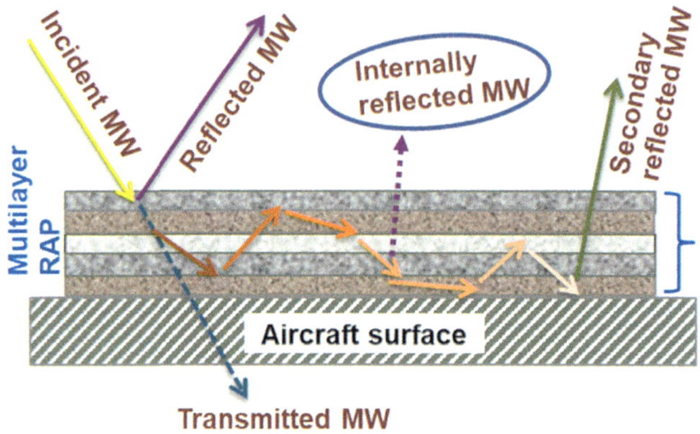

Figure 17.4 Schematic of the functioning of radar-absorbing paint and energy dissipation.

The Visby corvette was the first such ship to enter service. When designing a ship with a reduced radar signature, it is crucial to remove radar beams originating above the horizon. In addition, vertical shapes must be avoided, as they would perfectly reflect such a beam back to the emitter. In such cases, adaptive water curtain technology (AWCT) is used. Its main target is to deflect or scatter radar waves and, as a result, reduce the RCS of the ship. It uses high-conductive seawater to effectively create an angled radar-reflective curtain. The ship command can modulate the water curtain, to appear as "sea clutter." However, satellite data and local conditions must be taken into consideration, as trying to mimic sea clutter when satellite data show clear weather would be ineffective.

17.2.2 Infrared stealth

Infrared (IR) is one kind of electromagnetic wave having the wavelength of $0.76-1000\ \mu m$. IR radiation from objects is typically attenuated due to physical processes such as refraction, absorption, and scattering. However, IR rays of specific wave bands are capable of propagating through the atmosphere. This is termed as an atmospheric attenuation band or atmospheric IR window. Relatively little absorption of terrestrial thermal radiation by atmospheric gases occurs in this region of the IR spectrum. They lie in the region of wavelengths $0.76-2.6\ \mu m$ band, $3-5\ \mu m$ band, and $8-14\ \mu m$ band, with the latter band being the main window, as shown in Fig. 17.5. The IR signature from aircraft mainly comes from the radiation of the fuselage

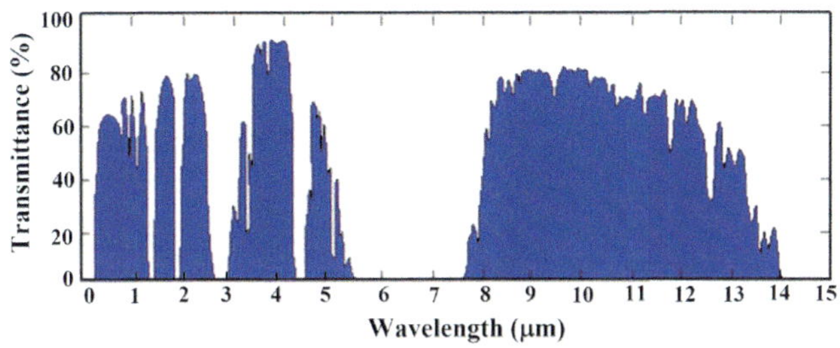

Figure 17.5 Electromagnetic spectral transmission window. From Wikipedia.

(due to aerodynamic heating of the surrounding atmospheric environment) and the radiation from the exhaust system. For effectively reducing contrast in the thermal image or an IR detection system, it is understood that reducing the IR emissivity of the target, controlling the surface temperature, and increasing the radiation energy of the background are important. Here, the first two factors are relatively feasible to achieve.

Not just radar, in modern warfare, IR search and tracking system is a major tool for monitoring enemy movements in air and water. The heat emitted from aircraft or warship engines is identified with the help of 'thermal vision' technology in this method. The heat energy is detected by the passive IR sensors. There are a few IR Signature Suppression Systems (IRSS) that could help with the IR problem. Black Hole Ocarina's (BHO) IRSS system is an optical blocking solution to obscure hot engine parts, the cooling exhaust duct, and the plume. The system is made up of a finned nozzle that has internal bends to shield hot internal exhaust surfaces from direct view. The whole system is designed to diffuse the exhaust plume and reduce the radiation of those plumes. The hot exhaust sucks in the cold air from the engine compartment, lowering outcoming gas temperatures. Film cooled tailpipe (FCT) is another IRSS system that consists of an improved nozzle, a flow wedge, and film cooling slots with cold air entries. This is a passive system that draws in ambient air by distributing pressure evenly along the tailpipe's length. The FCT system is specifically suited for helicopters and needs some modifications for installation.

IR-absorbing materials are also employed for this application. Polymers, organic and inorganic solids capable of absorbing in the near-infrared (NIR) region (1000 – 2000 nm), are emerging with great potential for use in telecommunications, for example, fused porphyrin arrays, semiquinones, mixed-valence binuclear metal complexes, metal fillers, inorganic

Table 17.1 Infrared (IR) transmittance of several organic binders.

Polymer	IR transmittance in the 8 – 14 μm window
Epoxy	0.35
Alkyd	0.39
Phenolic	0.5
High density polyethylene (HDPE)	0.5
Polyurethane (PU)	0.6
Ethylene propylene diene monomer (EPDM)	0.95

pigments, semiconducting fillers, phase change materials, etc. Organic NIR region materials that are electrically, optically, or thermally active at the telecommunication wavelengths and can be used in a device for optical attenuation, absorption, or antireflection. Their production cost is also relatively low. It is imperative to provide surface coatings that can lower the IR radiation of objects (such as weapons), which is an advancement of electronic warfare and IR detection technologies. The most extensively researched IR stealth coatings typically consist of a resin matrix, metal filler, and additives that significantly impact the IR reflection of the coatings. The commonly used organic polymeric binders with their transmittance in the 8 – 14 μm are given in Table 17.1 [4,5].

17.2.3 Visual stealth

Visual stealth is achieved by using paints or other materials to color and camouflage the image of a vehicle. Great examples of the use of this technology are the black-painted F-117 and B-2 Spirit designed to operate at night. The design of aircraft camouflage is complicated because an observer might be located above or under the airplane. Older camouflage schemes used countershading: light color underneath the plane and darker on top. Few other schemes take aircraft movement in battle into consideration, its orientation with respect to the ground, and apply a chosen pattern to the entire machine. The palette of colors in such a method is a selection of two or three dull, neutral shades. However, recently, with the usage of high-speed vehicles and heavy reliance on radar, the role of camouflage has decreased.

17.2.4 Sonar stealth

Sonar is an acronym for "sound navigation and ranging". This type of stealth technology focuses on submarine or ship sound reduction so that the unit remains undetected. One solution to this problem is using a device named Bathythermograph (B-thermograph), which records the temperature of the ocean at various depths. A B-thermograph can be placed on a ship or submarine. Sonar absorber is a submarine covering, able to absorb some of the sonar waves. This kind of solution presents some technical problems. The major problem is finding material that will have the requisite impedance and would be enough. What is more, it would need to cover a broad band of frequencies and boast high hydrostatic strength so that it could be used in deep water. The future of this technology is composite rigid syntactic foam combined with many fillers and ingredients. By dispersing appropriately sized elastomeric particles in a syntactic foam matrix, it can be used to make a truly broadband sonar absorber with a high hydrostatic strength.

17.2.5 Plasma stealth

The idea behind plasma stealth is to lower RCS by creating an ionized "layer" around the aircraft. Hazardous radar signatures are received in this quasiactive system, and they are absorbed/scattered by plasma that can absorb/spread a broad range of radar frequencies, angles, polarizations, and power densities. At higher frequencies, where the conductivity of the plasma allows it to interact strongly with the incoming radio waves, it is possible to use plasmas to control the reflected EM radiation from an object (a process known as plasma stealth). However, the wave can also be absorbed and converted into thermal energy instead of being reflected.

17.3 Development of stealth coatings

Paint coatings are one of the ways to address stealth requirements. A paint coating is a thin layered coating composed of polymers that provide superior adherence to the surface along with the desired functionalities. The role of the polymer here is primarily to act as a binder and hold the pigment or filler materials intact on the surface. The commonly used polymeric resins in the paints are epoxy, polyurethane, nitriles, polyalkenes, acrylic, and silicone systems. They are applied as multiple layers on the substrate. The base layer, or the undercoat, is applied to hide minor irregularities on the surface and provide a suitable surface for chemical anchoring of the subsequent top layers. The top layer, or top coat, is loaded with the fillers and pigments to render adequate strength, flexibility, and functionality. Fillers and pigments are generally organic/inorganic particles that are responsible for the performance characteristics.

Conducting polymers are another class of polymers that are used in the paint as a part of fillers to impart electrical conductivity. Polyaniline, polypyrrole, polyacetylene, polythiophene, polydopamine, etc., are such interesting materials, which are extensively used in stealth paints [6–9]. Generally, nanocomposites containing these conducting polymers and several filler materials such as ferrites, multiwalled carbon nanotubes (MWCNTs), graphene, and reduced graphene oxide show excellent MW absorption behavior. Further, these composites with resin also provide outstanding MW absorption properties.

In recent years, several polymeric coatings have been designed and subsequently developed, which have applications in both MW stealth and IR stealth. In the following sections, several polymer-based coatings for MW stealth and IR stealth applications are discussed.

17.3.1 MW stealth coatings

There have been several polymer-based coatings in the literature for MW absorption applications. Materials such as ferrites, perovskites, carbon materials, iron materials, conducting polymers, MXene, etc. are employed as fillers for the development of coatings. Some of the recent interesting studies are discussed here.

Spinel ferrites and hexaferrites in their nanostructured forms are employed in MW absorption applications since early era of the development of stealth coatings. $Mg_{0.5}Zn_{0.5}Fe_2O_4$ nanoparticle – polyurethane composite coatings with a thickness of ~100 μm on Al substrate were developed for the application of MW absorption in the 4 – 15 GHz frequency band [10]. The gel combustion method was used to make ferrite materials, and the coatings were prepared by the hand brush method. The optimized coating demonstrated significant broadband absorption properties with absorptivity greater than 88% in the frequency range of 4 – 15 GHz. In a report, $Ni_{0.40}Zn_{0.40}Cu_{0.20}Fe_2O_4$ (NZCF) nanoparticles were successfully incorporated into a poly(vinylidene fluoride) (PVDF) matrix in an effort to find a new material for MW absorption applications [11]. The NZCF-PVDF nanocomposite coating system emerged as a promising one for absorbing radiations in the microwave/GHz frequency range. Coating thicknesses were in the range of 0.18–0.25 mm. For coatings with 5 and 15 wt.% NZCF, the shielding effectiveness (SE_A) due to absorption was −39.5 dB and −34.9 dB at 11.5 GHz in the X-band region (8 − 12 GHz), respectively, whereas the coating with 10 wt.% NZCF was recorded to exhibit an SE_A value of −49.7 dB at 13.2 GHz in the Ku-band frequency band (12 − 18 GHz). In a work of Xu et al., lightweight renewable bio-based polyurethane foam (BPUF) and magnetic Fe_3O_4 composites for broadband MW absorption application were developed [12]. The findings demonstrate that the magnetic BPUF was multiinterfacial and had a porous hierarchical structure, which gave the Fe_3O_4 nanoparticles a matching layer and skeleton. The effective reflection loss (RL ~ 10 dB) of the composite varied from 4.16 to 18 GHz with 9 wt.% of Fe_3O_4 nanoparticles having a thickness of 1.5–5 mm. The surface density of the magnetic BPUF coating was less than 0.5 kg cm^{-2} , and the minimum reflection loss (RL_{min}) was −19.96 dB at 13.15 GHz for a thickness of 1.8 mm. Miszczyk prepared $Ni_xZn_{(1-x)}Fe_2O_4$ nano and micro ferrites (x = 0, 0.2, 0.4, 0.6, 0.8, and 1.0) and developed EM interference suppressing coatings with epoxy resin, wherein 10 vol.% ferrites were added as fillers [13]. Ferrite − epoxy coatings with 2 mm thickness displayed maximum attenuation in the range of 8–16 dB in the 6.5 − 15 GHz frequency band. Smitha et al. developed a thin radar-absorbing coating having $NiFe_2O_4$ (NF) and $Ni_{0.6}Zn_{0.4}Fe_2O_4$ (NZF) in an epoxy resin matrix [14]. They fabricated single-layer coatings with NF, NZF, and multilayer coatings with NF − NZF on aluminum sheets. For the NF sample, an RL_{min} of −14.3 dB was found at 8.2 GHz with a thickness of 1.5 mm. NZF with a thickness of 2.5 mm showed RL_{min} value of −33.2 dB at 9.8 GHz, and an effective absorption bandwidth of less than −10 dB was 3.0 GHz. It was found that the material combination of NF as the first layer, which had a thickness of 0.88 mm, and NZF as the second layer, which had a thickness of 0.84 mm, greatly increased RL to −45.0 dB at a relatively low total coating thickness of 1.72 mm. In another research work, MW-absorbing coatings with $Ni_{0.5}Zn_{0.5}Fe_2O_4$, carbon black in a PU matrix on cotton fabric were developed, wherein 30, 40, 50, 60, and 70 g of ferrite and 5 g of carbon black were mixed in each 100 mL PU [15]. The coated fabric, which was 1.86 mm thick and had a composition of 40% ferrite, 3% carbon, and 57% PU, provided 33.5% absorption, 26.5% transmission, and 39% reflection, respectively, in the X and Ku frequency bands. A reflection loss of over 7.5 dB in the entire Ku band and 22.5 dB at 13.5 GHz was reported. In a work of Atay and Çlik, barium hexaferrite − epoxy resin coatings on metal and glass were developed for radar-absorbing applications [16]. Barium hexaferrite was prepared by the sol-gel method; it was loaded in an epoxy matrix with different amounts, and coatings were fabricated. The radar-absorbing properties of the coatings were found to be substantially enhanced after loading of barium hexaferrite. RCS simulation for RAP on artillery shell (diameter: 30 mm and length: 130 mm) containing U-type hexaferrite magnetic material, $Ba_4(Co_{1-5x}P_{2x})_2Fe_{36}O_{60}$ (x = 0.05), and polyimide matrix was carried out employing multilevel fast multipole method (MLFMM) techniques [17]. Two RAPs with thicknesses of 1.43 and 1.83 mm were used for the simulation. Results of the RCS reduction at 10 GHz for the untreated shell and the coated RAP I and RAP II shells were compared. Average reductions on the order of 7–10 dB were reported for both paints at all aspect angles, with maximum reductions of 14 and 17 dB for RAP I and RAP II, respectively, observed at aspect angles of 165 degrees. The RAP II series performed relatively better, but as coating thickness grew, it added an additional weight penalty to the system. The literature described the preparation of MW-absorbing composite materials made of dendritic waterborne polyurethane-urea (WPU) polymer matrix with Ni – Zn ferrite filler [18]. The quick crosslinking of the dendritic structure led to a more efficient distribution of ferrite particles within the WPU matrix. It further ensured that the ferrite particles are trapped within a crosslinked structure. As a result, one coating generally exhibited broader absorption, and at some frequencies, another coating showed greater attenuation values in the 8.2 – 12.4 GHz frequency region.

In a recent work, a multilayer MW-absorbing coating consisting of iron powder, $MnFe_2O_4$, MWCNTs, and conducting polymer (polyaniline) in an epoxy resin, fabricated by a simple and cost-effective spray coating method, has been developed by Bera and coworkers [19]. The multilayer coating comprised three reflecting layers was designed keeping reflecting materials in descending order of their combined reflection from the substrate so that transmission, reflection, and absorption of MW radiation would occur inside the multilayer coating. EM simulation was carried out to predict the sequence, thickness, and concentration ratios of the above materials for possible layers. The multilayer coating with a thickness of ~2.1 mm and density of ~1.1 g cm^{-3} showed absorption of around 75 – 91% in the entire 8 – 12 GHz frequency range. Lakshmi et al. demonstrated enhanced MW absorption properties of polymethyl methacrylate (PMMA)-modified $MnFe_2O_4$–polyaniline (PMMF – PANI) nanocomposites [20]. They prepared $MnFe_2O_4$ by the solution combustion synthesis method, which is a very fast, economical, and simple method of preparation. When $MnFe_2O_4$ was chemically modified using polymethyl methacrylate, the 8–12 GHz frequency band experienced

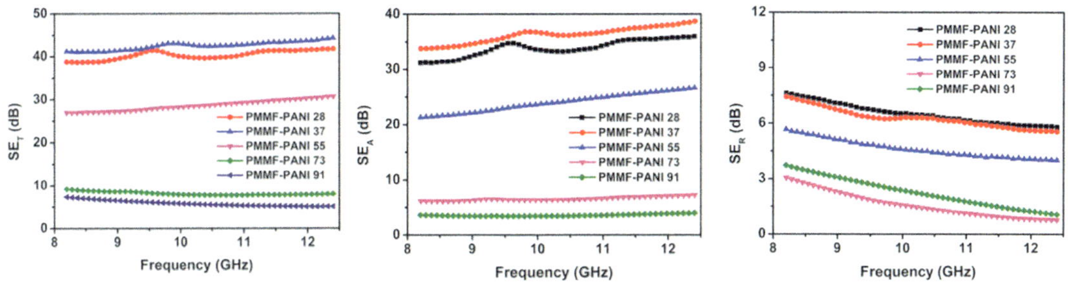

Figure 17.6 Plots of SE_T, SE_A, and SE_R of polymethyl methacrylate-modified $MnFe_2O_4$–polyaniline composites at various ratios as a function of frequency. Reproduced with permission from R.V. Lakshmi, P. Bera, R.P.S. Chakradhar, B. Choudhury, S.P. Pawar, S. Bose, et al., Enhanced microwave absorption properties of PMMA modified $MnFe_2O_4$-polyaniline nanocomposites. Physical Chemistry Chemical Physics 21 (9) (2019) 5068–5077. https://doi.org/10.1039/c8cp06943c, http://pubs.rsc.org/en/journals/journal/cp.

a 3.6 times increase in total EM interference shielding (SE_T). In the X-band region, the PMMF–PANI 37 composite (30 wt.% PMMF and 70 wt.% PANI) showed a SE_T value of 44 dB. The absorption component was the primary shielding mechanism in the PMMF–PANI optimized composites, according to a comparison of SE_A and SE_R values. SE_T, SE_A, and SE_R values of PMMF–PANI composites at various ratios as a function of frequency are presented in Fig. 17.6.

Fe- and Mn-based perovskite oxides are reported in the literature for their significant applications in MW absorption applications. Andriyanti et al. developed $BiFeO_3$ thin films on Kapton sheets for EM wave absorbing applications [21]. $BiFeO_3$ thin films were fabricated by DC sputtering under Ar gas with different times, and then their surfaces were modified by Ar ion implantation. Crystallite sizes and thicknesses of the films were in the range of 39 – 55 nm and 44 – 110 μm, respectively. EM wave absorption studies in the frequency range of 8 – 12 GHz demonstrated the reflection loss value of 40.6 dB at a frequency of 9.98 GHz, which is equivalent to 99.9% EM wave absorption. This indicated that $BiFeO_3$ film is a potential candidate as an EM absorber. In another work, $La_{0.7}Sr_{0.3}FeO_3/MoS_2$ with epoxy resin coating, prepared using a custom mold, exhibited the minimum reflection loss of –67.09 dB at 1.9 mm thickness and the maximum effective absorption bandwidth (EAB) of 7.28 GHz at 2.3 mm thickness [22]. Additionally, it demonstrated outstanding absorption capabilities for electromagnetic waves with several bands, spanning a broad spectrum of designated frequency ranges. Similarly, $La_{0.8}Sr_{0.2}MnO_3/Al_2O_3$–13%$TiO_2$ coatings were developed for MW absorption at 1000°C, having excellent oxidation resistance [23]. The coating material was prepared with polyvinyl alcohol with different mass ratios, and finally, the coatings were prepared by the plasma spray method. The coating with 30 wt.% $La_{0.8}Sr_{0.2}MnO_3$ and a thickness of 1.9 mm showed outstanding microwave absorption properties. The effective bandwidth was observed to be 1.7 to 2.4 GHz at temperatures below 773K. At a thickness of 1.7 mm and a temperature of 773K, the coating with 20 wt.% $La_{0.8}Sr_{0.2}MnO_3$ was reported to reach a maximum effective bandwidth of 1.6 GHz with a reflection loss of < -10 dB in the $8-12$ GHz frequency. Moreover, the bandwidth with RL< -5 dB could approach 2.4 GHz by keeping the thickness of 1.7 mm constant and raising the temperature to 973K. The preparation process and MW absorption properties of $La_{0.8}Sr_{0.2}MnO_3/Al_2O_3-13\%TiO_2$ coatings with different thicknesses and temperatures are presented in Fig. 17.7.

Iron particles and their alloys-based composites and coatings are used for the attenuation of MW absorption. Using an efficient and scalable approach, dual-shelled and hollow Fe@C microsphere composites were prepared by coating a ferric precursor on sulfonated polystyrene (PS) microspheres in a buffer solution and a thin layer of polydopamine (PDA) followed by annealing at 600°C to decompose the internal PS cores and carbonize the external PDA shell [24]. By combining the dielectric and magnetic losses, the carbon layers and the iron nanoparticles that are formed enhance the impedance matching, thus improving the MW absorption performances over a broad frequency range. The hollow microspheres cover showed a minimum RL of -54.4 dB at 8.8 GHz at a thickness of 4.5 mm in the whole X-band frequency range with an effective absorption bandwidth of 8.1 GHz at a thickness of 3 mm. In a work of Wang et al., a magnetorheological gel-based coating for EM wave absorption application in the $2-18$ GHz frequency band was developed [25]. Magnetorheological gel was prepared with flower-like carbonyl iron magnetic particles in a PU matrix. These magnetic particles move to form chains or columnar structures when placed in a magnetic field. This allows EM waves to pass through and scatter along the chains. MW absorption testing results showed the significant influence of magnetic particles on the MW absorption properties of the coatings. Enhancing the MW-absorbing capacity can be acheived by improving the loss ability and impedance matching of the coatings achieved with the right magnetic field. It was observed that the minimum RL values decreased first and then rose with an increase of magnetic field for different amounts of

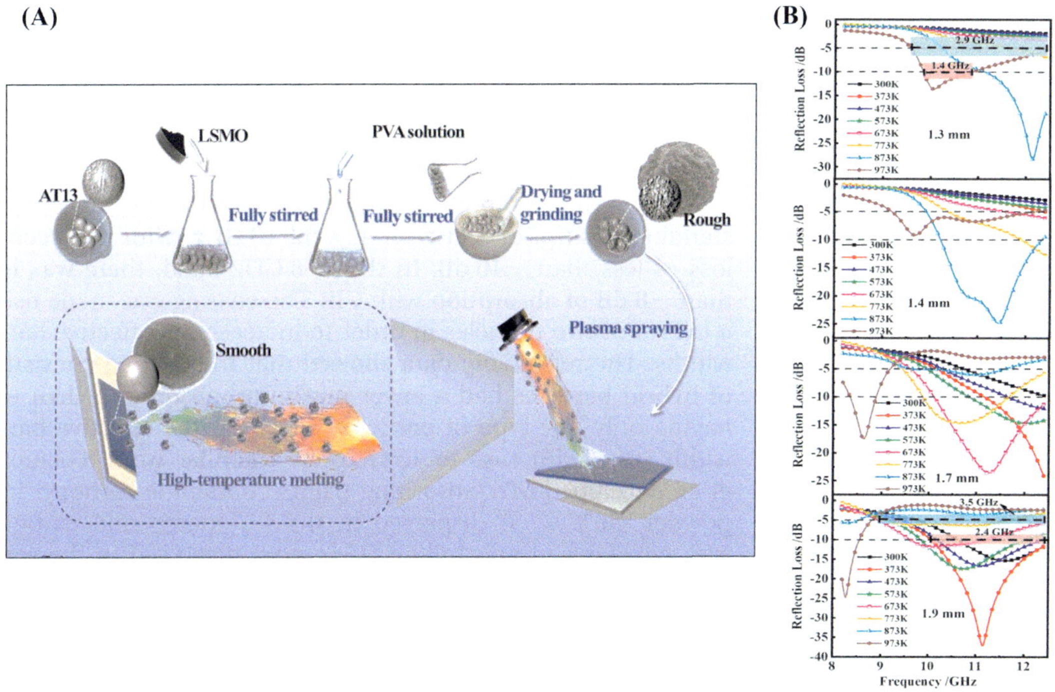

Figure 17.7 (A) Preparation process and (B) microwave absorption properties of $La_{0.8}Sr_{0.2}MnO_3/Al_2O_3-13\%TiO_2$ coating having $La_{0.8}Sr_{0.2}MnO_3$ to $Al_2O_3-13\%TiO_2$ mass ratio of 30 with different thicknesses and temperatures. Reproduced with permission from W. Wang, G. Liu, L. Wang, C. Ge, L. Wang, B. Wang, et al., Temperature-dependent dielectric and wide-temperature-range microwave absorption properties of $La_{0.8}Sr_{0.2}MnO_3/Al_2O_3-13\%TiO_2$ coatings, Ceramics International 49 (22) (2023) 34595–34602. https://doi.org/10.1016/j.ceramint.2023.08.106, https://www.journals.elsevier.com/ceramics-international.

carbonyl iron particles. In another work, novel PVDF/n-Fe composites were developed using an easy hot molding and bending process [26]. The composites with a filler loading of 20 vol.% and thicknesses of 1.90 mm showed EM interference (EMI) shielding effectiveness (SE) and reflection coefficient values of 40.21 and 20.97 dB (at 8.68 GHz), respectively. The thickness variation can be used to modify the observed absorption peak. Similarly, Zhou et al. developed carbon iron particles – silicone coating with different iron particle contents and thicknesses using the spray coating method for microwave absorption applications in the 2 – 18 GHz frequency range [27]. The observed RL value was observed below −5 dB in 5.76–18 GHz with 0.8 mm thickness. The frequency of the maximum absorption was shifted to a lower frequency with an increase in the thickness. It was also demonstrated that composite coating was suitable for MW absorption at 300°C. Ji et al.

developed MW-absorbing coatings containing Co ferrite and carbonyl iron particles (CIP) in a PU resin matrix in the $4-18$ GHz frequency region with a thickness of around 1.2 mm and a surface density of approximately 2.7 kg m^{-2} [28]. The weight fraction of Co-ferrite at 4–18 GHz could be adjusted to modify the electromagnetic and microwave-absorbing characteristics of the coatings. Co-ferrite particle coating showed outstanding absorption between 12 and 18 GHz, with a reflection loss of less than -10 dB. In the 4–8 GHz band, there was less than -8 dB of absorption with CIP. The coating was made using a hybrid of two particles in order to increase the effective bandwidths. The reflectivity data showed that the coating consisting of hybrid particles had a lower maximum absorption than one having only one type of particles. However, the effective bandwidth (< -8 dB) rose to 13 GHz. In a similar work, Folgueras et al. developed MW-absorbing paints containing carbonyl iron powder (90% w/w) dispersed in the PU matrix and carbonyl iron powder and polyaniline (15% w/w) dispersed in the same PU matrix on aluminium plates by the brush coating method [29]. Their thicknesses were 1.1 and 1.85 mm, respectively. They also prepared polyaniline-doped silicone sheets by molding with thicknesses of 2.8 and 4.4 mm. The incident EM wave was attenuated by exclusively contained carbonyl iron by approximately 7 dB, which is equivalent to 80% absorption of the EM energy. The paint that contained carbonyl iron and polyaniline reduced the wave's intensity by roughly 4 dB, corresponding to 60% absorption of the EM energy. Approximately 90% of the incident radiation was absorbed by the silicone rubber sheets made using a silicone matrix. Densities of all these materials were found to be less than 1 g cm^{-3}. In another research work, FeCoCr/silicone coatings were developed and subjected to examination for MW absorption behavior after heat treatment at 400°C [30]. The FeCoCr/silicone resin composite coatings were prepared using the spray coating process, with a 4:1 weight ratio between the two materials. The FeCoCr/silicone resin coatings were cured for 1 hour at 80°C and then for 3 hours at 180°C after spraying. A simulation of MW absorption properties of the coatings for the $2-18$ GHz frequency range was carried out. After 10 hours of heat treatment, the peak value of reflection loss of the coating with 1 mm thickness drops from -4.75 dB at room temperature to 6.18 dB, showing that the absorption performance of the coating was greatly improved and could withstand high temperatures. Simulated reflection loss values obtained with the coatings with different thicknesses indicate that coatings are potential candidates to work in low-frequency bands.

Several carbon-based nanomaterials are being employed for MW absorption applications. In recent years, different carbon nanostructured materials combined with fillers have emerged as potential candidates for this application. Hierarchically porous MW-absorbing materials (MAMs) containing magnetic crystallite-decorated hollow multicavity carbon nanosheet spheres were developed in a work of Liu and coworkers [31]. The materials were made by salt template-assisted spray drying, which involved a solution of magnetic metal salts and ammonium citrate, followed by thermal pyrolysis and template removal. This method is widely applicable and can be used to integrate different magnetic crystallites, including Fe_3O_4, Ni, FeCo, and $CoFe_2O_4$, into hollow multicavity carbon spheres. Excellent EM attenuation capability of −51.1 dB was achieved with an optimal Fe_3O_4-based absorber covering the entire X-band and with a thin thickness of 2.7 mm. It showed an effective absorption bandwidth of 7.7–13.4 GHz. Song et al. used biomass-derived porous carbon (BPC) for MW absorption [32]. Using BPC as the matrix and cobalt acetylacetonate as the raw material, a 3D porous BPC@Co/CoO nanocomposite was prepared through solvent heat and high-temperature calcination. Sample calcined at 500°C showed a minimum reflection loss (RL_{min}) value of −43.09 dB in the 2 − 18 GHz frequency region and an EAB of 6.75 GHz (11.25 − 18 GHz) with a matching thickness of 2.5 mm. The material exhibited outstanding EM wave absorption performance due to a variety of loss modes, including dipole polarization, interfacial polarization, conduction loss, and magnetic loss. Absorption also benefited from excellent impedance matching and a high attenuation coefficient, all of which were made possible by the unique microstructure. An Ag nanoparticles-decorated graphene coating on a flexible fabric substrate was developed for EM wave absorption [33]. To make a slurry with an acceptable viscosity, an adequate quantity of spherical graphene or Ag-decorated spherical graphene composite powders was homogeneously combined with PU resin. Subsequently, the slurry was stirred for 30 minutes while a small amount of silane coupling agent and antifoaming agent were also added. Then, the flexible fabric was evenly coated with the slurry using the blade coating method, and it was cured at 80°C. The 3 mm thick coating with 20 wt.% Ag has an EAB in the 9.08–11.52 GHz and a RL_{min} value of −15 dB. When combined with periodic frequency selective surface (FSS), the MW absorption performance was significantly improved. The attained RL_{min} value was -41.72 dB at 10.25 GHz, while the EAB was reached throughout the whole X-band frequency range. The elastic coating has an EAB that spans a broad angle range of 60–90 degrees. In another work, Ti_4O_7/carbon nanotubes (CNTs)/Al_2O_3 coatings were fabricated for MW absorption applications [34]. The first Al_2O_3 particles with different Ti_4O_7 and CNT concentrations were weighed and milled together for granulation. During the process, specific quantities of deionized water and polyvinyl alcohol (PVA) were added and heated to dissolve into glue. Then the coatings were developed using the plasma spray method. By varying the electrical conductivity, remarkable absorption and shielding capabilities of coatings could easily be achieved. With a thickness of 2.41 mm, the Ti_4O_7/CNTs/Al_2O_3 sample without CNTs had an ideal RL_{min} value of -55.7 dB at 8.9 GHz. The RL_{min} values of Ti_4O_7/CNTs/Al_2O_3 with 2.5 and 5 wt.% CNTs were only -7.7 dB at 8.2 GHz and -1.54 dB at 12.4 GHz, respectively. The carbon component enabled the coatings to exhibit reduced absorption of EM wave. A composite absorber comprising a pattern-impregnated honeycomb (PIH) structure with carbon black (CB) absorbent, a resistive FSS embedded in polypropylene (PP) foam, a flaky carbonyl iron (FCI) coating was designed [35]. FCI coating was prepared with epoxy resin and polyether amine curing agent. At the microscopic level, CB and FCI particles provided magnetic and electrical losses. With a thickness of 8.3 mm, the composite absorber achieved over 90% absorption from 2.06 to 18 GHz, which is equivalent to a fractional bandwidth of 158.9%. Joshi et al. prepared nanocomposite coatings using polyvinyl-pyrrolidone-stabilized silver nanoparticle-based graphene nanohybrid (Ag-PVP-CRG), chemically reduced graphene (CRG), and thermally reduced and exfoliated graphene (TRG) in a PU matrix [36]. It was observed that TRG-PU nanocomposite coating showed the best EMI shielding property with a total shielding effective value of -29.8 dB with 99.9% absorption efficiency compared to CRG-PU, Ag-PVP-CRG-PU, and PU coatings. MW-absorbing coatings, which comprised different concentrations of carbon fibers (CF) in a PU matrix, were developed in the work of Ban and coauthors [37]. The maximum reflection losses of the coatings were around -11.01 dB over the frequency range of 8–18 GHz when the CF content was 0.8 wt.% and the coating thickness was 1.2 mm. The coating surface density was 1.02 kg m^{-2}, and the bandwidth was less than $\sim$10 dB (almost 90% MW absorption) at roughly 4.2 GHz. Zeng et al. used a simple and inexpensive freeze-drying method to make porous MWCNT and WPU composites for EMI shielding applications in 8 – 12 GHz [38]. The EMI shielding performance of the lightweight composite was quite high, with SE of >50 dB. By simply changing the porosity, WPU matrix, and MWCNT percentage, the EMI shielding effectiveness (EMISE) might be regulated across a broad range. Porous MWCNT/WPU composites had densities as low as 20 mg cm^{-3}, and their SE values were greater than 20 dB, which meant that they reduced bulk composite weight by more than 98%. Specific shielding effectiveness (SSE), defined as SE divided by mass density, was also calculated in this work, which is an important criterion for evaluation of EMI SE. The SSE value was 1148 dB cm g^{-3}, which was very high as per literature was concerned. In another report, a melt blending process was employed for making electrically conductive polymer composites (ECPCs), which were based on poly(styrene-b-ethylene-ran-butylene-b-styrene) (SEBS) and expanded graphite (EG) or carbon black (CB) with thicknesses in the range of 1–5 mm [39]. Composites with 20 and 15 wt.% EG and CB showed −11.69 and −17.56 dB EMI shielding effectiveness values, respectively, wherein absorption was observed to be a dominant component in the shielding effectiveness [32]. Gupta et al. developed carbon black nanoparticles and PU-coated fabric for MW attenuation in the 8 – 18 GHz frequency range [40]. In X (8.2–12.4 GHz) and Ku (12–18 GHz) frequency bands, the coated fabric containing 6.51 wt.% carbon black and 93.49 wt.% PU with a thickness of 1.3 mm demonstrated approximately 46% absorption, 29.49% transmission, and 24.5% reflection. It was possible to produce a reflection loss of −31.39 dB at 17 GHz, more than −20 dB at 16–18 GHz, and more than −10 dB across the Ku-band. Additionally, −18 dB EMI shielding across the whole X- and Ku-band was obtained. In another report, Folgueras and coauthors employed the brush coating method for developing a radar-absorbing coating on a GF/PPS laminated composite using a nanostructured coating containing CNTs dispersed in a PU matrix [41]. Around 90% of the energy of the incident EM wave was found to be absorbed by the coating through the penetration of EM radiation, which was attributed to a relatively thin (0.15 mm) layer of the nanostructured coating with particular EM characteristics. Amorphous carbon nanotubes (ACNTs) in polyvinyl chloride with sizes ranging from 7 to 50 nm were used for making EM wave-absorbing composite films, as reported in the literature [42]. At the frequency range of 2–18 GHz, it was discovered that the reflection loss value of CNTs/PVC with a matching thickness of 2 mm increased from −13.20 to −25.02 dB. Once more, the absorption peak moved from 12.96 to 14.44 GHz, while the frequency bandwidth at −10 dB increased from 3.3 to 5.8 GHz.

MW absorption properties of five types of woven fabrics having different structural parameters and various fiber compositions with an electrically conductive coating containing the conjugated polymer system poly(3,4-ethylenedioxythiophene)-polystyrene sulfonate (PEDOT-PSS) were studied in the literature [43]. The coating samples were prepared by the screen printing method, and their MW absorption properties were investigated within the 2–18 GHz frequency range. It was observed that the better absorption ability was achieved within the 12–18 GHz range, and for fabrics coated on one side, better reflection reduction was obtained when the shielding effectiveness was below 20 dB. It became apparent that the electrical characteristics of the coated fabrics could be considerably regulated by adjusting the coating deposition on the fabric, which in turn affected the microwave characteristics of the coated fabrics. Different types of coated fabrics and their EMI shielding properties are displayed in Fig. 17.8.

Al/PANI composites were prepared using a straightforward insitu polymerization process [44]. The porous PANI was then coated onto the crystalline Al substrate with 20–30 μm thickness. At 14.56 GHz, the optimal reflection loss for Al/PANI was −46.09 dB, and the matching thickness was 1 mm. The composite coating's EAB was 1.76 GHz at this thickness. In a research work of Ren et al., a novel approach using chiral small molecules soft templates was put forth for the synthesis of superhelical chiral polypyrrole nanofibers [45]. These nanofibers have the potential to be developed into a high-performing, anticorrosive MW absorber. Because of its superior impedance matching and the EM wave attenuation ability by cross-polarization on EM waves, the optimized helical polypyrrole (HPPy) nanofibers displayed significant MW absorbing capacity with broad bandwidth and low packing ratio. When the packing ratio was only 6%, the ideal HPPy showed two minimum reflection loss peaks at 44.5 and 11.9 dB, respectively, for the effective bandwidth of 5.4 GHz. Reflection loss values are presented in Fig. 17.9.

Zhou et al. developed a conductive coated honeycomb metamaterial absorber composite with a conductive coating, and a photo-reactive acrylic resin reinforced with carbon fibers was used to 3D print the honeycomb skeleton [46]. Carbon fibers, resin, and MWCNTs made up the conductive covering. With a total thickness of 6 mm, the composite achieved less than 10 dB reflection loss in the 5.65–40 GHz range, indicating that it has a relative bandwidth of 152.9% and a relative thickness of 0.107.

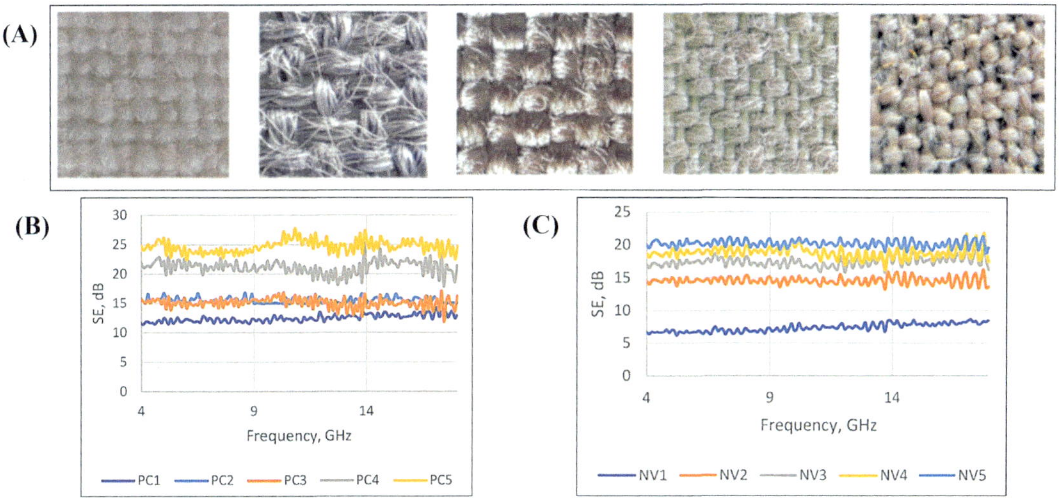

Figure 17.8 (A) Several PEDOT:PSS coated fabrics (cotton polyamide: PC and Nomex viscose: NV) (top panel) and (B and C) their electromagnetic interference shielding properties (bottom panel). Reproduced with permission from V. Rubežienė, S. Varnaitė-Žuravliova, A. Sankauskaitė, J. Pupeikė, P. Ragulis, A. Abraitienė, The impact of structural variations and coating techniques on the microwave properties of woven fabrics coated with PEDOT:PSS composition, Polymers 15 (21) (2023) 4224. https://doi.org/10.3390/polym15214224.

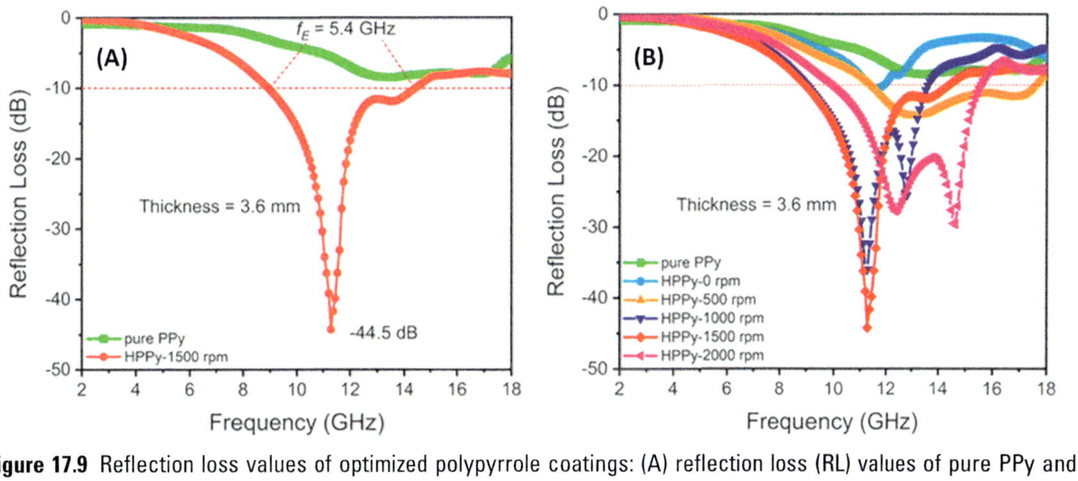

Figure 17.9 Reflection loss values of optimized polypyrrole coatings: (A) reflection loss (RL) values of pure PPy and HPPy-1500 rpm, (B) RL values of HPPy prepared at different polymerization speeds. Reproduced with permission from H. Ren, T. Li, H. Wang, Z. Guo, T. Chen, F. Meng, Two birds with one stone: superhelical chiral polypyrrole towards high-performance electromagnetic wave absorption and corrosion protection, Chemical Engineering Journal 427 (2022) 131582. https://doi.org/10.1016/j.cej.2021.131582.

This suggests that it has excellent EM absorption performance despite its small thickness and inherent lightweight. In another work, Bi et al. used sandwich nanoarchitectonics to develop honeycomb (HC) composites containing a carbon nanotube/carbon black/polyurethane (CNT/CB/PU) absorbing layer and an aramid HC core by immersion method [47]. The dielectric characteristics can be improved by modifying the carbon nanomaterial, and the impedance matching can be improved by using multilayer design. In the frequency range of 2.2–17.7 GHz, the developed three-layer sandwiched honeycomb composite achieves effective absorption (RL < −10 dB) with a minimum RL value of −26.4 dB.

Amudhu et al. developed a novel approach to enhance EM wave absorption employing glass fiber laminate composite (GFLC) with nanoparticles embedded nanofibers and FSS [48]. The FSS was coated with a mixture of PVDF and MWCNTs by electrospinning to enhance absorption. It was demonstrated that the fabricated composites increased EM absorption by 80% in comparison with metals over a wide frequency range at an overall thickness of only 3 mm. Comparing the experimental results to all other configurations, it was shown that RAS with 0.25 wt.% MWCNT/PVDF had an 87.5% bandwidth for over 90% absorption from 8.5 to 12 GHz. The customized composite material has great potential as RAS and has enormous potential for usage in the production of stealth aircraft.

MXene/PVA films by layer-by-layer blade coating method for EMI shielding application were reported in the literature recently [49]. It showed significant EMISE of 36.7 dB in the 8 − 12 GHz frequency region. In Fig. 17.10, the preparation and EMI shielding properties of MXene/PVA films are shown.

Deng et al. developed catecholamine-densified MXene coatings, which showed promising EMI shielding properties [50]. Here, the interfacial interaction between small MXene flakes and catecholamine molecules demonstrated the large-area fabrication of extremely densified and oriented MXene coatings by the blade coating method on conductive tapes and by brushing on various substrate materials, such as polar plastics, ceramics, fabric, and metal. By micro-crosslinking MXene nanosheets, the catecholamine molecules could greatly enhance the rheological characteristics of the coating. The coating showed outstanding EMI shielding performance in the 8 − 12 GHz frequency band.

17.3.2 IR stealth coatings

Similar to MW stealth, coatings with suitable fillers have also been employed for IR stealth applications. Metal powder, alloy

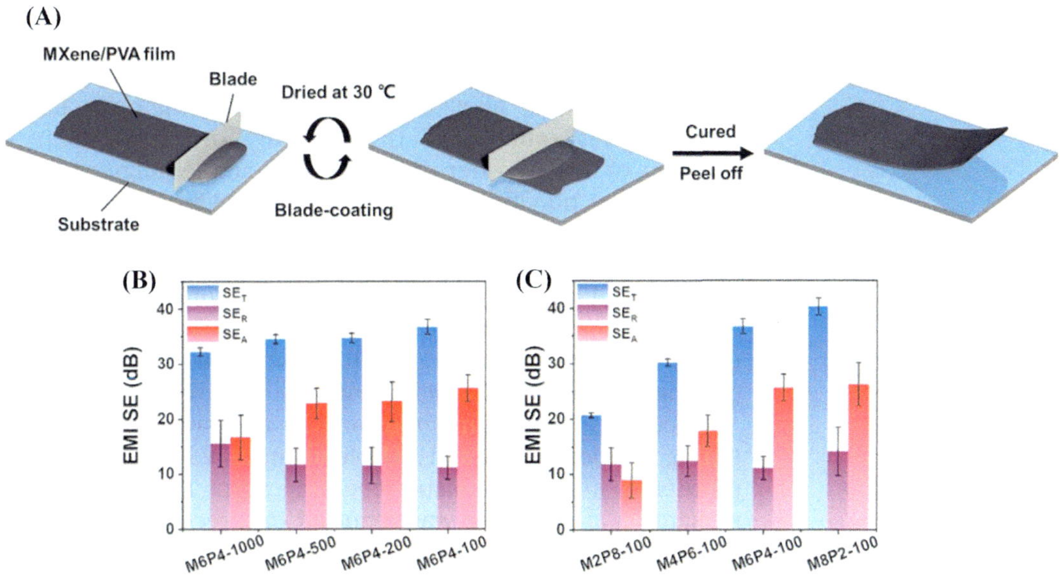

Figure 17.10 (A) Preparation of MXene/PVA coatings by blade coating method with different MXene (M) and PVA (P) mass ratios and scraper heights (100, 200, 500, 1000 μm), (B) EMI shielding performances of M6P4 composite films with different heights, and (C) EMI shielding performances of M2P8, M4P6, M6P4, and M8P2 composite films with 100 μm height. Reproduced with permission from J. Dong, Z. Li, C. Liu, B. Zhou, C. Liu, Y. Feng, Highly oriented MXene/polyvinyl alcohol films prepared by scalable layer-by-layer blade coating for efficient electromagnetic interference shielding and infrared stealth, Nano Research 17 (2024) 5651–5660. https://doi.org/10.1007/s12274-024-6486-8, https://www.springer.com/journal/12274.

powder, conjugated polymers, nanostructured oxide materials, carbon materials, hollow glass microspheres, etc., are prime fillers used for the development of IR stealth coatings. Materials, fabrication procedures, and IR emissivity properties of the polymer coatings have been discussed in this section.

In a recent review, research and development of fabrics used in IR stealth applications was addressed [5]. Authors discussed different types of fillers, such as metal-colored materials, semiconductors, phase change materials, thermal insulating materials, and several kinds of resins, such as epoxy, polyurethane, alkyd, phenolic resin, etc., in the review.

A low IR emissivity coating using Al powder, an IR functional pigment, and polysiloxane and glass powder as organic and inorganic binders was developed in a work of Qi and coworkers [51]. The findings indicated that the IR emissivity of the coatings was only 0.28 (between 7.5 and 14 μm) at 590°C because of the disintegration of polysiloxane and the uniform distribution of surface Al powder. Similarly, Ma et al. developed a novel aluminium/amorphous aluminium alloy/polyurethane composite coating with different concentrations of aluminium and amorphous aluminium alloy (AAA) and studied their mechanical, corrosion resistance, and IR emissivity properties [52]. The addition of AAA powder disrupted the parallel alignment of Al flakes in the polymer matrix. Al flakes moved toward the AAA particles and encircled them to form annular structures. It was observed that excess AAA powder resulted in lower adhesion strength and higher IR emissivity. An ideal concentration of 8.75 wt.% for Al-based alloy powder, of which 57 wt.% was AAA powder, showed the optimum overall performance of the composite coating; the details of which are given in Fig. 17.11.

Successful development of low IR emissivity coatings consisting of epoxy resin and PU blend modified resin, flake brass powder as a pigment, silane coupling agent KH560, and a dispersant as an adhesive was reported in a recent literature [53]. Authors comprehensively discussed epoxy and PU mass ratios, concentration of flake brass powder, KH560, and dispersant on the performance of the coating that were methodically examined. The findings indicated that when the mass ratio of epoxy resin to PU was 6:4 and the addition amount of brass powder was 50 wt.%, the coating exhibited emissivity and glossiness of 0.424 and 28.8, respectively. The composite coating made of a blend of epoxy and PU resin exhibits the best possible emissivity (0.332), glossiness (9.0), adhesion strength (grade 1), impact strength (50 kg cm), and flexibility (1 mm) at the same time.

The design and synthesis of three novel conjugated polymers containing Schiff base moieties by solution polycondensation was found in a work of Li and coauthors [54]. Three polymers, ATPTAL, ATPyTAL, and ATTPAL, were prepared using dialdehyde monomers such as p-phthalaldehyde, 2,6-pyridinedicarboxaldehyde, and 4,4'-(phenylazanediyl)dibenzaldehyde in amine-capped aniline trimer (ACAT) and dimethyl formamide, respectively. ATPTAL showed semicrystalline structures, and the nature of ATTPAL and ATPyAL was amorphous. IR emissivity of ATPTAL, ATPyAL, and ATTPAL coatings measured in the wavelength ranges of 8–14 and 2–22 μm at room temperature. At 8 – 14 and 2 – 22 μm wavelengths, the IR emissivity values ranged from 0.587 to 0.627 and 0.595 to 0.609, respectively. When compared to ATPTAL and ATPyAL, the IR emissivity of ATTPAL coatings was the lowest at 0.587 at 8–14 μm and 0.595 at 2–22 μm.

Core-shelled Cr_2O_3@stannic antimony oxide (ATO) structure by in situ precipitation method and fabrication of coatings with different Cr_2O_3/ATO mass ratios in silicone matrix by spray

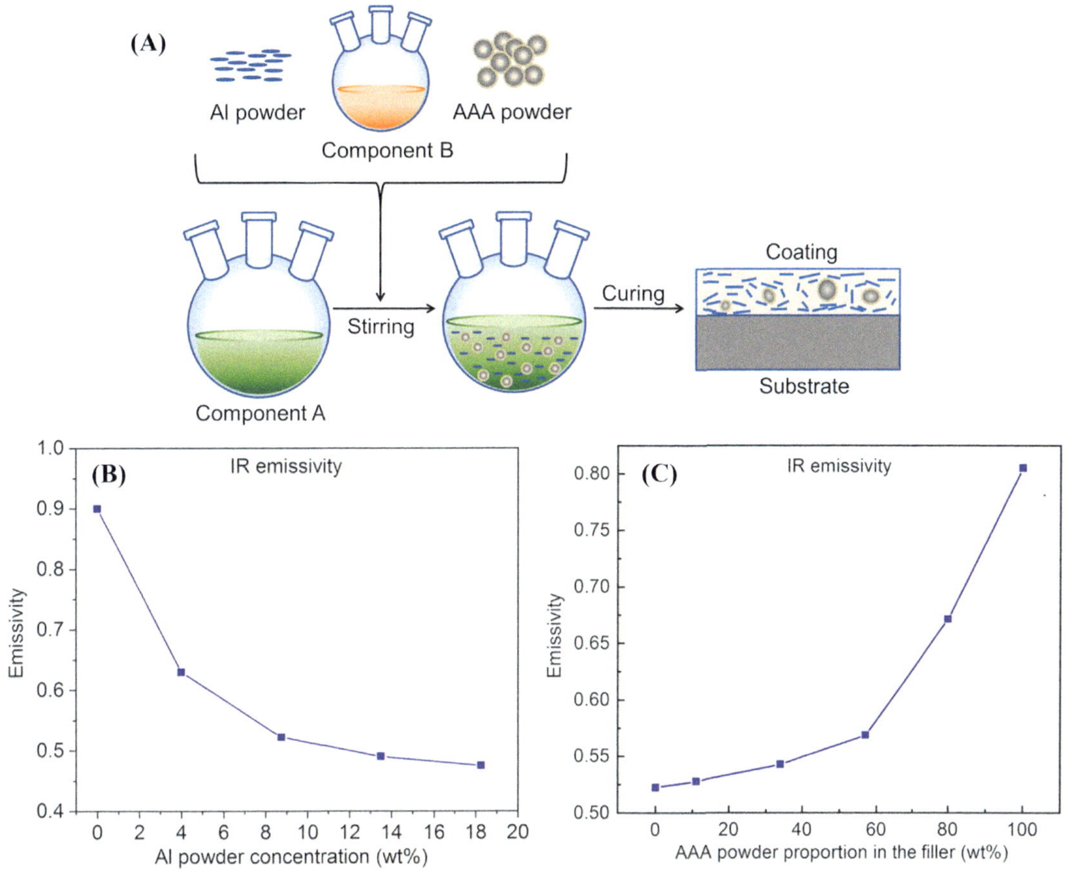

Figure 17.11 (A) Preparation of Al- and AAA-PU coatings, infrared emissivity values of (B) Al-PU coatings, and (C) AAA-PU coatings with different concentrations. Reproduced with permission from Z. Ma, Q. Li, J. Wei, C. Liang, T. Yang, G. Wang, et al., Effects of Al-based alloy powders on the mechanical behavior, corrosion resistance and infrared emissivity of polyurethane composite coatings, Colloids and Surfaces A: Physicochemical and Engineering Aspects 624 (2021) 126782. https://doi.org/10.1016/j.colsurfa.2021.126782.

coating method were reported by Chai and his group [55]. In this work, IR emissivity properties of the coatings and influence of temperature on the emissivity properties were studied. It was demonstrated that while the temperature increased from 50°C to 250°C, the IR emissivity of the optimized coating marginally reduced from 0.862 to 0.849, while that of the Cr_2O_3 coating remained at approximately 0.96. In another work, Chai et al. developed low emissivity coatings based on Al/M-Cr_2O_3/silicone resin using the spray coating method [56]. An ideal grass green IR stealth coating with an emissivity value of 0.525 was achieved by controlling the ratios of Al to M-Cr_2O_3. By coating the Al/M-Cr_2O_3-filled layer on the FeSiAl-based MW-absorbing coating, its absorption performance was enhanced by 41% compared to the FeSiAl MW-absorbing coating, which is attributed to the synergistic effect between the two layers. For greater thermal suppression efficacy and reduced IR emissivity, it is advantageous to increase the Al content of the mixed filler. Fig. 17.12 displays coating development and the corresponding IR emissivity values.

In a recent work, three-layer composite coatings were made using layered spraying method, in which a distinct paint recipe was sprayed on each layer [57]. The ethylene propylene diene monomer (EPDM) was used as the binder or matrix in each layer, providing low dielectric constant and loss. Three-layer composite coatings with varying aluminum powder contents and 4 wt.% carbon fiber (CF) were found to balance antistatic properties, low IR emissivity, and dielectric properties. The composite coatings showed an IR emissivity of 0.498. Youh et al. studied the IR emissivity of graphene-epoxy and Al-graphene epoxy coatings with various concentrations (0.01–1 wt.%) of graphene filler [58]. The coatings were developed using the spray coating method, and the thickness of the coating was 100 μm. In the 3–5 μm window, plain epoxy resin had an average emissivity of 0.75, while in the 8–12 μm window, it was 0.93. These values show that the epoxy coating has a high emissivity due to its excellent IR absorption. When 0.05 wt.% of graphene thermally reduced at 1100°C was added to the epoxy, the IR emissivity increased to 0.98 and 0.94 in 3–5 and 8–12 μm, respectively. These increases are caused by strong absorption of graphene in IR wavelengths, which is undesirable for IR camouflage. Fibrous membranes of PVDF and PVDF modified with single-wall carbon nanotubes (SWNT), designated as SWNT/PVDF, were developed using electrospinning technology [59]. PVDF and SWNT/PVDF membranes were subsequently decorated with Au nanoparticles, resulting in the Au-PVDF and Au-SWNT/PVDF. The IR emissivity of the films was measured between 2 and 22 μm wavelengths. For PVDF, SWNT/PVDF, Au-PVDF, and Au-SWNT/PVDF films, the corresponding IR emissivity values were 0.82, 0.77, 0.76, and 0.68, respectively. The outcome showed how well Au decoration reduced IR emissivity for fibrous membranes made of PVDF or SWNT/PVDF, suggesting a potential use for Au decoration in IR stealth camouflage.

Hollow glass microsphere (HGM) is also used in IR stealth applications. Low emissivity/thermal insulation IR stealth composites were fabricated, which were based on a combination of

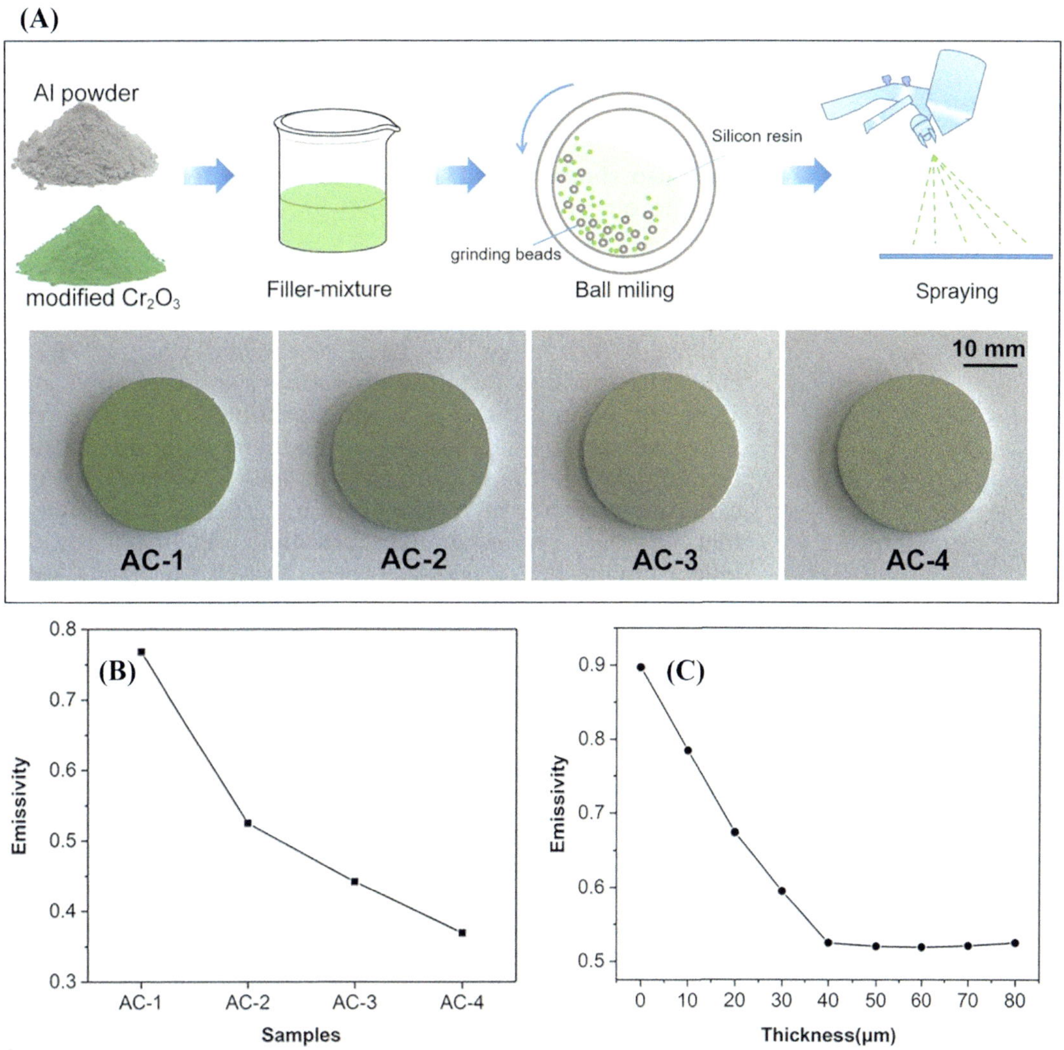

Figure 17.12 Development process of (A) Al/M-Cr_2O_3/silicone coatings (AC) and coating images, and infrared emissivity values of (B) different samples and (C) different thicknesses of AC-2 sample. Reproduced with permission from X. Chai, D. Zhu, Q. Chen, Y. Qing, K. Cao, F. Luo, et al., Tailored composition of low emissivity top layer for lightweight visible light-infrared-radar multiband compatible stealth coating, Advanced Composites and Hybrid Materials 5 (4) (2022) 3094–3103. https://doi.org/10.1007/s42114-022-00563-7, https://www.springer.com/journal/42114.

silver-plated HGM (HGM-Ag) and waterborne polyurethane through a one-step density-driven filler separation coupled with a freeze–drying method [60]. HGM-Ag particles with low emissivity were prepared by the electroless plating method. In the composites, HGM-Ag was concentrated on the top of the composites as a low emissivity layer, and the bottom layer with low thermal conductivity, about $0.044\ W\,m^{-1}\ K^{-1}$, served as the thermal insulation layer. The IR emissivity ($1-22\ \mu m$) of the composites can be reduced from 0.943 to 0.713, and the composites also showed good thermal insulation features because of their porous structure.

17.4 Conclusions and future perspectives

Stealth technology has been researched since WWII. This concept has widely been used in military and civil applications. There are several types of stealth technologies, such as radar, IR, visual, sonar, and plasma. Among them, radar and IR stealth technologies are the most important, and they have been under extensive research in recent years in terms of filler materials, matrices, easy processability, fabrication, weight penalty, durability, stability, etc.

Because of the flexible nature of polymers and easy application procedures, the technology of polymer coatings is expanding quickly, and these coatings can be used for a wide range of applications. In the last several years, polymer-based coatings have been used for MW and IR stealth applications. In general, polymers are used as a matrix and suitable nanofillers having MW-absorbing and low IR emissivity properties. Generally, epoxy, PU, PVDF, PVA, polydopamine, polyvinyl chloride, etc., are used as polymer matrices. Ferrites, perovskites, graphene, carbon black, and carbon fibers are used as filler materials. In recent years, use of new generation 1D and 2D materials such as CNTs, graphene, MXene, etc. for stealth coatings development has been observed. Because of extensive research in this area, several new-age materials are getting discovered, which will fulfill prime requirements for this technology. Therefore, further research and development regarding nanocoatings and nanotechnology for stealth are being augmented in several leading countries across the globe.

References

[1] A. Aytaç, H. İpek, K. Aztekin, B. Çanakçı, A review of the radar absorber material and structures, Scientific Journal of the Military University of Land Forces 198 (4) (2020) 931–946. Available from: https://doi.org/10.5604/01.3001.0014.6064.

[2] H. Ahmad, A. Tariq, A. Shehzad, M.S. Faheem, M. Shafiq, I.A. Rashid, et al., Stealth technology: methods and composite materials—a review, Polymer Composites 40 (12) (2019) 4457–4472. Available from: https://doi.org/10.1002/pc.25311, http://onlinelibrary.wiley.com/journal/10.1002/(ISSN)1548-0569.

[3] S.H. Kim, S.Y. Lee, Y. Zhang, S.J. Park, J. Gu, Carbon-based radar absorbing materials toward stealth technologies, Advanced Science 10 (32) (2023) 2303104. Available from: https://doi.org/10.1002/advs.202303104, http://onlinelibrary.wiley.com/journal/10.1002/(ISSN)2198-3844.

[4] H. Ishide, J. Yan, Study on infrared transmittance of Si-polymer hybrid structure press molded using a coupling agent, International Journal of Automotive Technology 13 (6) (2019) 817–824. Available from: https://doi.org/10.20965/ijat.2019.p0817, https://www.jstage.jst.go.jp/article/ijat/13/6/13_817/_pdf/-char/en.

[5] X. Zhou, B. Xin, Y. Liu, Research progress on infrared stealth fabric, Journal of Physics: Conference Series 1790 (1) (2021) 012058. Available from: https://doi.org/10.1088/1742-6596/1790/1/012058.

[6] R. Peymanfar, H. Dogari, E. Selseleh-Zakerin, M.H. Hedayatzadeh, S. Daneshvar, N. Amiri-Ramsheh, et al., Recent advances in microwave-absorbing materials fabricated using organic conductive polymers, Frontiers in Materials 10 (2023) 1133287. Available from: https://doi.org/10.3389/fmats.2023.1133287. journal.frontiersin.org/journal/materials.

[7] B.G. Soares, G.M.O. Barra, T. Indrusiak, Conducting polymeric composites based on intrinsically conducting polymers as electromagnetic interference shielding/microwave absorbing materials—a review, Journal of Composites Science 5 (7) (2021) 173. Available from: https://doi.org/10.3390/jcs5070173, https://www.mdpi.com/2504-477X/5/7/173/pdf.

[8] J. Yan, Y. Huang, X. Liu, X.X. Zhao, T. Li, Y. Zhao, et al., Polypyrrole-based composite materials for electromagnetic wave absorption, Polymer Reviews 61 (3) (2021) 646–687. Available from: https://doi.org/10.1080/15583724.2020.1870490, http://www.tandf.co.uk/journals/titles/15583724.asp.

[9] M. Zahid, R. Anum, S. Siddique, H.F. Shakir, Z.A. Rehan, Polyaniline-based nanocomposites for electromagnetic interference shielding applications: a review, Journal of Thermoplastic Composite Materials 36 (4) (2023) 1717–1761. Available from: https://doi.org/10.1177/08927057211022408, https://journals.sagepub.com/home/JTC.

[10] R. Bhattacharyya, S. Roy, O. Prakash, A.P. Singh, T.K. Bhattacharya, P. Maiti, et al., $Mg_{0.5}Zn_{0.5}Fe_2O_4$-polyurethane thin nanocomposite coating as broadband microwave absorber, SN Applied Sciences 1 (1) (2019) 38. Available from: https://doi.org/10.1007/s42452-018-0041-8.

[11] P. Saha, S. Das, S. Sutradhar, Influence of Ni-Zn-Cu-ferrite on electroactive β-phase in poly(vinylidene fluoride)-Ni-Zn-Cu-ferrite nanocomposite film: unique metamaterial for enhanced microwave absorption, Journal of Applied Physics 124 (4) (2018) 045303. Available from: https://doi.org/10.1063/1.5038134, http://scitation.aip.org/content/aip/journal/jap.

[12] X. Xu, X. Tian, G. Bo, X. Su, J. Yan, Y. Yan, Synthesis of lightweight renewable microwave-absorbing bio-polyurethane/Fe_3O_4 composite foam: structure analysis and absorption mechanism, International Journal of Molecular Sciences 23 (20) (2022) 12301. Available from: https://doi.org/10.3390/ijms232012301.

[13] A. Miszczyk, Protective and suppressing electromagnetic interference properties of epoxy coatings containing nano-sized NiZn ferrites, Frontiers

in Materials 7 (2020) 183. Available from: https://doi.org/10.3389/fmats.2020.00183.

[14] P. Smitha, I. Singh, M. Najim, R. Panwar, D. Singh, V. Agarwala, et al., Development of thin broad band radar absorbing materials using nanostructured spinel ferrites, Journal of Materials Science: Materials in Electronics 27 (8) (2016) 7731–7737. Available from: https://doi.org/10.1007/s10854-016-4760-6, https://rd.springer.com/journal/10854.

[15] K.K. Gupta, S.M. Abbas, T.H. Goswami, A.C. Abhyankar, Microwave absorption in X and Ku band frequency of cotton fabric coated with Ni–Zn ferrite and carbon formulation in polyurethane matrix, Journal of Magnetism and Magnetic Materials 362 (2014) 216–225. Available from: https://doi.org/10.1016/j.jmmm.2014.03.040.

[16] H.Y. Atay, E. Çelik, Barium hexaferrite reinforced polymeric dye composite coatings for radar absorbing applications, Polymer Composites 35 (3) (2014) 602–610. Available from: https://doi.org/10.1002/pc.22701.

[17] S. Padhy, S. Sanyal, R.S. Meena, R. Chatterjee, A. Bose, Characterization and performance evaluation of radar absorbing materials, Journal of Electromagnetic Waves and Applications 27 (2) (2013) 191–204. Available from: https://doi.org/10.1080/09205071.2013.743447.

[18] E. Açıkalın, O. Atıcı, A. Sayıntı, K. Çoban, H. Erkalfa, Preparation of dendritic waterborne polyurethane-urea/Ni–Zn ferrite composite coatings and investigation of their microwave absorption properties, Progress in Organic Coatings 76 (6) (2013) 972–978. Available from: https://doi.org/10.1016/j.porgcoat.2012.10.008.

[19] P. Bera, R.V. Lakshmi, R.P.S. Chakradhar, B. Choudhury, S. Jaikesh, R.U. Nair, H.C. Barshilia, A multilayer microwave radiation absorbing coating and a process for the preparation thereof, IN490560

[20] R.V. Lakshmi, P. Bera, R.P.S. Chakradhar, B. Choudhury, S.P. Pawar, S. Bose, et al., Enhanced microwave absorption properties of PMMA modified $MnFe_2O_4$-polyaniline nanocomposites, Physical Chemistry Chemical Physics 21 (9) (2019) 5068–5077. Available from: https://doi.org/10.1039/c8cp06943c, http://pubs.rsc.org/en/journals/journal/cp.

[21] W. Andriyanti, M.A.C.H. Nur, D.L. Puspitarum, T. Sujitno, H. Suprihatin, S. Purwanto, et al., Microstructures, magnetic properties and microwave absorption of ion-implanted bismuth ferrite thin films, Physica B: Condensed Matter 676 (2024) 415690. Available from: https://doi.org/10.1016/j.physb.2024.415690, https://www.sciencedirect.com/science/journal/09214526.

[22] A. Feng, D. Lan, J. Liu, G. Wu, Z. Jia, Dual strategy of A-site ion substitution and self-assembled MoS_2 wrapping to boost permittivity for reinforced microwave absorption performance, Journal of Materials Science & Technology 180 (2024) 1–11. Available from: https://doi.org/10.1016/j.jmst.2023.08.060, http://www.sciencedirect.com/science/journal/10050302.

[23] W. Wang, G. Liu, L. Wang, C. Ge, L. Wang, B. Wang, et al., Temperature-dependent dielectric and wide-temperature-range microwave absorption properties of $La_{0.8}Sr_{0.2}MnO_3/Al_2O_3-13\%TiO_2$ coatings, Ceramics International 49 (22) (2023) 34595–34602. Available from: https://doi.org/10.1016/j.ceramint.2023.08.106, https://www.journals.elsevier.com/ceramics-international.

[24] X.P. Li, Z. Deng, Yue Li, H.B. Zhang, S. Zhao, Y. Zhang, et al., Controllable synthesis of hollow microspheres with Fe@carbon dual-shells for broad bandwidth microwave absorption, Carbon 147 (2019) 172–181. Available

from: https://doi.org/10.1016/j.carbon.2019.02.073, http://www.journals.elsevier.com/carbon/.

[25] L.-R. Wang, M. Yu, P.-A. Yang, S. Qi, J. Fu, Synthesis of absorbing coating based on magnetorheological gel with controllable electromagnetic wave absorption properties, Smart Materials and Structures 28 (4) (2019) 044001. Available from: https://doi.org/10.1088/1361-665X/ab06e8, https://iopscience.iop.org/article/10.1088/1361-665X/ab06e8/pdf.

[26] H. Gargama, A.K. Thakur, S.K. Chaturvedi, Polyvinylidene fluoride/nanocrystalline iron composite materials for EMI shielding and absorption applications, Journal of Alloys and Compounds 654 (2016) 209–215. Available from: https://doi.org/10.1016/j.jallcom.2015.09.059.

[27] Y. Zhou, W. Zhou, Y. Qing, F. Luo, D. Zhu, Temperature dependence of the electromagnetic properties and microwave absorption of carbonyl iron particles/silicone resin composites, Journal of Magnetism and Magnetic Materials 374 (2015) 345–349. Available from: https://doi.org/10.1016/j.jmmm.2014.08.059.

[28] X. Ji, M. Lu, F. Ye, Q. Zhou, Preparation and research on the electromagnetic wave absorbing coating with Co-ferrite and carbonyl iron particles, Journal of Materials Science Research 2 (2) (2013). Available from: https://doi.org/10.5539/jmsr.v2n2p35.

[29] L.D.C. Folgueras, M.A. Alves, M.C. Rezende, Microwave absorbing paints and sheets based on carbonyl iron and polyaniline: measurement and simulation of their properties, Journal of Aerospace Technology and Management 2 (1) (2010) 63–70. Available from: https://doi.org/10.5028/jatm.2010.02016370, http://www.jatm.com.br/papers/vol2_n1/JATMv2n1_p63-70_Microwave_absorbing_paints_and_sheets_based_on_carbonyl_iron_and_polyaniline_measurement_and_simulation_of_their_properties.pdf.

[30] D. Yuan, Z. Chen, C. Cai, B. Yang, Y. Wang, Z. Chen, Microstructures, absorption and adhesion evolution of FeCoCr/silicone resin coatings at elevated temperature, Frontiers in Materials 10 (2023) 1168418. Available from: https://doi.org/10.3389/fmats.2023.1168418.

[31] Z. Liu, H. Tian, R. Xu, W. Men, T. Su, Y. Qu, et al., Magnetic crystallite-decorated hollow multi-cavity carbon nanosheet spheres for superior electromagnetic absorption, Carbon 205 (2023) 138–150. Available from: https://doi.org/10.1016/j.carbon.2023.01.031, http://www.journals.elsevier.com/carbon/.

[32] W. Song, Q. Zhao, Z. Wang, Magnetic biomass porous carbon@Co/CoO nanocomposite for highly efficient microwave absorption, Materials Research Bulletin 167 (2023)112371. Available from: https://doi.org/10.1016/j.materresbull.2023.112371.

[33] Y. Liu, S. Wang, Q. Wang, X. Su, Electromagnetic and microwave absorption properties of flexible fabric coatings containing Ag decorated spherical graphene powders with FSS incorporation, Ceramics International 50 (7) (2024) 11075–11084. Available from: https://doi.org/10.1016/j.ceramint.2024.01.008, https://www.sciencedirect.com/science/journal/02728842.

[34] Y. Li, Y. Qing, H. Yao, H. Xu, H. Wu, A novel plasma-sprayed Ti_4O_7/carbon nanotubes/Al_2O_3 coating with bifunctional microwave application, Journal of Colloid and Interface Science 645 (2023) 165–175. Available from: https://doi.org/10.1016/j.jcis.2023.04.145, http://www.elsevier.com/inca/publications/store/6/2/2/8/6/1/index.htt.

[35] Z. Ren, X. Liu, J. Su, Y.-Q. Liu, H. Zou, J. Tian, et al., Low-profile broadband microwave absorber based on magnetic coating and artificial electromagnetic structures, Chemical Engineering Journal 466 (2023) 143115. Available from: https://doi.org/10.1016/j.cej.2023.143115, http://www.elsevier.com/inca/publications/store/6/0/1/2/7/3/index.htt.
[36] M. Joshi, T. Bansala, S. Mukhopadhyay, Innovations in polymer nanocomposite coatings for EMI shielding applications, IOP Conference Series: Materials Science and Engineering 460 (1) (2018) 012012. Available from: https://doi.org/10.1088/1757-899x/460/1/012012.
[37] G.-dong Ban, Z.-hui Liu, S.-tian Ye, H.-bo Yang, R. Tao, P. Luo, Microwave absorption properties of carbon fiber radar absorbing coatings prepared by water-based technologies, RSC Advances 7 (43) (2017) 26658–26664. Available from: https://doi.org/10.1039/C7RA02631E.
[38] Z. Zeng, H. Jin, M. Chen, W. Li, L. Zhou, Z. Zhang, Lightweight and anisotropic porous MWCNT/WPU composites for ultrahigh performance electromagnetic interference shielding, Advanced Functional Materials 26 (2) (2016) 303–310. Available from: https://doi.org/10.1002/adfm.201503579, http://onlinelibrary.wiley.com/journal/10.1002/(ISSN)1616-3028.
[39] S. Kuester, C. Merlini, G.M.O. Barra, J.C. Ferreira Jr., A. Lucas, A.C. De Souza, et al., Processing and characterization of conductive composites based on poly(styrene-b-ethylene-ran-butylene-b-styrene) (SEBS) and carbon additives: a comparative study of expanded graphite and carbon black, Composites Part B: Engineering 84 (2016) 236–247. Available from: https://doi.org/10.1016/j.compositesb.2015.09.001.
[40] K.K. Gupta, S.M. Abbas, A.C. Abhyankar, Carbon black/ polyurethane nanocomposite-coated fabric for microwave attenuation in X & Ku-band (8–18 GHz) frequency range, Journal of Industrial Textiles 46 (2) (2016) 510–529. Available from: https://doi.org/10.1177/1528083715589752.
[41] L.C. Folgueras, M.A. Alves, M.C. Rezende, Evaluation of a nanostructured microwave absorbent coating applied to a glass fiber/polyphenylene sulfide laminated composite, Materials Research 17 (Supplement 1) (2014) 197–202. Available from: https://doi.org/10.1590/S1516-14392014005000009.
[42] T. Zhao, C. Hou, H. Zhang, R. Zhu, S. She, J. Wang, et al., Electromagnetic wave absorbing properties of amorphous carbon nanotubes, Scientific Reports 4 (2014) 5619. Available from: https://doi.org/10.1038/srep05619, http://www.nature.com/srep/index.html.
[43] V. Rubežienė, S. Varnaitė-Žuravliova, A. Sankauskaitė, J. Pupeikė, P. Ragulis, A. Abraitienė, The impact of structural variations and coating techniques on the microwave properties of woven fabrics coated with PEDOT:PSS composition, Polymers 15 (21) (2023) 4224. Available from: https://doi.org/10.3390/polym15214224.
[44] A. Zhu, H. Xing, Q. Fan, X. Ji, P. Yang, Conductive polyaniline coated on aluminum substrate as bi-functional materials with high-performance microwave absorption and low infrared emissivity, Synthetic Metals 271 (2021)116640. Available from: https://doi.org/10.1016/j.synthmet.2020.116640.
[45] H. Ren, T. Li, H. Wang, Z. Guo, T. Chen, F. Meng, Two birds with one stone: superhelical chiral polypyrrole towards high-performance electromagnetic wave absorption and corrosion protection, Chemical Engineering Journal 427 (2022) 131582. Available from: https://doi.org/10.1016/j.cej.2021.131582.

[46] Q. Zhou, C. Qi, T. Shi, Y. Li, W. Ren, S. Gu, et al., 3D printed carbon based all-dielectric honeycomb metastructure for thin and broadband electromagnetic absorption, Composites Part A: Applied Science and Manufacturing 169 (2023) 107541. Available from: https://doi.org/10.1016/j.compositesa.2023.107541.

[47] S. Bi, Y.Z. Song, G.-L. Hou, H. Li, Z.-H. Liu, Z.-L. Hou, et al., Sandwich nanoarchitectonics of heterogenous CB/CNTs honeycomb composite for impedance matching design and microwave absorption, Journal of Alloys and Compounds 943 (2023) 169154. Available from: https://doi.org/10.1016/j.jallcom.2023.169154, https://www.journals.elsevier.com/journal-of-alloys-and-compounds.

[48] L.B.T. Amudhu, R.V. Samsingh, S.E. Florence, B.S. Abirami, Novel radar absorbing material using resistive frequency selective surface based polymer composites for enhanced broadband microwave absorption, Proceedings of the Institution of Mechanical Engineers, Part L: Journal of Materials: Design and Applications 237 (10) (2023) 2115–2125. Available from: https://doi.org/10.1177/14644207231169362, https://journals.sagepub.com/home/PIL.

[49] J. Dong, Z. Li, C. Liu, B. Zhou, C. Liu, Y. Feng, Highly oriented MXene/polyvinyl alcohol films prepared by scalable layer-by-layer blade coating for efficient electromagnetic interference shielding and infrared stealth, Nano Research (2024) 5651–5660. Available from: https://doi.org/10.1007/s12274-024-6486-8, https://www.springer.com/journal/12274.

[50] Z. Deng, P. Jiang, Z. Wang, L. Xu, Z.-Z. Yu, H.-B. Zhang, Scalable production of catecholamine-densified MXene coatings for electromagnetic shielding and infrared stealth, Small 19 (46) (2023) 2304278. Available from: https://doi.org/10.1002/smll.202304278, http://onlinelibrary.wiley.com/journal/10.1002/(ISSN)1613-6829.

[51] L. Qi, Z. Xiang, X. Weng, Y. Wang, X. Wu, L. Yuan, Effect of high-temperature compositional and structural evolution of the polysiloxane coating with glass/Al powder on the infrared and mechanical properties, Polymer Degradation and Stability 215 (2023) 110461. Available from: https://doi.org/10.1016/j.polymdegradstab.2023.110461.

[52] Z. Ma, Q. Li, J. Wei, C. Liang, T. Yang, G. Wang, et al., Effects of Al-based alloy powders on the mechanical behavior, corrosion resistance and infrared emissivity of polyurethane composite coatings, Colloids and Surfaces A: Physicochemical and Engineering Aspects 624 (2021) 126782. Available from: https://doi.org/10.1016/j.colsurfa.2021.126782.

[53] J. Zhang, W. Zhang, Q. Guan, X. Li, D. Lv, Preparation and properties of epoxy resin and polyurethane blend resin-based low-infrared-emissivity coatings, Coatings 12 (11) (2022) 1708. Available from: https://doi.org/10.3390/coatings12111708.

[54] X.-Y. Li, L.-S. Zong, Y.-B. Wang, Q. Cao, J.-Y. Wang, X.-G. Jian, Construction of aniline trimer based conjugated polymers through Schiff base reaction and their use as feedstock for infrared stealth coatings, Chinese Journal of Polymer Science (England Edition) 41 (6) (2023) 851–858. Available from: https://doi.org/10.1007/s10118-022-2861-7, https://www.springer.com/journal/10118.

[55] X. Chai, D. Zhu, Y. Liu, Y. Qing, F. Luo, Z. Huang, et al., In-situ construction of Cr_2O_3@ATO hybrid pigment towards synergetic enhancement of visible light-infrared-radar compatible stealth, Journal of Colloid and Interface Science 645 (2023) 570–579. Available from: https://doi.org/10.1016/j.jcis.2023.04.175, http://www.elsevier.com/inca/publications/store/6/2/2/8/6/1/index.htt.

[56] X. Chai, D. Zhu, Q. Chen, Y. Qing, K. Cao, F. Luo, et al., Tailored composition of low emissivity top layer for lightweight visible light-infrared-radar multiband compatible stealth coating, Advanced Composites and Hybrid Materials 5 (4) (2022) 3094–3103. Available from: https://doi.org/10.1007/s42114-022-00563-7, https://www.springer.com/journal/42114.
[57] Y. Wang, M. Yu, Y. Gao, S. Liu, X. Yin, N. Pang, et al., Three-layer composite coatings with compatibility of low infrared emissivity and high wave transmittance, Journal of Alloys and Compounds 943 (2023) 169038. Available from: https://doi.org/10.1016/j.jallcom.2023.169038.
[58] M.-J. Youh, Y.-R. Huang, C.-H. Peng, M.-H. Lin, T.-Y. Chen, C.-Y. Chen, et al., Using graphene-based composite materials to boost anti-corrosion and infrared-stealth performance of epoxy coatings, Nanomaterials 11 (6) (2021). Available from: https://doi.org/10.3390/nano11061603, https://www.mdpi.com/2079-4991/11/6/1603/pdf.
[59] K.Y. Fang, F. Fang, Au-decorated SWNT/PVDF electrospun films with enhanced infrared stealth performance, Materials Letters 230 (2018) 279–282. Available from: https://doi.org/10.1016/j.matlet.2018.07.116.
[60] Y. Chang, Y. Wang, W. Wang, D. Yu, Highly efficient infrared stealth asymmetric-structure waterborne polyurethane composites prepared via one-step density-driven filler separation method, Colloids and Surfaces A: Physicochemical and Engineering Aspects 614 (2021) 126177. Available from: https://doi.org/10.1016/j.colsurfa.2021.126177.

18

Risk assessment of nanopaints coupled with environmental and economic feasibility

Neeta Shukla[1] **and Sumit Gupta**[2]

[1]Department of Physics, BGS College of Engineering and Technology (BGSCET), Bengaluru, Karnataka, India [2]Department of Mechanical Engineering, Amity School of Engineering and Technology, Amity University, Noida, Uttar Pradesh, India

18.1 Introduction

"Nanometer," referring to measurements on the order of billionths of a meter (10^{-9} m), was first coined by Nobel laureate Richard Zsigmondy in 1925. Derived from the Greek term for "dwarf," the prefix "nano" connotes genuinely tiny. Application of nanoparticles (NPs) in paints has, over the past two decades or so, entered, as it were, the big league of commercial automobile paint coatings. The National Nanotechnology Initiative in the United States defines nanotechnology as an interdisciplinary science, engineering, and technology endeavor at the nanoscale. In most definitions, it ranges from 1 to 100 nm. Some of the material's unique properties exhibit incredible applications coupled with chemistry, physics, biology, medicine, engineering, and electronics.

Interestingly, the histories of nanotechnology track back to ancient times [1]. Already in the fourth century CE, the Romans used NPs in their glass production; an example of their masterpiece is the famous Lycurgus Cup held in the British Museum. This marvelous artifact belongs to one of the earliest known uses of dichroic glass, changing its color depending on the lighting conditions. The three key categories that have dominated the marketplace in today's paint industry are decorative, industrial, and specialty coatings. Interior and exterior coatings for commercial, industrial, institutional, and residential buildings are needed—these serve not only to ornament the product but also make it more resilient and protective for products,

Nanopaints. DOI: https://doi.org/10.1016/B978-0-443-26512-9.00019-7

such as appliances, furniture, and transportation goods that are all compounded to endure certain conditions so they can function optimally in a variety of environments [2].

This paint market analysis emphasizes the growing environmental concern that has led to a shift from solvent-borne coatings emitting high levels of volatile organic compounds (VOCs) to more eco-friendly alternatives. These alternatives include waterborne, high solids, powder, and radiation-curable coatings. Waterborne coatings dominate due to their advantages, like less odor, lower VOC emissions, reduced viscosity, and ease of application. This trend reflects the industry's effort to reduce environmental impact while meeting regulatory and consumer demands for greener products. Silicon (Si), silver (Ag), titanium (Ti), and zinc (Zn) as nanomaterials (NMs) each bring unique properties that make them highly valuable in nanopaints. Here's a breakdown of their unique properties in nanopaint formulations.

Si-based nanopaints improve hydrophobicity, especially silica (SiO_2), and provide excellent water-repellent properties, making surfaces more resistant to moisture, stains, and mold. It also promotes self-cleaning, particularly in coatings like hydrophobic or photocatalytic surfaces, reducing dirt's adhesion and allowing rainwater to wash it away easily [3]. It is also helpful for generating ultraviolet light (UV) resistance, thermal stability, and low VOC emissions. Si NPs can help increase resistance to UV degradation, protecting the surface from fading, cracking, or becoming brittle over time. Si-based nanopaints perform well in high-temperature environments, retaining their structural integrity and protective qualities under thermal conditions. Si-based paints release fewer VOCs, making them more eco-friendly and safer for indoor environments.

Ag in nanopaints enhances antimicrobial properties, durability, and conductivity. AgNPs are well-known for their antimicrobial solid effects, which inhibit the growth of bacteria, fungi, and viruses on painted surfaces. This makes them ideal for healthcare, food processing, and other hygienic environments. Catalytic activity: AgNPs can act as catalysts, breaking down organic pollutants and providing self-cleaning or air-purifying benefits to surfaces coated with the paint. AgNPs improve the durability of coatings, making them more wear and tear-resistant while protecting against tarnishing or corrosion. AgNPs are excellent conductors of electricity, which is helpful for conductive coatings in electronics or other specialized applications.

Ti in nanopaints promotes photocatalytic activity. Titanium dioxide (TiO_2) NPs are widely used for their strong photocatalytic properties. Under UV or visible light, TiO_2 can break down organic pollutants, bacteria, and odors, making it ideal for self-cleaning and air-purifying paints. TiO_2 is highly effective at absorbing.

Zinc oxide (ZnO) nanopaints also offer UV protection, antimicrobial properties, antimicrobial properties, and unique characteristics, such as durability and scratch resistance inclusion, which increase the hardness and scratch resistance of coatings. This enhanced durability makes them suitable for high-traffic areas or surfaces exposed to physical wear, improving the overall lifespan and reducing maintenance needs.

Paint commonly refers to a liquid applied to any solid surface for coating, coloring, sealing, or shielding the surface. Nowadays, NMs are being implemented in paints to increase their adhesiveness effectiveness and to develop new textures and features. Paint commonly refers to a liquid applied to any solid surface for coating, coloring, sealing, or shielding the surface. Nowadays, NMs are being implemented in paints to increase their adhesiveness effectiveness and to develop new textures and features.

Nanopaint is a specialized coating capable of altering the properties of a surface or material based on predetermined parameters set by the user. Like conventional paint, nanopaint is initially applied as a liquid and solidifies. Within the liquid exists a suspension of minuscule particles known as nanotubes, which adjust their behavior in response to external conditions or upon receiving specific commands [4]. Currently, nanopaint is undergoing research and development stages. The functionality and characteristics of paints, such as their water and dirt-repelling properties, ease of cleaning, UV protection, antimicrobial resistance, and scratch resistance, are believed to be enriched using NMs. The preparation of nanopaints depends on the specific requirements of consumer or industrial organizations, such as radiation-free, decorative appearance, water-prone, microorganisms protective, fire protection, and thermal insulation. The performance and quality of any surface can be elongated with the application of engineered nano-based paints.

Nanopaints incorporate NPs (made up of distinct metals, metal oxides, polymers, or carbon-based materials like graphene) into their formulation. Nanopaints have gained significant interest and attention due to their unique properties and potential applications across various industries. These properties may include increased durability, improved scratch resistance, better surface adhesion, corrosion resistance, and enhanced UV protection. Additionally, depending on the type and mixture, nanopaints can offer functionalities, such as self-cleaning, antimicrobial properties, or even the ability to change color in response to environmental factors.

The functionality and characteristics of paints, such as their water and dirt-repelling properties, ease of cleaning, UV protection, antimicrobial resistance, and scratch resistance, are believed to be enriched using NMs. The preparation of nanopaints depends on the specific requirements of consumer or industrial organizations, such as radiation-free, decorative appearance, water-prone, microorganisms protective, fire protection, and thermal insulation. The performance and quality of any surface can be elongated with the application of engineered nano-based paints.

Despite having such important characteristics, NPs might be dangerous for the environment. The planet's ecosystem and human health may be affected by their existence in the air or water. Every day, millions of tons of potentially hazardous trash comprising nano- and microparticles are generated and disposed of due to human activities. Managing these particles necessitates a thorough understanding of the physics governing how NPs agglomerate and adhere to various surfaces. There is ongoing research and development in nanopaints to address these challenges and explore new applications. Researchers are investigating safer NPs, more efficient manufacturing processes, and novel functionalities to expand the utility of nanopaints across various industries. Today's most widely used photocatalysts are NMs, including TiO_2 and ZnO [5]. Their applications are diversified in nature. They are not only just applied as pigment in paints or coatings but also exhibit very vital antimicrobial properties. These become more important as public health concerns rise due to harmful microorganisms causing diseases. One major challenge is the photodegradation of organic components in paints, which begins to degrade the permanence and alter the color and appearance of paint over time. Potassium silicate, sodium silicate, lithium silicate, and phosphates are some inorganic binders promising alternatives that can offer a balanced approach toward making it both functional and long-lasting. This chapter reports on the improved photocatalytic performance of Ag-deposited ZnO (Ag/ZnO) and TiO_2 (Ag/TiO_2) NPs, where their performances in the degradation of one of the popularly used antibiotics, levofloxacin, are demonstrated under both UV and visible light.

The chapter provides an overview of understanding the benefits and distinct risks and assessment of nanopaints, which is also the current issue of investigation by researchers.

18.2 Literature review

Nanotechnology has revolutionized the construction process as it is now feasible to create highly advanced material products that reflect superior qualities and functionalities. Thus NMs boasting enhanced strength and durability, resistance to fire and thermal stability, self-cleaning properties, and photocatalytic action are being used in all types of cement, mortar, concrete, paints, coatings, insulation materials, and glass, among others. However, sustainability concerning these innovative materials is still in its infancy and needs to be understood extensively. Researchers have critically reviewed several antimicrobial additives across the board, from TiO_2, Ag, ZnO, and quaternary ammonium salts, with their performance in functional paints and coatings. Results have manifested that TiO_2, Ag, and ZnO NMs can act as high-performing antimicrobial agents [3]. They also represent potentially effective photocatalysts compatible with some inorganic binders used for paints and coatings, thus making them suitable for different coating applications. Among these, TiO_2 is especially attractive because of its inert chemical, nontoxic, and eco-friendly, which contribute unique advantages over nano-Ag and nano-ZnO. Nanoscale materials (1–100 nm) emphasize uniquely dimensioned material shapes with more surface area, translating to greater demand in the automotive, construction, health, and electronics sectors. Public awareness and debate in the early 2000s catapulted the first commercial applications of nanotechnology. While TiO_2 and silicon dioxide (SiO_2) stand at the fore of today's benefits appear to persist, other NMs, including ZnO, Ag, aluminum, cerium dioxide, copper oxide, and magnesium oxide, open new paths in research into their possible use. In this way, NMs can be selected to meet specific functional needs. Therefore it is available as free powders, stabilized suspensions, and even masterbatches and granules to ensure effective incorporation into the paint matrix. Here, under their structural features, NMs promise to give paints a leap of stature and usher in a new era in construction technology.

18.2.1 Manufactured nanomaterial

In current scenarios, the possible effects of manufactured nanomaterial (MNM) coatings (such as nano-TiO_2, nano-Ag, or nano-SiO_2) on the environment, human health, and safety should be considered holistically and prospectively throughout the entire life cycle. This concept is grounded in an innovative approach known as life cycle thinking, which integrates two complementary facets: a comprehensive analysis of the opportunities and risks across the entire product life cycle and an life cycle assessment (LCA) study by ISO 14040 standard ISO 2006a [6,7].

Research has identified TiO_2 as the ideal photocatalytic material for the paint industry, which can break and oxidize surface impurities and air pollutants like NOx into harmless gases [8,9]. However, recent studies have uncovered a sad limitation of this discovery—the photolysis of some VOCs is enhanced by the irradiation of photocatalytic paints, meaning that the latter may be formed in the environment rather than decreased. More detailed investigations revealed that TiO_2, even when exposed to visible light, can photogenerated carbonyl compounds, as indicated in many reports. Modifications of the NPs, including several coatings or optimization of their dispersion inside the organic matrix, might also be responsible for the overall modifications in photocatalytic efficiency to minimize degradation while maintaining optimal photocatalytic activity to give "safer-by-design" photocatalytic paints. Interestingly, several studies show that the TiO_2 NP shows minimal toxicity effects at lower concentrations. No significant harmful effects of TiO_2 NPs are observed in the set concentrations chosen for the study [10]. It is noteworthy, however, that such emissions from VOCs do not correlate with the degradative efficiency of the organic matrix in these paints [11]. This has brought forth the complexities of photocatalytic processes and underlines the need for relentless research efforts to balance efficacy and safety in using such paints containing TiO_2 NPs.

The manufacture of NMs and nanopaints significantly impacts the ecology. One possible advantage that could more than offset the more significant effect of producing nanopaints instead of conventional ones is the ability to prolong the paints' lifespan using NMs in the composition. The potential functions of coating NPs may influence the functionality of materials, such as composites, coatings, textiles, and energy production, excluding food and medical applications. The typical characterization of the NP types, including TiO_2 and SiO_2 in paints, consists typically of a diversity of powerful techniques for getting information on their morphologies, size, and chemical compositions. Such include X-ray diffraction, scanning electron microscopy (SEM), transmission mode SEM, energy dispersive X-ray spectroscopy, and scanning auger microscopy [12]. The principal driving force behind the growing addition of NMs to paints and lacquers is their potential to change the properties of surfaces and make, for example, surfaces self-cleaning or scratch-resistant [13].

More importantly, genotoxicity studies demonstrated that the paint or lacquer matrix contributes much more to DNA damage than NMs. In some samples, particular dust materials did not demonstrate a significant increase in DNA damage despite containing NPs within a paint or a lacquer matrix [14]. This fact demonstrates the importance of overall formulation when assessing risks associated with nanotechnology in coatings. Paints with MNM perform better in terms of the environment than paints without MNM [15]. The analysis demonstrated that several factors influence the results. Firstly, MNM cannot just be an extra ingredient; it must replace one of the original paint composition's active ingredients. Secondly, the new composition must increase paint longevity for a long time, reducing paint consumption during a building's life cycle. Thirdly, It is necessary to minimize MNM releases, especially by discarding leftover paint along with its packing.

18.2.2 Ecology of manufactured nanomaterial

The manufacture of NMs and nanopaints significantly impacts the ecology. One possible advantage that could more than offset the more significant effect of producing nanopaints as opposed to conventional ones is the ability to prolong the paints' lifespan by NMs in the composition. The manufacture of NMs and nanopaints significantly impacts the ecology. One possible advantage that could more than offset the more significant effect of producing nanopaints instead of conventional ones is the ability to prolong the paints' lifespan using NMs in the composition. Profound implications have been reported by distinct scientists on Ag-deposited ZnO (Ag/ZnO) and Ag-deposited TiO_2 (Ag/TiO_2) NPs [16]. These NPs enhance performance in degrading the antibiotic levofloxacin, particularly under UV and visible light irradiation. The presence of Ag on ZnO and TiO_2 surfaces typically improves their photocatalytic efficiency due to enhanced visible light absorption as Ag can absorb visible light more effectively, making the catalysts responsive not only to UV light but also to the visible spectrum, increased electron-hole separation as Ag acts as an electron trap, reducing the recombination of the photogenerated electron-hole pairs, which is crucial for photocatalysis, the enhanced surface plasmon resonance (SPR) causing the localized SPR effect of Ag contributes to better absorption and utilization of light, further enhancing the photocatalytic activity. In this sequence, enhanced photocatalytic properties make Ag/ZnO and Ag/TiO_2 effective in breaking down levofloxacin, an antibiotic that can persist in the environment.

NPs are incorporated into polymer matrices as fillers in conductive nanopaints. Metallic NPs and nanostructured carbons are the primary filler types used to formulate conductive nanopaints. The specific type of NM employed directly impacts the properties of the conductive nanopaint, such as electrical conductivity, corrosion resistance, and scratch resistance. [17]. The release of NMs is dispersed throughout every product's life cycle stage, from manufacturing and formulation through its use and eventual recycling or landfill placement. However, traditional research studies have been constrained by insufficient data in situations where previous datasets have become fragmented toward exposure estimation, such as for considerations of release, emission, or environmental fate [18]. Although there is considerable information on the release of NMs due to abrasion and weathering factors, the overall ecology of MNMs is still an emerging field. Science must continue research and, therefore move forward to understand the long-term ecological implications of these materials better, thus furthering safer nanotechnology development [19]. Such knowledge can significantly provide substantial support in further mitigating the risks associated with all stages of their life cycle.

18.2.3 Risks associated with nanopaints

The nanotechnology revolution heralds a new era in materials science, the golden light of innovation coming from this area is nanopaints. The addition of NPs to the formulations of paint could significantly enhance performance and improve durability, UV resistance, and antimicrobial activity. However, even such a benefit from using nanopaints must be matched against accumulated potential risks so that they can be used safely and responsibly. NPs have dualism in that they are good and bad. It seems that NPs can be responsible for some specific kinds of environmental degradation by their chemical interactions with specific environmental conditions. Some particles might be transformable and therefore would lead to the formation of toxic byproducts. Furthermore, because of their small size, NPs would enter and interact with biological systems differently compared to larger counterparts. Their persistence and bioavailability in the environment thus become a matter of concern [20].

On the contrary, NPs have tremendous advantages, such as aiding the water purification process through enhanced plant growth and easier production of greener chemicals derived from natural extracts instead of hazardous chemicals. This can potentially result in lower toxicity profiles in comparison to the traditional manufacturing procedure. Additionally, NPs find numerous applications in renewable energy technologies and drug delivery systems, among others [21].

While nanopaints offer a strong rationale for performance and sustainability, one must understand and balance this risk assessment to identify and manage the risks associated with their application. One of the most pressing risks associated with nanopaints is their vagrant behavior in the environment. Because NP material is unique, it creates unexpected interactions with other chemicals and biological systems. In some cases, the NPs will have varying reactivity compared to their bulk states, thereby leading to unforeseen toxicological effects. The effect of nanotoxicity can appear in different cytotoxic, genotoxic, and immunotoxic ways [22].

NPs can be seen to pass through cellular membranes and interact with organelles, which can be potential causes of oxidative stress and inflammatory response. Effects of such nature raise concerns over long-term exposure in manufacturing settings, consumers applying the paint, and living environment where nanopaint is applied. Therefore there is a need for rigorous testing and safety procedures to drive away such risks. Yet another significant matter related to nanopaints is bioaccumulation. Due to the small dimensions and unique features of NPs, they can readily enter biological systems and subsequently accumulate in organisms. Eventually, this bioaccumulation could cascade through food chains, in which toxic effects might be magnified as "NPs migrate up the trophic levels. A simple example is if aquatic organisms absorb NPs from contaminated water, then these NPs could enter the diets of larger predators, including humans [23].

It has been indicated that higher reactivity in NPs raises several questions regarding their fate in the environment. Once the NPs get into the environment, they react with soil, water, and air to result in complex chemical reactions. Certain of these particles may break down or change into something even harmful, which risks causing more danger to ecosystems and human health. This makes it mandatory to understand and study the entire life cycle of nanopaints from production and application to disposal. Nanotechnologies have advanced far quicker than the regulatory frameworks assuring their safe and effective utilization. Most of the regulations, which were developed currently, did not entirely consider the specificity of NPs and some associated hazards. Thus it poses the risks of being overly complex, too expensive, or both for the manufacturers and consumers alike when complying with the regulation.

Manufacturers must find their way in a regulatory patchwork, which by country varies and even sometimes is vague about specific nanomaterials. This is also so on the definite regulatory control, where the market introduction of harmful products is possible without adequate data on their safety. Consumers, being unaware, may expose themselves to risks relating to nanopaints. Public perception plays a very significant role in the acceptance of nanotechnology and its applications of nanopaints. The fear of nanopaints among the public may overcome several other aspects where nanopaints outperform their variants due to issues related to safety and environmental impact [24]. Misinformation or misperception of nanotechnology compounds the problem, thereby making the public skeptical toward such paints. Another relevant dimension of nanopaints is economic consequences. Often, the costs of compliance, testing, and public relations efforts pose a heavy burden on manufacturers' budgets. These costs may filter their way to the consumers and thus limit market prevalence of nanopaints, especially in areas sensitive to prices. For instance, though strong perceptions from the public and acceptance would raise demand and create new business opportunities for companies offering safe and open products, strong perceptions would increase market demand while weak perceptions would result in lower demand and fewer business opportunities for companies selling safe and open products.

As the scenario of nanotechnology is constantly in transition and a new risk is to be seen emerging each day during research, continuous monitoring and assessment of nanopaints become inevitable to stay ahead of the potential hazards. Further collaboration by researchers, regulatory bodies, and industry stakeholders may further evolve ways for best practices and safety guidelines in relation to reducing risks. Proper handling and safety measures need to be in place at each stage of nanopaint from manufacturing to application and disposal. Secondly, there is also the need for continuous research into interactions between NPs with biological systems as well as their long-term environmental impact [25].

18.2.4 Risk assessment coupled with environment

Engineered NMs (ENMs) risk assessment is unlike any traditional risk assessment because the material properties are at the nanoscale. While the general components of a risk assessment remain feasible—that includes hazard identification, hazard characterization, exposure assessment, and risk characterization—the characteristics of the ENMs add new complexity. Characterizing materials at the nanoscale requires specific techniques, and existing toxicity tests may require adaptation to assess the risks associated with new materials accurately. NPs have been included in most lines of business due to their unique physical, chemical, and biological properties. However, their lifecycle emissions are challenging to trace. Various methodologies have been utilized to trace these emissions, especially in nanopaints [26].

In nanopaint formulations, NPs improved the properties of durability, UV resistance, self-cleaning, and bonding. Despite the wide variety of benefits, these NMs' lifecycle threatens human health and the environment. Therefore it is necessary to carry out thorough risk assessments so as not to compromise safety while utilizing the advantages of nanotechnology in paints and coatings. NPs, because of their size, can travel long distances in air and water. Many NPs, mainly metal oxides (such as ZnO and TiO_2), are highly persistent in the environment. They do not degrade quickly, meaning they can remain in ecosystems for long periods, potentially causing long-term harm. This increases their potential to spread widely and impact distant ecosystems. Organisms can take up NPs, especially in aquatic environments, where they can bioaccumulate through food chains. This can lead to higher concentrations of toxins in organisms, including fish, birds, and even humans, causing long-term ecological and health issues. While NMs can enhance paint performance, their high surface area and reactivity can make them more toxic to aquatic life and microorganisms than their bulk counterparts.

Environmental risk assessments for **Si**, **Zn**, and **Ti**-based nanopaints focus on the unique behavior and characteristics of these NMs in the environment. The primary environmental concerns arise from the potential toxicity, bioaccumulation, and persistence of NPs, which can affect ecosystems and various forms of wildlife. Understanding the entire lifecycle of nanopaints, from raw material extraction to disposal, becomes essential to minimizing their environmental and health impacts. LCA not only helps identify potential hazards and exposure pathways but also provides insights into mitigating these risks. By addressing the entire lifecycle, industries can develop safer, more sustainable nanopaints that offer the benefits of nanotechnology while protecting ecosystems and human health.

The lifecycle of nanopaints is critical and involves necessary steps, such as the **production and sourcing of NMs**, manufacturing and formulation, application, and end of the life cycle, that is, disposal and recycling. Several case studies indicate that TiO_2 is frequently used in nanopaints for its UV-blocking and self-cleaning properties, but studies have shown that TiO_2 NPs can be toxic to aquatic life when released into water systems [27]. However, a potential solution can be found by evaluating risk mitigation through LCA. For instance, implementing filtration systems in water treatment plants to capture NPs before they enter water bodies and ensuring proper disposal of paint residues can help minimize environmental contamination. It was determined through abrasion tests of composites or coatings exhibiting NPs that it is improbable that free NMs are released. Environmental issues also arise in scenarios involving the destiny of composite particles released due to the abrasion effect and their potential to liberate NMs upon subsequent exposure to environmental factors like weathering.

An enhanced understanding of NMs' environmental behavior, fate, and impacts has significantly advanced over the past decade. As a result of insights gained from testing NMs, it has become evident that the unique characteristics of testing and evaluating these materials in the environment should be considered when developing testing and assessment strategies. Consequently, conducting environmental risk assessments of NMs is feasible. However, uncertainties persist in the robustness of such assessments due to the absence of standardized test methods, guidance for hazard assessment, updated data requirements, adjusted default values, and monitoring data.

Elevated understanding of the environmental behavior, fate, and impacts of NMs has significantly advanced over the past decade. As a result of insights gained from testing NMs, it has become evident that the unique characteristics of testing and evaluating these materials in the environment should be considered when developing testing and assessment strategies. Consequently, conducting environmental risk assessments of NMs is feasible. However, uncertainties persist in the robustness of such assessments due to the absence of standardized test methods, guidance for hazard assessment, updated data requirements, adjusted default values, and monitoring data.

The risk assessment of ENM is unique compared to the traditional framework due to their property of being nanoscale in size, which provides different complexities. While the general practice of risk assessment, which includes hazard identification and exposure evaluation, applies to ENMs also, they demand more peculiar approaches than others in assessing the potential risk since they have a high surface area and reactivity that calls for some changes in the available toxicity tests, which are more significant because of their vast probable application in most sectors of industries [28]. Nanopaints are yet another important application of ENMs that also hold promises in enhanced hardness, Ultraviolet light (UV) resistance, and self-cleaning properties. Nanopaints' lifecycle risks human health and the environment, from production to disposal. As NPs are highly sized and reactive, they can travel through air and water and accumulate in ecosystems, such as soil and biota. The best illustration would be that of metal oxides, such as ZnO and TiO_2 used in nanopaints; they are not biodegradable, thus remaining persistent in the environment with a tendency for bioaccumulation in the food chain, affecting predators, such as fish and birds and posing long-term ecological risks [29].

The behavior of NPs, such as bioavailability and bioaccumulation, must be considered in evaluating environmental risks. Bioaccumulation was reported to affect the higher toxicity of ZnO and TiO_2 NPs in marine organisms. Such unique behaviors included in methodologies assessing risk can guide the development of a regulatory framework. The exposure pathways for nanopaints from extraction up to disposal need to be evaluated to identify stages for mitigation [30]. Minimizing the environmental impact of nanopaints starts with recycling. Reusing NPs to form novel products can reduce environmental leakage and support a circular economy. This would reduce the need for raw materials, reducing waste generation. Closing the lifecycle loop for industries reduces the environmental burden and that of material.

LCA methodologies can be beneficial in finding risks and promoting safety interventions that concern the safe handling of NMs. NPs in wastewater may be filtered before finally entering aquatic ecosystems. More importantly, proper disposal and recycling should be ensured so NPs do not spread into the environment. Another potential strategy for sustainability is to recycle waste from nanopaints by reclaiming them to be integrated into production lines [31,32].

Although initial tests suggest that NPs are less likely to leach out of coatings under normal conditions, further research must ensure long-term behavior under changing conditions. Continuous improvement in recycling technologies and considerable improvement in regulation aspects help balance innovation with environmental sustainability for the industries using NMs [33].

Traditional risk assessment may not aptly portray the behavior of NPs in complex ecosystems. New protocols that account for NPs' impact on microbial communities and aquatic life, even at trace concentrations, need to be established. In the future, evaluations must employ new NP detection methods in environmental matrices like water and soil to ensure qualitative assessments. Recycling programs must also lay down precautions against environmental contamination. The NM has also shown this application increase, highlighting the challenges and issues associated with the regulatory and societal levels [34]. Most of the frameworks lack the issues behind the pace of the technology, as a gap is left for oversight purposes. In that case, there is, therefore a need for interdisciplinary collaboration between scientists, regulators, and industry players to come up with adequate risk assessments. Recycling regulations should also be included in the described frameworks to ensure that products like paints are recycled.

Due to nanotechnology's global impact, harmonizing international regulations for NMs will also be an urgent issue. Public engagement and transparency might be the only ways to inculcate confidence in the technology and promote prudent development. Also, public participation in recycling nanopaints could minimize environmental risks [35,36].

Risk assessment methodologies should be updated with new scientific findings as NM research progresses. Although there is currently a lack of known long-term effects from releasing NMs into the environment, continued research on bioaccumulation, recycling processes, and ecological impacts further helps reduce these uncertainties. Interdisciplinary research is crucial in developing safer NMs and reducing their ecological footprint.

18.2.5 Risk assessment and economic implications for nanopaints

The process of risk assessment combined with economic evaluation for nanopaints is complex and requires in-depth looks at the potential risks associated with using NPs while weighing economic implications throughout the product lifecycle. Nanopaints use advanced materials like TiO_2, ZnO, and SiO_2, improving the performance characteristics to exhibit a higher UV resistance, antimicrobial properties, and general durability. These advantages make nanoscopic paints appealing in automotive, construction, and electronics markets. Nonetheless, using nanotechnology also involves unique risks and costs that must be weighed against these economic benefits. Careful considerations of market demand and consumer adoption, production and supply chain risks, the cost of complying with regulations, and environmental and broader economic implications must be considered and reviewed carefully [37].

The biggest challenge in the nanopaints field is that NMs are used in production. Generally, their manufacturing processes consume much more energy than traditional varieties. This can consequently lead to high production costs, which are often too high for manufacturers to absorb through adjustment of prices. In addition, supply chain effects, mainly sourcing raw materials for the creation of NPs, may influence the price volatility of these nanopaints. For instance, the availability and cost may be drastically affected by geopolitical tensions, trade policy changes, and environmental factors regarding some specific raw materials [38]. The economics of nanopaints may be vulnerable based on the sensitivity of prices in various areas. Although these paints may offer good performance and longevity, higher initial prices may serve as a limitation to market access, especially by consumers who are constrained by budget restraints and industries.

However, the long-term benefits of nanopaints might be improved for less frequent maintenance services because of increased durability. For example, durable paints with robust protection for the automotive and construction industries could gradually reduce the total cost of ownership over time. Thirdly, the regulatory environment for NMs is becoming increasingly onerous, and strong environmental and safety legislation is controlling them. Such compliance requires large financial outlays on testing, safety precautions, and compliance certifications. Failure of such companies to comply with the regulatory standards exposes them to severe financial sanctions in the form of fines and physical ramifications in the form of legal implications and severe damage to their reputation. Environmental contamination due to improper handling and disposal of nanopaints poses the threat of high remediation costs and depletion of industries reliant on a clean environment [39].

This leads to new markets and economic benefits through sustainability certifications and the development of eco-friendly nanopaints. Sustainability in products is increasingly becoming a source of consumption for environmentally conscious consumers, thus offering companies an investment opportunity in nanotechnology at the cost of being an innovative leader in their field. These firms can obtain an advantage in overseas markets, which translates to revenue through patents, market dominance, and enhanced product performance. Although the cost of development and installation might be high in the case of nanopaints, the long-term economic benefits cannot be overlooked.

Economic analysis should adopt risk assessment for the safe and sustainable use nanopaints. This is why companies can avert risks and spur long-term economic growth by balancing the higher costs of nanotechnology with the gains made on innovation, performance improvements, and conformity with regulations. Failure to consider the risks of nanopaints for their producers, buyers, and other stakeholders might lead to substantial financial loss. The unsanitary nature of nanopaints could result in a costly mistake with product recalls, legal battles, and regulatory fines. A good risk analysis involves consideration of events that could lead to some loss of assets, investment failures, or loan defaults. Therefore companies, governments, and investors must comprehensively conduct detailed risk assessments before entering new ventures in the nanotechnology domain [40].

18.3 Discussion

Nanoscale Si and TiO_2 are the desired NMs of interest for the paint industry. However, parallel research on Ag, ZnO, aluminum oxide, cerium dioxide, copper oxide, and magnesium oxide is ongoing. To enhance the safety and effectiveness of future photocatalytic nanopaints, widely used in the industry, we have discussed the morphological structure, chemical composition, structural parameters, and properties of TiO_2 NPs. In contrast, nano-based paints can effectively reduce VOCs, such as xylene; the degradation of the organic matrix results in the release of harmful carbonyl compounds like formaldehyde. Abrasion experiments were carried out to examine how photocatalytic paints endure weathering and aging, uncovering the dispersion of TiO_2 NPs into the air. The effectiveness of TiO_2-infused paints in ameliorating air quality is of concern. In response, we advocate for a "safer-by-design" strategy, suggesting various enhancements to these paints before their introduction to the market.

Nowadays, research has demonstrated that Nano SiC paint exhibits exceptional efficacy in enhancing the abrasion resistance, hardness, and adhesion of paint films on wood flooring. Moreover, this paint film boasts increased durability and weather resistance, providing superior protection for engineered wood flooring. Through comparison of various groups with different ratios of nano SiC powder additions, the optimal mass fraction is determined to be 2%. Similarly, advancements in nanotechnology applied to paints enable the manipulation of radio frequencies, allowing for the blocking or passage of specific wavelengths with nanopaints. Additionally, there is a growing demand for conductive nanopaints and coatings in the automotive industry to enhance substrates' structural, thermal, and electrical properties while achieving a chrome-like appearance.

Despite the numerous benefits of engineered nanopaint materials, challenges persist in managing the disposal or treatment of NMs. While previous studies have discussed appropriate models for risk assessment, insufficient data availability has been noted in some cases. The characteristics of NMs, including size and shape, significantly influence optical, magnetic, electrical, physical strength, surface area, sensitivity, stability, and reactivity. Therefore employing accurate models to assess the toxicity of specific classes of NMs is advisable for more effective risk assessment outcomes. Moreover, nanotechnology has opened up new opportunities for creating advanced materials with unique properties and enhanced functionality in the construction industry. NMs, such as TiO_2, Ag, SiO_2, and ZnO have significantly improved the performance of construction materials like cement, concrete, paints, and coatings. These NMs offer benefits like increased durability, fire resistance, self-cleaning, and photocatalytic abilities, making them highly effective in various applications.

However, the sustainability of nanotechnology in construction is still in its early stages, and more in-depth research is needed to understand these materials' environmental and health impacts fully. While NMs like TiO_2 are recognized for their eco-friendly and nontoxic properties, others, such as nano-Ag and nano-ZnO, may pose environmental risks, particularly regarding bioaccumulation and potential toxicity. On the other hand, Si-based paints provide excellent durability, water resistance, and eco-friendliness, but their higher cost and complexity of application can be limitations for some users. As the use of NMs in construction grows, it is essential to adopt a life cycle approach that considers both the benefits and risks of these materials, ensuring their safe and responsible use. Ongoing research, safety regulations, and environmental risk assessments are critical in balancing the technological advantages of nanotechnology with its potential ecological and health-related risks.

18.4 Conclusion

Nanopaints may release NPs into the air during sanding, spraying, and other application processes or through composites in the air or water. Proper ventilation and personal protective equipment are necessary to minimize exposure risks to workers and bystanders. NPs in nanopaints could exhibit higher toxicity than larger particles of the same material due to their increased surface area and reactivity. This heightened toxicity may harm human health and the environment if proper precautions are not taken during production, application, and disposal. NPs in nanopaints may agglomerate over time, affecting the desired properties and performance of the paint. Strategies to prevent or mitigate NP agglomeration are essential to maintain product efficacy. Despite extensive research, the long-term effects of NP exposure in nanopaints are not fully understood. Continuous monitoring and research are necessary to assess potential health and environmental impacts over time. The chapter offers guidance on supporting aspiring researchers in academic or nonacademic realms to advance their careers and cultivate a fresh cohort of dynamic, adaptable, entrepreneurial, and inventive early-stage researchers. This is achieved through hands-on experience in independent and interdisciplinary research, participation in local and network-wide training initiatives, and engaging in intersectoral and international placements. By examining the advantages and potential pitfalls of early-stage innovation, minimizing risks and capitalizing on the opportunities presented by nanopaints is possible. Given the evolution of nanotechnology, it is essential to remain vigilant, as new risks associated with nanopaints may arise. Thus continual monitoring and research are imperative for proactively identifying and addressing these emerging risks.

References

[1] About Nanotechnology. National Nanotechnology Initiative. Available from: https://www.nano.gov/about-nanotechnology (accessed 20.08.24).

[2] L. Pokrajac, et al., Nanotechnology for a Sustainable Future: Addressing Global Challenges With the International Network4Sustainable Nanotechnology, ACS Publications, Washington, DC, 2021, pp. 18608–18623.

[3] M.C. Chen, P.W. Koh, V.K. Ponnusamy, S.L. Lee, Titanium dioxide and other nanomaterials based antimicrobial additives in functional paints and coatings, Progress in Organic Coatings 163 (2022) 106660.

[4] T. Nguyen, et al., Impact of UV irradiation on multiwall carbon nanotubes in nanocomposites: formation of entangled surface layer and mechanisms of release resistance, Carbon 116 (2017) 191–200.

[5] F. Van Broekhuizen, P. Van Broekhuizen, Nano-Products in the European Construction Industry, European Federation of Building and Wood Workers and European Construction Industry Federation, Amsterdam, 2009, pp. 1–109.

[6] A.R. Köhler, C. Som, A. Helland, F. Gottschalk, Studying the potential release of carbon nanotubes throughout the application life cycle, Journal of Cleaner Production 16 (8–9) (2008) 927–937.

[7] B. Nowack, Is anything out there?: What life cycle perspectives of nano-products can tell us about nanoparticles in the environment, Nano Today 4 (1) (2009) 11–12.

[8] T. Salthammer, F. Fuhrmann, Photocatalytic surface reactions on indoor wall paint, Environmental Science & Technology 41 (18) (2007) 6573–6578.

[9] J. Auvinen, L. Wirtanen, The influence of photocatalytic interior paints on indoor air quality, Atmospheric Environment 42 (18) (2008) 4101–4112.
[10] M. Horie, S. Sugino, H. Kato, Y. Tabei, A. Nakamura, Y. Yoshida, Does photocatalytic activity of TiO_2 nanoparticles correspond to photo-cytotoxicity? Cellular uptake of TiO_2 nanoparticles is important in their photo-cytotoxicity, Toxicology Mechanisms and Methods 26 (4) (2016) 284–294.
[11] D. Truffier-Boutry, et al., Characterization of photocatalytic paints: a relationship between the photocatalytic properties–release of nanoparticles and volatile organic compounds, Environmental Science: Nano 4 (10) (2017) 1998–2009.
[12] A. Mohammadipour, et al., The effects of exposure to titanium dioxide nanoparticles during lactation period on learning and memory of rat offspring, Toxicology and Industrial Health 32 (2) (2016) 221–228.
[13] S. Rades, et al., High-resolution imaging with SEM/T-SEM, EDX and SAM as a combined methodical approach for morphological and elemental analyses of single engineered nanoparticles, RSC Advances 4 (91) (2014) 49577–49587.
[14] S. Sepeur, European Coatings Tech Files. Nanotoxicology: Technical Basics and Applications, *Vincentz Network*, Hannover, 2008, pp. 5–168.
[15] A.T. Saber, et al., Inflammatory and genotoxic effects of sanding dust generated from nanoparticle-containing paints and lacquers, Nanotoxicology 6 (7) (2012) 776–788.
[16] F. Jandaghian, A.E. Pirbazari, O. Tavakoli, N. Asasian-Kolur, S. Sharifian, Comparison of the performance of Ag-deposited ZnO and TiO_2 nanoparticles in levofloxacin degradation under UV/visible radiation, Journal of Hazardous Materials Advances 9 (2023) 100240.
[17] C. Som, et al., LICARA Guidelines for the Sustainable Competitiveness of Nanoproducts, Zeist, Dübendorf, St. Gallen, 2014, pp. 1–44.
[18] M.N. de Queiroz, et al., Conductive nanopaints: a remarkable coating, Nanotechnology in the Automotive Industry, Elsevier, 2022, pp. 429–449.
[19] B. Giese, et al., Risks, release and concentrations of engineered nanomaterial in the environment, Scientific Reports 8 (1) (2018) 1565.
[20] R. Aitken, K. Creely, C. Tran, Nanoparticles: An Occupational Hygiene Review, HSE Books, London, 2004.
[21] A. Spoială, et al., Zinc oxide nanoparticles for water purification, Materials 14 (16) (2021) 4747.
[22] C. Egbuna, et al., Toxicity of nanoparticles in biomedical application: nanotoxicology, Journal of Toxicology 2021 (1) (2021) 9954443.
[23] C. Lopez-Chaves, J. Soto-Alvaredo, M. Montes-Bayon, J. Bettmer, J. Llopis, C. Sanchez-Gonzalez, Gold nanoparticles: distribution, bioaccumulation and toxicity. In vitro and in vivo studies, Nanomedicine: Nanotechnology, Biology and Medicine 14 (1) (2018) 1–12.
[24] V. Subramanian, et al., Assessing the social impacts of nano-enabled products through the life cycle: the case of nano-enabled biocidal paint, The International Journal of Life Cycle Assessment 23 (2) (2018) 348–356. Available from: https://doi.org/10.1007/s11367-017-1324-9.
[25] C.A. Gosselin. Nanotechnology in the World of Paints and Coatings. https://www.paint.org/coatingstech-magazine/articles/nanotechnology-in-the-world-of-paints-and-coatings/ (accessed 20.09.24).
[26] K. Savolainen, H. Alenius, H. Norppa, L. Pylkkänen, T. Tuomi, G. Kasper, Risk assessment of engineered nanomaterials and nanotechnologies—a review, Toxicology 269 (2–3) (2010) 92–104.

[27] R. Arora, T. Roy, P. Adak, A review of the impact of nanoparticles on environmental processes, BIO Web of Conferences 86 (2024) 01001.
[28] J.R. Lead, et al., Nanomaterials in the environment: behavior, fate, bioavailability, and effects—an updated review, Environmental Toxicology and Chemistry 37 (8) (2018) 2029–2063.
[29] B. Nowack, T.D. Bucheli, Occurrence, behavior and effects of nanoparticles in the environment, Environmental Pollution 150 (1) (2007) 5–22.
[30] V.K. Sharma, J. Filip, R. Zboril, R.S. Varma, Natural inorganic nanoparticles–formation, fate, and toxicity in the environment, Chemical Society Reviews 44 (23) (2015) 8410–8423.
[31] S.F. Hansen, E.S. Michelson, A. Kamper, P. Borling, F. Stuer-Lauridsen, A. Baun, Categorization framework to aid exposure assessment of nanomaterials in consumer products, Ecotoxicology 17 (2008) 438–447.
[32] A.B. Boxall, K. Tiede, Q. Chaudhry, Engineered nanomaterials in soils and water: how do they behave and could they pose a risk to human health? Nanomedicine Nanotechnology, Biology, and Medicine 2 (6) (2007) 919–927.
[33] T.Y. Sun, N. Bornhöft, K. Hungerbühler, B. Nowack, Dynamic probabilistic modeling of environmental emissions of engineered nanomaterials, Environmental Science & Technology 50 (2016) 4701–4711.
[34] E. Asmatulu, J. Twomey, M. Overcash, Life cycle and nano-products: end-of-life assessment, Journal of Nanoparticle Research 14 (3) (2012) 720. Available from: https://doi.org/10.1007/s11051-012-0720-0.
[35] S.K. Yadav, Z.A. Khan, B. Mishra, Impact of nanotechnology on socio-economic aspects: an overview, Reviews in Nanoscience and Nanotechnology 2 (2) (2013) 127–142.
[36] D.E. Babatunde, I.H. Denwigwe, O.M. Babatunde, S.L. Gbadamosi, I.P. Babalola, O. Agboola, Environmental and societal impact of nanotechnology, IEEE Access 8 (2019) 4640–4667.
[37] E. Kabir, V. Kumar, K.-H. Kim, A.C. Yip, J. Sohn, Environmental impacts of nanomaterials, Journal of Environmental Management 225 (2018) 261–271.
[38] J. Kekäläinen, Assessment of Environmental Impacts With Life Cycle Methods in Nanotechnology Industry, Master's thesis, AtoZ Nano (2013).
[39] K. Srinivas, Nanotechnology in the Paint Industry, International Journal of TechnoChem Research 4 27–39.
[40] R.V.S.K. Jakkula, P. Sethuramalingam, "Analysis of coatings based on carbon-based nanomaterials for paint industries – a review, Australian Journal of Mechanical Engineering 21 (3) (2023) 1008–1036.

19

Gaps, challenges, and future prospects of nanopaints

Vijay Mannari[1] and Prakash Chander Thapliyal[2]

[1]Coatings Research Institute, Eastern Michigan University, Ypsilanti, MI, United States [2]Advanced Concrete, Steel and Composites Group, CSIR-Central Building Research Institute, Roorkee, Uttarakhand, India

19.1 Introduction

Nanotechnology-based materials are providing pioneering solutions in construction sector acting in response to unmet needs. However, suitable standardized techniques need to be accessible for worth and protection evaluation of these pioneering products prior to their rendition into the ground and for observing their performance while manufacturing practice is altered. The matter which technological solutions are presently accessible in the scientific society to carry the demanded characterization of nanotechnology-based materials and which procedural progresses should be prioritized to sustain product developers in their regulatory evaluation [1].

Nanopaint is a variety of coating matter that has nanoparticles, minute particles with nanometer proportions. These nanoparticles give the paint only one of its kind properties for instance better strength, color preservation, etc. As a result, nanopaint is used in a variety of purposes, together with automotive, aerospace, and building industries for superior act in addition to long-lasting finishes. Due to inclusion of nanoparticles, nanopaints are proficient in conveying better act and functionality contrast to long-established paints, for example, self-cleaning, thermal-insulating, antimicrobial functionalities, etc., very much required in a range of industries. Despite the fact that nanopaints tender plentiful earnings, their ecological and wellbeing forces are still not entirely implicit. As a result, more research and appraisal are required to make certain the protected and accountable development of these pioneering coatings.

Nanopaints and coatings marketplace is anticipated to amplify at a compound annual growth rate (CAGR) of 7.53% from 2024 to

Nanopaints. DOI: https://doi.org/10.1016/B978-0-443-26512-9.00020-3

2031, from an estimation of USD 1.8 billion in 2024 to USD 19.73 billion by 2031. Nanoparticles are more and more integrated in floor coatings as well as inside paints to get better abrasion and microbial stability. One meticulous apprehension of this function is discharge of nanoparticles from these surfaces due to recurring mechanical stresses acted upon them, such as mopping, walking, cleaning, furniture moving, and scratching. Such discharge of the nanosize matter is potentially detrimental to inhabitants, mostly kids who usually feel greater publicity to substances deposited at floor level [2].

There has been a sudden increase in the quantity of nano-based goods in modern times. Investigations in this field have fascinated an enormous scientific curiosity in electronic, biomedical along with coating fields. This is an advance that presents mutual openings and tests. Even as exploitation of materials at the atomic degree has unlocked gate to likely new-fangled advances, it has, in addition, increased a few grave wellbeing worries. One such apprehension is over the fitness outcomes of breathable nanoparticles found in many end goods, such as paints. There are also apprehensions that release of a few of these nanoparticles into waterways along with rivulets could be injurious to gill membrane of fish and crustaceans.

Preparation of nanopaints was done by inserting zinc oxide (ZnO) and Fe_2O_3 nanoparticles in linseed alkyd resin by stirring with soya lecithin and turpentine oil [3]. By using these diverse nanopaints formulations, fouling properties were bargained with greater than before antibacterial properties based on which coatings were scrutinized for the corrosion behavior. Expansion of antifouling nanopaints with nanoparticles as biocides will considerably advance efficacy of antifouling nanopaints. The results put forward that with an augment in concentration of nanoparticles, there was an enhancement in antifouling and anticorrosion.

Learning from our empirical study and applying existing outline guides us to post a few key procedural involvements. Latest nanopaint know-how has technical hitches in challenging other technologies and as a consequence, it is vital to form protected space. Research and development at the nanoscale entail a great level of amalgamation, from union of research regulations to innovative associations between start-ups, regional actors, and research facilities.

19.2 Gaps

Nanopaint doles out a changing resolution in field of material science as well as coatings. Its rationale is many-sided, as this sophisticated category of paint tenders extraordinary properties mainly consequent from its nanoscale particles. Nanopaint's tiny particles permit paint to launch a consistent surface by way of better adhesion and robustness than usual counterparts. Planned to rally a variety of user-specific necessities, nanopaint can be engineered to display properties, such as electrical conductivity, heat resistance, or hydrophobicity, among others. Accordingly, this cutting-edge expertise is swiftly gaining thrust in more than a few sectors, such as the construction, automotive, etc. Thanks to its potential for attractive durability, performance, and esthetics.

For instance, self-cleaning nanopaint is used in architectural surroundings to lessen repair costs and improve a building's look. Hydrophobic behavior of this paint repels water along with dust, making certain nominal clean-up labors. In automotive business, heat-resistant and anticorrosion nanopaints assist in engine endurance and automobile element protection, in due course ensuring safe, cost-effective, and long-lasting performance. Likewise, aeronautical segment has observed noteworthy return with acceptance of lightweight nanopaints that give to oil efficiency and reduce on the whole load of airliner. As a consequence, nanopaint is more and more sighted as a game-changer in materials and coatings sphere for its pioneering applications, ground-breaking properties, and eco-friendly uniqueness.

Paints having biodegradable biocides could go down on their shielding impact, as biocides get cleaned out or inactivated by physical/chemical sources. For that reason paint production units are focusing on diverse plans to trounce this trouble. The substitute of organic-supported, degradable biocides by nanomaterials is one of special approaches, which paint production units have taken in deliberation [4]. Dry blasting harmful nanopaints can tranquil direct to higher discharge of SiO_2 and new compounds. Contact to higher quantities of respirable silica dioxide can escort to silicosis and cancer [5,6]. Nanosilica can in accumulation provoke necrosis at elevated quantities when nanosilica-doped nanopaints are in touch with human tissue [7]. Presently, the only helpful way of getting rid of nanoparticles from water is in course of application of energy-exhaustive methods, such as reverse osmosis using semipermeable film.

Nanopaints have been used for biofouling deterrence. Prior job assessed antifouling effects of nano as well as micro size exterior outlines which are motivated by environment [8–12]. One such illustration of biomimetic antifouling nanopaint is one motivated by microtopography of shark hide [13].

C_{18} nanopaints with higher intensities of nano roughness were found to restrain settlement of spores along with their antifouling outcome was revealed to surpass that of commercially existing fouling release coating [14].

Scardino et al. [15] observed superhydrophobic coatings with nano-scaled coarseness barred settlement of key micro as well as macro fouling variety. On another hand, engineered nanopaints have widespread applications in renewable energies, pharmaceuticals, electronics, along with antifouling coatings, etc. [16]. Of late, antifouling as well as antialgal behaviors of nanopaints having ZnO nanorods and TiO_2 nanoparticles have been deliberated [17–20].

In view of impending long-term societal as well as fiscal profits of fresh nano-enabled products, there is an apparent requirement for exploration and advancement to target closing space in nanomaterials safety. Following guidelines for upcoming investigate works can help out in satisfying the research gaps on the theme areas covered [21,22]:

(1) Initiating a blend of diverse analytical methods for determining nanomaterial mass concentration, particle concentration, and morphological information, (2) carrying out toxicity revisions using surroundings pertinent to environmental revelation of nanomaterials for individuals to get dose-response information, (3) acquiring applicable statistics to build up quantitative nanostructure–toxicity relationships, and (4) carrying out efforts for devising guidelines for regulating occurrence of nanomaterials in diverse environmental media.

19.3 Challenges

Metal-based nanomaterials generally release metallic ions which are looked to be mainly noxious nanomaterials amid all. Bioavailability and toxicity of these metal-based nanoparticles are very much reliant on quantity of adsorbing particles in water column [23]. Nanoparticles bound to other compounds forming bigger complexes are not as much bioavailable and as a result less deadly. Nanoparticles released into surface water may result in toxicological troubles [24]. Fortunately, quantities of nanometals (nanosilver, nano-ZnO, and nanotitanium dioxide [TiO_2]) established in surface waters are in array of ng L^{-1}, which is still small [25].

Qian et al. [26] investigated the outcome of bridge painting activities, such as coating thickness, coating process, etc., and meteorology situations, that is, temperature, moisture, etc., in contact with ketone, aromatic, ester, etc., using mixed-effect models. Park et al. [27] studied coating works at several construction workplaces covering residential buildings, office buildings, logistics centers, and a swimming pool. Based on this work, they concluded painters are rendering to different solvents, containing carcinogens and reproductive contaminants, and intensity of total organic volatile contents concentration in countless painting jobs surpassed exposure restrictions. Construction workers must be saved from harmful chemicals throughout their painting jobs with special shielding devices and/or work practice procedures.

Zuin et al. [28] studied the discharge of TiO_2, silver, and silica nanoparticles from three unlike paints by means of water immersion experiment. Fiber-cement boards were coated with paints and afterward exposed to ultraviolet beam. After static water immersion investigation, a very low discharge of Ti (4–8 $\mu g\ L^{-1}$) was observed, while the Ag calculated in leachates was underneath detection limit (0.1 $\mu g\ L^{-1}$). Similarly, minute discharge of Si was calculated in leachates, with 73 mg L^{-1} of Si released from paints containing SiO_2 after 120 hours of water immersion.

Jorgensen et al. [29] studied emissions from four interior nanopaints. Emission was considered for both bases as well as full-pigmented nanopaints, that is, water-borne acrylic paints and solvent-borne alkyd paint. Every trial was carried out two times in a test chamber in standardized settings. Discharges during the paint-drying were considered by assessing number concentration of nanoparticles and dimension distribution in the nanorange. The outcomes from the solvent-borne paint gave the peak concentration with the least particle size distribution. Outcomes from this experimentation demonstrated that exposure to nanoparticles for inhabitants assessed over a week or 28 day phase was little and that interior paints are most likely not awfully vital when it comes to recognizing products that release nanoparticles into interior environments.

Industries that comprehensively use nanoparticles, such as paints and coating segments set to greatly profit from cost-efficient know-how for eliminating nanoparticles from used water. Use of nanotechnology in nanopaints region is anticipated to rise exponentially and for vehicle makers, it will be obligatory to utilize antiscratch paints and coats on their vehicles. A few studies are available regarding discharge of nanomaterials from goods and processes used by businesses and clients. The quantity and features of the discharged particles establish the potential environmental exposure. Kattan et al. [30] examined discharge of titanium from paints having pigment—TiO_2 and nano-TiO_2 by exposing the coated panels to virtual weathering having sunlight and rain in environment chambers. Small-scale leaching tests were also carried out to reveal the effect of different factors on discharge-, such as composition of water, nature of support, and UV light suggesting that paints holding nano-TiO_2 may perhaps discharge barely partial quantities of materials into environment.

Construction painters have not been considered well in provisions of their vulnerabilities exposure. Park et al. [27] in their study tried to estimate exposure intensities of entire volatile organic compounds for painters in the construction trade for waterproofing activities and three painting activities by summing up fifteen targeted materials expressed in exposure-index (EI) values. The maximum EI for painting was 5.61 for interior epoxy primer task, followed by interior epoxy-resin task (2.03), outside oil-based-spray-paint task (1.65), outside water-based-paint task (0.66), and interior oil-based-paint task (0.15). This work acknowledged that construction painters are exposed to stages of volatile organic compounds (VOC) concentration in lots of painting jobs exceeding exposure restrictions. Construction workers need to be sheltered from chemical means during their painting works by using personal protective devices and/or work practice measures.

Nanotechnology is frequently well thought-out capable technology realm in terms of trade prospects in a range of industries, particularly in the background of grave societal challenges relating to energy, health care, clean water, and climate change. However, nanotechnology metrics are not there. Better metrics are necessary to keep an eye on and yardstick nanotechnology for potential policy requirements.

A lot of nanopaints for interior and open-air applications include biocides and additives to safeguard against microbial, physical, as well as chemical weakening. The biocides should stay lively as long as they are integrated into the nanopaints. Once the biocides are released they should degrade within a small point in time so that there are no build-up in surroundings. The paint manufacturing is not only targeting their investigations in creating superior paint formulations through degradable biocides. They as well believe in using nanomaterials, such as nanosilver, nanocopper, nano-ZnO, photocatalytic active nanotitania, and nanosilica as additives for the shielding of nanopaints, against microbial degradation as well as physico-chemical deterioration. At the time there is no certification that nanomaterials in nanopaints and fascias will accomplish their assignment in long run since there are no long-term studies present. From nanosilver-doped nanopaints, it is known that Ag is simply washed out by rainwater. Photo-active nano-TiO_2 adsorbs UV light and creates hydroxyl radicals, which not only hinder microbial growth but can also start or speed up the photocatalytic degradation of paint medium. Thus it is still unknown if it makes sense to add nanomaterials into paints. Intensive study and advance are still needed to unearth the answers [31].

Of late, a raise in the making of intelligent nanomaterials has offered extensive functionalities where TiO_2 and ZnO nanoparticles were produced in a green chemistry approach. Afterward, commercial enamel paint was customized by adding 2–5 wt./vol.% of nanoparticles. These nanopaints were subsequently applied onto diverse surfaces, such as carbon steel sheets, wooden sheets, and aluminum disks. The anticorrosion results have shown that nanopaints offer a hurdle due to decline of agglomerates on the exterior keeping away from porosity. These results showed promise for modification of paints with TiO_2 and ZnO nanoparticles, and application on solid surfaces for the construction and textile areas [32].

There is at present no all-inclusive information accessible about the types of nanomaterials in commercial coatings in the market. To offer simplicity about nanomaterial containing coatings in the value chain and the clients, at least those products for which the discharge of nanomaterials across their entire life cycle cannot be ruled out, should be reported. Clients should be educated about the likely environmental profits and hazards of nanopaints and on the subject of particular handling needs. Researchers, product designers, clients, and decision-makers should persist to distribute information to guard the environment. Further R&D must target exposure appraisal of other risks for painters including construction workers, to make out high-risk jobs and to get better risky occupational environments.

Additional research and development requirements take place with reverence to exposure to environmental conduct of nanopaints, their toxicological effects on individuals as well as surroundings, and the sustainability of nanopaints. From an ecological standpoint, there are for those reasons wide-ranging requirements in terms of research and development. Hence it is required to:

- Build up and adjust appropriate standardized testing and analytical techniques for assessing exposure to nanopaints in a range of ecological medium (water, soil, and air) taking into version that nanomaterials are classically not on the loose into the surroundings as one-off nanoparticles other than may be rooted in or close to nanopaints.

- Build up and adjust test procedures that guarantee comparability of investigate conclusions concerning environmental impacts and conduct of nanomaterials.
- Build-up techniques that perceive the discharge of nanomaterials all through total life cycle of goods that are coated with nanomaterials counting their exploits and dumping; carry out personal case studies of nanomaterials, resolve their strength in the coatings and their outcomes, and conduct in the surroundings, for example, after weathering or abrasion.
- Examine the biopersistence (in particular in the lungs), bioavailability, and (eco)toxicity of nanopaints.
- Carry out life cycle scrutinies assessing the impacts of nanopaints on the surroundings and individual wellbeing right through the full product life cycle to be able to appraise their benefits and weigh against them to usual goods.
- Revise dumping of nanopaints, such as incineration or recycling, since there is only slight data on hand about the conduct and discharge of nanomaterials and their influence on these processes.

19.4 Conclusions

Customers show concern about the existence of nanoparticles in paints along with associated hazards of exposure. As a result, in paint industry most of the time declares that no nanoparticles are at large in the interior surroundings without documentation. There is a need to undertake a comprehensive study to guess cytotoxic amounts of diverse nanoparticles (nano-TiO_2 and nanosilica dioxide) which might be used by paint traders in vicinity of opportunity to get better properties of paints as nanoparticles might be on the loose from nanopaints and get shifted into surface water. More studies are needed of the hour on exposure evaluation of new risks for construction workers so that high-risk jobs can be identified and hazardous occupational environments may be improved.

Nanomaterials have revolutionized functionalities achievable by various products across numerous industries. Within the construction segment, nanopaints and nanocoatings integrate a more and more quantity of engineered nanoparticles (ENPs). These augment resistance to aging and put in properties associated to biocidal action, air purification, thermal insulation, and self-cleaning. Public is exposed to nanomaterials in huge quantities and on surfaces inside their habitats. Making out probable contact levels and hazards is consequently vital to individual and environmental wellbeing.

Researchers explored normally used nanomaterials in nanocoatings and nanopaints comprising silver, TiO_2, and silicon dioxide. Studies illustrated that very few ENPs are out from paints even when subjected to hard abrasion and leaching. Under inconsiderate conditions, they are principally at large still rooted in the nanopaint medium. Weathering outcomes in the discharge of only a small amount of nanomaterials. Nanomaterials discharge is affected mainly by the binder used, meaning that new formulations could lessen release throughout the life cycle, including in a landfill. Further, there was no significant discharge of nanomaterials into the surroundings under landfill circumstances since geomembrane lining systems in landfills obstruct discharge and there was no confirmation of nanoparticles release in fumes in fire [32].

Subsequent to the precautionary theory, fresh applications should normally be tested for wellbeing with regard to individual health and the environment before they are promoted. A hazard appraisal is to leave out apprehension or explain hazard management measures with which the hazard can be abridged to a good enough stage. In our view, the environmental compatibility of nanomaterials and their applications are a significant characteristic when talking about the opportunities and hazards they cause. This is predominantly spot on when nanomaterials come into getting in touch with individuals or are at large into the environment during their life cycle. Although there is a little information existing about the toxicological properties of the nanomaterials used, this information is frequently deficient and not analogous making concluding appraisal of the hazards unfeasible. Hence, the crucial opinion that nanomaterial in nanopaints is safer for the surroundings and individual health the more strongly they are rooted in the coatings should be well thought-out in their manufacture. Several environmental profits are to be likely from nanotechnology-based nanopaints, such as thinner nanopaint coats, minimization or switch of solvents along with noxious compounds used in nanopaints. In addition, self-cleaning or "easy-to-clean" surfaces may curtail the need for cleaning along with diminishing the utilization of energy, as well as cleaning agents, and extending the life extent of the coated objects.

Nevertheless, to make certain the safety of nanopaints that contain nanomaterials, studies should be carried out at an early point of their expansion not only at the environmental benefits but also at the risks posed by nanomaterials and their applications. It is recommended that improvement and standardization of appropriate measuring and analytical methods allocate an enhanced detection and specific exposure assessment. Prior to marketing, nanopaints should be tested with respect to the probable discharge of nanomaterials across their full life cycle. Additionally, their ecological sustainability with particular deliberation of material flows, energy consumption, waste, and emissions should be studied.

Nanotechnology has provided a gigantic prospective for fresh solutions to many of the existing challenges society is facing. On the other hand, nanotechnology is also a part of product improvement that is on the rise more rapidly than regulatory frameworks. This is owing to soaring intricacy of many nanomaterials, need of a worldwide harmonised regulatory explanation, and diverse scopes of directive at worldwide stage [33].

Even though there has been a noteworthy improvement linked to analytical approaches as well as to the expansion of fitting test guiding principles for nanomaterials and their safety evaluation, there is a still a call for larger comprehensive alliance and consent in the regulatory field.

Acknowledgments

Author (PCT) wishes to show gratitude toward Director, CSIR-Central Building Research Institute, Roorkee for their constant encouragement. Also both authors state that they do not have any conflict of interest.

References

[1] B.H. Kenzaoui, R.J. Vandebriel, A. Howarth, M. Siccardi, C.A.W. David, N.J. Liptrott, et al., Methodological needs in the quality and safety characterisation of nanotechnology-based health products: priorities for method development and standardisation, Journal of Controlled Release 336 (2021) 192–206.

[2] L. Sung, T. Nguyen, NIST Technical Note 1835 – Nanoparticle Released from Consumer Products: Flooring Nanocoatings and Interior Nanopaints, National Institute of Standards and Technology, United States, 2014, p. 68.

[3] M.V.K. Mohan, T.V.K.B. Prakash, A. Mukherjee, Antifouling paint formulations with ZnO and Fe_2O_3 nano-paints for marine applications, Sustainable Chemistry and Pharmacy 30 (12) (2022) 100858.

[4] I.K. Koponen, K.A. Jensen, T. Schneider, Comparison of dust released from sanding conventional and nanoparticle-doped wall and wood coatings, Journal of Exposure Science & Environmental Epidemiology 21 (2011) 408–418.

[5] R. Kaegi, A. Ulrich, B. Sinnet, R. Vonbank, A. Wichser, S. Zuleeg, et al., Synthetic TiO_2 nanoparticle emission from exterior facades into the aquatic environment, Environmental Pollution 156 (2008) 233–239.
[6] J. Locker, A. Carstens, High pressure injection of silica-based paint into a finger, The New Zealand Medical Journal 123 (2010) 104–106.
[7] J.P. Kaiser, S. Zuin, P. Wick, Is nanotechnology revolutionizing the paint and lacquer Industry. A critical opinion, The Science of the Total Environment 442 (1) (2013) 282–289.
[8] M.V. Graham, N.C. Cady, Nano and microscale topographies for the prevention of bacterial surface fouling, Coatings 4 (2014) 37. Available from: https://doi.org/10.3390/coatings4010037.
[9] F.W.Y. Myan, J. Walker, O. Paramor, The interaction of marine fouling organisms with topography of varied scale and geometry: a review, Biointerphases 8 (2013) 30. Available from: https://doi.org/10.1186/1559-4106-8-30.
[10] C. Richards, A. Slaimi, N.E. O'Connor, A. Barrett, S. Kwiatkowska, F. Regan, Bio-inspired surface texture modification as a viable feature of future aquatic antifouling strategies: a review, International Journal of Molecular Sciences 21 (2020) 5063. Available from: https://doi.org/10.3390/ijms21145063.
[11] A.J. Scardino, R. De Nys, Mini review: biomimetic models and bioinspired surfaces for fouling control, Biofouling 27 (2011) 73–86. Available from: https://doi.org/10.1080/08927014.2010.536837.
[12] G.D. Bixler, B. Bhushan, Biofouling: lessons from nature, Philosophical Transactions of the Royal Society A 370 (2012) 2381–2417. Available from: https://doi.org/10.1098/rsta.2011.0502.
[13] D.W. Bechert, M. Bruse, W. Hage, Experiments with three dimensional riblets as an idealized model of shark skin, Experiments in Fluids 28 (2000) 403–412. Available from: https://doi.org/10.1007/s003480050400.
[14] P. Majumdar, E. Lee, N. Patel, K. Ward, S.J. Stafslien, J. Daniels, et al., Combinatorial materials research applied to the development of new surface coatings IX: an investigation of novel antifouling/fouling-release coatings containing quaternary ammonium salt groups, Biofouling 24 (2008) 185–200. Available from: https://doi.org/10.1080/08927010801894660.
[15] A.J. Scardino, H. Zhang, D.J. Cookson, R.N. Lamb, R.D. Nys, The role of nano-roughness in antifouling, Biofouling 25 (2009) 757–767. Available from: https://doi.org/10.1080/08927010903165936.
[16] E.R. Bandala, M. Berli, Engineered nanomaterials (ENMs) and their role at the nexus of food, energy, and water, Materials Science for Energy Technologies 2 (2019) 29–40.
[17] M. Al-Fori, S. Dobretsov, M.T.Z. Myint, J. Dutta, Antifouling properties of zinc oxide nanorod coatings, Biofouling 30 (2014) 871–882. Available from: https://doi.org/10.1080/08927014.2014.942297.
[18] V.K. Yemmireddy, Y.C. Hung, Photocatalytic TiO_2 coating of plastic cutting board to prevent microbial cross-contamination, Food Control 77 (2017) 88–95. Available from: https://doi.org/10.1016/j.foodcont.2017.01.025.
[19] Z. Zhu, H. Cai, D.W. Sun, Titanium dioxide (TiO_2) photocatalysis technology for nonthermal inactivation of microorganisms in foods, Trends in Food Science and Technology 75 (2018) 23–35. Available from: https://doi.org/10.1016/j.tifs.2018.02.018.
[20] Bhattacharya, S., Missing gaps in indian nanotechnology development: exploring effective policy interventions for innovation and commercialisation, Proceedings of the International Seminar on "Creation

and Diffusion of Technology" organized by Knowledge Forum (http:// http;//fgks.in), hosted by Indian Institute of Technology Bombay in partnership with TATA Trusts, pp. 205–217.

[21] A. Kumar, P. Kumar, A. Anandan, T.F. Fernandes, G.A. Ayoko, G. Biskos, Engineered nanomaterials: knowledge gaps in fate, exposure, toxicity, and future directions, Journal of Nanomaterials (2014) 16. 130198.

[22] I. Furxhi, F. Murphy, C.A. Poland, M. Cunneen, M. Mullins, Precaution as a risk in data gaps and sustainable nanotechnology decision support systems: a case study of nano-enabled textiles production, Nanoethics 15 (2021) 245–270.

[23] L. Kiaune, N. Singhasemanon, Pesticidal copper (I) oxide: environmental fate and aquatic toxicity, Reviews of Environmental Contamination and Toxicology 213 (2011) 1–26.

[24] S.W.P. Wijnhoven, W.J. Peijnenburg, C.A. Herberts, W.I. Hagens, A.G. Oomen, E.H. Heugens, et al., Nano-silver: a review of available data and knowledge, Nanotox 3 (2009) 109–138.

[25] F. Gottschalk, T. Sonderer, R.W. Scholz, B. Nowack, Modeled environmental concentrations of engineered nanomaterials (TiO_2, ZnO, Ag, CNT, fullerenes) for different Regions,, Environmental Science & Technology 43 (2009) 9216–9222.

[26] H. Qian, N. Fiedler, D.F. Moore, C.P. Weisel, Occupational exposure to organic solvents during bridge painting, The Annals of Occupational Hygiene 54 (4) (2010) 417–426.

[27] H. Park, H.D. Park, J.K. Jang, Exposure characteristics of construction painters to organic solvents, Safety and Health at Work 7 (2016) 63–71.

[28] S. Zuin, L. Golanski, M. Gaiani, A. Ferrari, Leaching of nanoparticles from experimental water-borne paints under laboratory test conditions, Journal of Nanoparticle Research 16 (2014) 2185.

[29] R.B. Jorgensen, I.G. Hveding, K. Solheim, Nanosized emission from commercially available paints used for indoor surfaces during drying, Chemosphere 189 (2017) 153–160.

[30] A.A. Kattan, A. Ulrich, A. Wichser, S. Zuin, R. Vonbank, S. Brunner, et al., Release of TiO_2 from paints containing pigment-TiO_2 or nano-TiO_2 by weathering, Environmental Science: Processes & Impacts 15 (2013) 2186–2193.

[31] J.P. Kaiser, L. Diener, P. Wick, Nanoparticles in paints: a new strategy to protect façades and Surfaces, Journal of Physics: Conference Series 429 (2013) 012036.

[32] R. Solano, D.P. Ruiz, A. Herrera, Preparation of modified paints with nano-structured additives and its potential applications, Nanomaterials and Nanotechnology 10 (2020) 1–17.

[33] J. Allan, S. Belz, A. Hoeveler, M. Hugas, H. Okuda, done Patri, et al., Regulatory landscape of nanotechnology and nanoplastics from a global perspective, Regulatory Toxicology and Pharmacology 122 (2021) 104885.

Index

Note: Page numbers followed by "*f*" and "*t*" refer to figures and tables, respectively.

U

V

W

X

Y

Z

9780443265129